MUSIC
SPEECH
HIGH-FIDELITY

SECOND EDITION

WILLIAM J. STRONG
GEORGE R. PLITNIK

SOUNDPRINT

Published by SOUNDPRINT
Typesetting and layout by Aimee Holdaway, Liberty Press
Printed and bound by Liberty Press
Printed in the United States of America

ISBN 0-9611938-0-8

Contents

III. The Ear and Hearing

IV. Listening Environments

V. The Human Voice and Speech

VII. Electronic Reproduction of Music

Appendices

Bibliography

Index

Preface to the Second Edition

This second edition of **Music Speech High-Fidelity** represents our continuing effort to provide reading materials and associated exercises for student use in descriptive acoustics courses. We have incorporated suggestions from students and colleagues who used the first edition. The second edition includes many photographs, many new figures, extended coverage of many topics, as well as the indexing of new terms. We welcome comments from interested readers and will incorporate their suggestions in such future versions of the book as might be warranted.

Most acoustics textbooks available at the descriptive level concentrate on one or two selected areas of acoustics to the partial or complete exclusion of others. This may be desirable when depth of treatment, at the necessary expense of a broad overview, is the aim. However, we feel that once some of the basic conceptual tools have been developed it is not only possible, but highly desirable, to treat several areas of acoustics at the descriptive level. In particular, many common ideas are applicable to both musical acoustics and speech acoustics. In **Music Speech High-Fidelity** we have attempted to develop the relevant physical concepts and to apply these to acoustical aspects of music, speech, high-fidelity, room acoustics, and reproduction of sound. The approach is, of necessity, "broad-brush" in nature, and many of the more subtle points are only alluded to. In keeping with our aim that the book be primarily descriptive, we have had to sacrifice scientific accuracy in some instances in order to present the material in a simple manner. We hope the student will make use of the resource materials listed at the end of each chapter to explore areas of interest in greater depth and also to enlarge upon other topics only partly developed in the book.

We have attempted to emphasize the application of physical principles in explaining and describing many diverse acoustical phenomena. We hope that the student will give serious consideration to the exercises and demonstrations provided at the end of each chapter in exploring the application of physical principles to the world of sound. Most of the exercises are self-contained and can be completed with the material supplied in the text. The demonstrations are intended to be carried out by the instructor to illustrate physical principles. However, many of the "demonstrations" can be made most meaningful when carried out by the student as laboratory activities. There is no substitute for the firsthand experience of using laboratory instruments to observe physical phenomena and collect data.

We hope that the student and instructor alike will find the listing of resource materials useful. The listing includes books and journal articles, audio tapes, films, and video tapes. Many of these items will be available to the student in the school library or learning resource center. Books that provide useful reference material for several chapters are listed in a bibliography and are referred to by chapter number. Journal articles and books with more restricted content are referenced in full at the end of each chapter. A bibliography of publications and audio-visual suppliers appears at the end of the book.

More than enough material for a one-semester course is included in this book. After Parts I and II (the physical principles) have been completed, the other chapters can be used more or less independently. It is possible to structure a course involving hearing and music (parts III and VI), hearing and speech (parts III and V), hearing and hi-fi (parts III and VII), or other combinations. In some instances it may be desirable to initially skip Parts I and II and then refer to them as needed while studying the other material.

We acknowledge our indebtedness to many workers in the various fields of acoustics. In order to make our presentation most readable, we have not often given credit in the running text to those whose research is described. However, the references at the end of each chapter indicate the sources of our material. Many of our figures have been taken or adapted from journal articles and books as noted in the figure captions; others have been supplied by colleagues from their unpublished material. Many companies have supplied photographs and other materials.

The books on musical acoustics by John Backus and Arthur Benade, the book on speech by Peter Denes and Elliot Pinson, and the books on hearing by Reinier Plomp and Juan Roederer have particularly stimulated our thinking in the preparation of this book.

We are indebted to many colleagues for fruitful discussions, useful information, and other assistance. In particular, we must mention Donald Allen, John Backus, Irvin Bassett, Arthur Benade, John Coltman, Duane Dudley, Neville Fletcher, Carleen Hutchins, Edward Jones, Norman Kinnaugh, Alan Melby, George Orvis, Paul Palmer, Michael Piper, Thomas Rossing, William Savage, Del Scott, Stephen Stewart, Ingo Titze, and Glenn Williams.

We appreciate the support and encouragement given by the Brigham Young University Department of Physics and Astronomy in the preparation of the manuscript. We are especially grateful to several individuals who played major roles in the preparation of the book. Shemay Reddish typed many revisions of the manuscript. Kurt Haug at Communitect edited the manuscript. Stephen Strong designed and generated many of the figures. McRay Magleby and Brent Burch designed the cover.

Finally, we wish to thank our wives, Charlene Strong and Gail Plitnik, who gave us encouragement and sacrificed time and resources that might have been spent otherwise.

I. Fundamentals of Physics and Vibration

Courtesy of Brigham Young University Archives. Photo by L. W. Taylor.

1. Science and Acoustics

We live in a sea of sound surrounded by such things as familiar voices, strains of music, and intruding noises. Each sound has its own characteristics to which we respond—sometimes positively, sometimes negatively, sometimes unconsciously.

What makes the sounds we hear? The familiar sound of a voice is produced almost unconsciously through skillful control of a human vocal system. Skilled musicians are able to create subtle musical sounds with instruments. And yet, the noises we hear are produced in a manner that requires no apparent skill.

The production of desirable sound has two aspects: that involving a skilled performer controlling a sound producing device, and that involving the physical mechanisms underlying the working of the device. The first aspect might be categorized as art and the second as science. Each has its own special description of sound production, and there are benefits and insights to be gained from considering sound production from each vantage point.

Since we will be concerned primarily with a scientific view of sound, we will begin by establishing the philosophical point of view from which science proceeds. **Science** can be defined as knowledge of the operation of general laws. Physical science is knowledge gained through study of the physical world; it is knowledge about the operation of general laws governing the physical world. The task of the scientist is to discover and state the general relationships connecting observed facts by using methods of observation, experimentation, and generalization.

Several comments are in order regarding the nature of science. Scientific knowledge is organized through the process of generalization by abstraction from particular observations. Science is an everchanging, dynamic enterprise based on continuing observations and new generalizations. The inherent limitation of science lies in the fact that it uses methods based upon observation and thus, anything that is beyond sensory observation is outside the boundaries of science. It is not relevant to consider whether there are physical laws that man can never hope to discover. The ultimate authority in science is observation. If questions can be answered, at least in principle, by observation, they may be considered scientific questions. Science is not limited to observation alone, but its superstructure must finally stand or fall on the basis of observed facts.

Science is a human activity performed by humans for the benefit of humans and because of its intellectual challenge. "Scientists are people of very dissimilar temperaments doing very different things in very different ways. Among scientists are collectors, classifiers, and compulsive tidiers-up; many are detectives by temperament and many explorers. Some are artists and others are artisans. There are poet-scientists and philosopher-scientists and even a few mystics" (P.B. Medawar, **The Art of the Soluble**, Barnes and Noble, 1967).

Characteristics of Science

To what extent do the various sciences have common characteristics which distinguish them from other disciplines such as the humanities? The answer, of course, is not as simple as the question. The best we can do is to mention a few points for consideration, stressing four characteristics which, when combined, make the sciences unique: generalization, classification, quantification, and experimentation.

Generalization is the expression of particular situations in more general terms. For example, a person may begin to sneeze and cough on repeated occasions when petting the family cat and therefore draw the general conclusion that he or she is allergic to cats. (In science the process of abstraction does not usually consist of casually drawn inferences, as in this example, but is based upon detailed and careful observation.)

If you ask physicists what the weight of a moose would be on top of a tower 6400 km high, they will answer that its weight will be about one-fourth what it is on the earth's surface. How can the physicist know this? No one has ever built such a tower nor is it likely that one ever will be built. The answer is based on a generalization of science which states that an object's weight decreases in proportion to the inverse square of its distance from the earth's center.

The sciences make use of generalization in varying degrees. We can picture a chain of natural sciences in this order: physics, chemistry, geology, biology. The current extent to which generalization has been found possible and fruitful decreases as we go along this chain. However, virtually all other fields of human knowledge also employ generalization in varying degrees.

A second feature of science is **classification,** or the sorting of objects and phenomena into different classes or categories. Classification catalogs information into a more easily accessible form and has been a mark of science since the days of the Greeks. It is particularly important for biology and geology.

Quantification is a third scientific characteristic that involves the attaching of numerical values to generalizations. While this could be done in any field (e.g., attaching numerical "weights" to various authors of English literature), if quantification is to be meaningful it must be used to express unique relationships in mathematical form—as, for example, in an algebraic equation.

Generalization, classification, and quantification, however, are still not sufficient to unequivocally set the sciences apart from other fields. Business accounting, for example, has all these characteristics. Science, however, possesses another important characteristic—experimentation. **Experimentation** is a procedure whereby generalizations are tested.

The importance of experimentation to science cannot be overemphasized. Past attempts at establishing a science without experimentation have produced false conclusions. The famous Greek, Aristotle (who lived in the fourth century B.C.) concluded that the heavier an object, the faster it would fall toward the earth. This seems like a reasonable generalization, for we have all observed that if a stone and a feather are dropped simultaneously, the stone reaches the ground first. But, many centuries later, Galileo observed that all objects, regardless of weight, fall toward the earth with the same acceleration. Any differences are due to air resistance. Aristotle's seemingly reasonable generalization was discarded when adequate experimentation was finally performed.

Scientific Method

The so-called "scientific method" is not one, specific, detailed method of procedure used by all scientists but is any systematic procedure used by scientists in their work. There are probably as many different methods used by scientists as there are scientists; nevertheless, the general way in which most scientists tackle a problem can be considered as "scientific method."

First, when a scientist attempts to start from a specific observation and arrive at a generalization, he is using **inductive reasoning**. When he then attempts to use the generalization to search for evidence of it in scientific observations, he is using **deductive reasoning**. Both types of reasoning are important in science, and each, for different reasons, must ultimately refer back to experimentation. After induction we must constantly devise new experiments to ascertain that we have not overextended our generalization. In deduction we must rely on experimentation to verify the validity of our initial assumptions, because the conclusions we draw will be valid only if the initial assumption is valid. Abraham Lincoln once asked, "If you call a tail a leg, how many legs does a dog have?" If we take the first clause as a postulate, the deductive process tells us that the dog has five legs. But it remains an unalterable fact that a dog has only four legs. Nothing is wrong with the deductive process here, it is only that we have a faulty postulate. As Lincoln observed, "Calling a tail a leg does not make it a leg."

Scientific method usually consists of several steps. The problem to be solved is stated as precisely as possible. Published material is searched to find information pertinent to the problem. Experimentation is conducted to observe and collect data applicable to the problem. The resulting data are organized by means of tabulation, graphs, and so on. A generalization based on relationships among the data is formulated to provide an answer to the problem. Predictions are deduced from the generalization. Additional experiments are then performed to see if the predictions are verified. If the predictions are not verified, modifications may be made to the generalization and additional attempts made at achieving experimental verification. Finally, the findings are published so that others may learn the results, which accounts for much of the success in the ongoing processes of science

As an example of the application of the above steps assume that you have different pieces of brass tubing that are identical except for length. Assume further that each piece of tubing is fitted with an identical clarinet mouthpiece. The problem involves finding the relationship between the tubing length and the pitch produced when the tube is blown. When you experiment by blowing into tubes of different lengths, you observe that there are different pitches associated with each. Through organization of your data, you also discover that the longer lengths of tubing produce the lower pitches. You then make the generalization that the pitch is inversely proportional to the length of the tube. You then make some predictions about the pitches that will be produced by some of the tubes that you have not yet blown. You blow into these tubes and measure the pitches and compare the measured values with the predicted values in an attempt to verify your generalization. You find that your generalization is approximately verified, but that you get more accurate results if you include the length of the mouthpiece along with the length of the tube. This leads to some modifications of your original generalization and may be followed by new attempts at experimental verification of additional predictions.

Application of the scientific method to physical phenomena may enable you after some practice to obtain new knowledge, structure knowledge so that it is more easily accessible, apply knowledge to new situations, and test the validity of new knowledge.

Hypotheses, Laws, and Theories

A physical scientist generally proceeds on the assumption that the phenomena he observes in his laboratory are governed by natural laws. His task is to discover these physical laws and to express and explain them in a clear manner. If his interpretation of the laws is accurate, relationships among any experimental data he gathers can be accommodated within his statement of these laws.

The experimental observations or data measurements that a scientist makes constitute the **facts** of science. Certain controls and conditions are usually associated with the measurements of data, and these in effect become a part of the facts. Furthermore, scientific facts or measurement must be repeatable by different observers in different parts of the world at different times, or at least be equally accessible to all comparably equipped scientists in the case of nonrepeatable phenomena.

Once a sufficient number of facts are available the process of generalization is used to express relationships among the facts. A **hypothesis** is a tentative statement of relationships among observed facts. It implies an insufficiency of experimental evidence and is often intended only as a tentative assumption to be modified and expanded with further experimentation.

As a hypothesis is tested and refined it may evolve into a law or a theory. A **law** usually refers to scientific generalizations and implies definite relationships among natural phenomena under certain prescribed conditions. When using a scientific law we must be aware of its limitations and the conditions under which it applies.

Some laws appear to be universal, but sooner or later a situation may arise in which these laws will be found to be inaccurate or too limited. However, within the limits prescribed when the law was discovered, we expect the law to hold because it is based on the previously observed facts.

When a law is supplemented by new information, it does not always mean that the old law must be discarded. Typically, new information can be expressed in the form of additional conditions on the law. So long as we remember the limitations of a law and do not exceed them, the old form of a law may be very useful and in some cases easier to apply than the newer form.

A **theory** is a conceptual scheme or model that provides a description of natural phenomena and may incorporate diverse laws of science. Both laws and theories imply greater refinement and a greater range of experimental evidence than does a hypothesis. Inasmuch as empirical laws are often incorporated within the framework of a theory (e.g., Newton's laws of motion within the theory of classical mechanics) theories are more general then empirical laws. For a theory to be most useful, a scientist should be able to make predictions based on the theory that can be tested experimentally. Ideally a theory should correlate many separate and seemingly unrelated facts into a logical, easily grasped structure. A good theory will make it possible to predict specific new observable phenomena and relate all previously observed phenomena.

Obviously, scientists do not expect a theory to go through the ages unchanged. In fact, an important aspect of science is trying to disprove existing theories. A theory can never be proven but only disproven; no number of observations or experiments will prove a theory to be true. Rather, we speak of observations and experiments as verifying theories. The most beautiful and elaborate of theories, however, are subject to modification or discard if they cannot explain observable facts.

Science and Technology

Scientific activities may be expressed as an interrelated threefold process, as shown in Figure 1.1. First, science utilizes experimentation; that is, scientists are involved in observing nature and gathering data. Second, science is theoretical; scientists discover and state laws and theories which generalize, predict, interpret, and otherwise make data more manageable and understandable. Thirdly science can be applied to transform our environment. Whereas theoretical, or ''pure'', science is concerned with experimentation and the formulation of new laws and theories, ''applied'' science seeks to find useful applications of these laws and theories in the matters of everyday living.

The three components of science are interrelated as shown in Figure 1.1. In the figure each circle is connected with each of the others by two oppositely directed arrows, which indicate that the processes of science go in either direction. The arrows between the circles representing experiment and theory, for instance, show that experimentation leads to theory and that theory often influences the type of experiment performed and what one expects to observe from the experiment. The double arrows between the circles representing theory and application indicate that theory is important in determining the possibility, as well as the probable direction, of some useful application. Conversely, an enrichment of theory is often the result of a struggle with new problems arising from an attempt to apply the theory. Finally, the circles representing experiment and application are also connected, showing that an experiment may directly suggest an application, or that technological developments may suggest new areas for gathering data.

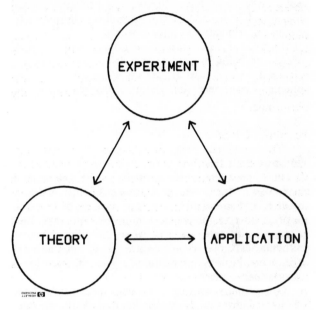

Figure 1.1. The threefold nature of science.

The area of applied science generally goes by the all inclusive title of **technology**. Throughout most of recorded history, however, technology developed by trial and error, independent of science. (This is evidenced in the Greek root techne, which means ''skill'' or ''art.'') Prior to this century, most technological developments were based more on the extrapolation of existing technology than on existing science. For example, it is well known that the technological development of musical instruments advanced to a high level of sophistication before there was very much scientific understanding of their operation. In recent decades, however, science has done much in the way of giving us an understanding of the operation of musical instruments and has been used in some instances to aid in the design or modification of musical instruments. During the past century, science and technology have become more intertwined and close cooperation between them has permitted the development of modern systems such as digital audio recording.

Several comments are in order regarding the nature of technology as opposed to science. The goal of science is to obtain knowledge about our physical environment; its ultimate products are the concepts, laws and theories which describe nature. The goal of technology is to apply scientific knowledge and skills to the solution of technological problems; its ultimate products are new and better things. Technology, like science, is a human activity and is engaged in by humans in order to provide products that will benefit humanity. Those engaging in

5

technology wish to reap intellectual and material rewards for meeting challenges in innovative ways.

Much of our modern technology is based upon prior scientific discoveries. "Pure" scientific research, which often appears to be irrelevant, provides a vast reservoir of knowledge from which new technological advances may come. The laser is an excellent example of this. When first invented it was regarded as a toy with no useful applications; but ten years later it was being utilized in numerous ways. Even now, not all the useful applications have been discovered. Modern science relies very heavily upon technological developments to provide experimental tools for research.

In summary, science and technology reinforce each other, each being mutually dependent upon the other. Science functions most effectively when its three components—experimentation, theory and application—are properly balanced.

Nature of Acoustics

In its broadest sense, **acoustics** is the science and technology of the production, transmission, reception, and effects of sound and vibration. Its areas of research are often interdisciplinary, crossing the boundary lines of many traditional academic areas including physics, mathematics, electrical engineering, mechanical engineering, biology, psychology, medicine, audiology, speech science, music, and architecture.

Our primary concern is to describe and illustrate the physical laws and principles that underly the production and perception of music and speech and the electronic reproduction of sound. From this point of view acoustics may be regarded as a subdivision of physics—other subdivisions being mechanics, optics, thermodynamics, electromagnetism, and modern physics. Physics in turn is a subdivision of physical science—other subdivisions being astronomy, chemistry, and geology. Physical science is one of three subdivisions of science—the others being the biological and social sciences.

Although acoustics is an extensive discipline, it may be divided rather conveniently into two main aspects: where the role played by a human is as an experimenter attempting to determine physical aspects of sound, and where the human perceives sound through direct interaction with sound. The first aspect where there is little direct human involvement in the processes studied is referred to as **physical acoustics**. The following subdivisions may be included: ultrasoncis, infrasoncis, atmospheric sound, mechanical vibration and shock, bioacoustics, underwater sound, seismic waves, noise production and control, sound production by musical instruments, speech production, and architectural acoustics.

The second aspect is referred to as **perceptual acoustics** since the human is directly involved as a perceiver. The following subdivisions may be included: physiological acoustics, psychoacoustics, effects of noise, perception of musical sounds, speech perception, perception of sound in enclosures, and electroacoustics.

Underlying and unifying all aspects of acoustics is a mechanical vibration in some material. The main divisions of acoustics, physical acoustics and perceptual acoustics, are determined by the role of a human. The subdivisions of acoustics may fall in one or more of the categories of physical science, engineering, life science, and fine arts, which illustrates the enormous range of applications of acoustics. There is virtually no aspect of human experience which is not influenced by some facet of acoustics. Truly, acoustics is an interdisciplinary subject without parallel.

Brief History of Acoustics

Sound has always been very significant to man as a means of communication, as a source of pleasure, and even as a source of annoyance. Yet, acoustics, is relatively new. Prior to the seventeenth century the study of acoustics was concerned primarily with musical sounds and architecture. The ancient Greek philosophers had little interest in the scientific study of sound, but they had considerable interest in music. In the sixth century B.C., Pythagoras discovered the relationship between consonant musical intervals and string lengths. In the first century B.C. the Roman architectural engineer, Vitruvius, had some understanding of the wave theory of sound. In his books on architecture, he discussed the acoustical characteristics of Greek theaters and demonstrated some knowledge of echoes and reverberation.

The first scientifically valid studies of sound were made about 1600 by Galileo, who extended the observations of Pythagoras with carefully constructed experiments, thus laying the foundation of experimental acoustics. It was Galileo who first stated the relationship between pitch and frequency, the laws of musical harmony and dissonance, and the dependence of a stretched string's natural frequency upon its length, mass and tension. At about the same time, Mersenne measured the return time of an echo and thus obtained a value for the speed of sound in air which was within 10% of the currently accepted value. In 1660 Robert Boyle suspended a bell in a glass enclosed vacuum and showed that although the clapper could be made to strike the bell by means of an attached string, no sound was heard outside the enclosure. Newton studied theoretical aspects of sound and calculated, from theoretical considerations alone, a value for the speed of sound in air that was somewhat lower than the experimental values.

The 18th century was primarily a period of development in theoretical acoustics. Chladni made many notable discoveries concerning vibrating strings and rods. LaPlace changed Newton's formula and calculated a speed of sound which agreed closely with experimental values. New mathematical tools in the hands of creative mathematicians such as Lagrange, Bernoulli, and Euler brought increased understanding to such diverse subjects as pitch and tone quality, and the nature of sound transmission in liquids.

In the 19th century the mathematics necessary for analyzing complicated waves was developed by Fourier and applied to sound waves. The first accurate measurement of the speed of sound in water was made, and the fundamental law governing the dependence of wave velocity upon the density and elasticity of a medium was formulated. Three devices which proved to be useful in the study of sound—the siren, the stethoscope, and the stroboscope—were invented. Much attention was given to the standardization of pitch, and tuning forks of exceptional accuracy were manufactured. The two most famous 19th century scientists who worked in acoustics were H.L.F. Helmholtz and Lord Rayleigh. Helmholtz, who was both a physicist and a physician, published his scientific masterpiece entitled **On the Sensations of Tone**, in which he summarized his theories and experimental investigations of musical tone, vowels, and the hearing process. Lord Rayleigh, a scientific synthesist, summarized in his tome **The Theory of Sound**, virtually everything that was known about acoustical theory. By the end of the century, the telephone, the microphone, and the phonograph had been invented and were destined to have a powerful influence on the science and technology of sound.

The age of electronics was born in the twentieth century, and a host of new instruments for the production and reception of sound became available. It became possible to make simple and accurate quantitative measurements of sound. The rapid advances in acoustics during the 20th century are evident in radio, talking motion pictures, hi-fi systems, and public address systems. Improved instrumentation for the study of speech and hearing has stimulated the creation of new branches of physiology and psychology. Ultrasonic devices have become tools for medical diagnosis and therapy. Music is being influenced by technological devices such as the computer and the synthesizer. There is virtually no area of human activity which is not being influenced by the science and technology of sound.

Further Reading

Hunt, F.V. 1978. **Origins in Acoustics** (Yale University Press).

Lindsay, R.B. 1976. **Acoustics: Historical and Philosophical Development** (Dowden, Hutchinson, and Ross).

Miller, D.C. 1935. **Anecdotal History of the Science of Sound** (Macmillan).

2. Basic Physical Quantities and Laws

Musicians, audiologists, and audiophiles all have their own specialized terminology. Once you have had sufficient exposure to the terminology of a particular discipline, you are able to understand and use it in meaningful ways. For our study of music, speech, and high fidelity from the point of view of acoustics, we need to have some familiarity with the terminology and concepts involved. In this chapter we consider some fundamental physical quantities and laws that form the foundation upon which we build our physical description of musical instruments, the human speech mechanism, and hi-fi systems.

Operational Definitions

Suppose you ask three people to measure the length of a given piece of string. The first person reports a length of 0.25 meters, the second 25.4 centimeters, and the third 10 inches. You notice that they all perform the measurement in a similar way, even though they all get different numerical answers. It strikes you that they all give an answer in some unit of length (meters, inches, etc.). Furthermore, all obtained their answers in the same manner, by comparing the piece of string to a standard (although different in each case) measuring device. The important point is the fact that they each had the same concept of length, namely that it is something obtained by measurement.

Length is defined by the operation of making a measurement. This is what is meant by an **operational definition**. Does an operational definition tell us what length "really" is? No, it only tells us how to measure it, and for physical science that is all that is necessary. In physics, any concept which is not ultimately defined by operations is also meaningless, and any question which cannot be answered by making a measurement (even if we don't have the technical ability at present to perform the task) is meaningless. While a question like "Does a table cease to exist when not being observed?" may be an interesting topic of discussion for a philosophy class, it is meaningless for physics.

Four Fundamental Physical Quantities

All the terms which are used in science must be defined carefully in order to avoid possible confusion and misunderstanding. As you will see, once a term has been defined, it is used to define still other terms, and in this manner the scientific jargon is formulated. Somewhere, however, there must be a starting point; that is, there will have to be several basic quantities which are used to start the hierarchy of definitions. For the physics which we will use, four fundamental quantities are needed. Although the choice of which physical quantities are taken as fundamental is somewhat arbitrary, the four which are usually chosen are length, time, mass, and electric charge. Since these quantities are used to define all the other physical quantities which will follow, they can only be defined operationally; that is, we cannot define length independent of measurement since length is "defined" by the act of making a measurement. Likewise, time, mass, and charge have meaning in physics only in terms of their measurement.

Length (ℓ), the first fundamental quantity, can be defined as the spatial distance between two points. The measurement is accomplished by comparing the unknown distance to some standard length. The international standard of length is the **standard meter** (m), defined as the distance between two lines engraved on gold plugs near the ends of a certain platinum-iridium bar kept at the International Bureau of Weights and Measures at Sevres, near Paris. (Since 1960 the standard has been defined by comparison to one particular color of light emitted by a certain isotope of krypton.) Convenient units of length for many acoustical measurements are the centimeter and the millimeter, defined as one one-hundredth and one one-thousandth of a meter, respectively.

Time (t) is the second fundamental quantity to be considered. As with length, we have an intuitive feeling for time even though we would probably have great difficulty in defining it. We may think of time in terms of the duration of events, and of the measurement of time as the comparing of the duration of an unknown event with that of a standard event. If we define a solar day as the time it takes the earth to make one complete rotation on its axis, we have a standard event which can be used to measure time. Careful measurement, however, has shown that the solar day varies slightly in time during the course of the year; so a mean solar day was defined. This is just the "time length" of a solar day averaged over a year. The mean solar day can then be subdivided into twenty-four hours, each hour into sixty minutes, and each minute into sixty seconds. Thus there are 24 × 60 × 60 = 86,400 seconds in a mean solar day, and the **standard second** (s) can be defined as 1/86,400 of a mean solar day. (Recent astronomical measurements have shown that the rotating earth can no longer be regarded as a satisfactory clock because there are slight irregularities in its rotation. Just as the meter can be defined more accurately in terms of a certain color of light, time can be measured more accurately by using the vibration properties of atoms or molecules.)

Mass (m) is the third fundamental quantity to be considered. The concept of mass is not as intuitively obvious as length and time, but we can try to develop some feeling for this concept through the following experiment. Suppose you are walking barefoot along a country lane and you see an old tin can. You kick the can as hard as you are able and it sails down the street. Farther down the lane you encounter a similar tin can, which you also kick. This time, however, someone has previously filled the can with lead. When you kick the can it barely moves, but you experience considerable pain in your big toe. One way to describe the difference between the two cans is to say that the second can (because of all the lead) had a greater mass than the first can. The mass of an object is related to the difficulty of changing the state of motion of the object. To determine the mass of any object it is only necessary to compare the unknown to a standard mass by use of a balance. The international standard of mass is the **standard kilogram** (kg), standardized as the mass of a particular platinum-iridium cylinder kept with the standard meter at the International Bureau of Weights and Measures. The gram, defined as one one-thousandth of a kilogram, is often used to express mass.

All objects are composed of particles possessing **electric charge** (q), the fourth fundamental quantity. Friction can "rub off" some of these particles so that the effect of isolated charges can be demonstrated, even if the charge itself is not directly observable. For example, an ordinary hard rubber comb, when drawn through your hair, becomes charged and will attract small pieces of paper. Experiments show that when two materials are charged by being rubbed together, the resulting charge of each material is fundamentally different from the other. In order to distinguish between these two types of charge, one is labeled negative as is the charge found on the rubber comb, while the other is called positive. (These labels are arbitrary, albeit conventional. The charges could just as easily have been called red and green.) It is now known that electrons all possess an identical negative charge, hence a negatively charged comb is a comb with an excess of electrons. Like the other fundamental quantities, charge is defined operationally by comparison to a standard. A **coulomb** (C) of electric charge is defined as the total charge of 6.24×10^{18} electrons, that is, about six and a quarter billion billion electrons.

Derived Quantities

The following quantities are defined in terms of the four fundamental quantities. The dimensions associated with each quantity signify the manner in which it was obtained from the fundamental quantities.

When an object changes position, it is said to undergo a **displacement** and the magnitude of that displacement is the length or distance it has moved. Distance alone, however, is not sufficient to determine displacement. Displacement must also include the direction in which the object is moved. In many practical applications displacement is measured from some natural or rest position, as, for example, when considering the displacement of a point on a guitar string. To completely specify the displacement, we must specify not only the distance the string has moved, but also the direction of motion from the rest position.

If we enclose a rectangular region on a flat table top by drawing two pairs of parallel, but mutually perpendicular, lines we can define the **surface area** (S) of the enclosed region as the product of the length of two of the perpendicular sides. The unit used to measure area is the square meter, or something comparable. (Area can be defined for non-rectangular regions on a flat surface, but the definition becomes more complicated.) If we enclose a "rectangular" space with three sets of mutually perpendicular walls, we can define the **volume** (V) of the space as the product of the area of one of the walls and the distance to its parallel wall. Volume has the dimensions of area times length or length cubed and the unit used to measure volume is the cubic meter, or something comparable.

Often we are more concerned with how fast something is moving than with the actual distance traveled. While you are driving on an interstate highway you may be constantly monitoring your speedometer so that you don't drive faster than 55 mph, while little thought is given to the total distance traveled. The average rate at which distance is traveled is defined as the **average speed**. That is,

$$\text{average speed} = \frac{\text{distance traveled}}{\text{elapsed time}}$$

As an example, suppose it takes you one-half hour to travel from one town to another town twenty kilometers away. Your average speed for the trip is calculated as follows:

$$\frac{20 \text{ kilometers}}{0.5 \text{ hour}} = 40 \text{ km/hr}$$

The average speed, however, does not specify how the speed varied during your trip. For instance, part of the trip was on a freeway where you traveled at 60 km/hr and part of the trip was through town where your speed varied from 25 km/hr on some roads to zero km/hr while you waited for a traffic light. To take into account these variations of speed, another concept, that of instantaneous speed, has been developed. **Instantaneous speed** is just the speed at any instant of time. This is the speed which shows on your speedometer at any given moment.

Although the words are often used interchangeably, in science we draw an important distinction between speed and velocity. **Average velocity** is defined as the time rate of change of displacement. That is,

$$\text{average velocity} = \frac{\text{change of displacement}}{\text{elapsed time}}$$

Velocity, then, includes speed (the magnitude of the velocity) and direction. To accurately state a velocity you must give the speed (20 km/hr) and the direction (due north). As an example, suppose you travel 20 kilometers due west in half an hour. Your average velocity would be

$$\frac{20 \text{ km due west}}{0.5 \text{ hour}} = 40 \text{ km/hr due west}$$

Suppose you then return to your starting point by a different route, but your average speed is again 40 km/hr. Now, your average speed for the entire trip is 40 km/hr; but your average velocity for the trip is zero km/hr because the net displacement is zero, since you returned to your starting point. **Instantaneous velocity** (v) is the velocity at any instant of time, just as instantaneous speed is the speed at any instant of time. In discussing acoustical systems, instantaneous velocity is more often of interest than is average velocity.

An object experiences an **acceleration** (a) when its velocity changes, i.e., it speeds up or slows down. More precisely,

$$\text{acceleration} = \frac{\text{change in velocity}}{\text{elapsed time}}$$

Since velocity includes both speed and direction, an acceleration can be obtained by either a change of speed or a change of direction, or a change of both speed and direction.

Laws of Motion

Almost everyone has, while standing at rest, been pushed or shoved, so we have an intuitive grasp of the physical concept of **force** (F). If you push on a boulder you exert a force on the boulder and may cause it to move from its state of rest. Over three hundred years ago Isaac Newton wrestled with the concepts of force and acceleration to formulate his three laws of motion. These important laws give us a physical definition of force.

The **first law** states that a body in a state of uniform motion (or at rest) will remain in that state of uniform motion (or rest) unless acted upon by an outside force. The first law is a qualitative definition of force, for while it does not tell us what a force is, it does tell us when a force is not acting. If we observe an object at rest, we know there is no net force acting on the object. If you take this book out to the edge of the solar system and heave it as hard as you can off into space, it will travel forever in a straight line at a constant speed (unless, of course, it is influenced by another object).

While Newton's first law tells us when a force is or is not acting Newton's **second law** tells us what happens when a force acts. When an unbalanced force acts on an object with mass, the object is accelerated. The acceleration is directly proportional to the magnitude of the force and in the same direction as the force. When a force acts there is a change in the motion of an object. The force which caused the change is proportional to both the mass being accelerated and the resultant acceleration. We can represent this information symbolically as

force = mass × acceleration or F = ma.

Newton's second law then is a quantitative definition of force. The law tells us that force is the product of mass and acceleration. The unit of force is a **newton** (N), which is defined as the force which causes a mass of 1.0 kg to accelerate at a rate of 1.0 m/s^2.

As an example of the second law, suppose that you wish to accelerate two different objects—a bicycle and an automobile—at the same rate. You must exert a much larger force on the automobile because of its much larger mass. When the small mass of the bicycle is multiplied by the acceleration we see that a small force is required. On the other hand, multiplying the large mass of the automobile by the acceleration results in a much larger force.

Newton's **third law** is probably the best known and yet least understood of the three laws. For every action (force) there is an equal and opposite reaction (force). The law tells us that forces do not occur singly in nature, but in pairs. When two objects interact and object A exerts a force on object B, then object B exerts an equal and opposite force on object A. However, only one force in the pair acts on each object, and it is this force that we must consider when applying the second law of motion to determine whether an object will be accelerated. For instance, when a violin string is bowed, the bow exerts a force on the string and the string exerts an equal and opposite force on the bow; but for determining the acceleration of the string, we consider only the force exerted on the string by the bow and not the force exerted on the bow by the string.

Additional Derived Quantities

It is easy to confuse the concepts of mass and weight, and people often use these words as synonyms. In science, however, weight and mass are defined differently, even though they are related concepts. Mass as defined earlier is a measure of the inertia or sluggishness a body exhibits when any attempts are made to change its motion in any way. **Weight** (w) is the force due to gravitational attraction acting on a mass. Since your mass depends on the quantity of matter of which you are composed, it does not change if you visit the moon or Mars. Your weight on Mars, however, would be only 40% of your weight on earth, due to the smaller gravitational attraction on Mars. Since weight is a force, it is measured in newtons (or pounds), just like any other force.

Which weighs more, a pound of lead or a pound of feathers? Don't be misled by confusing weight and density. A pound of lead and a pound of feathers have the same weight (and the same mass), but the pound of feathers will certainly occupy a much greater volume. Lead is a material in which massive atoms are very com-

11

pactly grouped, while the less massive molecules which make up feathers are not very compactly grouped. The relative molecular masses and "degree of compactness" of different materials is expressed by the concept of density. The **density** (D) of a material is the mass (e.g., grams) occupied by a standard volume (e.g., cm³) of the material:

$$\text{density} = \frac{\text{mass}}{\text{volume}}$$

The density of water is 1.0 gm/cm³, while lead has a density of 11.3 gm/cm³; air has the very low density of 0.0012 gm/cm³.

A balloon filled with air will fly around a room when it is released. A metal can which has the air pumped out of it will be crushed by invisible forces. To better understand the interaction between gases and container walls, the concept of pressure is useful. **Pressure** (p) is defined as the force per unit area acting on the surface of an object. Thus,

$$\text{pressure} = \frac{\text{force}}{\text{area}}$$

The unit for measuring pressure is the **pascal** (Pa) defined as one N/m². As an example of the difference between pressure and force, consider a brick resting on its broad side on a table top. The pressure exerted on the table by the brick is the weight of the brick divided by the area touching the table. If the brick is turned on edge, the force (weight of the brick) remains the same, but because a smaller area is in contact with the table, the pressure is increased. If we up-end the brick so its smallest face rests on the table, the pressure is even greater; the weight, of course, remains unchanged.

To understand why a metal can is crushed when the air is pumped out of it, consider that we are living at the bottom of an ocean of air. In much the same way as the water behind a dam exerts a pressure on the dam, the atmosphere above us exerts its pressure (called atmospheric pressure) on all of us. Although the pressure is considerable (10^5 Pa at sea level), it usually goes unnoticed since we are immersed in it. When the pressure is removed (as for the evacuated can), or even imbalanced slightly the effects are noticed immediately. Who has not experienced an uncomfortable feeling in the ears when driving down a mountain, or when changing altitude suddenly in an airplane? The slight change of atmospheric pressure produces an unbalanced force on the eardrum which can only be relieved by equalizing the pressure.

Derived Electrical Quantities

Electric charge is an intrinsic property of electrons, and as electrons move, their charge moves with them. We can picture a stream of electrons and their associated electric charges moving from one location to another. A flow of electric charge is called an **electric current** (i) which is defined as the amount of charge passing a given point per unit time, that is

$$\text{current} = \frac{\text{charge}}{\text{time}}$$

The unit of current is the **ampere** (A) defined as a coulomb per second. To maintain a continuous current requires a continuous supply of electrons on one end of a wire and a place for the electrons to go on the other end. These two requirements can be met by a battery. The flow of electrons from one end of the battery through the wire to the other terminal of the battery is analogous to the flow of water through a pipe. To obtain a continuous flow of water through a pipe, a pump and a water supply must be provided at one end, and an opening for the water to escape at the other end. The resultant flow is measured in cubic meters per second just as electric current is measured in coulombs per second. The total water available is the number of cubic meters in the reservoir, just as the total charge available would be the number of coulombs.

A pressure difference (supplied by a pump) must exist between two points in a pipe if water is to flow between them. In an analogous way, an **electrical potential difference** (V) supplied by a battery must exist between two points in a wire if electrical current is to flow between them. The unit of potential difference is the **volt** (V) which is somewhat analogous to the pascal, the unit of pressure. We will see in the next chapter that these analogies can be very useful for developing relationships among electrical quantities when we know the relationships among mechanical quantities—and vice versa.

Just as two objects with mass exert gravitational forces on each other, two objects with electrical charges exert electrical forces on each other. When the two objects have like charges they repel each other; when they carry opposite charges they attract each other. The deflection of an electron stream in a TV set or a laboratory oscilloscope provides a practical example of electrical forces.

In addition to electric forces between charges at rest there are magnetic forces between moving charges. We will not define magnetism but will state some qualitative features about magnetic forces. When an electric current exists in a straight wire or in a loop of wire, a magnetic field is produced. A magnetic field is capable of exerting a magnetic force on any moving charge in its vicinity. (A permanent magnet may be thought of as due to many tiny molecular current loops that have all been oriented in the same direction to produce an effective "internal" current loop in the magnet.)

When a wire coil carrying a current is placed in a magnetic field it may be caused to move because of the magnetic force on its moving electrons; the greater the current the larger the force. Conversely, when a wire coil in a magnetic field is moved, current may be produced because of the magnetic force on the coil's electrons; the larger the motion, the greater the current. Electric motors, dynamic loudspeakers, and dynamic microphones are based on this principle.

When a wire loop is placed in a changing magnetic field a current is induced in the coil. This is the principle on which an electric generator is based.

Exercises

1. Determine which of the following statements and questions are operationally meaningless: Are there natural laws which man can never hope to discover? All matter and space are permeated with an undetectable substance. The entire universe is expanding so that everything in it has all its linear dimensions double every month. Is the sensation which I experience when I see green the same as that which you experience when you see green?

2. Describe experiments which would demonstrate that there are two and only two types of electrical charge.

3. According to Newton's first law, once an object is started in motion it will continue in a straight line motion forever if no forces act on it. Explain, then, why objects we start in motion (by sliding, or throwing, etc.) always come to a stop.

4. A 50 kg boy and a 100 kg man stand in identical carts on a level surface. Each is pushed with equal forces. Explain what happens and why.

5. Explain how a moose can pull a cart in one direction when the force of the cart on the moose is in the opposite direction. Be sure to consider all force pairs acting, but gravity may be ignored.

6. Suppose you are in an orbiting satellite where everything is "weightless." If you have two identical cans, one empty and one filled with lead, what experiments could you perform to determine which has the greater mass?

7. Compute the average velocity in each of the following cases. An auto moves 60 m west in 100 s. A bicycle moves 1 cm in 1 s. A person walking moves -0.1 cm in 10^{-3} s.

8. Compute the average acceleration for each of the following cases. The speed of an auto changes 6000 cm/s in 10 s. The speed of a bicycle changes -1.0 cm/s in 10^{-2} s.

9. Instantaneous velocity can be determined by taking a very small change in displacement (respresented as Δd) and dividing it by the very small elapsed time (Δt) to give $v = \Delta d/\Delta t$. If a vibrating string moves a distance of 0.001 cm in 0.0001 s, calculate the instantaneous velocity. (The Δ notation is used to indicate a small change of a quantity such as time, distance, etc.)

10. The relationship among instantaneous velocity, displacement, and time can be written $v = \Delta d/\Delta t$. If $\Delta d = 0.50$ cm and $\Delta t = 0.01$ s, find the instantaneous velocity. If $v = 100$ cm/s and $\Delta t = .05$ s, find Δd. If $v = 100$ cm/s and $\Delta d = 0.20$ cm, find Δt.

11. The instantaneous acceleration of an object can be written $a = \Delta v/\Delta t$. When $\Delta v = 0.10$ cm/s and $\Delta t = 0.05$ s, find a. When $a = 6.0$ cm/s^2 and $\Delta t = 0.05$ s, find Δv. When $a = 5.0$ cm/s^2 and $\Delta v = 0.10$ cm/s, find Δt.

12. Imagine that you are driving your car on a perfectly straight highway. Calculate your acceleration for each of the following situations: You increase your speed from 20 km/hr to 40 km/hr in 10 s. You decrease your speed from 40 km/hr to 20 km/hr in 5 s. You remain at a constant speed of 30 km/hr for 20 s.

13. Explain (in terms of electrons) why charging objects by rubbing together (static electricity) always yields equal amounts of positive and negative charge.

14. Distinguish between 10 N and 10 N/m^2.

15. Explain why spiked heels will puncture a hard floor while a normal heel will not. As an example, compare the pressure exerted by a 50 kg woman wearing 1 cm^2 heels to that of a 100 kg man wearing 50 cm^2 heels.

16. How would the air density at the earth's surface compare to the air density at the bottom of a deep mine? On top of a high mountain? Explain.

Demonstrations

1. Atmospheric pressure—Watch a can get crushed by the atmosphere as it is evacuated. Try to pull apart the two halves of an evacuated spherical shell.

Further Reading

Ashford, Chapter 3
Backus, Chapters 1, 15
Ballif and Dibble, Chapters 2, 3
Krauskopf and Beiser, Chapters 1, 2
Rossing, Chapters 1, 18
Sears, Zemansky, and Young, Chapter 1

Audiovisual

1. **Laws of Motion** (12 min, color, 1952, EBE)
2. **Newton's First and Second Laws** (Loop 80-273, EFL)
3. **Newton's Third Law** (Loop 80-274, EFL)
4. **Electricity** (13 min, color, 1971, WD)
5. **Force on a Current in a Magnetic Field** (Loop 80-169, EFL)
6. **Faraday's Law of Induction** (Loop 80-4179, EFL)

3. Work, Energy, Power, and Intensity

A detailed application of the concepts discussed in Chapter 2 may be impossible or at best too laborious to be useful. The concepts of work, energy, and power may in some ways be regarded as providing an alternate description for most things we find of interest in the physical aspects of music, speech, and high fidelity. These new quantities can be defined in terms of the previously defined quantities, but in many ways they are handier to use.

Work

In our everyday life we often evaluate the difficulty of a task in terms of the work involved. If a friend lifts a heavy object for you, he may complain that he is working too hard. If you push with all your strength against an immovable wall you would probably claim that you are working equally hard. An observing scientist, although sympathetic to the effort you are exerting, would contend that your friend is working, and you are not. The scientific definition of work takes into account the force exerted and the distance an object moves when the force is applied. The force must cause displacement of the object or the work done is zero. The **work** (E) done by a force is the product of the force times the displacement (in the direction of the force) caused by the force. Symbolically,

work = force × displacement (in direction of force).

Since you exerted a force on a wall which did not move, no work was performed. When your friend lifted the heavy object, however, the force he exerted caused the object to be displaced upward and work was done. If we double the load that your friend has to lift, we double the work he must do. Likewise, if he must lift his load twice as high, the work is doubled. The unit of measurement in which work is expressed is the product of distance and force units. In the metric system, the unit of work is the **joule** (J) defined as a newton-meter. One joule (rhymes with pool) of work results when a force of one newton displaces an object a distance of one meter in the direction of the force.

Mechanical Energy

"Energy" is perhaps the most fundamental unifying concept of all science. Yet, despite the prevalence of the term, it is an abstract concept which cannot be simply defined. When you start a car, you use "human energy" to turn the ignition key. The ignition key engages the battery which converts chemical energy to electrical energy. The electrical energy from the battery is used in the starter to produce mechanical energy which starts the engine. The engine then converts chemical energy from fuel into heat energy and then into mechanical energy which is used to propel the car. If it is dark you turn on the lights, which are converting electrical energy from the battery into light energy. If an animal runs in front of your car you honk the horn (converting electrical energy into sound energy) and step on the brakes (converting mechanical energy into heat energy). The basic points of this hypothetical exercise are that (1) energy in some form is involved in all our activities; (2) energy can appear in many different forms; and (3) energy can be changed from one form to another. We will begin our discussion by considering mechanical energy, one of the most recognizable forms of energy.

Consider a simple frictionless pendulum (or swing) as shown in Figure 3.1. If we do work by applying a force to the pendulum we can move it from its rest position B to position A. If the pendulum is now released from position A it will gain speed until it reaches B and then lose speed until it reaches C and then move to B and to A. If the pendulum is truly frictionless the motion will repeat itself indefinitely, reaching the same height each time at A and C and having the same speed at B. There is an implication here that something is conserved, but the something can be neither speed nor height because both change throughout the motion. We define the conserved quantity as **total mechanical energy** which consists of energy due to motion and energy due to position. The energy due to motion is termed **kinetic energy**, expressed as $KE = 1/2\ mv^2$, where m is the pendulum mass and v is its velocity magnitude (speed) at any instant. The energy due to position is termed **gravitational potential energy** expressed as $GPE = wh$, where w is the weight of the pendulum and h its height above some rest position. The total mechanical energy of the pendulum is equal to the sum of KE and GPE and is conserved, or in other words, it is constant. As KE increases GPE decreases (and vice versa) in such a way that their sum is always the same. As the pendulum goes from position A through position B to position C and back again, there is a continual transfer of energy from all potential (at A and C because the pendulum is not moving there) to all kinetic (at B where the speed is greatest). At the in-between points the energy is a combination of potential and kinetic, but at any point in the motion the sum of the potential and kinetic energy is exactly the same as at any other point.

Potential energy may also be associated with a stretched rubber band, a compressed spring, or a stret-

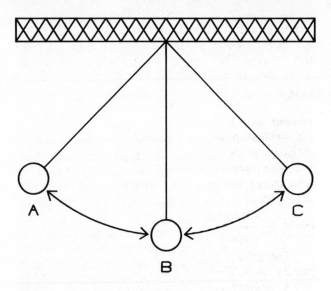

Figure 3.1. Vibrating pendulum.

ched drumhead. Potential energy associated with stretching or compressing objects is more relevant to our studies than is gravitational potential energy. For example, the **potential energy** of a spring is defined as PE = 1/2 sd² where s is a "stiffness" associated with the spring and d is the displacement magnitude of the spring from some rest position.

The greater the amount of work done to raise a pendulum or to compress a spring the greater its potential energy. We then surmise that energy must be expressed in the same units as work, namely the joule.

Whenever a given physical quantity remains constant in a changing situation we generalize the result as a law. In this case the law states that the total mechanical energy (the sum of kinetic and potential energies) of a system remains constant when no frictional forces are present. In any real system there will always be some friction present and so this law is only an approximation of reality. However, it is still useful for analyzing vibrating systems, so long as its limitations are not forgotten.

Other Forms of Energy

When friction is present, what becomes of mechanical energy? Consider the kinetic energy produced by a girl scout rubbing two sticks together. Where does this energy go? It turns into a different form of energy, one which we observe as heat in the sticks. Whenever mechanical energy disappears due to friction, **heat energy** appears. If we accurately measured the total heat energy produced and the total mechanical energy that disappears in a given situation, we would discover that the two amounts were equal; the mechanical energy lost equals the heat energy gained. Likewise, heat energy can be changed to mechanical energy, as in a steam locomotive. But heat energy alone does not account for all possible energy transformations that take place. Suppose that you hammer a nail into a piece of wood. Some

of the mechanical energy of the swinging hammer is converted into heat (the nail gets warm), but what becomes of the remainder of the hammer's kinetic energy? It takes energy to deform the wood when the nail is driven in, and a sound is produced every time the hammer collides with the nail. Deformation and sound are other forms of energy which may appear when mechanical energy disappears.

Further experiments involving the transformation of say, mechanical energy to electrical energy (by means of a generator) or electrical energy to mechanical energy (by a motor) convince us that whenever one form of energy "disappears," equivalent amounts of some other energy forms "appear" in its place. The study of numerous transformations of energy from one form into another form has led to one of the great generalizations of science, the **law of conservation of energy**: energy cannot be created nor destroyed, but it may be transformed from one form into another; the total amount of energy, however, never changes. The word "conservation" as used here does not mean to save for the future, but rather signifies that total energy always remains constant. Energy never appears or disappears, it is merely transformed.

The concept of energy conservation is very important in speech production, musical systems, control of the energy level in an auditorium, and so on. Speech production and production of musical sounds require the performer to supply energy to the vocal mechanism or to the musical instrument. Part of the energy supplied by the performer is converted into acoustical energy that is emitted by the instrument as useful vocal or musical energy; part of the energy is lost internally due to frictional heating. In the design of listening rooms, we are often concerned with purposely converting sound to other forms of energy in order to reduce loudness, or so that the sound does not reverberate too long.

Bernoulli's Law

A force important in speech production as well as sound production in musical instruments can be derived from conservation of energy in a moving fluid. When a fluid moves slowly its pressure is high and when it moves rapidly its pressure is low. This relationship accounts for the lift of airplane wings and the loss of roofs from houses in tornadoes. It is also the basis for a driving mechanism of the vocal folds, the lips, and mechanical reeds.

Consider the law of conservation of energy as it applies to water flowing through a pipe. When the water arrives at a constriction in the pipe, as shown in Figure 3.2, it must move faster to avoid piling up. (This effect would be analogous to three lanes of bumper-to-bumper traffic moving at 20 km/hr and merging into one lane of traffic. If the traffic is to keep moving, so that a delay is avoided, each car must speed up to 60 km/hr at the constriction. If this unsafe method of merging traffic could be carried out, it would avoid further congestion.)

16

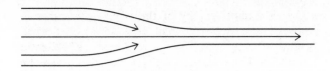

Figure 3.2.   Water flow in a constricted pipe.

When a flowing fluid reaches a smooth constriction, it increases its speed and the smooth flow continues with no congestion. But when the fluid gains speed, it also gains kinetic energy. Where does this energy come from? In the eighteenth century a Swiss scientist, Daniel Bernoulli, reasoned that the kinetic energy was acquired at the expense of decreased potential energy. One form of Bernoulli's Law states that where kinetic energy in a flowing fluid is large, potential energy is small (and vice versa) so that the total energy remains constant. Figure 3.3 illustrates this law for a pipe with a constriction. Notice that the total energy in the pipe remains constant, but as the pipe is constricted kinetic energy increases at the expense of potential energy. As the pipe expands, the fluid moves more slowly, and potential energy is gained at the expense of kinetic energy.

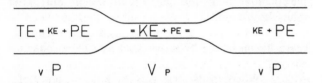

Figure 3.3.   Illustration of Bernouli's Law for fluid flow in a constricted pipe. Kinetic energy and velocity are large in the constriction. Potential energy and pressure are large in the unconstricted pipe.

A more common form of Bernoulli's Law states that where the speed of a moving fluid is large the fluid pressure is small and vice versa, as indicated by the labels at the bottom of Figure 3.3. [Kinetic energy of a fluid is defined in terms of the square of the velocity of the fluid. Potential energy of a fluid must somehow be defined in terms of a compressing or stretching of the fluid in a manner analogous to that of a spring. Potential energy in a fluid can be defined in terms of the square of its displacement amplitude times a "fluid stiffness," but it is more commonly defined in terms of the square of the pressure. Usually, when considering energy in fluids we talk about kinetic energy density as mass per unit volume times velocity squared and potential energy density as pressure squared times a fluid constant. The crucial relationships for our purposes are the dependence of KE on velocity squared and that of PE on pressure squared.]

Power

In our discussion of energy transformations we did not discuss the important aspect of how rapidly energy can be transformed. One gallon of gasoline contains a certain amount of chemical energy, but the power we are able to produce will depend on how rapidly we use this energy. One gallon of gasoline in a power lawnmower yields 10 horsepower for 3 hours while the same gallon of gasoline in an automobile may produce 60 horsepower for one-half hour. If the same gallon of gasoline were burned in a jet plane it could produce 100,000 horsepower for 1 second! The difference is in the power of the engine, not in the energy content of the fuel. The concept of **power** (P), then, includes both the total energy expended and the time involved in expending it. More precisely,

$$power = \frac{expended\ energy}{elapsed\ time}$$

Energy is expressed in the same units as work, namely joules. The unit of power is the **watt** (W) defined as a joule (energy) per second (time). This unit may be used for either electrical power or mechanical power. (Power may also be expressed in the unit of a horsepower but such a unit is not convenient for our uses.)

[Just as energy is expressed in units of joules equal to newtons times meters, power may be expressed in units of watts equal to newtons times meters divided by seconds: $W = Nm/s$. But meters divided by seconds is a unit for velocity so power may be expressed in units of force times velocity. Acoustical power can be expressed as pressure times "volume velocity." Electrical power can be expressed as voltage times current. The unit of power is the watt in all of these cases.]

As an illustration of the difference between energy and power, consider the following example. Suppose you must exert a force of 400 newtons to climb 10 meters up to the fourth floor of a certain building. One time you dash up the stairs in 10 seconds. Another day you walk up the stairs in 100 seconds. After one particularly rough day you crawl up the stairs in 1000 seconds. What was different in each of these cases? The work done by you was the same in each case because your weight and the distance traveled was identical. The work done (or energy supplied) by you would be $E = F \times d = 400\,N \times 10\,m = 4000\,J$. But the time it took you to do this amount of work was different in each case, so the power you had to supply was also different. The power exerted in the first case was $P = E/t = 4000\,J/10\,s = 400\,J/s = 400\,W$. For the second case, $P = 4000\,J/100\,s = 40\,W$, and for the third case, $P = 4000\,J/1000\,s = 4\,W$. Hence, the longer it took you to travel up the stairs, the less the power required. The total work or energy in each case was the same because, even though less power was supplied, it was being applied for a longer time.

The concept of power has many important applications in the science of acoustics. When we discuss hearing acuity, we will be concerned with the sound power perceived by an ear. When noise problems are discussed, they are rated by their over-all sound power. Musical instruments are judged to be loud or soft based on their power output, and hi-fi amplifiers are rated by their power output in watts.

The wattage rating of electric light bulbs is a power rating—a 200-watt bulb requires twice as much power, or twice as much energy per second, as a 100-watt bulb. The power we encounter when dealing with sounds is very much smaller than typical electrical powers—which testifies to the remarkable sensitivity of the ear. But even so, the same principles apply and can be used to rate in watts the sound output of various sources. Consider, for example, a 100-watt light bulb. While perfectly adequate to provide light for an average-sized room, it is not a very powerful source of light. Yet those familiar with high fidelity equipment will recognize that a 100-watt amplifier would rock the entire house! As a matter of fact, a 10-watt amplifier is capable of producing more sound than the average housewife will tolerate.

Intensity

How are we to account for the fact that as we move farther from a sound source such as a loudspeaker, it sounds less loud? Even though a loudspeaker produces constant sound power we perceive the sound as less loud when we move away from the loudspeaker—especially out doors where there are no walls to reflect the sound. A similar effect occurs with a 100 watt light. The power output of the light bulb remains the same but somehow we perceive it as less bright as we move away.

Imagine 100 watts of sound power spreading out in all directions in an ever enlarging sphere. Although the total sound power remains constant on the surface of the sphere, it is spread over a larger area as the size of the sphere increases. The **intensity** (I) of a spherical sound wave is the power per unit area; that is, the total sound power divided by the surface area of the sphere. The unit of intensity is watt per square meter (W/m²). Figure 3.4 is a representation of sound power being spread over larger and larger areas as it travels away from the source. As we move away from the source, the sound becomes less intense (i.e., the same power spread over a larger area) and is perceived as softer. In Figure 3.4 we can see that if a certain amount of sound power falls on a 1 m² surface located at a distance of 10 m from the source, the same sound power would be spread over $2 \times 2 = 4$ m² at a distance of 20 m and $3 \times 3 = 9$ m² at a distance of 30 m. The intensity, then, decreases as the inverse square of the distance from the source; when the distance from the source is doubled, the sound intensity falls to one- fourth its previous value. (This assumes of course that there are no reflecting walls and that the sound spreads out uniformly in all directions.)

In addition to the decrease of sound intensity due to "spreading out," sound energy and intensity may be decreased because of the absorption of sound energy when a sound wave contacts an object, and because of the transformation of sound energy into heat energy as a sound wave moves through air. These causes of energy loss are important for rooms and will be discussed in

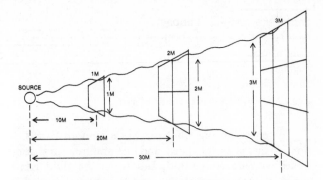

Figure 3.4. Diagram illustrating the decrease of intensity with increasing distance from the source. The sound power is spread over a larger area at greater distance from the source.

Part IV. The absorption of sound energy at the walls of wind instruments can be very significant, often consuming in excess of 90 percent of the energy supplied by a player. Energy is also lost from musical instruments by radiation from strings, membranes, or tone holes.

Exercises

1. A pumpkin held above the ground has potential energy. If the pumpkin is released, what becomes of the potential energy just before it hits the ground? After it hits the ground?

2. Why does a swinging pendulum eventually cease its motion? What becomes of the mechanical energy?

3. Where in its motion is the kinetic energy of a pendulum bob at a miminum? Where is kinetic energy at a maximum? When the potential energy is half of its maximum value, what is the kinetic energy?

4. Why does your car tend to lurch toward the left when you pass an oncoming truck on the highway? As your relative speeds increase, will the effect increase or decrease?

5. When you drive a car along a road at a constant speed, the engine is continuously burning fuel and doing work on the car, and yet the car does not gain energy. Explain.

6. When you raise a book above a table you do work on the book. When the book is at rest one meter above the table top, in what form is the energy and from where did it come? When you drop the book, in what form is the energy the instant before the book hits the table? When the book hits the table and comes to rest, what happens to the energy?

7. Why do automobile brakes get hot when stopping the car?

8. Tell for each of the following examples what type of energy the electrical energy is being changed to: a light bulb, an electric motor, an oven.

9. Would it be correct to state that all changes in the physical world involve energy transformation of some kind? Explain.

10. Would it be possible to hurl a snowball against a wall at such speed that the snowball would

completely melt upon impact? Explain.

11. A device has been proposed which has a grinding wheel hooked to an electric motor which also runs a generator, which in turn generates the electricity to turn the motor. Without knowing much about generators and motors, why would you conclude that such a device will not work?

12. Why does an electric motor require more current when it is started than when it is running continuously?

13. An oboe is found to be only 3% efficient in converting the player's input energy into useful acoustic energy that is radiated as sound. What happens to the rest of the energy? Where is it lost?

14. Two excellent musicians (an oboist and a trumpeter), with comparable lung capacities, have a contest to see who can blow their instrument for the longer time. Since both performers are able to sustain a continuous tone for the same time period, the result of the contest is a tie. However, people who listen to the sound output observe that the trumpet tone is consistently louder than the oboe tone. Does this observation offer any clues as to the relative efficiencies of the two instruments? Explain.

Demonstrations

1. Energy conservation—Make a pendulum by suspending a large mass on a light string. Set it into motion. Observe the time required for the motion to cease. What becomes of the energy?

2. Repeat Demonstration 1 for a mass on a spring.

3. As a demonstration of Bernoulli's Law, place a ping-pong ball inside a funnel through which air is streaming upward through the narrow end. The ball will not be blown up and out of the funnel. Now invert the funnel and the ball will be held in position rather than blown away by the air. The air flow around the ball is constricted and the pressure is consequently less than the atmospheric pressure below the ball. The ball is actually pushed into the region of lower pressure above it by the greater pressure below it, so that it remains supported in midair.

4. The Bernoulli force can be demonstrated by placing a small card (3 × 3) on a table. Place a thread spool on end and centered on the card. Use a tack through the center of the card to keep it centered on the spool. Now, as you blow through the spool, lift it away from the table and the card will come too.

Further Reading
Ashford, Chapter 3
Backus, Chapters 1, 15
Ballif and Dibble, Chapter 6
Krauskopf and Beiser, Chapter 4
Rossing, Chapters 1, 15
Sears, Zemansky and Young, Chapter 6

Audiovisual
1. **Energy and Power** (30 min, color, 1957, EBE)
2. **Energy Conversion** (Loop No. 80-3437, EFL)
3. **Conservation of Energy** (Loop No. 80-276, EFL)

4. Properties of Simple Vibrators

What do a wagging dog's tail, a swinging clock pendulum, and a sounding trumpet have in common? Vibration is the common element and the study of vibration will engage our attention throughout the remainder of this book, for all sounds ultimately result from and consist of mechanical vibrations. At first we will consider only **periodic motion**, motion that repeats itself over and over again in equal time intervals. Later, as our studies lead us to more and more complex vibrating systems we will broaden our perspective to include non-repeating vibratory motions. Periodic motion may consist either of a simple repetitive pattern or a complex repetitive pattern. In this chapter we will consider the simplest of all periodic motions, simple harmonic motion.

Simple Harmonic Motion

Suppose we consider a swinging pendulum as an example of simple periodic motion. The pendulum hangs straight down and is "centered" when at rest. As it oscillates or swings, it moves from left to right and back again. If we label the center as the "rest position," then we may speak of the "displacement from the rest position" as the distance left or right from the center to the position of the pendulum at any given time. This information would be more useful, however, if it could be graphically represented. Suppose we attach a small paint brush to the bottom of the pendulum and pull a strip of paper beneath the oscillating paint brush so that a curve is traced on the paper as indicated in Figure 4.1. Examine the curve produced by the swinging pendulum. Distance along the paper strip represents time. The back and forth distribution of paint represents the left or right displacement of the pendulum from its rest position. Although this curve was produced for one particular motion, experimentation with any similar repetitive motion would produce a similar characteristic curve.

Any object which is capable of vibrating must have a rest position where it remains while not vibrating; such a position is called the **equilibrium position** because the sum of all forces acting on the object at this point is zero. Any disturbance of the object will cause it to move away from this position of equilibrium to some new position in which the forces on the object no longer add up to zero. The object then experiences a "restoring force" which pulls it back toward its equilibrium position. This restoring force may be due to gravity, to internal stiffness (elasticity) of a material, or to an externally applied tension. As the restoring force pulls the object back toward its rest position, the object eventual-

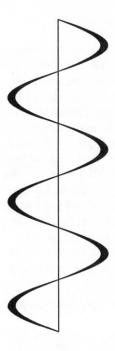

Figure 4.1. Idealized "paint brush curve" traced out on a moving strip of paper by a swinging pendulum.

ly returns to the rest position, but in doing so it acquires speed which causes it to overshoot and travel some distance to the other side of its equilibrium position before coming to rest. The object now experiences a new force which again pulls it back toward the rest position; again it overshoots the rest position. This back and forth motion continues to repeat itself. Motion of this general type is called an oscillation or vibration. It is characterized by the fact that the farther the object moves from the equilibrium position, the greater the restoring force which acts to pull the object back. If the restoring force is exactly proportional to the distance from the rest position, we have the special type of oscillation known as **simple harmonic motion** (abbreviated SHM).

Simple harmonic motion is characterized by five special attributes.

(1) The motion continually repeats itself in a definite interval of time, called the period; the **period** (T) is the amount of time required for one complete oscillation.

(2) The motion is symmetric about the equilibrium position. The maximum displacement to either side of the rest position is identical.

(3) The displacement curve traced out in time by an object undergoing SHM is a special curve called a **sinusoid**.

21

(4) The velocity of the object varies continuously throughout the oscillation. The velocity must be zero at the positions of extreme displacement where the object momentarily stops, whereas it passes through the rest position with maximum speed. The velocity varies in a symmetric, repetitive manner and can also be described by a sinusoidal curve.

(5) The acceleration of the object also varies sinusoidally.

As an illustration of SHM, consider a mass which is permanently attached to a spring. The spring is placed in a horizontal position on a frictionless table top (e.g., an air table), with one end fastened to the table. Figure 4.2 is a top view of this arrangement and shows that the mass is free to slide on the table. We can start the mass oscillating by first displacing it (which stretches the spring) directly away from the table support and then releasing the mass. (We are confining our attention to up and down vibration in one dimension. You can imagine other vibrations with the mass swinging back and forth like a pendulum or twisting the spring to produce torsional motion. In many practical situations we can describe a vibrator adequately in terms of one dimensional motion. We will follow this approach unless stated otherwise.)

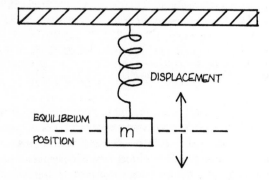

Figure 4.2. Mass and spring vibrator.

Suppose that we mount a movie camera directly above the oscillating mass of Figure 4.2 and film the motion at a rate of 10 frames per second. The movie film provides us with a detailed record of the location of the mass relative to its rest position at time intervals of 0.1 s. To obtain the displacement of the mass from its rest position at each time interval we need only measure the appropriate distance as recorded in each frame of the film. Some of these measured data are recorded in the displacement column of Table 4.1. The initial frame (labeled 0) was arbitrarily chosen from one of the frames where the mass was passing through its rest position. The displacement data from Table 4.1 can be used to construct the upper graph of Figure 4.3 which shows how displacement varies with time. The displacement has both positive and negative values because it is measured either "above" or "below" the rest position. After the data are plotted the individual points are connected by a smooth curve (the solid line in the graph).

The graph clearly shows how the sinusoidal curve is produced by plotting the displacement of the mass versus time.

Table 4.1.   Values of displacement and velocity for a mass and spring vibrator.

Frame	Time (s)	d (cm)	v (cm/s)
0	0	0.00	6.3
1	0.1	0.59	5.1
2	0.2	0.95	1.95
3	0.3	0.95	-1.95
4	0.4	0.59	-5.1
5	0.5	0.00	-6.3
6	0.6	-0.59	-5.1
7	0.7	-0.95	-1.95
8	0.8	-0.95	1.95
9	0.9	-0.59	5.1
10	1.0	0.00	6.3
11	1.1	0.59	5.1
12	1.2	0.95	1.95
13	1.3	0.95	-1.95
14	1.4	0.59	-5.1
15	1.5	0.00	-6.3
16	1.6	-0.59	-5.1
17	1.7	-0.95	-1.95
18	1.8	-0.95	1.95
19	1.9	-0.59	5.1
20	2.0	0.00	6.3

The column of velocities in Table 4.1 was computed from the frames of the movie film in order to obtain the instantaneous velocity of the mass. A positive velocity represents motion of the mass upward even if the mass is below the equilibrium position. A negative velocity represents downward motion of the mass. A plot of the velocity versus time is shown in the lower graph of Figure 4.3. You will note that the curve of velocity versus time is also a sinusoid, but it is shifted with respect to the displacement curve. When the displacement is zero, the velocity is at a maximum, i.e., as the object passes through the equilibrium position (displacement zero) it is traveling at its maximum speed. On the other hand, when the displacement is at a maximum (either positive or negative), the object stops momentarily and

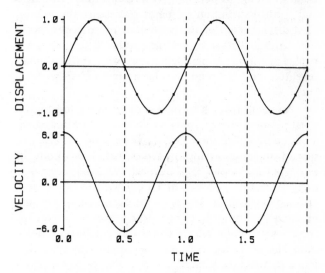

Figure 4.3. Displacement and velocity for the mass and spring vibrator.

thus, velocity is zero.

In summary, when a mass is undergoing SHM, the curves representing the manner in which the displacement and the velocity change with time are sinusoids.

Comparing Simple Harmonic Motions

The foregoing discussion has considered features that are common to all simple harmonic oscillators. However, not all simple harmonic motions are identical; they may differ in size, repetition rate, and relative starting times.

The maximum displacement in either direction from the rest position is known as the **displacement amplitude**. The total displacement from the maximum on one side of the rest position to the maximum on the other side of the rest position is thus twice the amplitude. In general, for any sinusoid, the amplitude

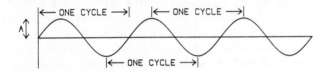

Figure 4.4. The amplitude and several ways of defining a cycle for SHM.

(represented by the letter A in Figure 4.4) is the distance from the rest position to a peak of the sinusoid. The **velocity amplitude** refers to the maximum velocity, which occurs when the object passes through the rest position.

One cycle of a vibration is defined as one complete excursion of the mass from the rest position over to one extremity, back through the rest position to the other extremity, and back again to the rest position. One cycle of a sinusoid can be used to represent one complete oscillation. It is not mandatory that a cycle be counted from the equilibrium position. Any point of the sinusoid is a valid starting point, but one cycle always means that you have come to an identical point on the sinusoid.

When considering a vibrating object, it is often more convenient to speak of how frequently it repeats its motion than to speak of the time required for completion of one cycle which time we defined previously as the period. The term **frequency** (f) is used to express the number of cycles completed in each second. The unit of frequency is the **hertz** (abbreviated Hz and representing a cycle per second), named after a 19th century scientist who did much to advance our understanding of wave motion. Since frequency is the number of cycles occurring each second and the period is the number of seconds required for one cycle, it can be seen that they are reciprocally or inversely related; when one increases, the other must decrease. For example, if a vibrating mass on a spring has a frequency of 10 Hz, its period would be 1/10 s or 0.1 s; if the frequency of vibration is decreased

to 2 Hz, the period is increased to 0.5 s.

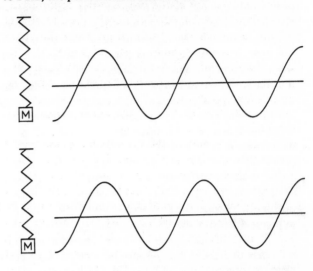

Figure 4.5. Two identical mass and spring systems vibrating in phase.

If we displace two identical mass and spring systems the same distance and release them together (Figure 4.5), they will oscillate exactly "in step." Their motions are identical in every way: they have the same amplitude and the same frequency. If we displace the masses by the same amount, but in opposite directions before releasing them, they will still oscillate with the same frequency and the same amplitude (Figure 4.6), but now they will oscillate exactly "out of step"—when one is moving upward, the other is moving downward. The difference between the motions illustrated in Figures 4.5 and 4.6 is expressed in terms of phase. **Phase** is the fraction of a cycle covered between some reference time and any point on a sinusoid. **Relative phase** is the relative time relationship between two sinusoids. When two masses are moving in step (Figure 4.5) they are said to be **in phase** (zero phase difference), while two masses

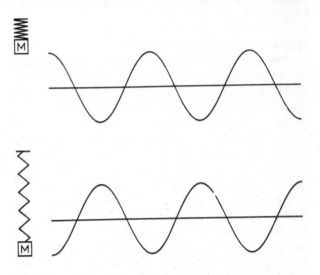

Figure 4.6. Two identical mass and spring systems vibrating out of phase.

23

moving opposite each other (Figure 4.6) are **out of phase** (half cycle phase difference). The displacements are always exactly opposite each other for out-of-phase motion, although the amplitude and periods may be identical. One way to express the phase difference between the two sinusoids is in fractions of a cycle. Note that the lower curve of Figure 4.6 could be made identical to the upper curve if it were merely shifted to the left by one-half cycle. The relative phase difference between these two curves is one half of a cycle. All phase differences may be expressed in fractions of a cycle. A phase difference of zero means the oscillations are in phase; a phase difference of 1/4 would mean the motions are one-quarter of a cycle out of phase. The phase difference between the displacement and velocity curves in Figure 4.3 is one-quarter of a cycle.

Phase differences may also be expressed by reference to a point moving around a circle. One complete rotation of the point is a cycle. Since there are 360° or 2π radians in one rotation around a circle, these quantities are applicable to a cycle. Hence, a phase of zero cycles is equivalent to 0° or 0 radians; a phase of one half cycle is equivalent to 180° or π radians.

[It is possible to write an expression for the instantaneous displacement of a simple vibrator in terms of a sinusoid as $d = A \sin (t/T + \phi)$ where A is displacement amplitude, t is elapsed time, T is the period, ϕ is the phase when time is zero, and sin is the sine function. The quantity $(t/T + \phi)$ forms a dimensionless argument for the sine function, values of which are tabulated in Appendix 6. Note that everytime t increases by T this argument increases by one cycle. The above expression can be written in a more common form by realizing that $f = 1/T$ so that $d = A \sin (ft + \phi)$. The argument $(ft + \phi)$ can be interpreted as number of cycles where ft and ϕ are each expressed in cycles. It is more common to write $d = A \sin (360 ft + \phi)$ with the argument expressed in degrees or $d = A \sin (2\pi ft + \phi)$ with the argument expressed in radians. Equivalent arguments in cycles, degrees, and radians are listed in Appendix 6.]

Effect of Mass and Stiffness on SHM

Small objects generally vibrate rapidly while large objects tend to vibrate more slowly. The wings of a mosquito oscillate rapidly enough to produce a high pitched hum, whereas an earthquake may jolt the earth and cause undulations having a period of one hour or longer. What is the explanation for the generally different vibration rates of small and large bodies? The explanation does not depend upon size, but rather upon mass; the more massive the vibrator, the slower its natural vibration. Consider again a mass attached to a spring and vibrating with SHM. If we double the mass, the same spring force must change the motion of twice as much mass. Because the increased mass has more resistance to changes of motion (inertia) these changes take place more slowly which results in a longer time to complete one oscillation. Thus, increasing the mass of a vibrator increases the period of vibration and decreases the frequency. Conversely, reducing the mass of a vibrator reduces the period and increases the frequency.

What would happen if, instead of changing the mass, we exchanged the spring for a stiffer one? The stiffer spring, when displaced from equilibrium, will exert a greater restoring force on the mass, thus producing more rapid changes of motion; the mass completes its motion more rapidly, the period is reduced, and the frequency increased. Increasing the stiffness of a simple harmonic oscillator results in an increased frequency, while decreasing the stiffness gives a corresponding decrease in frequency. The **stiffness** (s) of a spring is measured by the spring constant. The greater the force required to stretch a spring 1 cm, the greater the **spring constant**. A stiff spring has a large spring constant, i.e., large stiffness, while a "soft" spring has a small spring constant.

The frequency with which a simple mass-spring system (having a particular spring constant and a particular mass) vibrates is called its **natural frequency**. Natural frequency is directly proportional to the square root of the spring stiffness and inversely proportional to the square root of the mass. When a simple mass-spring system vibrates at its natural frequency its form or pattern of motion is called its **natural mode**.

Damped and Forced Oscillation

In our discussion so far we have been implicitly assuming that once we start an object vibrating, it continues oscillating indefinitely. Such, of course, is not the case. If, for example, the freely vibrating mass and spring of Figure 4.2 is placed on a table where there is even a small amount of friction, the vibration amplitude will decrease slowly as illustrated in the upper part of Figure 4.7. When the system is subjected to a very large resistance, such as would result from a rough surface, the amplitude decreases more rapidly, as can be seen in the lower part of Figure 4.7. **Damped oscillation** results when the amplitude of an oscillator decreases with time. All oscillators will eventually come to rest if no means for sustaining their oscillation is provided, because oscillatory energy is lost to the various forms of friction which are always present. In a mechanical system, such as the mass on a spring, friction converts mechanical energy into heat. In musical instruments acoustical energy is converted into heat by several forms of friction. In all cases, the vibration energy decreases with time; systems with more friction lose their energy more rapidly. A convenient way to express the rapidity with which amplitude decays, is the time required for the amplitude to decrease to one-half its initial value. This is referred to as the **damping time.** An interesting characteristic of the damping time is that it is independent of the initial amplitude; the amplitude decreases by equal fractions in equal time intervals. If the initial amplitude, A, decreases to A/2 during the first second, the amplitude will decrease to 1/2 of A/2 or A/4 during the second second, and to A/8 during the third second.

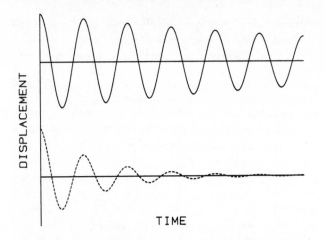

Figure 4.7. Displacement versus time for a free vibrator with small resistance (solid line) and a free vibrator with large resistance (dashed line).

This is illustrated in Figure 4.8. The initial amplitude for this case is 10 cm, and the damping time is 2.0 seconds. Note that every two seconds the amplitude is one-half its previous value. These concepts are particularly important to the understanding of percussion instruments as discussed in Part VI.

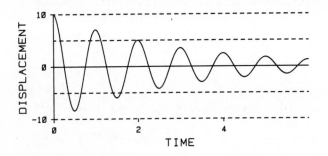

Figure 4.8. Illustration of equal fractional decrease in amplitude in equal time intervals for a free vibrator.

For the strings and membranes used in most percussion instruments the damping time is sufficiently long so that once set into motion the vibration continues long enough for the instrument's intended purpose. Examples are a piano string, a guitar string, and a drumhead. In non-percussion instruments, however, the oscillations would die out too rapidly for intended musical purposes if energy were not being supplied continuously by the performer. We distinguish then between a **free vibrator**, which is any vibrator set into oscillation and then allowed to decay, and a **forced vibrator** to which energy is continuously supplied. In free vibration the object vibrates at its own natural frequency, which remains constant even though the amplitude decreases continuously. In forced vibration, when a vibrator is driven by a continuous sinusoidal force the vibrator will vibrate at a frequency the same as that of the driving force, but generally with a small amplitude. For instance, if a sounding tuning fork is held over the opening of a bottle a feeble sound of the

same pitch as the fork will be heard from the air in the bottle. The vibration of the air is forced by the periodic sound waves emitted by the tuning fork. Blowing across the mouth of the bottle will result in a much louder sound of different frequency, as this corresponds to the free vibration of the air in the bottle.

Suppose a sinusoidal force having a frequency of 0.5 Hz is applied to a mass on a spring having a natural frequency of 1.0 Hz; after an initial period of instability, the vibrator will settle down to vibrating at the driving frequency of 0.5 Hz, but with a relatively small amplitude. The displacement and the driving force will be in phase—when the force pushes upward, the mass moves up. If we now raise the frequency of the driving force to 2.0 Hz, the mass will vibrate with a frequency of 2.0 Hz and with a small amplitude, but with one important difference; the displacement and the force are now one half cycle out of phase! The force pushes up while the vibrator is moving down and vice-versa. In fact, it is a general characteristic of driven vibrating systems with small damping that when the driving frequency is below the natural frequency, the displacement is in phase with the driving force, while above the natural frequency they are out of phase.

Resonance

Now, suppose that the frequency of the driving force is started well below the natural frequency, and increased to some frequency well above the natural frequency. We discover that the amplitude of vibration gradually increases to its maximum value at the natural frequency, and then decreases at higher frequencies as shown in Figure 4.9. **Resonance** is the condition of large amplitude vibration which occurs when a simple vibrator is driven at a frequency equal to or very near its natural frequency. In addition, the phase of the driving force is one quarter of a cycle ahead of the displacement so that the driving force is in phase with the velocity. When a system resonates, the vibrator achieves its greatest amplitude as the frequency of the driver becomes equal to the natural frequency.

As an example of the principle of resonance, suppose that a bell ringer wishes to ring a heavy church bell, but the bell is too large to be set ringing by a single pull on the rope. However, the ringer pulls the rope as hard as he can and then releases it. The bell then begins to swing, which causes the rope to move up and down. After the bell has performed a complete swing, the rope is back at its original position but it is moving downward. If the ringer now pulls down on the rope again the amplitude of the bell's swing will be increased with only slight additional effort. With repetition of the process the amplitude of the swing increases until the bell is ringing with as much vigor as is desired. Physically, the force exerted by the ringer is a periodic force with the same period as that of the free oscillations of the bell (and with the correct phase) so the amplitude of the vibration builds to a large value.

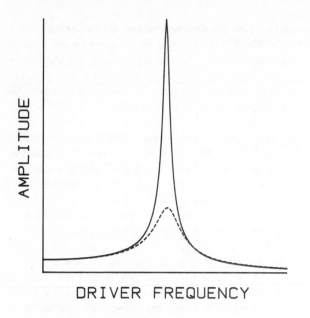

AMPLITUDE

DRIVER FREQUENCY

Figure 4.9. Vibration amplitude versus driving frequency for a mass and spring vibrator with small resistance (solid line) and a vibrator with large resistance (dashed line).

A struck tuning fork emits a sound having a definite frequency. The sound itself, however, is rather feeble. If the tuning fork is placed on a wooden box with one face open, a much louder sound is heard. In this situation, the driving force (the tuning fork) causes resonance in the box and the large increase of amplitude results in a louder sound (the hollow box is consequently called a resonance box). Suppose now that we have two identical tuning forks mounted on resonance boxes. If the open end of each box is faced toward the other, then striking one fork will cause the other fork to vibrate by resonance. If the frequency of one fork is now changed slightly by placing a small piece of clay on one prong, very little energy will be transferred by resonance, even though the actual frequency change of the fork has been slight. In this case, the resonance is said to be a **narrow resonance** and is highly selective.

Not all resonators exhibit such narrow resonances. If you blow across the mouth of an empty soda bottle, a sound at the natural frequency of the bottle is produced. If we hold a tuning fork of the same frequency over the bottle, a loud sound is heard due to resonance. If the frequency of the tuning fork is changed slightly by a small lump of clay, very little change in loudness is perceived; other frequencies close to the resonance frequency also elicit a fairly large response from the bottle. As a matter of fact, as the frequency is slowly varied from its resonance value, the response of the bottle falls off rather slowly and we have to make a fairly large change in the frequency before the bottle's response decreases appreciably. We describe this situation by saying that the resonance is a **broad resonance**. Some objects, such as the tuning fork on a resonance box, exhibit narrow resonances while other objects, such as the sounding board of a piano, display broad resonances. Most musical instruments have resonance

characteristics which are somewhere between these two extremes. Brass instruments have fairly narrow resonances, but they are sufficiently broad that the player can miss the exact resonance frequency by a small amount and still get a good response. If the resonance is too broad, however, the player will not be able to hold the note on pitch.

Why do different vibrating systems exhibit different resonance characteristics? The difference can be attributed to differing amounts of damping (resistance) and sound radiation which remove mechanical energy from the vibrating system. The more rapidly energy is removed, the more rapidly the amplitude of a free vibrator decays and the broader its resonance.

The relative amount of resistance in a vibrating system determines whether the system will have narrow resonance or broad resonance. Again we consider the mass on a spring and assume that it is being driven by a small forcing device of variable frequency. If we plot the amplitude of vibration as a function of frequency, we get a response curve like the one shown by the solid line in Figure 4.9. If we now let the mass vibrate on a rough table top, friction will have a considerable effect on the amplitude of vibration. Plotting the amplitude for different driving frequencies, as was done before, we get the dashed curve shown in Figure 4.9. Note that with the increased resistance, the originally narrow response curve has been appreciably broadened. Note also that the amplitude achieved at resonance is much smaller for the larger resistance than for the original case. Systems with a narrow resonance respond strongly to a limited, narrow range of frequencies; systems with a broad resonance respond slightly to a greater range of frequencies. Finally, close examination of Figure 4.9 shows that not only does resistance affect the amplitude and thus the width of resonance curves, but it also influences the resonance frequency itself. When little resistance is present the resonance frequency of a driven system occurs very nearly at the natural frequency of the vibrator. As damping is increased, however, the resonance frequency moves to lower values.

We can summarize the effects of resistance as follows: (1) A large resistance causes free vibrations to decay rapidly and produces a broad resonance (or response curve) for a forced vibration. (2) A small resistance results in a long-lasting free vibration and a narrow resonance for a forced vibration. (3) A narrow resonance has a high amplitude, while a broad resonance has a low amplitude. (4) As the resistance of a driven oscillator is increased, the resonance frequency is lowered.

Exercises

1. If a suspended mass is pulled down by stretching the spring, what direction will it move when released? What provides the force to accelerate it? When the mass reaches the rest position, will it stop or keep moving? What role does inertia play in the behavior of the mass? What forces are acting on the

mass when it is in the rest position?

2. For SHM, what is the phase difference between the acceleration curve and the velocity curve?

3. Consider a mass vibrating on a spring. Ignoring friction, at what position is the kinetic energy of the mass greatest? At what position is the potential energy greatest? How does the sum of the kinetic and potential energies compare at these various points? What happens to the total energy if friction is present?

4. Two identical tuning forks are struck identical blows and caused to vibrate. One fork is held in the air, the other has its base pressed against a table. Which tuning fork will sound louder? Why? Which fork will vibrate for a longer time? Why?

5. Name some musical instruments which are free vibrators. Name some musical instruments which are forced vibrators. Are the vocal folds free or forced?

6. What is the effect of resistance on an oscillatory system? What causes the resistance?

7. Make a rough graph of the displacement-versus-time curve for a free vibrator when resistance is present to a moderate degree.

8. List several examples of resonance.

9. Will a narrow or a broad resonance allow the greater displacement of a forced oscillator? Which type of resonance is more frequency selective?

10. Describe the mass, the restoring force, and the resistance for each of the following three simple vibrators: (a) a ball on a massless spring, (b) a ball on a stretched massless string, (c) an air-filled bottle with a narrow neck. The relation $f \approx 0.16 \ \sqrt{(s/m)}$, where m is the mass of the vibrator and s is its stiffness, gives the approximate frequency of vibration for these vibrators. An air-filled soft drink bottle can be viewed as a simple vibrator with the air in the bottle providing the stiffness and the air that moves in the neck of the bottle providing the mass.

11. A mass of 100 gm hung on a spring was set into motion. A movie camera with a framing rate of 20 frames per second was used to photograph the mass. The values for the displacement of the mass obtained by reading each frame of the film are listed in Table 4.2. How were the velocities determined? How were the accelerations determined? Does a constant force act on the mass? What produces the force?

12. Graph the values for displacement, velocity, and acceleration from Table 4.2. (Notice the relationships among these three quantities and that each can be represented with a sinusoid.)

13. Construct diagrams representing the displacement, velocity and acceleration for SHM. If the phase of the displacement sinusoid is 0, what is the phase of the velocity sinusoid? Of the acceleration sinusoid? What is the phase of acceleration relative to velocity?

14. Use the equation from the text to plot the displacements for one cycle of SHM where T = 10 s, A = 2.0 cm, and t = 0, 2, 4, 6, 8, 10 s. Use starting phases of 0, 1/4, and 1/2 cycle.

Table 4.2. Values of displacement, velocity, and acceleration for the mass and spring vibrator of Exercise 11.

t(s)	d(cm)	v(cm/s)	a(cm/s²)
0.0	0.00	6.3	0
0.05	0.31	6.0	-12.2
0.10	0.59	5.1	-23.3
0.15	0.81	3.7	-32.0
0.20	0.95	1.95	-37.5
0.25	1.00	0	-39.2
0.30	0.95	-1.95	-37.5
0.35	0.81	-3.7	-32.0
0.40	0.59	-5.1	-23.3
0.45	0.31	-6.0	-12.2
0.50	0.00	-6.3	0
0.55	-0.31	-6.0	12.2
0.60	-0.59	-5.1	23.3
0.65	-0.81	-3.7	32.0
0.70	-0.95	-1.95	37.5
0.75	-1.00	0	39.2
0.80	-0.95	1.95	37.5
0.85	-0.81	3.7	32.0
0.90	-0.59	5.1	23.3
0.95	-0.31	6.0	12.2
1.00	.00	6.3	0

15. Explain by using Newton's second law, why period increases as mass increases. How can Newton's second law explain the reduction in period which results from increased stiffness?

16. Explain how you could utilize resonance phenomena in order to determine the frequency of a tuning fork. If the frequency of the fork were too high, how could it be lowered to the correct frequency?

17. Observe and describe natural vibrators such as trampolines, automobiles, beds, floors, and so on. Identify spring elements, mass elements, and sources of friction.

18. Give examples of simple vibrators.

Demonstrations

1. "Air Cart Mass and Spring," Freier and Anderson, page M-60

2. "Graphical Display of Mechanical Resonance, "Freier and Anderson, page M-60.

3. "Helmholtz Resonators," Freier and Anderson, page S-12

4. "Musical Bottles," Freier and Anderson, page S-13.

5. "Tuning Forks with Resonators," Freier and Anderson, page S-14

6. Simple vibrators and forced vibration—Suspend a mass from a spring and set the system in oscillation. Measure the period. (You could, for example, time 10 oscillations and this period divided by 10 would be the period for one oscillation.) Give the value of the period in seconds and fractions of a second. Give the value for frequency in Hz. Increase the mass appreciably (e.g., double it). Repeat the above. What happens to the period? Does it increase, decrease, or is it unchanged? What happens to the frequency? Give the value of T and f. Now suspend the original mass on a

stiffer spring. Repeat the above. What happens to the period? What happens to the frequency? Give values for T and f. Attach a driving apparatus (your finger if nothing better) to the mass and spring. Adjust the driving frequency until the maximum amplitude of oscillation is obtained. This is called the resonance condition. Measure T and f under this condition. Are they approximately the same period and frequency as those obtained for the natural oscillation? Why?

7.   Experiment with a large pendulum such as a mass on the end of a rope or a child in a swing. How do you adjust your pushing frequency relative to the natural frequency to achieve large amplitude? How do you adjust the phase of your pushing force to achieve this result? When the swing is at its position of maximum displacement, the "high point" where it momentarily stops, you exert no force. As the swing begins to move you gradually increase the force (by pushing harder) until you are exerting your maximum effort at the equilibrium position. Your driving force must always be one-quarter of a cycle ahead of the displacement amplitude in order to realize resonance. That is, when the displacement is at a maximum, the force is zero and when the displacement is zero (at equilibrium) the force is at a maximum. Having the right frequency, then, is only part of the resonance story; the driving force must also have the correct phase. If you wish to

stop the swing, how do you adjust the frequency and phase of your force?

8.   Blow on various bottles and observe their resonance frequencies. Identify the "mass" and "stiffness" of the air in the bottles. Sing into a wash basin while varying your pitch.

9.   Measure the half amplitude damping time of a mass and spring vibrator. Add a "tail" to the mass and let the tail move in water or light oil. How does the damping time change? Drive the two systems so that they resonate. Compare their amplitudes.

10.   Measurement of displacement, velocity, and acceleration of a moving vibrator—Attach a mass to the end of a suspended spring. Place a Polaroid camera and a stroboscopic light about 1 meter away from the mass. Set the strobe to flashing about 20 times per second. Set the mass in motion. Open the camera shutter for about 2 seconds and pan the camera. Then close the shutter and develop the print. Construct a table having the form of Table 4.1. Fill in the table by measuring the displacements on the print and then calculating the velocities and accelerations. If you do not have a usable print, Figure 4.10 may be employed for these measurements. At which of the tabulated times was the velocity greatest? Smallest? How can you tell? At which of the tabulated times was the acceleration greatest? Smallest?

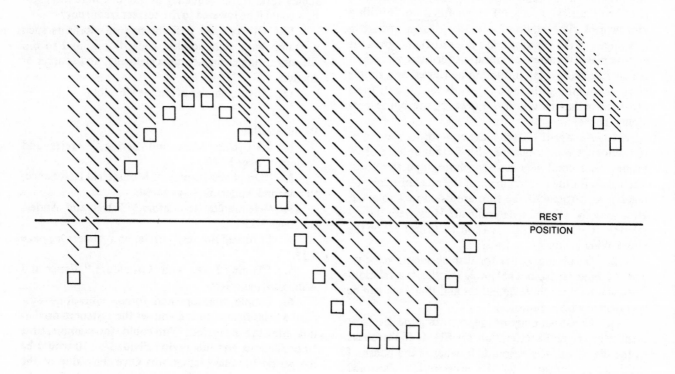

Figure 4.10.   Simulated motion of a mass and spring vibrator.

Further Reading

Backus, Chapters 2, 4
Benade, Chapter 10
Crawford, Chapters 1, 3
Denes, Chapter 3
French, Chapters 3, 4
Rossing, Chapters 2, 4
Sears, Zemansky, and Young, Chapter 11

Audiovisual

1. **Vibrations** (14 min, color, 1961, EBE)

2. **Velocity and Acceleration in Simple Harmonic Motion** (Loop No. 80-225, EFL)

3. **Tacoma Narrows Bridge Collapse** (Loop No. 80-2181, EFL)

4. **Simple Harmonic Motion** (Loop No. 80-3098, EFL)

5. **Damped Oscillators** (Loop No. 80-166, EFL)

6. **Damped and Driven Oscillators — Resonance** (REB)

7. **Glass Breaking with Sound Resonance** (REB)

5. Properties of Compound Vibrators

Most vibrating systems, including musical instruments and the voice, are not simple vibrators, nor can they be described solely in terms of the definitions obtained for simple vibrating systems. On the other hand, analysis has shown that almost any complicated vibrating system can be represented as a combination of many different simple vibrators. In this chapter we will consider compound vibrators and find that they also may execute periodic motions. We will find that the concept of a single natural mode as applied to a simple vibrator must be extended to include several natural modes for a compound vibrator. We will end the chapter with a discussion of vibration recipes (or spectra) which tell us about the motion of a compound vibrator.

Natural Modes

In Chapter 4 we considered two examples of simple vibrators: a simple pendulum and a mass attached to a spring vibrating on a frictionless table top. Let us now consider a more complicated vibrator: two masses attached with springs, as shown in Figure 5.1. The ends of the outer springs are attached to supports so they do not move. This compound vibrator has all the same ingredients as a simple vibrator—mass, restoring force, and resistance. Note that instead of one mass and a restoring force, we now have two masses and restoring forces.

Figure 5.1.   Vibrator consisting of two masses attached to springs.

(The two mass system shown in Figure 5.1 can be approximated by taking two "long" pendulums, connecting them to end supports with springs and coupling them to each other with a spring. If the motions are kept small, the masses change their heights only slightly and thus restoring forces due to gravity can be neglected. The only forces to be considered are due to the expansion and compression of the springs.)

There are many possible ways in which the two mass systems of Figure 5.1 can vibrate depending on which springs are compressed or expanded. Most of the possible vibration patterns will not result in periodic motion. We look for the simplest possible periodic vibration patterns for the system, considering only motion back and forth in the direction of the springs. (We do not consider up and down motion or motion

crossways relative to the direction of the springs.) Suppose we displace both masses by the same amount in the same direction. When they are released they move back and forth like two simple mass and spring vibrators because the spring coupling them is neither compressed nor expanded and does not influence the motion. This vibration pattern, illustrated in Figure 5.2A, is termed the first natural mode, because the frequency of oscillation is the lowest possible frequency for this system.

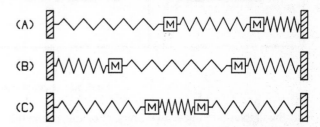

Figure 5.2.   (A) First natural mode of two mass vibrator of Figure 5.1. (B) Second natural mode of two mass vibrator in one phase of oscillation. (C) Second natural mode in another phase of oscillation.

Now suppose one mass is displaced to the left while the other is moved to the right by the same amount. When released, the masses will move in opposite directions and the coupling spring will alternately be expanded and compressed. Once again, there is only one frequency of oscillation, but due to the additional restoring force provided by the coupling spring, the frequency will be higher than in the first natural mode. Two phases of this motion, illustrated in Figures 5.2B and 5.2C, is known as the second natural mode of this system. If the system starts vibrating in one natural mode it will continue to vibrate in that mode. Any complicated type of vibration can be constructed from a combination of these two natural modes. Note that the one-mass system has one natural mode and one natural frequency. The two-mass system has two natural modes and two natural frequencies.

Consider now a set of vibrators consisting of blocks attached to a string, as shown in Figure 5.3. The one-block system has one natural mode, which consists of the block vibrating up and down. The two natural modes of the two-block system are shown in the second row of Figure 5.3. Note that in the first mode the masses always move together, while in the second mode the blocks move in opposition to each other. A three-mass system would have three natural modes, as shown in the third row of Figure 5.3. The lowest-frequency mode (first natural mode) again has all the masses moving together as a unit. The higher frequency second mode

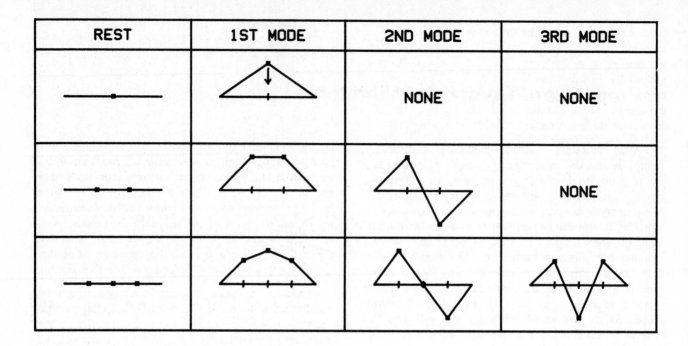

REST	1ST MODE	2ND MODE	3RD MODE
		NONE	NONE
			NONE

Figure 5.3. Natural modes of blocks on a string.

has the masses moving as two units. (For this mode the center mass does not move at all.) Finally, in the highest frequency mode (third natural mode) the masses move as three units, each moving in the direction opposite that of its nearest neighbor. Note that in the three-mass system the frequency of the second mode is higher than that of the first mode, and the frequency of the third mode is higher than that of either the first or second mode.

Let us now generalize our observations for one dimensional oscillations: (1) Multi-mass systems can oscillate in different ways. (2) There are as many natural modes as there are masses in a system. (3) The higher modes have higher frequencies than the lower modes. (4) In each mode all masses move with the same frequency.

Suppose now that the number of blocks on the string is increased from three to ten. There will now be ten natural modes. Let us consider the first three natural modes and compare them to the first three natural modes for the three-block system as shown in Figure 5.4. Note the similarities of the first, second, and third modes. Also note that when we increase the number of blocks the string shape of vibration smooths out; that is, the sharp angles present in the three-block system are replaced by smoother curves. Imagine now what would happen if we increase the number of blocks to 100, or 1000, or to a virtually infinite number—in which case a perfectly smooth curve would be produced. The form of the curve would then be a portion of a sinusoid.

You may have noticed that for several of the modes of vibration one block may remain at rest and not vibrate at all (e.g., the center block in the second mode of the three-block system). When a block (or a point of a continuous system like a string) always remains at rest,

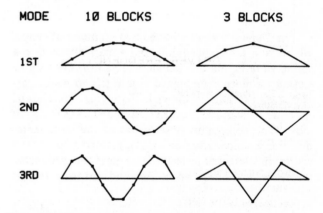

Figure 5.4. Ten-block and three-block vibrators illustrating correspondence of their first three natural modes.

it is termed a **node**. A point of maximum vibration is termed an **antinode**. Note that for any of the vibrating systems shown the number of antinodes present is equal to the mode number.

Resonances of Compound Vibrators

Forced vibration has the same meaning for compound vibrators as for simple vibrators. However, in a two-mass system there will be two different frequencies (as opposed to one in a simple vibrator) to which the compound vibrator responds. When a compound vibrator is forced to vibrate at one of its natural frequencies, a large displacement of the vibrator results.

By driving the simple vibrator of Figure 4.2 with a variable frequency driving force a response curve like that shown in Figure 4.9 is obtained. It is characterized by a single peak whose height and width depend on the amount of resistance present. By driving the compound vibrator of Figure 5.1 with a variable frequency driving

32

force a response curve like that shown in Figure 5.5 is obtained. The response curve for the compound vibrator exhibits two peaks corresponding to the two natural frequencies associated with each of its natural modes. Resistance plays the same role in this case as in the case of simple vibrators. The resistance may be different for different natural modes, but generally the resistance is greater for higher frequency modes because of their greater number of back-and-forth motions. The greater the resistance associated with any mode the more rapidly it decays during free vibration, and the broader its response peak when undergoing forced vibration.

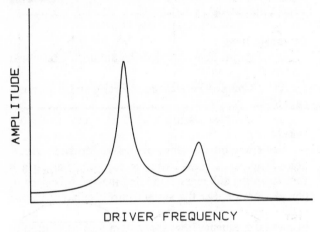

Figure 5.5. Vibration amplitude versus driving frequency for a two mass vibrator.

The response curves of a multi-mass vibrator will exhibit multiple peaks, each peak occurring at one of the natural frequencies of the vibrator. In principle, vibrators with a very large number of masses will have a large number of natural modes and natural frequencies. In practice, only the first dozen or so modes are of much practical importance.

Motions of Compound Vibrators

Obviously there are many ways in which a set of blocks on a string could vibrate besides the natural modes we have discussed. From careful analysis of a set of blocks vibrating in some complicated manner we discover that any oscillation irregardless of its apparent complexity, can be represented as a composite of its natural modes. As an example consider the three-block system shown in Figure 5.6. Suppose each of the three possible natural modes vibrate with the amplitudes and phases shown in (A) through (C). Now suppose that the system vibrates in a complex manner, such as shown in (D). This complex motion is just the second mode added to the first mode (having an amplitude of 1.0 unit). To obtain this complex vibration we merely superpose the second mode upon the first mode and the combination gives the initial displacement shown. This system vibrates with two different frequencies, the frequency of the first mode having an amplitude of one unit and the frequency of the second mode having an amplitude of

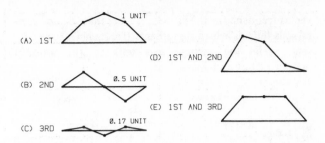

Figure 5.6. Three blocks on a string vibrator: (A) first mode, (B) second mode, (C) third mode, (D) first and second modes combined, and (E) first and third modes combined.

one-half unit. The resulting complex motion will not generally be periodic. However, when the second natural frequency of a system is exactly twice the first natural frequency the complex motion will be periodic. This occurs because the second mode completes exactly two cycles of vibration while the first mode completes one. For a few masses on a string the second natural frequency usually is not exactly twice that of the first so the complex motion does not repeat. We can see that the starting motion shown in Figure 5.6E is the sum of the first and third natural modes in a manner similar to that described for Figure 5.6D. Again, the system vibrates with two frequencies, the lower-frequency first mode having a one-unit amplitude and the higher-frequency third mode having an amplitue of 0.17 unit. (There are some slight complications in specifying the amplitudes of some higher modes because largest displacements above and below rest position may not be equal. This is the case in mode 3 where we have taken the amplitude to be the displacement of any block—the middle one in this case. In most cases of interest we will not be concerned with this much detail.)

If we know how much of each mode is to be added, the addition can be done graphically. For example, we can take a ruler and measure the displacement of the left block in (A) and add its displacement in (B) to get its resulting displacement in (D). Doing the same for the middle block just gives the displacement from (A) because the middle block has zero displacement in (B). The resulting displacement of the right block in (D) is smaller than that in (A) because its displacement in (B) is negative and so substracts. Similar graphical additions of (A) and (C) can be done to obtain (E).

Vibration Recipes

In the creation of a "culinary delight" we have many ingredients at our disposal, and a recipe tells us which ingredients to use and how much of each is to be used. Similarly, information on the ingredients present in a complex vibration can be conveniently presented in a "**vibration recipe.**" The vibration recipe, as with a culinary recipe, lists the various ingredients present and the amount of each. For vibrating systems, the ingredients are the natural modes or natural frequencies and the amount of each is represented by the corresponding amplitude. A vibration recipe is termed a **spectrum** and is conveniently presented on a bar graph of amplitude versus frequency as in Figure 5.7. Note that the horizontal axis tells us which ingredients (frequencies) are present, while the vertical axis tells us the amount (amplitude) of each ingredient.

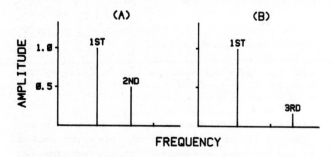

Figure 5.7. (A) Spectrum for combined modes of Figure 5.6D. (B) Spectrum for combined modes of Figure 5.6E.

The spectrum in Fig. 5.7A for the complex vibration shown in Fig. 5.6D has a bar one unit long located at the frequency of the first mode. This represents the amount (amplitude) of the first mode present. Another bar 0.5 unit long at the frequency position of the second mode represents the amount (amplitude) of the second mode. A similar description can be given for Figure 5.7B, the spectrum of the complex vibration shown in Fig. 5.6E.

Exercises

1. How many natural modes are there in each of the following cases? A two-mass vibrator; three masses on a string; four masses on a string; twenty masses on a string; N masses on a string; two connected air volumes.

2. Make charts similar to Figure 5.3 for four blocks on a string and five blocks on a string.

3. Sketch as a bar graph a recipe showing the following ingredients: 2 cups flour, 1 cup sugar, 1/2 cup eggs, and 1/2 cup butter.

4. Sketch the following spectrum as a bar graph: 1.0 unit of frequency f_1, 0.6 units of frequency f_2, and 0.3 units of frequency f_3.

5. Repeat Exercise 4 for the three natural frequencies given in Figure 5.6A, B, C.

6. Construct bar graphs and draw vibration patterns for two blocks on a string in each of the following cases: (a) 1 unit of mode one; (b) 1 unit of mode two; (c) 1 unit of mode one plus 1 unit of mode two.

7. Show the vibration pattern and construct bar graph spectra for the following combinations of the modes in Figure 5.6: (a) Add the modes in B and C. (b) Add the modes in A, B, and C.

8. Plot spectra (in bar graph form) for the two complex vibrations of two masses on a string as shown in Figure 5.8. Refer to Figure 5.3 (middle) to get modal patterns.

9. Repeat Exercise 8 for the complex vibrations of a three-mass system shown in Figure 5.9.

10. Sketch initial mass positions for various com-

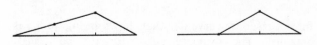

Figure 5.8. Two combined mode vibration patterns for a two mass system.

Figure 5.9. Two combined mode vibration patterns for a three mass system.

binations of modes 1 and 2 for the two mass and spring system. (See Figures 5.1 and 5.2)

Demonstrations

1. "Adjustable Coupled Pendulum," Freier and Anderson, page M-61.

2. "Coupled Pendulums," Freier and Anderson, page M-62.

3. "Coupled Oscillators," Freier and Anderson, page M-62.

4. Compound vibrators and forced vibration—Suspend a mass from a spring. Then attach another spring and mass to the first mass to form a compound vibrator. Set the system into oscillation in one natural mode. What is the period of this mode? What is its natural frequency? Set the system into oscillation in the other natural mode. What are the period and frequency of this mode? Attach a driving apparatus (such as your finger) to one of the masses or springs. Adjust the driving frequency until a resonance is obtained. Measure the period for this condition. Are these approximately the same period and frequency as those obtained for one of the natural modes? Why? Adjust the driving frequency until the other resonance is found.

5. Repeat Demonstration 4 for two masses on a string. (Attach fluorescent clay masses to a string and use a loudspeaker driver or other appropriate device to drive the string.)

6. Repeat Demonstration 4 for two connected air volumes.

7. Repeat Demonstration 4 for two coupled pendulums.

Further Reading

Backus, Chapters 3, 4.
Benade, Chapters 6, 7.
Crawford, Chapters 2, 3.
French, Chapter 5.
Hall, Chapter 8.
Rossing, Chapters 2, 4.
Edge, R.D. 1981. "Coupled and Forced Oscillations," Physics Teacher 19, 485-488.

Audiovisual

1. **Coupled Oscillators: Part III, Normal Modes** (Loop No. 80-2694, EFL)

2. "Coupled Oscillations" (REB)

6. Measurement of Vibrating Systems

Many of the phenomena of vibrating systems are not directly observable because the oscillations are taking place too rapidly to be visible. For instance, the strings of a violin move much too rapidly for us to make any successful direct measurements of their displacements. And of course the sounds themselves are invisible. In order to make accurate measurements of these invisible waves, we need some means for turning them into some visible form. In this chapter we will be concerned with some of the instruments which aid us in our study and measurement of vibrations and sounds.

Transducers

A **transducer** is any device which transforms one form of energy into another. The transducers most useful to us are those which involve acoustical, mechanical or electrical energy. An **acoustical generator** is a transducer that produces acoustical disturbances in matter, while an **acoustical receiver** detects acoustical disturbances. The human voice is an acoustical generator, since it converts an input of "nerve energy" into acoustical energy (speech), while the ear serves as an acoustical receiver, converting acoustical energy (sound) into nerve impulses. Two particularly significant categories of transducers (with respect to sound) are microphones and loudspeakers, which will be discussed in the next section. Other transducing devices and instruments useful in acoustical applications will be discussed in following sections.

There are several different physical mechanisms whereby energy may be transformed from one form to another. One such mechanism is the **piezoelectric effect** employed in ultrasonic devices and microphones. When certain crystals or ceramic materials are compressed or distorted they acquire an electric charge. If output leads are attached to the crystal or ceramic an electrical output signal is obtained which is proportional to the mechanical input, in this case, distortion. Conversely, when an electrical voltage is applied to a piezoelectric crystal or ceramic it will change its shape, thus transforming electrical energy into mechanical energy.

A second transduction mechanism is the **capacitive effect** employed in condenser microphones. When two parallel metal plates are charged (one plate positively, the other plate negatively) the voltage between the plates will vary as the separation of the plates changes. Hence, mechanical motion of the plates may be transformed into a varying electrical signal.

A third method of transduction is based on the effect of magnetic forces on electrons moving in a magnetic field. When a current flows through a coil in a magnetic field, the electrons and the coil containing them experience a magnetic force. This transduction mechanism is used in dynamic loudspeakers to convert electrical energy into acoustical energy. When a coil is mechanically moved in a magnetic field, its electrons can experience a force that causes them to move as a current. This transduction mechanism is used in dynamic microphones to convert acoustical energy into electrical energy.

Not all transducers are sensitive to the same input amplitudes. Some respond to small amplitude inputs and others only to large amplitude inputs. For a transducer to be most useful, its output should bear a linear relationship to its input; i.e., doubling the size of the input should double the size of the output. The **dynamic range** of a transducer is the range of amplitudes over which its output response is nearly linear.

Not all transducers are sensitive to the same frequencies. The human ear responds to a frequency range of approximately 20 to 20,000 Hz, while a dog's ear responds roughly to the range of 15 to 50,000 Hz. Since the dog can hear much higher frequencies than a human, the dog will respond to high-frequency dog whistles which we cannot hear. A good recording microphone may respond to the frequency range of 20 to 20,000 Hz, while the transducers in the telephone respond only to the more limited frequency range of about 300 to 3000 Hz.

Microphones and Loudspeakers

A **microphone** is a transducing device which converts acoustical energy into electrical energy in the form of a time-varying voltage. The voltage should fluctuate so as to accurately represent the vibration of the air. The basic operation of a microphone can be represented as shown in Figure 6.1. Sound impinging on a diaphragm causes the diaphragm to vibrate. The mechanical motion of the diaphragm is coupled to the transducer, which produces the varying voltage. The original acoustic vibrations are thus preserved in the form of a time-varying voltage.

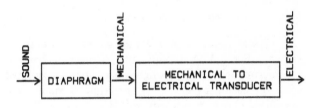

Figure 6.1. Basic processes in a microphone.

Microphones can be classified as one of two types—pressure sensing or velocity sensing—and combinations of these two types. In the pressure sensing type sound is allowed to strike only one side of the diaphragm. Thus the diaphragm is sensitive to pressure changes. Pressure microphones are said to be omnidirectional or non-directional since they respond to sound coming from any direction. In velocity microphones both sides of the diaphragm are exposed to the sound. Velocity microphones respond mostly to sounds striking one side or the other of the diaphragm in a head-on fashion. When sound approaches the edge of the diaphragm, it exerts equal but oppositely directed pressures on each side of the diaphragm so that no motion is produced. A velocity microphone is said to be bidirectional. Directional properties of microphones will be considered in more detail in Part VII.

The piezoelectric effect is the transducing principle employed in **crystal microphones** and **ceramic microphones**. Ceramic materials are more commonly used because they are less sensitive to mechanical shock, humidity, and temperature than are crystal materials. Typical elements of a ceramic microphone are shown in Figure 6.2. One end of the ceramic is held fixed to the microphone casing and the other end is coupled to the diaphragm by a drive pin. Conducting material is deposited on the two surfaces of the ceramic. When the diaphragm moves, the ceramic element is bent and a positive or negative (depending on the direction of bending) voltage appears on one conductor. This voltage is amplified by a small preamplifier mounted in the casing. Ceramic microphones are used in portable sound equipment, tape recorders, and hearing aids. They have a wide, but somewhat uneven, frequency response.

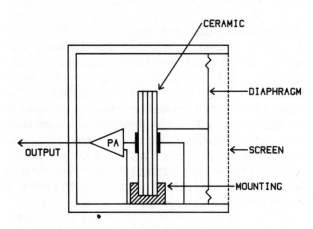

Figure 6.2.   Elements of a ceramic microphone.

Typical elements of a **condenser microphone** are shown in Fig. 6.3. One conducting plate is held fixed to the microphone casing, though it is electrically isolated from the casing by insulating material. A metal or a metallized plastic diaphragm is used and acts as the second (and movable) conducting plate of the capacitor. When the diaphragm vibrates an oscillating voltage is produced on the fixed plate. This voltage is amplified by

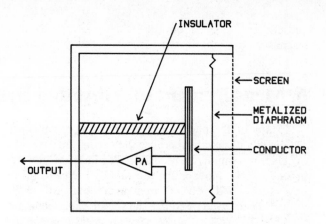

Figure 6.3.   Elements of a condenser microphone.

a small preamplifier mounted in the casing. Condenser microphones are used in applications requiring high precision and wide, uniform frequency response. Formerly a major disadvantage of condenser microphones has been their requirement for a high voltage source. This disadvantage has been eliminated in modern electret condenser microphones with only a very modest decline in performance. Electret condenser microphones are used in portable sound equipment and in hearing aids.

A **dynamic microphone** has a coil of wire attached to its diaphragm as shown in Figure 6.4. The transducing element is a coil that is free to move between the poles of a magnet. The electrons in the coil move with the coil and experience a force (as described by the magnetic force law) which produces a varying electric voltage across the ends of the coil. Dynamic microphones are capable of a relatively high power output, are rugged, and are capable of a broad frequency response over a wide dynamic range. Since they are able to withstand the high intensity sound levels often associated with popular music they are widely used for live performances and for recording sessions.

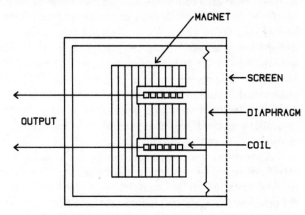

Figure 6.4.   Elements of a dynamic microphone.

A **loudspeaker** is a transducing device which converts electrical energy into acoustical energy. One common type of loudspeaker, the direct radiator dynamic loudspeaker, operates on the same principle as the dynamic mike—only in the opposite sense. Figure 6.4 may be regarded as an illustration of a dynamic loudspeaker as well as a dynamic microphone. A voltage source is connected to a coil of wire situated in the magnetic field of a permanent magnet. A varying voltage produces a varying current in the coil. The coil experiences a force (due to the presence of the magnetic field) which causes the coil to vibrate. Since the coil is attached to a diaphragm, the motion of the coil "drives" the diaphragm, which causes the air to vibrate. Because of the simplicity of construction, the small space requirements, and the fairly uniform frequency response, this type of speaker is the most widely used. Other types and aspects of loudspeakers will be considered in Part VII.

Other Transducing Devices

Devices termed "pickups" (various types of which are used in record players, electric guitars, and so on) are similar to microphones but do not have diaphragms. Such pickups are typically used to convert mechanical energy to electrical energy. The transducing element is coupled to the object of interest via a "drive pin" or by direct contact. In **piezoelectric pickups** used in inexpensive record players the needle serves to couple mechanical motion to a piezoelectric transducer. In **dynamic pickups** the needle couples mechanical motion to a coil in a magnetic field. Dynamic pickups typically have a better frequency response and are used in high fidelity systems.

Specially designed piezoelectric pickups can be used to measure the force that a string is exerting on the bridge of a stringed instrument—from which the string motion can be inferred. The output voltage from the pickup can be amplified and played through a loudspeaker to produce a much more intense sound than is otherwise possible. Certain electric guitars and electric pianos use piezoelectric pickups in this fashion. **Capacitive pickups** are used to transform the mechanical vibrations of struck chimes to electrical signals which can be amplified and played outdoors. The metal body of the chime becomes one plate of the capacitor, while the second plate is mounted loosely to the chime by means of a flexible material.

A type of **magnetic pickup** useful for measuring the velocity of vibrating steel strings consists of a coil with an adjustable iron pole piece in its center. The pickup is placed below, but not touching, a string. The vibrating steel string varies the magnetic flux through the coil which induces a proportionately varying voltage in the coil. From information on the velocity of the string, other parameters of the motion, such as displacement and acceleration can be inferred. The string motion can thus be described in terms of the voltage generated by the magnetic pickup.

Another transducer which has proven to be useful in acoustical research is the **accelerometer**—a device which measures acceleration. Small accelerometers are useful in measuring the acceleration of the tubes of wind instruments, the bodies of stringed instruments, the acceleration of the human nose (for determining relative amounts of nasal energy), and so on. In general, they need to be much smaller than the object whose acceleration they are measuring so that they do not interfere with the motion of the vibrating object. The output from a typical accelerometer is in the form of a varying voltage that is proportional to the varying acceleration. Velocity and displacement also can be obtained from this information.

Other Instruments

The **oscilloscope** is a very useful instrument which is used in a wide range of applications in laboratories, electronic shops, hospitals, and so on. With an oscilloscope it is possible to display a time-varying voltage on a phosphorescent screen so that it can be seen. The main components of an oscilloscope are a cathode ray tube (similar to a TV picture tube) and electrical control circuitry. The cathode ray tube contains an "electron gun" that shoots out a continuous stream of electrons—much as a garden hose shoots out a stream of water. The electron stream is fired at a screen which has been treated with special material that phosphoresces and gives off light when struck by the electrons. The cathode ray tube has a deflecting mechanism that causes the electron stream to move from left to right on the screen so as to draw a "line of light" across the screen. The rate at which the electron stream is moved across the screen is regulated by the control circuitry. The control circuitry is adjusted via various knobs and switches on the front panel of the oscilloscope. When the electron stream reaches the right-hand side of the screen, the stream is momentarily turned off while the electron gun is abruptly moved back to aim again at the left-hand side of the screen. An additional deflecting mechanism is also provided in the tube to produce deflections in the vertical direction. The vertical deflections are controlled by an input voltage. A positive voltage causes an upward displacement of the beam, while negative voltage causes a downward displacement. A fluctuating voltage (originating from a microphone or other device) will cause the electron beam to move up and down repeatedly. If the beam is being swept horizontally at the same time, the result is a "drawing" of a waveform on the screen. The vertical axis is proportional to amplitude and the horizontal axis represents time. The three sketches in Figure 6.5 show a side view of a cathode ray tube and the electron stream, a front view of the oscilloscope displaying a straight line (when no input voltage is present to control the vertical deflections), and another front view of the screen with a waveform displayed. Figure 6.6 shows a typical laboratory oscilloscope. Figure 6.7 shows several different waveforms as displayed on an oscilloscope.

A **spectrum analyzer** converts time-varying voltages

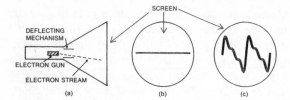

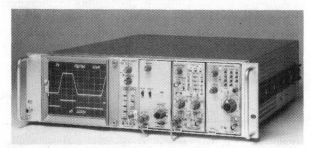

Figure 6.5. Elements of an oscilloscope: (a) side view, (b) front view of screen with no signal present, and (c) front view with signal present.

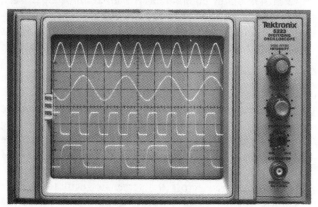

Figure 6.6. A rack mounted laboratory oscilloscope. (Courtesy of Tektronix, Inc.)

Figure 6.7. Oscilloscope screen showing waveform displays. (Courtesy of Tektronix, Inc.)

into their corresponding spectra (vibration recipes) so that the amount of energy present at each of many different frequencies can be determined. Figure 6.8 illustrates the basic process of spectral analysis. Note that the input signal is in the form of voltage versus time and that the spectrum is in the form of voltage versus frequency.

WAVEFORM   SPECTRUM   SPECTRUM
(VOLTAGE VS. TIME)   ANALYZER   (VOLTAGE VS. FREQUENCY)

Figure 6.8. Basic processes of a spectrum analyzer.

There are several types of spectrum analyzers. One type uses a tunable filter that only allows a single frequency—the one to which it is tuned—to pass through at any one time. A second type of analyzer has multiple filters, each tuned to a different frequency, so that more than one frequency can be analyzed at a time. Figure 6.9 illustrates a multiple filter analyzer that provides fine frequency resolution by using many filters (or by simulating many filters through analog or digital

methods).

High resolution spectrum analyzers typically provide spectral information at equally spaced frequencies (Figure 6.12). Some multiple filter analyzers do not

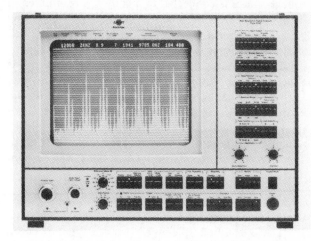

Figure 6.9. High-resolution real-time spectrum analyzer. (Courtesy of B & K Instruments.)

display spectra at equal frequency intervals, but display spectra based on divisions of frequency into fractions of octaves. Figure 6.10 illustrates a multiple filter analyzer in which there are three filters in each octave and each filter covers a frequency range of one third of the octave. An octave represents a frequency doubling; thus, third octave filters span smaller frequency ranges at low frequencies and larger ranges at higher frequencies. The audio frequencies used for third octave analyzers are (expressed in Hz): 25, 31.5, 40, 50, 63, 80, 100, 125, 160,

Figure 6.10. One-third octave real-time spectrum analyzer. (Courtesy of Ivie Electronics, Inc.)

200, 250, 315, 400, 500, 630, 800 1000, 1250, 1600, 2000, 2500, 3150, 4000, 5000 6300, 8000 10000, 12500, 16000, 20000. Third octave spectrum analyzers provide spectral information at non- uniformly spaced frequencies (Figure 6.12). High resolution and third octave analyzers are said to be real time analyzers in that they are capable of displaying a varying spectrum as it changes. More will be said in later chapters about the application of spectrum analyzers for the determination of the frequencies present in sounds produced by the human voice and musical instruments.

A **function generator** (Figure 6.11) is a device that provides a variable frequency oscillating voltage corresponding to one or more of the following wave types: sine, sawtooth, square, triangular, and pulse; typical waveforms are shown in the left column of Figure 6.12.

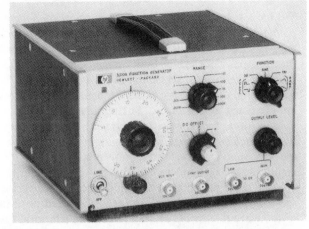

Figure 6.11. A function generator with sine, square, triangular, pulse, and ramp functions. (Courtesy of Hewlett-Packard.)

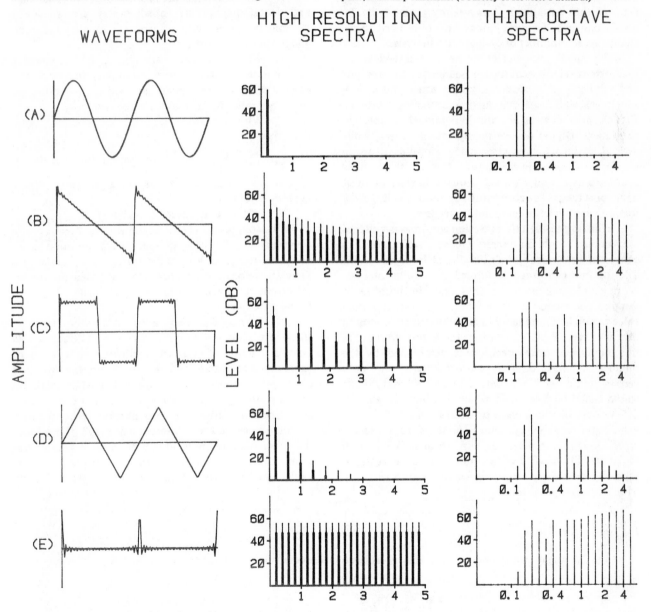

Figure 6.12. Various waveforms and their high resolution and one-third octave spectra: (A) sine, (B) sawtooth, (C) square, (D) triangular, and (E) pulse.

We have seen that any complex vibration of a compound oscillator can be constructed from its various natural modes. Likewise, the waveforms produced by a function generator can be built up from "natural vibrations" in the form of sinusoids. In Figure 6.12 the spectra (vibration recipes) for each of the above waveforms are presented in two forms. The middle column shows the spectrum of each wave as analyzed by a high resolution spectrum analyzer, while the right column shows the spectrum from a third octave analyzer. The low amplitude background results mostly from noise in the system. The finite width of the peaks occurs because the spectrum analysis was performed on a limited portion of the wave (about ten cycles). When the temporal extent of a wave is decreased, the widths of the frequency components of its spectrum are increased. (This is one form of the well-known "uncertainty principle.") To obtain narrower frequency lines we would need to include a greater number of cycles in the analysis.

The digital computer and other digital devices are used extensively in acoustical research for the analysis and synthesis of musical instrument tones and speech sounds as well as for modeling acoustical systems. Digital audio devices are also becoming available for consumers. Digital devices manipulate numbers, while acoustical signals, even after being transduced, are in the form of continuous voltages. If we wish to input sounds to a computer (or other digital device) we must have some means for converting continuous voltages into numbers which the computer can accept.

In **analog-to-digital conversion** continuous voltages are sampled to obtain discrete voltages which are then converted into numbers as illustrated in Figure 6.13. The continuous voltage is sampled at uniform intervals of time. (A sampling rule says that the number of samples per second must be equal to or greater than twice the highest frequency in the signal.) An analog to digital converter (ADC) converts these discrete samples into numbers whose precision is determined by the number of binary digits in the converter. Some converters (and other digital devices) have 16 binary digits (often called bits), each of which can have a value of zero or one. (Decimal digits by contrast can each have a value from zero through nine.) Numbers from 0 to 2^{16} = 65536 can be represented with 16 bits. (Very often the abbreviation ADC will be used to refer to both the sampling and the converting shown in Figure 6.13.)

In **digital-to-analog conversion** a digital to analog converter (DAC) converts numbers into discrete voltages which are smoothed into continuous voltages as illustrated in Figure 6.14. The "smoother" shown in Figure 6.14 is actually a filter used to smooth the waveform and eliminate spurious high frequencies.

Exercises

1. Why were hi-fi and stereo impractical before the 1940s?

2. In what sense does an auto engine act as a transducer?

3. Name some items useful in everyday life that depend on transducers.

4. What role do transducers play in the application of the "scientific method"?

5. An X-ray machine is a useful tool for medical and dental diagnostics. In what way can an X-ray machine be considered a transducer? Identify the input and output energy forms.

6. Why is it necessary for a hi-fi recording microphone to respond over essentially the same frequency range as the human ear, while a more limited frequency range is acceptable for the telephone?

7. Why is a DAC necessary to produce "computer music"?

8. Complete Table 6.1.

9. Which of the instruments in Exercise 8 involve transducers? Which are acoustical generators? Which are acoustical receivers?

10. Name and describe instruments appropriate to do the following: (1) Show the varying pressure of a clarinet tone. (2) Show the spectrum of an oboe tone. (3) Show the spectra of different speech sounds. (4) Show the velocity of a guitar string. (5) Measure the acceleration of the wall of a sounding trumpet.

Demonstrations

1. Transducers and measuring instruments—Connect a microphone to an oscilloscope. Draw a diagram of your setup showing what forms of energy exist at different points. Speak into the mike and observe the patterns on the scope. What do the peaks on the scope display represent? What do the valleys represent? Speak a steady vowel into the mike. What features do you observe in the oscilloscope display? "Hiss" into the mike. What features do you observe? Speak a nasal sound into the mike. What features do you observe? Whistle into the mike. What features are present? Snap

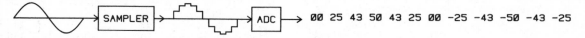

Figure 6.13. Analog-to-digital conversion showing how a continuous voltage waveform is sampled and converted into a string of numbers.

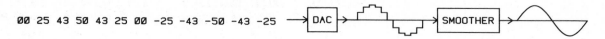

Figure 6.14. Digital-to-analog conversion showing how a string of numbers is converted into discrete samples and smoothed into a continuous waveform.

Table 6.1. Various "instruments" with their corresponding input energies and output energies.

"Instrument"	"Input energy"	"Output energy"
movie camera	light	photo
human ear	acoustical	nerve
human voice	nerve	acoustical
oscilloscope		
DAC		
ADC		
TV camera		
TV receiver		
microphone		
loudspeaker		
accelerometer		
magnetic pickup		
piezoelectric pickup		
spectrum analyzer		

your fingers or clap your hands by the mike. What features do you observe in these waveforms?

2. Function generator and measuring instruments—Connect a function generator to a loudspeaker and to an oscilloscope. Draw a diagram of this setup. Turn the function generator control to produce sine waves. Describe the appearance and sound of the sine waves. Repeat for triangular waves and for square waves. Connect the function generator to a spectrum analyzer and observe the spectra of different waves. (If you don't have a spectrum analyzer available, use the spectra of Figure 6.12.) Describe the pertinent features of each spectrum you observe.

3. Connect an electric guitar to an oscilloscope, a loudspeaker, and a spectrum analyzer. Repeat pertinent items from Demonstration 1.

4. Get an inexpensive ceramic phono cartridge and connect it to an oscilloscope and spectrum analyzer. Drag it across a "smooth" surface and observe the results.

5. Use your finger tip as a transducer. Place it on your Adam's apple while singing or speaking and while increasing and decreasing pitch and loudness. What do you observe?

6. Calibrate your finger tip as a transducer by gently touching the bass strings on a piano as the correct key is struck. At what frequency does your finger no longer sense the vibrations?

Further Reading

Backus, Chapter 15
Olson, Chapters 9, 10
Rossing, Chapters 10, 20-22, 28, 33
Kamperman, G.W. 1977. "Sound and Vibration Measuring Equipment," Sound & Vibration 11 (Jan), 8-9.

Audiovisual

The Oscilloscope (Loop 89-3966, EFL)

II. Characteristics of Sound Waves

Courtesy of Brigham Young University Archives. Photo by S. Worton.

7. Waves in Matter

Up until now, our concern has been with simple and complex vibrations that stayed at one place and didn't travel. Many vibrations, including those we hear, travel or propagate as waves from a vibrating source to other places. A **wave** is a disturbance travelling outward from a vibrating source. A **medium** is the substance or material through which a wave travels. A wave may carry energy from one place to another, but a wave does not carry the medium in which it travels from one place to another. We can understand this by imagining flipping a rope or dropping a rock in a pond of water. We can visualize the disturbance travelling in the medium (rope or water), but the medium itself does not travel from one place to another. In this chapter we will consider how waves travel through solids, liquids, and gases. We also will consider how these waves can be represented graphically.

Models of Matter and Wave Propagation

An **impulse** is a burst of energy or a disturbance of short time duration. It is often useful to "impulse" a system and observe how the system responds to the impulse. In this manner some of the interesting features of the system can be observed. Suppose, for example, that we bang a piece of iron once with a hammer—that is, we impulse the iron. If it were possible to observe the individual atoms of the iron we would notice that the impulse which we imposed on the system does not remain stationary but travels from the point of impact outward. To learn how the impulse travels we must consider the atomic structure of the solid. Figure 7.1 is a two-dimensional representation of the atoms in a solid. (In an actual solid the structure would be three-dimensional.) The structure is characterized by an orderly arrangement of atoms which are held in place by electrical forces, represented by springs. The electrical force law governs the forces that bind together the atoms or molecules in a solid. Each atom is composed of a positively charged nucleus surrounded by negatively charged electrons. The positively charged nuclei of different atoms repel each other, but they are attracted by the negatively charged electrons located between them and thus are also bound to each other. The electrical binding forces are analogous to springs, as shown in the diagram.

When one molecule of a solid is displaced from its normal position the forces represented by the spring return it to its normal position. Suppose, for example, that one molecule is displaced to the right. The "spring" to the right of the molecule is compressed while the "spring" to the left is expanded. There are, then, two forces acting to return the molecule to its normal position: the force due to the compressed "spring," which tends to push the molecule back into place, and the

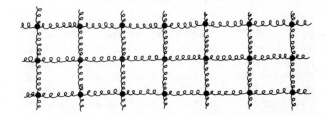

Figure 7.1. Two-dimensional diagram of molecules in a solid. The electrical binding forces are represented by springs.

force due to the expanded "spring," which tends to pull the molecule back into place. Each molecule has an equilibrium position and may undergo back and forth excursions, or oscillations, in a manner quite similar to that of a mass on a spring. When the system is disturbed by an impulse from a hammer blow, several molecules are displaced and several "springs" are correspondingly compressed and expanded. As the forces move the molecules back toward their equilibrium positions, the molecules oscillate for a short time. Since each end of a "spring" is attached to a molecule and each molecule has several "springs" attached, displacement of one molecule will soon cause more molecules, which are attached to the other ends of the "springs," to be displaced. This process will continue from each molecule to its neighbors and the impulse, in addition to causing oscillation at one point in the solid, will cause a disturbance that travels throughout the solid.

Disturbances may also propagate through liquids and gases. The propagation mechanism for liquids and gases is somewhat different than it is for solids. Figure 7.2 is a representation of the molecules in a gas. Again, there are electrical forces represented by the springs, but the forces are not strong enough to keep the molecules in a nice, orderly arrangement. The forces act only when the molecules are very close, as in a collision. For this reason the "springs" are shown attached to only one molecule. When two molecules get sufficiently close, the "springs" cause the molecules to repel; otherwise they do not affect each other. Suppose now that we impulse a gas by bumping some of the molecules on the left and thus displace them toward the right. Since the molecules are not attached, there is no restoring force, but still the molecules bumped toward the right encounter other molecules and bump them toward the right. The process will continue through the entire region where the gas is enclosed; so an impulse can be propagated in a gas as well as in a solid or liquid.

The structural models of solids and gases given here are greatly simplified to make evident some of the more salient properties of wave propagation and its dependence on the properties of the matter through which the wave passes. It is perhaps worth noting that

Figure 7.2. Two-dimensional diagram of molecules in a gas. The electrical forces of repulsion are represented by springs.

we view solids as composed of closely packed molecules that are bound in place, liquids as composed of closely packed molecules that are free to slide past each other, and gases as composed of widely separated molecules that are in a constant state of random motion and collision with one another. The strong intermolecular forces ("springs") of solids cause a solid to maintain its shape, whereas liquids and gases typically take on the shape of their containers because of the weaker intermolecular binding forces.

Wave Types

Often, a continuously recurring vibration, rather than a single impulse, disturbs a material medium. The medium is then disturbed in a continuous manner and the disturbance travels as a continuous wave. The particles which compose the medium vibrate or "wave" back and forth as the disturbance travels through the medium.

When a jump rope is pulled tight and flipped, a disturbance is caused which travels along the rope. The individual parts of the rope move up and down as the disturbance (wave) passes. A wave of this type, in which the particles of the medium move transverse (perpendicular) to the direction in which the wave travels, is called a **transverse wave** or **shear wave**. If the molecules at the left end of the solid represented in Figure 7.1 are "jiggled" up and down (as shown in Figure 7.3a), the disturbance travels through the solid toward the right because the molecules are bound to each other. If one molecule of a gas at the left (as in Figure 7.2) were jiggled up and down, the disturbance would not travel to the right because the molecules are not bound together. Transverse waves can propagate in solids but not in gases and liquids because gases and liquids have no restoring force which would act to return a displaced molecule to its former position.

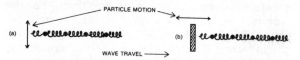

Figure 7.3. Two wave types: (a) tranverse and (b) longitudinal.

If the left wall of a gas container is jiggled (as in Figure 7.3b), however, this disturbance does travel through the gas. The wall bumps the nearest molecules, thus exerting a force which is propagated to other molecules. The forces between molecules occur only during collisions, at which time the "springs" are momentarily compressed. The disturbance, which consists of molecules shaking to and fro along the direction the disturbance propagates, spreads because the jiggling molecules bump into their neighbors. A wave in which the particles of the medium move parallel to the direc-

tion the wave travels is called a **longitudinal wave** or **compression wave**. If a solid is jiggled in and out on its left side this back and forth disturbance can also travel along the solid to the right. Longitudinal waves can therefore propagate in solids, liquids, and gases.

Finally, there are **surface waves**, such as ocean waves. These waves are essentially transverse waves, even though they are associated with liquids. They are possible because the surface tension of the liquid provides the restoring force necessary to transmit a transverse disturbance on the surface of the liquid.

Representation of Waves

Waves on a string are transverse displacement waves; the particles of the string move up-and-down as the waves travel from left to right (see Figure 7.4). Sound waves in air are longitudinal waves; the air molecules are displaced back and forth in the direction of wave travel as a sound wave passes. For sound waves in air, however, it is more convenient to speak of pressure changes produced rather than the displacement of the air molecules. When the air molecules are forced closer together than normal by the impinging disturbance, the resulting excess of air molecules produces a slight increase of pressure called a **condensation**. An increase of pressure is relative to and above normal atmospheric pressure. When the air molecules are pulled slightly farther apart than normal there is a slight decrease in pressure, called a **rarefaction**. We will refer to pressures greater than atmospheric pressure as positive (+) pressures and pressures less than atmospheric as negative (-) pressures.

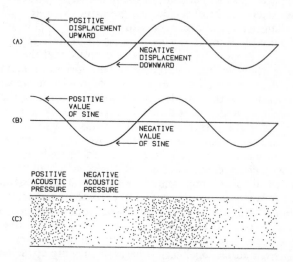

Figure 7.4. Representations of waves: (a) displacement waves on a string, (c) pressure waves in a tube, and (b) sinusoid used to represent either displacement waves or pressure waves.

Transverse waves on a string can be represented by sinusoids if we adopt the convention that the positive and negative portions of the sinusoid represent the upward and downward displacement of the string, respectively. Likewise, the longitudinal pressure waves in a tube of air can be represented by sinusoids if we adopt the convention that the positive part of the sinusoid represents pressure above atmospheric (+) and the negative part represents pressure below atmospheric (-).

As an example, imagine a vibrating piston, such as a loudspeaker diaphragm, mounted on the left end of a long tube filled with air, as shown in Figure 7.4C. As the piston moves in and out at a certain frequency it alternately produces condensations and rarefactions. The disturbance created by the vibrating piston then moves from left to right in the tube with a certain speed. The wavelength of the wave train thus produced is the distance from one condensation to the next. The series of condensations and rarefactions in the tube at any instant of time can be represented by a sinusoid as shown in Figure 7.4B.

Now consider a long coiled spring through which a longitudinal wave is passing. When the spring is compressed we have an increased "pressure" (or compression) and when the spring is expanded we have a decreased "pressure" (or expansion). The spring with a set of compressions and expansions is shown in Figure 7.5. Note that the points of maximum compression and expansion of the spring do not correspond to the points of maximum displacement of the coil. For instance, at a point of maximum compression, the displacement is zero. That is, at the center of a compression the spring is pulled in from either side, so that the spring does not move at the center. Likewise, at the center of an expansion the spring moves outward from either side and there is no displacement. The displacement of the spring in the positive direction (toward the right in Figure 7.5) is at a maximum at a point halfway between a "rarefaction" and a "condensation" (such as point A in the figure). The displacement of the spring in the negative direction (toward the left in the figure) is at a maximum at a point halfway between a "condensation" and a "rarefaction" (such as point B in Figure 7.5). Note that the pressure wave is a sinusoid, but it is displaced from the corresponding displacement wave by one-quarter of a cycle. This generally holds for pressure waves—the pressure wave differs in phase from the displacement wave by one-quarter cycle.

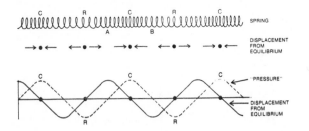

Figure 7.5. "Pressure" and "displacement" waves in a spring illustrating the relative phases between the two.

Exercises

1. Describe how an impulse travels through a solid and how the "internal" forces of a solid help to transmit the impulse. What keeps the solid from coming apart?

2. Repeat Exercise 1 for a gas.

3. Name several disturbances that travel in a gas; in a liquid; in a solid.

4. What materials (solid, liquid, gas) will transmit longitudinal waves? Why does a gas not transmit both kinds of waves?

5. If a transverse wave cannot be propagated through liquids, how do you explain water waves? What is the medium for these waves?

6. What are displacement, velocity, and acceleration for waves on a string?

7. Repeat Exercise 6 for waves in a gas.

8. Complete Table 7.1

Table 7.1. Wave features in different vibrating systems. Fill in table with appropriate features for each "instrument."

"Instrument"	Free or forced vibration?	Longitudinal or transverse wave?	Wave travel in solid or gas?
clarinet reed			
vocal folds			
vocal tract			
piano string			
violin string			
oboe tube			
drum head			
chime			

Demonstrations

1. "Sound in Vacuum," AAPT, pages 162, 243
2. "Wave Machine," Meiners, Section 18-2
3. "Wave Models," Meiners, Section 18-8
4. "Wave Speed," Meiners, Section 18-3
5. "Wave Pulse on a Rope," Freier and Anderson, page S-3
6. "Waves on an Air Track," Freier and Anderson, page S-7
7. Observe longitudinal and transverse waves in a horizontally suspended slinky. Waves can also be seen in a slinky lying on a table.
8. Observe longitudinal and transverse waves in a one-dimensional array of suspended spring-coupled masses
9. Observe longitudinal waves in a one-dimensional array of suspended uncoupled masses. Note that transverse waves cannot be produced in the uncoupled masses.
10. Generate waves in a pan of water and watch them propagate. One drop from an eye dropper produces an impulse.

Further Reading

Backus, Chapter 3
Denes and Pinson, Chapter 3
Krauskopf and Beiser, Chapter 9
Rossing, Chapter 3
Sears, Zemansky, and Young, Chapter 21

Audiovisual

1. **Discovering Where Sounds Travel** (11 min, color, 1965, PAROX).
2. **Sounds and How They Travel** (11 min, color, 1965, PAROX).
3. **Sound Waves and Their Sources**, 2nd ed.(10 min, 1950, EBE).
4. **Waves and Energy** (11 min, color, 1961, EBE)
5. "Bell in Vacuum" (REB)

8. Wave Properties and Phenomena

All waves have several common features. Waves travel with certain speeds governed by the medium in which they travel. Periodic waves have a well-defined frequency and wavelength. The wave phenomena of reflection, refraction, diffraction, and the Doppler effect are common to all waves. However, some phenomena are more apparent with sound than with light. Interference, another wave phenomenon, is so important in acoustics that the entirety of Chapter 9 is devoted to it.

Wave Properties

The medium through which sound waves travel is air; the medium for transverse waves on a taut string is the string itself. The **wave speed** is the speed at which a wave travels through a medium as determined by the physical properties of that medium. If we pluck a stretched string, we create a transverse wave which travels at a constant speed. The wave speed is increased when the string is stretched more tightly. The wave speed is decreased when the density of the string is increased. This information can be summarized by the relation $v = \sqrt{F/D}$ where v is the wave speed, F is the stretching force (or tension) applied to the string, and D is the string's density (given as mass per unit length).

The speed of sound in air is obtained from the relation $v = \sqrt{1.4p/D}$ where p is the atmospheric air pressure and D is the density of the gas (given as mass per unit volume). Note that this expression is quite similar to the equation for the string, with pressure analogous to force. A change in the temperature of the air will change both the atmospheric pressure, the density of the air, and the molecular speeds. The changes are such that as temperature increases, the speed of sound increases in direct proportion. The relation $v = 331.7 + 0.6 \times$ temperature may be used to determine the speed of sound in air (in m/s) at any temperature expressed in degrees centigrade. At room temperature the speed of sound is approximately 343 m/s.

A periodic wave, in which the wave motion repeats, has a frequency which is identical to the frequency of the vibrating source producing the wave. A periodic wave also has a wavelength associated with it. An important relationship among speed, frequency, and wavelength for a periodic wave will now be derived. Consider a vibrating string with a periodic disturbance caused by an oscillator (which moves up and down sinusoidally) attached to the left end. The wave motion travels to the right, as shown in Figure 8.1. The frequency of the oscillator is represented by f (the number of vibrations per second). The period of vibration (the time for one cycle of vibration) is related to the frequency by the expression T = 1/f. The **wavelength**, represented by λ in Figure 8.1, is the distance any part of the wave travels during a time equal to one period. Wavelength can also be defined as the distance between nearest like points on a periodic wave. The definition of wave speed is v = d/t, where d is the distance the wave travels in a time t. In a time equal to T (one period) a point on a wave will travel a distance equal to λ (one wavelength). Putting this information in the definition of speed, we obtain v = λ /T. Since 1/T = f, the expression for speed can be written as v = λ f. Hence, the speed of a wave is equal to the product of the frequency of vibration and the wavelength. Since the speed of a wave is a constant determined by the physical properties of the medium, if the frequency of the oscillator changes the wavelength changes in such a manner that the product of frequency and wavelength is constant. For instance, if the frequency is doubled, the wavelength will be halved, and vice versa.

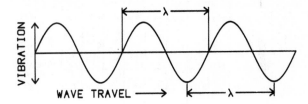

Figure 8.1.   Wave motion in a vibrating string.

Reflection

One wave phenomenon is **reflection,** an abrupt change in direction of travel when a wave encounters a change of medium. When a wave encounters a surface, or any other discontinuity in the medium, it is either reflected or absorbed—or a combination of the two. Reflections may be one of two types, regular or diffuse. A **regular reflection** occurs when a wave encounters a hard, smooth surface. All of the wave then bounces off the surface in the same direction and the wave merely changes direction. For example, a mirror reflects light waves in a regular manner, as shown in Figure 8.2a. A **diffuse reflection** occurs when the waves encounter a rough surface, as shown in Figure 8.2b. The reflected waves no longer travel in one direction, but in many different directions. This type of reflection occurs when light encounters a wall. The light reflects, but not in a regular manner as from a mirror. When sound waves are reflected in a relatively regular manner, we hear an echo if the reflecting surface is sufficiently far away. When the sound is diffusely reflected by walls we may perceive a continuation of the sound called reverberation.

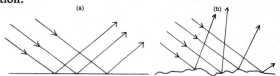

Figure 8.2.   (a) Regular reflections from a smooth surface. (b) Diffuse reflections from a rough surface.

Buildings should be properly designed to distribute reflected sound energy more or less uniformly over the entire listening area. Generally, concave surfaces are undesirable because they concentrate waves and produce focusing as shown in Figure 8.3. The vertical dashed lines represent plane waves (no curvature of the wave fronts) travelling from left to right. The horizontal dashed arrows show the direction of travel. The plane waves are reflected from the concave surface as "curved" waves travelling away from the surface. The arrows pointing away from the surface show the directions different parts of the reflected waves travel. It can be seen that these arrows all converge at a common point, the focal point for this concave parabolic surface. Waves striking this particular surface will tend to be focused when they are reflected. (We will often use arrows to show the direction waves are travelling. We will also use straight and curved lines to show wavefronts as they travel to different places in a medium.)

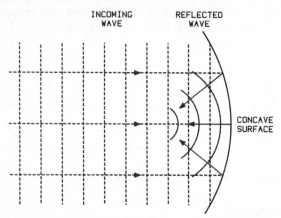

INCOMING WAVE   REFLECTED WAVE

CONCAVE SURFACE

Figure 8.3. Reflections from a concave surface. Incoming plane waves and their direction of travel are shown by dashed lines and dashed arrows, respectively. Reflected waves and their directions of travel are shown by solid curved lines and solid arrows, respectively.

The Mormon Tabernacle in Salt Lake City is an example of a building where the sound is focused because of the approximately elliptical shape of the building. When a sound, such as a pin being dropped, is produced at one focal point of the ellipse (near one end of the building) the sound energy is focused at the other focal point (near the other end of the building), so that even a soft sound is easily heard. The paths the sound takes are shown in Figure 8.4a. This focusing effect, while interesting, is highly undesirable for an auditorium because the sound is concentrated in one small region of the room instead of being distributed uniformly over the entire audience.

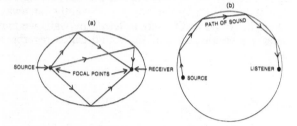

(a)   (b)

PATH OF SOUND

SOURCE   FOCAL POINTS   RECEIVER

SOURCE   LISTENER

Figure 8.4. Effects of regular reflections of sound: (a) focusing of sound in an elliptical enclosure and (b) creeping of sound around a circular enclosure.

A similar phenomenon is that of the so-called "whispering gallery." The whispering gallery is a circularly shaped enclosure. When a person stands by one wall and whispers, the sound is clearly heard on the opposite side of the enclosure, as well as at many other places along the wall. The sound is being reflected along the wall, so that it "creeps" from side to side as shown in Figure 8.4b.

Wavefronts of direct and reflected waves for a circular wave produced by a spark source in an auditorium are shown in Figure 8.5. Note the complexity of the reflection pattern as the waves reflect from the many interior surfaces. The picture is further complicated by diffraction effects, which will be discussed later.

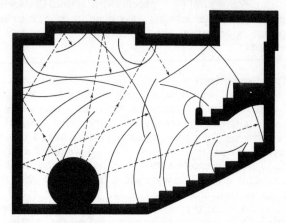

Figure 8.5. Curved wavefronts of direct and reflected sound from spark source located behind circular object. Solid curved lines show wave fronts and dashed arrows show directions of travel for some waves. Note the reflections from various structures in the hall. (After Davis and Kaye, 1927.)

Refraction

The bending of waves when they pass from one medium into a medium having a different wave speed is called **refraction**. Light, for example, bends when it passes from air to water, since light travels more slowly in water than in air as illustrated in Figure 8.6. Refraction of sound is most commonly observed when overlying layers of air are either warmer or cooler than underlying layers. The air near snow covered ground is cooler than the overlying air and consequently the sound speed is lower in the air near the ground. In such circumstances a sound wave will be bent downward,

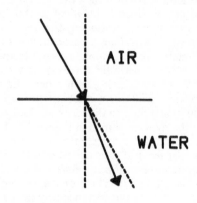

AIR

WATER

Figure 8.6. Refraction of light at air-water interface. Arrows show directions of travel. Dashed lines provide references.

because sound travels more slowly closer to the ground, where the air is cooler, and the wave is bent toward the direction of lower speed, as illustrated at the left in Figure 8.7. Under such conditions the sound will travel great distances along the ground, an effect you may have noticed during a quiet nocturnal walk through the snow. In the summer the air is generally cooler higher above the surface, and sound is consequently bent upward as illustrated at the right in Figure 8.7.

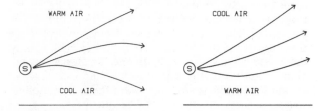

Figure 8.7. Refraction of sound due to cooler and warmer layers of air.

Diffraction

Diffraction is the bending of waves around obstacles or through openings. Figure 8.8 illustrates the bending of sound waves around a corner. We are able to hear around corners even though we cannot see around them. The amount of bending is large when the wavelength is equal to or larger than the size of the object. Sound wavelengths are large, usually measured in terms of meters. Wavelengths of light are of a smaller order measured in millionths of a meter. Sound is strongly diffracted around meter sized objects while light is not.

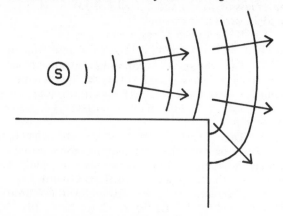

Figure 8.8. Diffraction of sound around a corner.

Diffraction also occurs when waves pass through openings. Consider a wave incident upon an opening from the left, as shown in Figure 8.9, the crest of each wave being represented by a solid line. If the opening is quite large compared to the wavelength, as portrayed in Figure 8.9a, the wave passes through the opening and continues much as before. When the opening is smaller, however, as in Figure 8.9b, some bending of the wave is observed at the edges. When the opening is smaller than the wavelength, the diffraction of the waves is so substantial that the opening becomes essentially a new source of waves, as represented in Figure 8.9c. Note that diffraction always occurs, but to be noticeable, the opening must be about the same size as, or smaller than,

a wavelength. Furthermore, the smaller the opening, the greater the diffraction. Figure 8.9d shows that for a longer wavelength and a larger opening the same amount of diffraction is observed; i.e., the diffraction depends on the ratio between the size of the opening and the wavelength. Since sound waves can be conveniently measured in meters, diffraction of sound waves through openings smaller than a meter is an important effect which is easily observed. You may have noticed that whenever someone in your dormitory is playing their stereo loudly with the door open, you hear mostly bass sounds which are the long waves. The shorter wavelength treble sounds do not diffract much through doorways and around corners and consequently are not prominently heard in the hallway.

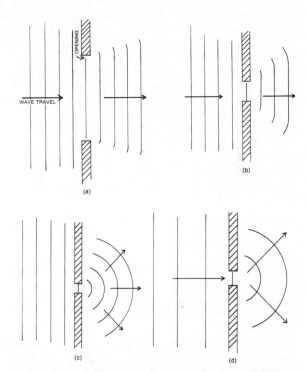

Figure 8.9. Diffraction effects as waves pass through openings: (a) wavelength small compared to size of opening, (b) wavelength and opening about the same size, (c) wavelength large compareed to size of opening, and (d) longer wavelength and larger opening.

Doppler effect

You have probably experienced the Doppler effect (at least for sound) even if you didn't understand it. As the cars on a road zoom past, you may have noticed a change in the sound's pitch as the cars approach and then pass you. This is an example of the Doppler effect. The **Doppler effect** is the change in the apparent frequency of a sound due to a relative motion between the sound source and the listener. Since we perceive frequency primarily as pitch, we hear a pitch change as the source of sound passes us. For example, when the horn on an approaching car is sounded, the pitch is higher than when the car is at rest. After the car passes us, the pitch of the horn is lower than normal.

The cause of such a pitch (or frequency) change can be illustrated by a still pool of water into which you throw your pet bug, Bugsy. If Bugsy remained stationary in the water he would create a concentric set of waves every time he splashed with his legs, as shown in Figure 8.10a. However, as he swims toward the right he creates new waves farther to the right; so as the waves move outward, they no longer form concentric circles but rather look like those shown in Figure 8.10b. Note that the waves are closer together in the direction in which Bugsy is moving and farther apart in the direction away from which he is moving. Since the wavelength is the distance between waves, we can see that Bugsy's motion through the water shortens the wavelength of the waves in the direction toward which he moves and lengthens those in the direction away from which he moves. If you are sitting in the water at a location behind Bugsy you experience a longer wavelength when Bugsy is moving (Figure 8.10b) than when he is at rest in the water (Figure 8.10a). If you are located in front of Bugsy you experience a shorter wavelength when Bugsy is moving (Figure 8.10b) than when he is at rest (Figure 8.10a). Recall, however, that wavelength and frequency are inversely related—the longer the wavelength, the lower the frequency and vice versa. Hence, at a position where the source is moving away from you you receive a lower frequency, while at a position where the source is approaching you you receive a higher frequency.

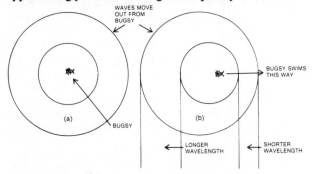

Figure 8.10.   (a) Concentric wavefronts for a vibrator at rest in the medium. (b) Displaced circular wavefronts for a vibrator moving in the medium.

The Doppler effect applies to sound waves as well as to the water waves just discussed. Suppose one of your musician friends plays a C on his trumpet while riding in an open car. When the car is not moving you perceive a pitch appropriate for C. When the car is moving toward you at a fairly high speed you perceive a C sharp even though the trumpeter is sounding a C. When the auto moves away you hear a B. The change of pitch is due to the Doppler effect.

The Leslie loudspeaker system, often used with electronic organs and by popular music groups, uses the Doppler effect to produce a frequency vibrato. In one version of the system the loudspeaker is rotated so that it has relative motion toward and away from a listener. In another version a baffle is rotated to produce apparent motion of the loudspeaker. As the loudspeaker moves toward the listener the apparent frequency is higher and as it recedes the apparent frequency is lower. The rotation rate of the loudspeaker determines the fluctuation rate of the frequency.

As an analogy to the Doppler effect, suppose that two people work at opposite ends of a conveyor belt which moves at constant speed. One person (the source) places cartons on the conveyor belt at the rate of one per second. The second person (the receiver) removes the cartons at the other end of the belt. One day the source who always places one carton per second on the belt begins walking toward the receiver while continuing to put cartons on the belt at a rate of one carton per second. The receiver, who is accustomed to receiving exactly one carton per second is now confused because the frequency of cartons has increased to two per second. Hence, motion of the source of cartons toward the receiver results in an increased frequency of cartons, even though the source is outputting the cartons at the same frequency of one per second. When the source stops his excursions the frequency returns to one carton per second. The receiver however, decides to investigate the cause of the disturbance. He begins to walk toward the source while still picking up one carton per second. Soon he notices that he has to pick up more cartons per second because he is moving toward the now stationary source. Hence, an increase of frequency also results from a stationary source and a moving receiver. Similarly, relative motion between the source and the receiver so that they are moving apart results in a decrease in frequency, regardless of whether the source or the receiver moves.

Exercises

1.   Assuming that a wave has a frequency of 500 Hz and a wavelength of 0.01 m, what is the speed of the wave?

2.   Given a wave frequency of 100 Hz and a speed of 1.0 m/s, compute the wavelength.

3.   Given a speed of 10.0 m/s and a wavelength of 0.10 m, find the frequency.

4.   If the tension in a string is 0.1 N, and the string has a mass of 10^{-5} kg and a length of 1.0 m, what is the speed of waves in the string?

5.   If the ambient pressure in air is 10^5 Pa and the density of air is 1.3 kg/m³, calculate the speed of sound.

6.   Take the speed of sound in air to be 340 m/s. What is the wavelength in air if f = 340 Hz? What is the frequency when the wavelength is 0.10 m? What is the wave speed in helium if f = 1,000 Hz and the wavelength is 0.97 m?

7.   The velocity of sound in air is about 340 m/s. If the space between the earth and the moon were filled with air, how long would it take sound to travel from the moon to the earth (a distance of 4×10^6 m)?

8.   Calculate the speed of sound at the following temperatures? (a) 70° C; (b) 32°C; (c) 12°C; (d) 0°C; (e) 20°C.

9.   What happens to the wavelength when a wave travels from a less dense to a more dense medium?

10.   Explain the difference between refraction and diffraction. Imagine running on a treadmill with different track speeds for each foot to illustrate refraction.

11.   Why can we hear around corners but not see around corners?

12.   Is there any practical significance to the fact that sound waves can be refracted?

13.   If you can clap your hands and hear an echo 0.2 s later, how far away is the reflecting surface?

14.   Fill in Table 8.1 for the frequencies heard by a listener for various possibilities of relative motion between the source of sound and the listener. Use the appropriate equation from Appendix 5. Take the speed of

sound to be 340 m/s and assume the speeds in the table are in m/s. Negative speeds in the table mean that the listener (or source) moves away from the source (or listener).

Table 8.1. Doppler changes of frequency. Fill in frequencies listener receives on basis of frequencies source emits and relative velocities of source and listener. Frequencies are in Hz and velocities are in m/s. Negative velocities indicate motion away from the source or the listener.

f(source)	v(listener)	v(source)	f(listener)
100	34	0	
1000	-68	0	
500	0	17	
300	0	-34	
200	34	-34	

15. Assume that on summer days air near the ground is warmer than air higher above the ground, but that the opposite is true at night. Will an outdoor concert band be heard over greater distances during daytime or nighttime practices?

Demonstrations

1. "Velocity of Sound," AAPT, pages 57, 123
2. "Velocity of Sound," Meiners, Section 19-2
3. "Diffraction of Sound," Meiners, Section 19-7
4. "Doppler Effect," Meiners, Section 19-6
5. "Refraction of Sound," AAPT, pages 139-140
6. "Refraction of Sound," Meiners, Section 19-8
7. "Ripple Tank Wave Phenomena," Meiners, Section 18-6
8. "Wave Pulse on a Rope," Freier and Anderson, page S-3
9. Demonstrate the Doppler effect with a small loudspeaker connected with a strong cable to an oscillator so it can be whirled about.
10. Measure sound speed in air by clapping your hands about 20-30 meters away from a plain, flat-walled building. Use a rhythm so that each clap coincides with the reflection of the previous clap. Measure the time between claps and the distance to the building and calculate the sound speed.
11. Find the frequencies of a few tuning forks by comparing them with the output of a sine wave generator; that is, adjust the frequency of the generator until it matches that of the tuning fork you are using. How do the labeled frequencies on the tuning forks compare with the sine wave generator setting? How do you account for discrepancies?
12. Measurement of wavelength and sound velocity—Set up the apparatus shown in Figure 8.11. Two sinusoids will be seen on the screen of the oscilloscope, one from the direct connection and one coming via the microphone. Try changing the distance between the speaker and the microphone. You will notice that one of the sinusoids moves relative to the other. Set the microphone close to the speaker and note the relative phases of the two sinusoids on the scope. Slowly move the microphone away from the speaker until the traces again have the same relative phase. As you do this the microphone moves by a wavelength. What is the wavelength in meters? What is the frequency reading

on the sine wave generator? Compute the sound speed in air from the relationship $v = f\lambda$. How does your result agree with the accepted value?

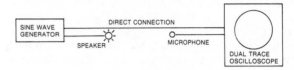

Figure 8.11. Apparatus set up for measurement of wavelength.

13. Fill a flat glass cake pan with lightly colored water and light it from below. Drop a small object, such as a marble, into it at different places relative to the sides of the pan and observe the reflected wave patterns. (Dropping water from an eye dropper may work better.)
14. Use the setup of Exercise 13, but form barriers of aluminum foil in various shapes and observe the results.
15. Doppler effect—Take a small tape recorder and record the sounding of an auto horn. Next, place the recorder beside a relatively straight stretch of road. Have the auto get up to a speed of 10-20 m/s (30-50 mph) and sound its horn as it approaches and then passes the recorder. Also, note the speedometer reading. Play the recorder into a spectrum analyzer and measure the most significant frequencies for the auto at rest, for the approaching auto, and for the receding auto. Using the frequency formula for the Doppler effect (see Appendix 5), calculate the approach speed of the auto from the measured frequencies of the horn for the auto at rest and approaching. Compute the receding velocity of the auto in a similar manner. How closely do the calculated speeds compare with each other and with the speedometer reading?
16. Fabricate two "person size" parabolic reflectors. Place them facing each other across a room. Have two people stand at the reflector focal points and carry on a conversation across the room.

Further Reading

Backus, Chapter 3
Denes and Pinson, Chapter 3
Hall, Chapter 4
Krauskopf and Beiser, Chapter 9
Olson, Chapter 1
Rossing, Chapter 3
Sears, Zemansky, and Young, Chapter 23
Stevens and Marshofsky, Chapter 1
Davis, A.H., and G.W.C. Kays. 1927. **The Acoustics of Buildings** (G. Bell and Sons)

Audiovisual

1. **Diffraction and Scattering of Waves Around Obstacles** (RT-16, MLA)
2. **Doppler Effect in a Ripple Tank** (Loop 80-2371, EFL)
3. **Reflection of Circular Waves from Various Barriers** (RT-2, MLA)
4. **Reflection of Waves from Concave Barriers** (Loop 80-233, EFL)
5. **Refraction of Waves** (Loop 80-234, EFL)
6. "Speed of Sound" (REB)
7. "Reflection from Concave Surfaces" (REB)

8. ''Refraction — Sound Lens'' (REB)
9. ''Diffraction'' (REB)
10. ''Doppler Effect'' (REB)

9. Interference and Standing Waves

In prior chapters we have discussed waves travelling through different media. However, we assumed that only single waves were present. And yet, there are many examples in which two or more waves are present at the same position or in the same region of a medium at the same time. Think of dropping two rocks into a pond at different places. The waves moving outward from each will eventually be at the same place on the water surface. The sound reaching our ears when a band plays is actually the combined sound of the sounds produced by the individual instruments.

When two or more waves occupy the same place in a medium they superpose on each other to produce a wave which is the sum of the individual waves. The **principle of linear superposition** states that when two or more waves simultaneously occupy the same position in a medium the resultant wave is the algebraic sum of the individual waves. (Algebraic sum means that positive and negative values are taken into account in the addition process.) Waves on strings of most musical instruments and pressures waves in air are generally of small amplitude so that linear superposition holds. Other cases in which wave amplitudes are so large that linear superposition does not hold usually will not concern us.

In this chapter we will find that two waves can add up to produce a larger resultant wave, or they can add in such a way as to produce a smaller resultant wave; or they can add to produce any number of intermediate cases.

Interference of Waves

Interference is the manner in which two or more waves add at given points in a medium. **Constructive interference** occurs when the waves add together in phase and create a larger wave, as shown in Figure 9.1. If the waves add out of phase a smaller wave is created, as shown in Figure 9.2, resulting in **destructive interference**.

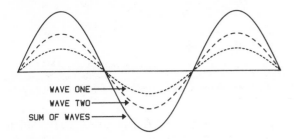

Figure 9.1. Constructive interference resulting from two waves adding in phase.

As an example of constructive interference, consider two identical positive pulses moving toward each other on a string. Because of the principle of linear superposition, the waves interfere as they occupy the same part of the string. Also, each wave maintains its

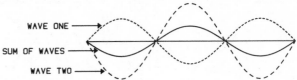

Figure 9.2. Destructive interference resulting from two waves adding out of phase.

original identity after the waves "pass through" each other. As the pulses pass through each other a larger positive pulse results, as shown in Figure 9.3A. On the other hand, if one of the pulses is negative and the pulses pass through each other, a momentary cancelation occurs, as shown in Figure 9.3B. The first case is an example of constructive interference because the waves add together and produce a larger wave. The second case illustrates destructive interference because the waves momentarily add to zero.

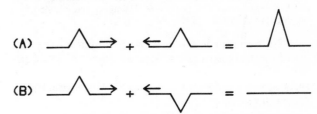

Figure 9.3. Interference of two pulses on a string: (A) constructive when pulses have same polarity and (B) destructive when pulses have opposite polarity.

Sound waves consist of regions of condensation and rarefaction. However, they can be represented as a series of peaks and valleys as shown above. Constructive and destructive interference apply equally to pressure waves in a gas. When two pressure waves of equal frequency, amplitude and phase coexist in the same medium, constructive interference results. When two pressure waves of equal frequency and amplitude but of opposite phase are traveling through the same medium, destructive interference results. This does not mean that energy is destroyed; rather the energy is redistributed.

Beats

When two sinusoids of slightly different frequencies are sounded together the resulting wave fluctuates between large and small amplitudes because the sinusoids alternately interfere constructively and destructively. The periodic fluctuations in amplitude are termed **beats**; the number of beats per second is equal to the difference in frequency between the two sinusoids. The number of amplitude fluctuations per second is equal to the number of beats. Figure 9.4 shows how two sinusoids alternately interfere constructively and destructively to produce beats.

As it applies to sound waves, the phenomenon of beats results in a fluctuation in the loudness. The sound is loud when the sinusoids add constructively and soft when they add destructively. If two sinusoids having frequencies of 200 Hz and 210 Hz are combined they will be in phase and interfere constructively 10 times per second; they will also be out of phase and interfere destructively 10 time per second. The resultant sound will flucuate from loud to soft to loud 10 times per second.

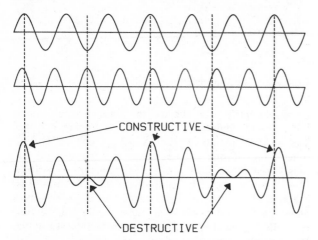

Figure 9.4. Production of beats by the alternate constructive and destructive interference of two waves. The upper and middle waves add together to form the lower wave.

The phenomenon of beats is often used for the tuning of musical instruments. A piano tuner tunes one string of three to a tuning fork by eliminating the beats. He then tunes the other two strings to the first string by the same process.

Reflected Waves and Interference

In the first sections of this chapter we considered interference of two waves created in the same medium. Another case of equal importance for most musical instruments occurs when a single wave is created on a string or in an air filled tube and is reflected from the end of the string or tube. The reflected wave travels in the opposite direction from the original wave and can interfere with the original wave. We must first determine the nature of the reflected wave in terms of the original wave and the reflecting end of the string or tube. We can then determine the nature of the interference between the original and reflected waves.

For waves on a string, the end of the string provides an abrupt change of medium for the wave. When a wave encounters a change of medium, part of the wave (in some cases all of the wave) is reflected back in the direction from which it came. If the incident (incoming) wave is positive (a crest), the reflected wave may be either positive or negative (a trough), depending on how the medium changes. Consider a pulse (Figure 9.5A) on a string. Assume that the pulse travels to the end of the string, which is fastened in place so it cannot move. As the pulse encounters the fixed end of the string, it attempts to pull the string upward. The string, however, cannot move. A reflected pulse is then created in such a way that the reflected pulse and the incident pulse add together at the fixed end of the string so there is no displacement of the string. The reflected pulse must be the inverse of the incident pulse so that when the two are added together at the fixed end they produce zero

displacement by destructive interference, as shown in Figure 9.5.

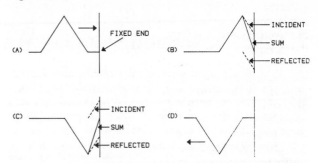

Figure 9.5. Wave reflection from the fixed end of a string: (A) incident wave, (B) wave encounters fixed end and reflection begins, (C) reflection almost complete, and (D) reflected wave has opposite polarity from incident wave.

If a pulse encounters a free end of a string, a different kind of reflection occurs. When the end of the string is free to move, an upward motion of the string occurs as the wave reaches the end of the string. Since there is no force constraining the string, a reflected pulse is produced which is in phase with the incident pulse; that is, the reflected pulse adds to the upward displacement of the string. The reflected pulse then appears identical to the incident pulse at the free end of the string, as shown in Figure 9.6. Note in Figure 9.6 that because of the constructive interference of the incident and reflected pulses the displacement of the end of the string is greater than it would have been had the pulse just passed through. This additional displacement of the end of the string causes the familiar effect of snapping a whip. When the pulse is reflected at the free end of the whip the large displacement causes the tip of the lash to move faster than the speed of sound in air, thus causing the familiar "crack" sound.

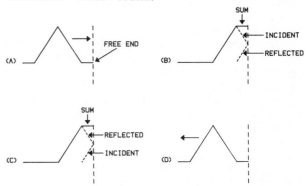

Figure 9.6. Wave reflection from the free end of a string: (A) incident wave, (B) wave encounters free end and reflection begins, (C) reflection almost complete, and (D) reflected wave has same polarity as incident wave.

Pressure waves in an air filled tube can be treated in a way analogous to displacement waves on a string. At the open end of a tube much of the original wave is reflected back into the tube because of the drastic change in propagation conditions when the wave is no longer confined by the tube. The pressure at the open end of a tube will be approximately zero because the surrounding atmosphere tends to cancel either condensations or rarefactions. (Obviously, the pressure is not exactly zero because we hear some sound. However, for the sake of determining the nature of the reflected wave we can assume the pressure is zero.) Pressure waves are reflected at the open end of a tube in the same way displacement waves are at the fixed end of a string; a positive pressure is reflected as a negative pressure. At the closed end of a tube condensations and rarefactions can occur because the closed end prevents the atmosphere from leaking in or out. Pressure waves are reflected at the closed end of a tube in the same way displacement waves are at the free end of a string; a positive pressure is reflected as a positive pressure from the closed end of a tube.

Suppose we generate a continuous triangular wave starting at the left end of a string and moving toward the right as shown in Figure 9.7A. The right end of the string is fixed so that it cannot move. As the wave encounters the boundary, it is reflected out of phase. The positive part of the wave, which encounters the boundary first, is reflected as a negative part as shown in Figure 9.7B. The negative part which follows the positive part in the incident wave is reflected as a positive part, which now follows the negative part in the reflected wave, as shown in Figure 9.7C. The result is that the reflected wave has the same appearance as the incident wave (Fig. 9.7C) because two reversals took place: one due to reflection from a fixed end, and the other due to the fact that the first pulse to encounter the fixed end is the first pulse reflected.

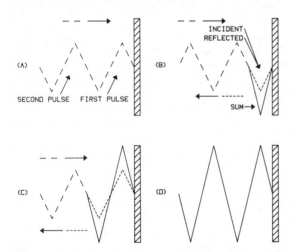

Figure 9.7. Interference of incident and reflected waves to form a standing wave: (A) incident wave, (B) positive pulse reflects as negative pulse and combines with following negative pulse to produce negative pulse of double amplitude, (C) double amplitude negative and positive pulses appear after reflection of one wavelength, and (D) standing wave after reflection of two wavelengths.

The wave pattern that results from combining the incident and reflected waves on the string in Figure 9.7 does not travel back and forth along the string as do the waves from which it is composed, but oscillates up and down while "standing still." A **standing wave** is a wave pattern that results from the alternate constructive and destructive interference of two or more travelling waves; it has points of minimum amplitude vibration called **nodes** and points of maximum amplitude vibration called **antinodes**. The positions of the nodes and antinodes are stationary or "standing" and do not move along the string.

A standing wave pattern on a string is illustrated in

Figure 9.8. Labels A, C, and E are at nodes where there is no string motion. Labels B and D are at antinodes where the vibration amplitude is at a maximum. The string vibrates from the configuration shown by the solid line to that shown by the dashed line and back again repeatedly.

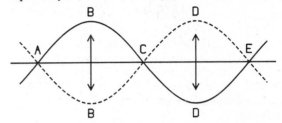

Figure 9.8. Standing wave pattern on a string illustrating nodes at points A, C, and E and antinodes at points B and D. The solid and dashed curves show the string at two different times.

When the reflection from only one end of a string or a tube is considered, as we have done here, standing waves are produced for all frequencies of vibration. However, when we consider reflections from both ends of a string or tube, as are typical of musical instruments, we will find that standing waves are produced only for certain frequencies, the natural frequencies of the system. These standing wave patterns are the natural modes of the system and will be discussed in the next chapter.

Exercises

1. Consider two waves having the same speed and wavelength approaching each other from opposite ends of a string, as shown in Figure 9.9. Sketch the resulting waves for the following conditions: Points A & AA coincide; B & BB coincide; C & CC coincide; D & DD coincide; and E & EE coincide.

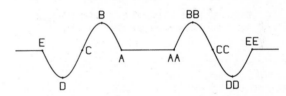

Figure 9.9. Interference of waves travelling in opposite directions.

2. Describe a situation comparable to Exercise 1 for two pressure waves in a tube.

3. Compute the beat frequencies for the tone pairs in Table 9.1.

4. A tuning fork has a frequency of 440 Hz. When sounded with another tuning fork, 5 beats per second are heard. What is the frequency of the second fork? How could you tell whether the frequency of the second fork is greater or less than the first fork?

5. Consider a positive pressure pulse in a tube which is open at the far end. Sketch the incident and reflected pulses. For an open tube the results are only approximate. Why? Consider a positive pressure pulse in a tube which is closed at the far end. Sketch the incident and reflected pulses.

6. Suppose the pulses in Exercise 5 were negative instead of positive. Sketch the incident and reflected impulses for the open and for the closed tubes in this case.

Table 9.1 Beat frequencies of tone pairs. Fill in beat frequency for each tone pair.

Tone pair	f_1 (Hz)	f_2 (Hz)	f_{beat} (Hz)
1	100	101	
2	300	306	
3	400	392	
4	1,000	998	

7. The results in Exercises 5 and 6 seem to be just opposite from what we might expect intuitively from the examples of displacement waves on a string as found in the text. Why? Would we get different results if we considered displacement waves in the tube rather than pressure waves?

8. Suppose a triangular wave generator is attached to the far left end of the string in Figure 9.10. The right end of the string is fixed, so that it cannot move. Suppose the sketches in Figure 9.10 show just enough of the right portion of the string so that one wavelength of the original wave train is visible. Sketch the resulting reflected wave and then the sum of the original and reflected waves.

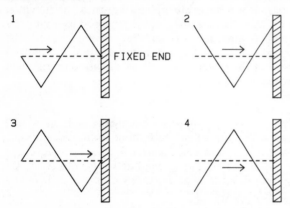

Figure 9.10. Triangular wave incident at the fixed end of a string. Frames 1, 2, 3, and 4 show the wave at successively later times.

9. Sketch the incident, reflected, and sum of the waves as in Exercise 8 for the free end of a string.

10. Repeat Exercise 8 for pressure waves at the open end of a tube.

11. Repeat Exercise 8 for pressure waves at the closed end of a tube.

Demonstrations

1. "Acoustical interference," AAPT, page 65
2. "Interference of Waves," Meiners, Section 18-4
3. "Wave Machine," Meiners, Section 18-2
4. "Wave Reflections at a Discontinuity," Freier and Anderson, page S-5
5. "Interference of Sound Waves," Freier and Anderson, page S-26
6. Interference of sinusoids as beats—Connect a sine wave generator and a function generator to a mixer. Connect the mixer to an oscilloscope and to a loudspeaker. This will enable you to both see and hear the results. Set the function generator to produce sine waves and adjust its frequency to about 200 Hz. Set the amplitudes of the two generators to be about equal. (Compare it with Figure 9.4.) Adjust the frequency of the sine wave generator so that you get about one beat per second. Describe the waveform that you see on the scope. Adjust the sine wave generator so that you get two or three beats per second. How does the result differ.
7. Set up standing waves in a wave machine, a jump rope, or a pan of water by getting the phase of the incident wave adjusted properly relative to that of the reflected wave.

Further Reading

Backus, Chapter 3
Hall, Chapter 4
Olson, Chapter 1
Rossing, Chapter 3
Sears, Zemansky, and Young, Chapter 22

Audiovisual

1. **Standing Waves in a Gas** (Loop 80-3874, EFL)
2. **Standing Waves on a String** (Loop 80-3866, EFL)
3. **Superposition of Pulses** (Loop 80-2397, EFL)
4. **Vibrations of a Drum** (Loop 80-3924, EFL)
5. **Vibrations of a Metal Plate** (Loop 80-3932, EFL)
6. **Vibration of a Rubber Hose** (Loop 80-3890, EFL)
7. "Interference" (REB)
8. "Beats" (REB)

10. Standing Waves

As noted in the previous chapter, large standing waves can occur at any frequency on a "long" string where the relationships between incident and reflected waves need be considered at only one end of the string. In this chapter we will consider the conditions that must be satisfied to produce standing waves on "short" strings and in "short" tubes where relationships between incident and reflected waves must be considered at both ends of the system. In particular, the boundary conditions at each end of a string or tube determine the natural modes and natural frequencies for standing waves. The same concepts are extended to standing waves in two and three dimensional systems.

Standing Waves on Strings

Strings used on musical instruments such as guitars, violins, or pianos, are fixed at both ends. Any disturbance produced on such a string will travel to the two ends of the string where it will be reflected in accordance with the rules discussed in Chapter 9. The various waves travelling back and forth along the string will interfere with each other. Nodes always exist at the two fixed ends of the string but may or may not exist at other points on the string. If we vibrate the string at just the right frequencies, we can produce standing waves where some parts of the string are stationary and where other parts move with maximum amplitude. The standing waves are examples of the natural modes of compound systems discussed in Chapter 5. We must now explore these natural modes in more detail.

The simplest standing wave pattern which can be produced on a string fixed at both ends is that shown in Figure 10.1A. It consists of a node at each end of the string and an antinode (point of maximum vibration) in the center. The length of the string in this case is one-half wavelength. The frequency corresponding to this wavelength is called the **fundamental frequency** of the vibrating string; so called because it is the frequency associated with the first natural mode which can form a standing wave pattern on the string. If the string is vibrated more rapidly (thus increasing the frequency) another standing wave pattern eventually results, as shown in Figure 10.1B. This pattern consists of a node at each end and one in the center. In this case the length of the string is equal to one wavelength. The wavelength is half as long and the frequency twice as great as in the preceding case because the wave speed is the same for all frequencies. The next standing wave pattern which can be produced is shown in Figure 10.1C; the frequency in this case is three times the fundamental frequency. These frequencies are the first three natural mode frequencies of the string and the wave pattern associated with each frequency is called a natural mode of vibration.

The natural frequencies can also be calculated from the relation $v = f\lambda$, where the wave speed is determined from the string's density and tension. The wavelength

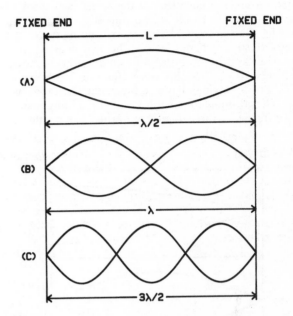

Figure 10.1. Natural modes of a string fixed at both ends: (A) first mode with $\lambda = 2L$, (B) second mode with $\lambda = L$, and (C) third mode with $\lambda = 2L/3$.

can be expressed in terms of the length of the string. For the first mode half a wavelength is equal to the length of the string as can be seen in Figure 10.1A. The wavelength of the first mode is then twice the length of the string or $\lambda_1 = 2\ell$. This results in $f_1 = v/\lambda_1 = v/2\ell$ for the fundamental frequency of a string fixed at both ends. Similarly, for the second mode the wavelength is equal to the string length or $\lambda_2 = \ell$. This results in a second mode frequency of $f_2 = v/\lambda_2 = v/\ell$ which is twice the frequency of the first mode. The third mode has a wavelength one third that of the first mode and a frequency three times that of the first mode, and so on for the higher modes.

From the foregoing we see that the frequency of each mode is the mode number multiplied by the fundamental frequency for a string fixed at both ends. If the fundamental frequency is 100 Hz, the second mode has a frequency of 200 Hz, the third 300 Hz, and so on. We can summarize by saying that the natural frequencies of a "fixed-fixed" string, that is, a string fixed at both ends, are given by $f_n = n \times f_1$, where n = 1,2,3 . ... The natural frequencies of the fixed-fixed string are termed **harmonics** because they are integer (whole number) multiples of the fundamental frequency. More will be said in Chapter 16 about harmonics and their role in musical harmony.

Consider now the physically unrealistic, but instructive, situation of a string which is fixed at one end and free at the other. The fixed end of the string must always be a node, since it is fastened in place. Similarly, the free end of the string will always be an antinode, since there is no force to hold it in place. Figure 10.2A shows the smallest part of a wave which can "fit" on the string and still satisfy the endpoint conditions. Clearly this is one-quarter of a wavelength, so the fundamental wavelength is four times the length of the string. Now try to fit a standing wave on the string that is equal to one-half of a wavelength. Such a wave would require a node at each end of the string, as shown in Figure 10.2B, but this is not possible since the free end must be at an antinode. Hence there is no mode for the fixed-free string where $f = 2f_1$. The next smallest fraction of a wavelength which "fits" on the string is shown in Figure 10.2C. For this second natural mode three-fourths of a wavelength fits on the string; this mode has a wavelength equal to one-third that of the fundamental and a frequency three times that of the fundamental, as shown in Figure 10.2C.

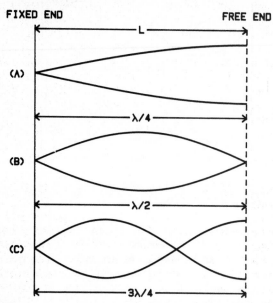

FIXED END   **FREE END**

Figure 10.2. Natural modes of a string fixed at one end and free at the other: (A) first mode with $\lambda = 4L$, (B) unallowed mode, and (C) second mode with $\lambda = 4L/3$.

The modal frequencies of a fixed-free string can be calculated in a manner similar to that used for a fixed-fixed string. Referring to Figure 10.2A, we see that the wavelength of the lowest mode is one quarter the length of the string or $\lambda_1 = 4\ell$. This results in $f_1 = v/4\ell$ as the frequency of the fundamental mode. For the second natural mode (Figure 10.2C) three fourths of a wavelength is equal to the string length or $\lambda_2 = 4\ell/3$; this results in $f_2 = v/\lambda_2 = 3v/4\ell$ for the frequency of the second mode, which we note is three times the frequency of the first mode. By studying higher modes we discover that the higher natural mode frequencies are odd integer multiples of the fundamental frequency. We can summarize by saying that the natural frequencies of a fixed-free string are given by $f_n = (2n - 1) \times f_1$, where $n = 1, 2, 3 \ldots$.

Two interesting features become apparent when we compare a fixed-fixed string with a fixed-free string. First of all, for fixed-fixed and fixed-free strings of equal length, ℓ, and with the same wave speed, v, the fundamental frequency of the fixed-fixed string ($f_1 = v/2\ell$) is twice that of the fundamental frequency of the fixed-free string ($f_1 = v/4\ell$). Secondly, the fixed-fixed string has natural frequencies that are integer multiples

of its fundamental frequency ($f_n = n \times f_1$), whereas the fixed-free string has natural frequencies that are odd integer multiples of its fundamental frequency ($f_n = (2n - 1) \times f_1$).

Standing Waves in Tubes

While musically useful strings are usually fixed at both ends, musically useful air tubes can be open at both ends, as with the flute, or closed at one end and open at the other, as with the clarinet. Still, we can use the same basic techniques to explore natural modes for air columns as we used for strings.

For tubes whose diameters are small compared to the wavelength of waves travelling in them, the pressure waves in the air in the tube exhibit many of the same features as the standing waves in a string—except that pressure is approximately zero at the open end of a tube. That is, a pressure node exists at an open end. This is because a pressure wave is reversed upon being reflected at the open end of a tube, much as a displacement wave on a string reflects from a fixed end of a string. Hence, at an open end the incident and reflected waves add together destructively, thus producing a pressure node. At the closed end of a tube, a pressure wave reflects with the same phase as the incident wave. The incident and reflected waves then add together constructively, thus producing a pressure antinode at the closed end of a tube.

In many ways, pressure waves in an open-open tube are analogous to displacement waves on a fixed-fixed string, and pressure waves in an open-closed tube are analogous to displacement waves on a fixed-free string. A general rule is that when the endpoint conditions are the same at the two ends of a string or a tube the higher mode frequencies are integer multiples of the fundamental; when the end point conditions are different at each of the two ends of a string or a tube, the higher mode frequencies are odd-integer multiples of the fundamental.

The following method can be used to determine experimentally the natural frequencies for a tube closed at one end and open at the other. A variable frequency sine wave oscillator is used to drive a small loudspeaker that is placed in the closed end of the tube, as shown in Figure 10.3. A microphone also is inserted in the closed end to measure pressure, and the corresponding microphone voltage output is displayed on an oscilloscope. The frequency of the oscillator is varied from low to high, and a graph representing the pressure (the microphone output) at the closed end of a tube 170 mm long is made as shown in Figure 10.4. A means (not shown in Figure 10.3) is provided to cause the loudspeaker to move constant amounts of air at all frequencies so that variations in the microphone output as

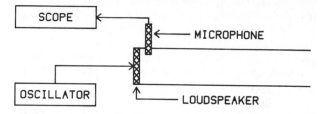

Figure 10.3. Apparatus for measuring the input impedance response curve of a tube.

the frequency changes are due to wave interference in the tube.

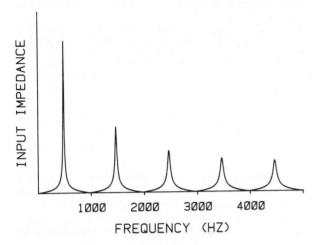

Figure 10.4. Response curve of a cylindrical tube.

Two basic modifications occur as the waves travel in the tube. Waves travelling back and forth along the tube interfere and produce large standing waves at the natural frequencies of the tube as indicated by the peaks in the curve. Waves in the tube lose part of their energy because of frictional forces at the walls of the tube. Losses at the walls are greater at higher frequencies as indicated by the smaller peaks at higher frequencies in the curve.

Measured at different frequencies, the pressure response of a tube gives a response curve of input impedance plotted versus frequency. The **input impedance** of a tube is the ratio of pressure to volume of air flow in a sound wave where both pressure and air flow are measured at the input end of the tube. The response curve shows us how the tube responds at many different frequencies. High values of input impedance occur for frequencies at which strong standing waves are produced and low values occur for weak standing waves. Note in Figure 10.4 that the frequencies at which the input impedance is largest are the resonance frequencies of a closed-open tube and the higher resonance frequencies are odd-integer multiples of the lowest resonance frequency. The frequencies of the first three peaks are about 500, 1500, and 2500 Hz.

Standing Waves on Membranes

Waves moving from one end of a string to the other end and back again are one-dimensional waves; that is, they are confined to move in one dimension back and forth along the string. When a stretched flexible membrane, such as a drumhead, is hit with a mallet, a pulse is produced. The pulse can travel in two dimensions because the membrane has extent in two dimensions. The pulse travels outward from the point of impact until a fixed boundary is encountered. At each boundary encounter the wave is reflected, and standing waves can be set up on the membrane, in a manner similar to that on a vibrating string. The nodes of a vibrating membrane are not points, as for the string, but are lines on the surface, called **nodal lines**. The boundaries of the membrane are obviously one set of nodal lines. Each mode of vibration has a definite frequency, the lowest fre-

quency mode being the fundamental. Although the frequency of each higher mode may be expressed in terms of the fundamental, we will see that higher mode frequencies may or may not be integer multiples of the fundamental.

The lowest mode of a rectangular membrane fixed along all of its edges is one in which all parts of the membrane move together; an antinode exists in the center while nodes exist along the edges. The vibration frequency of the lowest mode is given by $f_{1,1} = (v/2) \sqrt{((1/\ell 1)^2 + (1/\ell 2)^2)}$, where $f_{1,1}$ is the frequency of the lowest mode, v is the wave speed in the membrane and $\ell 1$ and $\ell 2$ are the lengths of the membrane in each of its two dimensions. The wave speed v depends on the mass per unit area of the membrane and on the tension per unit length along the edge of the membrane. For comparison recall that the lowest mode frequency of a string fixed at both ends is $f_1 = v/2\ell$, which is analogous to the above expression. Note that the natural mode frequency for the membrane has two subscripts. Each subscript refers to the conditions on the membrane in one of its two dimensions. The fact that each subscript is a "1" indicates that the membrane vibrates as one unit in that direction. The vibration frequencies of other modes of the membrane are given by $f_{n1,n2} = (v/2) \sqrt{((n1/\ell 1)^2 + (n2/\ell 2)^2)}$, where $f_{n1,n2}$ is the vibration frequency of the n1, n2 mode, v is the wave speed in the medium, $\ell 1$ and $\ell 2$ are the lengths of the membrane in each of its two dimensions, and n1 and n2 are integers indicating the mode. When n1 = 1 and n2 = 1 we have the first mode discussed above. With n1 = 1 and n2 = 2 we have a mode in which the membrane vibrates in one part along the first dimension and in two parts along the second dimension with a nodal line separating the two parts. The frequencies of the higher modes will not all be integer multiples of the lowest mode frequency, although some higher mode frequencies may fulfill this condition depending on the particular proportions of the rectangular membrane. Figure 10.5 illustrates the fundamental (1,1) mode and the (2,1) mode on a rectangular membrane.

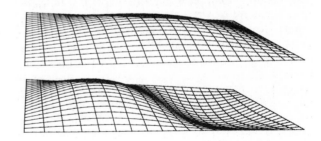

Figure 10.5. Two natural modes of a rectangular membrane: $f_{1,1}$ mode (upper) and $f_{2,1}$ mode (lower).

As examples of the first few natural frequencies, consider a membrane with $v = 40$ m/s, $\ell 1 = 0.1$m, and $\ell 2 = 0.2$m. For this membrane, $f_{1,1} = 223.6$ Hz, $f_{1,2} = 283$ Hz, $f_{1,3} = 360$ Hz, $f_{1,4} = 447$ Hz, and $f_{2,2} = 447$ Hz $= 2f_{1,1}$.

Sound Waves in Rooms

Sound waves moving back and forth between opposite walls or between floor and ceiling in rectangular rooms are also one-dimensional waves. However, two-dimensional waves (striking four surfaces) or three-dimensional waves (striking six surfaces) can also exist in rooms. As a result, very complex standing wave patterns can exist in rooms because of the interference of waves reflected from room surfaces or from objects in the room.

Consider three-dimensional pressure waves in a hard-walled rectangular room. Pressure antinodes will exist at the walls. The vibration frequencies of the various modes of the room are given by $f_{n1,n2,n3} = (v/2) \sqrt{((n1/\ell1)^2 + (n2/\ell2)^2 + (n3/\ell3)^2)}$, where v is the wave speed in air, $\ell1$, $\ell2$, and $\ell3$ are the lengths of the room in each of its three dimensions, and n1, n2, and n3 are integers indicating the mode. With only one of the n's not equal to zero, one-dimensional modes exist; with two of the n's nonzero, two-dimensional waves exist; and with all three n's nonzero, three-dimensional waves exist. In general, most of the higher mode frequencies are not integer multiples of the fundamental, although some may be.

As examples of the first few three-dimensional modes, consider a room with $\ell1 = 4.0$ m, $\ell2 = 5.0$ m, $\ell3 = 6.0$m, and v = 340 m/s. For this room, $f_{1,1,1} = 61$ Hz, $f_{2,1,1} = 96$ Hz, $f_{1,2,1} = 85$ Hz, $f_{1,1,2} = 79$ and $f_{2,2,2} = 122$ Hz $= 2f_{1,1,1}$.

Hard-walled rooms having parallel sides give rise to very strong standing waves, which means that the pressures measured at different points in the room may be quite different, depending on whether the measurement is made at a node, an antinode, or an intermediate spot. So-called "dead spots" occur at nodal positions in rooms and are especially prevalent when strong standing waves exist. Anything that can be done to reduce the standing wave and to make the sound pressures in the room more uniform will help to alleviate the problem of dead spots. Standing waves can be reduced by placing absorbing materials in the room and by making the room more irregular.

Natural modes exist on membranes and in rooms that are other than rectangular in shape. It is not possible, however, to write simple expressions for natural mode frequencies in these cases. More will be said in later chapters about such modes.

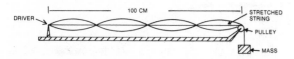

Figure 10.6. Apparatus to demonstrate standing waves in a wire.

Exercises

1. Sketch the third, fourth, and fifth modes for a fixed-fixed string. Express the frequency of each mode in terms of the fundamental frequency, f_1.

2. Sketch the third, fourth, and fifth modes for a fixed-free string. Express the frequency of each mode in terms of the fundamental frequency, f_1.

3. Draw the first four modes for pressure waves in an open-open tube.

4. Draw the first five modes that might be considered pressure waves in an open-closed tube. Place an

X through any mode that is not possible, such as the mode shown in Figure 10.2B.

5. Fill in Table 10.1.

Table 10.1 Features of string and tube systems. Fill in the table with an expression for f_n for each system. List other features of each system.

System	f_1	f_n	Features of system
fixed-fixed string	200		
fixed-free string	100		
open-open tube	200		
open-closed tube	100		

6. Calculate the frequencies of (1,1), (1,2), (2,1), and (2,2) modes of a square membrane with a side length 0.5 m if the wave speed is 30 m/s. What special relationship exists between the (1,2) and (2,1) modes that does not hold for rectangular membranes in general?

7. What gives rise to "dead spots" in rooms?

8. Describe how treating walls with absorbing materials, making a room of irregular shape, and placing objects in a room help to alleviate "dead spots" in the room.

Demonstrations

1. "Resonance Tube," Freier and Anderson, page S-12

2. "Hoot Tube," Freier and Anderson, page S-13

3. "Frequency Tube," Freier and Anderson, page S-13

4. "Flaming Tube," Freier and Anderson, page S-8

5. "Organ Pipe," Freier and Anderson, page S-14

6. "Sonometer," Freier and Anderson, page S-21

7. "Mode of String Oscillations," Freier and Anderson, page S-21

8. "Two Dimensional Birthday Cake," Freier and Anderson, page S-9

9. "Chladni Plates," Freier and Anderson, page S-9

10. "Standing waves in strings—The apparatus shown in Figure 10.6 is set up to produce standing waves in a wire. A 60-Hz driver is attached to the left end. The density of the wire is D = 0.000642 kg/m. The length of the wire is 1.0 m. The tension applied to the wire can be controlled by adding mass. When a mass of 0.3 kg is suspended from the wire, the fourth mode is produced as shown. What is the wave speed for this mode? What is the wavelength? What mass must be suspended to produce the second mode? The first mode?

11. A loudspeaker is set in the corner of a hard-walled rectangular room with dimensions of 3.0 × 4.0 × 5.0 m. A sine wave generator is used to drive it. A sound level meter is placed in a diagonally opposite corner to measure sound pressure levels. A maximum sound pressure level is observed to occur at a frequency of 78 Hz. What room mode is being excited? What is the approximate sound speed in the room? For what mode and at what next higher frequency would another sound pressure level maximum occur?

Further Reading
AAPT, pages 83-85
Backus, Chapters 4, 9
Benade, Chapters 7, 9
Hall, Chapters 9, 12
Olson, Chapters 1, 8
Roederer, Chapter 4
Sears, Zemansky, and Young, Chapter 22
Stevens and Warshofsky, Chapter 4

Audiovisual
1. **Standing Waves in a Gas** (Loop 80-3874, EFL)
2. **Standing Waves on a String** (Loop 80-3866, EFL)
3. **Vibration of a Rubber Hose** (Loop 80-3890, EFL)
4. **Vibrations of a Drum** (Loop 80-3924, EFL)
5. **Vibrations of a Metal Plate** (Loop 80-3932, EFL)
6. "Shive Wave Machine - Standing Waves" (REB)
7. "Standing Sound Waves" (REB)
8. "Reflection of Pulses" (REB)
9. "Standing Waves in a String" (REB)
10. "Mersenne's Law" (REB)
11. "Standing Waves in air Columns" (REB)
12. "Chladni Plates" (REB)
13. "Resonance Curves" (REB)

11. Complex Waves

Almost every sound we hear is caused by a complex wave. A **complex wave** is any wave other than a sinusoid. The fact that many waves can be superposed to form a single complex wave makes possible the recording of a full symphony orchestra in a single record groove or on a single track of tape. When a record is played, the complex waveform recorded in its groove is picked up by the needle, eventually causing a loudspeaker to vibrate and reproduce the orchestra sound. Some understanding of complex waves is also important in understanding sound waves produced by the human voice and by musical instruments. Any musical sound, no matter how complicated, can be represented by a complex wave.

Any complex waveform can be constructed from sinusoidal components. The study of complex waves can be approached from complementary directions: by analysis, which is the breaking down of a complex wave into its components, or by synthesis, which is the building up of a complex wave from its components. We will first consider the analysis of sound waves, then their synthesis, and finally the role of a wave's spectrum in determining tone color.

Analysis of Complex Waves

The manner in which complex waves are related to the sinusoids discussed in previous chapters is rather interesting. In Chapter 5 we found that any complicated motion of a compound vibrator can be formed by adding various amounts of its natural modes. The superposition principle discussed in Chapter 9 states that when two or more waves simultaneously occupy the same region of a medium, the resultant wave is the point-by-point sum of all individual waves. We already know that the natural modes of waves on strings or in tubes are sinusoids with frequencies which are multiples of the fundamental frequency. By applying the superposition principle we see that complex waves can be constructed by the addition of the natural mode sinusoids. This result was obtained (by mathematical means) in 1822 by a French mathematician, Joseph Fourier. In nonmathematical terms, **Fourier's theorem** states that any repetitive wave pattern, no matter how complex, may be broken down into constituent sinusoids of different amplitudes, frequencies, and phases. If a complex wave exists on a string or in a tube, it can be shown that the sinusoidal components are in fact the natural modes of the system.

Each sinusoid helping to make up a complex wave is called a **partial**. The first partial, the lowest-frequency sinusoid, is known as the **fundamental**. The higher partials (second, third, etc.) of the complex tone can be either one of two different types: harmonic or inharmonic. Higher partials whose frequencies are integer multiples of the fundamental frequency are called **harmonic partials**, or **harmonics**. When the frequencies of the higher partials are not integer multiples of the fun-

damental frequency they are termed **inharmonic partials**. The term **overtone** is sometimes used to refer to those harmonics having frequencies higher than the fundamental. Within this system the second harmonic is known as the first overtone, the third harmonic as the second overtone, and so on. Because discussion of the overtone system is not necessary here, and because it often leads to confusion, we will not consider it further.

As noted above, each ingredient of a complex wave is a sinusoid having a different frequency. The spectrum for a particular wave tells us the amount of each ingredient present. The analysis of a complex wave, then, means determining the spectrum of the wave. There are several common methods used to analyze complex waves in order to determine spectra, including electronic filtering and digital Fourier analysis. Since the several spectral analysis methods produce similar results, the conceptually simpler filter method will be discussed.

A **filter** is a device which allows certain frequencies to pass through unchanged, while other frequencies are eliminated. You may think of a set of filters as having a function analogous to a set of screens, each with a different mesh size. The screens are used to sort a conglomerate of gravel into coarse, medium, and fine components. Likewise, a set of three filters (each having different characteristics) could be used to separate a complex wave into low, medium, and high frequency components. Figure 11.1 shows how a complex wave could be resolved into its components by passing it through a set of filters. The high resolution spectrum analyzer described in Chapter 6 can be thought of as being composed of such a set of filters. However, there are about 200 to 400 filters in a typical spectrum analyzer, as opposed to the three filters illustrated. The greater the number of filters, the more finely the frequency components can be resolved during a spectrum analysis; that is, with more filters smaller differences of frequency may be detected.

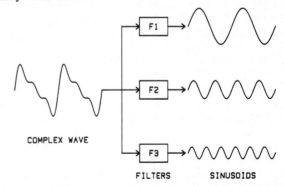

Figure 11.1.   The analysis of a complex wave with a set of filters tuned to different frequencies.

As an example of the spectral analysis of complex waves, consider the three waveforms shown in the left column of Figure 11.2. The waveforms shown are the pictures obtained when three different "stylized" instruments are sounded: (a) a tuning fork, (b) a clarinet, and (c) a trumpet. Because of the complex nature of these waveforms, little information can be gleaned by observing them. If the waveforms are analyzed with a spectrum analyzer to obtain their spectra, as shown in the right column of Figure 11.2, the pertinent information is more apparent. We see, for instance, that the tuning fork waveform consists of only one component, the fundamental, while the clarinet displays predominantly odd harmonics. The trumpet is seen to have all harmonics, but the second harmonic has the greatest amplitude.

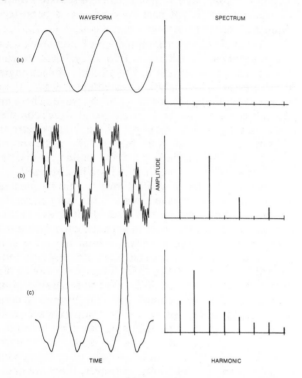

Figure 11.2. Idealized waveforms and spectra of several "musical instruments": (a) tuning fork, (b) clarinet, and (c) trumpet.

Synthesis of Complex Waves

In Chapter 9 we used the superposition principle to add waves having identical or only slightly different frequencies. The same principle can be used when waves of quite different frequencies are added together. The synthesis of a complex wave implies that the resultant wave is constructed by adding together simple sinusoids, "piece by piece." A repetitive waveform of any shape can be constructed by this method if enough components are added together. As a simple example of a graphical method used to add sinusoids consider the following situation. A mass is attached to a spring which is mounted on a frame in a boat. When the mass is set into vibration it repeats its motion every second when observed from inside the boat. The boat rocks up and down once every four seconds due to passing waves. When the motion of the mass is observed from the dock, we see a resultant motion that is the sum of two motions. If we plot the motion of the mass as seen from the boat we obtain Figure 11.3A. Figure 11.3B is a plot of the motion of the boat as seen from the dock. Using a synthesis approach and adding these two yields Figure

11.3C which is the motion of the mass as seen from the dock.

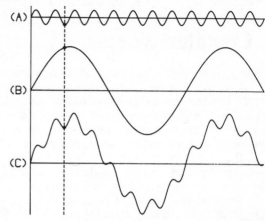

Figure 11.3. (A) Vibrating mass as seen from boat. (B) Motion of boat as seen from dock. (C) Motion of mass as seen from dock.

The procedure used to add two waves graphically is illustrated in Figure 11.3. A point along the time axis of the two waves (Figures 11.3A and 11.3B) is marked by the dashed vertical line. A ruler is used to measure the displacement of each wave at this point in time along the vertical line. The two measurements are added keeping in mind that displacements above the axis are positive numbers while those below the axis are negative. A mark is made in Figure 11.3C representing the result of the addition. The procedure is repeated for the remaining points on the waveforms. Connecting the marks provides a smooth curve.

An equation representing pressure waves can be written as $p = A \sin(ft + \phi)$, where p is the quantity you would plot on a graph of pressure versus time, A is the pressure amplitude of the sinusoid, f is the frequency, t is time, and ϕ is the phase in fractions of a cycle. As an example, we can obtain the relative amplitudes for the first two non-zero partials of a square wave as 1 and 1/3 from Chapter 6. We do not get information on the relative phases which some experimentation would show should be 0 in both cases. If the frequencies for these partials are 10 Hz and 30 Hz, we can write an expression for our square wave as $p = p_1 + p_2$, where $p_1 = 1.0 \sin(10 t)$ and $p_2 = 0.33 \sin(30 t)$. We now construct Tables 11.1 and 11.2 which give values for p_1 and p_2 at different instants of time. Adding p_1 and p_2 for each instant of time gives p. In order to produce smoother curves the graphs showing p_1, p_2, and p, in Figure 11.4 were plotted at much shorter time intervals than those given in Tables 11.1 and 11.2. The data points from the tables are also shown in the figure. The lower graph is a sum of the upper two. The total time interval from 0 to 0.1 s was chosen to cover one cycle of vibration.

Now consider a different relative phase for p_2 so that $p_2 = 0.33 \sin(30 t + 0.5)$. We construct Table 11.3 for p_2 with a phase of 0.5. Adding the original p_1 and the p_2 with the new phase gives us the complex wave in Figure 11.5. The spectra of the complex waves in Figures 11.4 and 11.5 are the same, but the wave shapes differ because of different relative phases. If the two were continuously repeated, they would sound almost identical.

66

Table 11.1. Steps in calculation of $p_1 = \sin(10t + 0)$. Time t is in seconds and arguments for sine function are in cycles.

t	10t	(10t + 0)	sin (10t + 0)	p_1
0	0	0	0	0
0.01	0.1	0.1	0.59	0.59
0.02	0.2	0.2	0.95	0.95
0.03	0.3	0.3	0.95	0.95
0.04	0.4	0.4	0.59	0.59
0.05	0.5	0.5	0	0
0.06	0.6	0.6	-0.59	-0.59
0.07	0.7	0.7	-0.95	-0.95
0.08	0.8	0.8	-0.95	-0.95
0.09	0.9	0.9	-0.59	-0.59
0.10	1.0	1.0	0	0

Table 11.2. Steps in calculation of $p_2 = 0.33 \sin (30t + 0)$. The starting phase is 0 cycle.

t	30t	(30t + 0)	sin (30t + 0)	p_2
0	0	0	0	0
0.01	0.3	0.3	0.95	.31
0.02	0.6	0.6	-0.59	-0.19
0.03	0.9	0.9	-0.59	-0.19
0.04	1.2	1.2	0.95	.31
0.05	1.5	1.5	0	0
0.06	1.8	1.8	-0.95	-0.31
0.07	2.1	2.1	0.59	0.19
0.08	2.4	2.4	0.59	0.19
0.09	2.7	2.7	-0.95	-0.31
0.10	3.0	3.0	0	0

Table 11.3. Steps in calculation of $p_2 = 0.33 \sin (30t + 0.5)$. The starting phase is 0.5 cycle.

t	30t	(30t + 0.5)	sin (30t + 0.5)	p_2
0	0	0.5	0	0
0.01	0.3	0.8	-0.95	-0.31
0.02	0.6	1.1	0.59	0.19
0.03	0.9	1.4	0.59	0.19
0.04	1.2	1.7	-0.95	-0.31
0.05	1.5	2.0	0	0
0.06	1.8	2.3	0.95	0.31
0.07	2.1	2.6	-0.59	-0.19
0.08	2.4	2.9	-0.59	-0.19
0.09	2.7	3.2	0.95	0.31
0.1	3.0	3.5	0	0

As a final example, consider the synthesis of a pulse-like wave formed by adding the following three sinusoids: $p_1 = 1.0 \sin (10t + 0.75)$; $p_2 = 1.0 \sin (20t + 0.25)$; and $p_3 = 1.0 \sin (30t + 0.75)$. The values of p_1, p_2, and p_3 at various times t, as well as the sum $p = p_1 + p_2 + p_3$, are given in Table 11.4. The waveforms for p_1, p_2, and p_3, are graphed in Figure 11.6. The lower part of the figure shows their sum. Note that this complex wave could be synthesized by adding the numerical values of Table 11.4 directly or by adding the waves of Figure 11.6 graphically.

Tone Color and Spectrum

We have mentioned previously that the pitch of a sound is related to its frequency while the loudness is related to the wave amplitude. Obviously, more than just pitch and loudness information is needed to describe different sounds. We can imagine a situation in which a flute and violin each play the same note (same

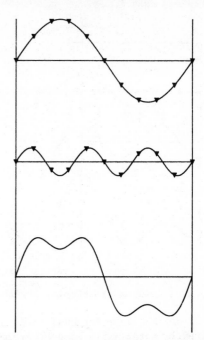

Figure 11.4. Fundamental (upper), third harmonic with a phase of 0 relative to the fundamental (middle), and sum of the fundamental and third harmonic (lower).

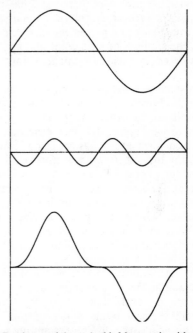

Figure 11.5. Fundamental (upper), third harmonic with a phase of 0.5 cycle relative to the fundamental (middle), and sum of the fundamental and third harmonic (lower).

Table 11.4 Tabulation of p_1, p_2, p_3, and their sum as described in text. Time t is in seconds.

t	p_1	p_2	p_3	$p = p_1 + p_2 + p_3$
0.0	-1.0	1.0	-1.0	-1.0
0.01	-0.81	0.31	0.31	-0.19
0.02	-0.31	-0.81	0.81	-0.31
0.03	0.31	-0.81	-0.81	-1.31
0.04	0.81	0.31	-0.31	0.81
0.05	1.0	1.0	1.0	3.0
0.06	0.81	0.31	-0.31	0.81
0.07	0.31	-0.81	-0.81	-1.31
0.08	-0.31	-0.81	0.81	-0.31
0.09	-0.81	0.31	0.31	-0.19
0.10	-1.0	-1.0	1.0	-1.0

67

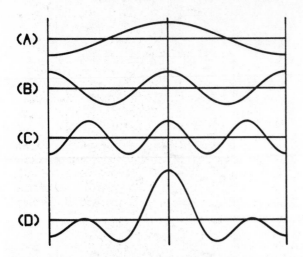

Figure 11.6. (A) Fundamental with phase of 0.75 cycle. (B) Second harmonic with phase of 0.25 cycle. (C) Third harmonic with phase of 0.75 cycle. (D) Pulse-like wave resulting from the addition of (A), (B), and (C).

pitch) with the same loudness. We are able to distinguish between these two sounds. We say they each have a different tone color. The tone color of a complex sound is related to the spectrum of that sound; different spectra produce sounds of differing tonal qualities. The spectrum is a measurable physical characteristic of a sound; tone color is a perceptual characteristic of that sound. The subjects of tone color and tone quality will be considered in greater detail in Chapter 14. Other characteristics of musical tones which are important for distinguishing various instruments will be discussed in Chapter 31. For now, let it suffice to say that different tone colors are due to different spectra. By considering the spectra of Figure 11.2, for example, we see that although each waveform has the same fundamental frequency, the three different waveforms will each have a different tone color because of their different spectra. The tuning fork has a rather simple spectrum of only one component; thus the sound is considered "pure." The clarinet, with a spectrum consisting predominantly of odd harmonics, has a tone which can be described as "hollow"; presumably because of the missing even harmonics. Finally, the trumpet has a spectrum consisting of many harmonics, the second harmonic having the greatest amplitude. The trumpet tone is often referred to as "bright" and "full." In light of the spectrum shown, this seems to be an apt description.

In summary, the differences among tone colors of complex sounds depends primarily, but not entirely, upon the presence, number, and relative amplitudes of the partials. The relative phases of the partials have little effect upon the resultant tone color (except in extreme cases), even though phase differences may alter the shape of a waveform drastically.

Exercises

1. Is a free or a forced system more likely to produce harmonic partials? Which system is more likely to produce inharmonic partials? Why? (Hint: forced systems give rise to periodic waves. Free systems most often give rise to nonperiodic waves.)

2. Consider two sinusoids given by sin (ft) and sin (ft + ϕ), where ϕ is the phase. Plot these two sinusoids for f = 2Hz and ϕ = 0.25 over an interval of 1 second.

3. What is the amplitude of each wave in Exercise 2? The frequency? The phase?

4. Do all musical notes that have the same fundamental frequency and the same number of higher partials sound the same? What besides the number of higher partials would have to be the same for identical sound to result?

5. What determines the tone color of a musical sound? Are inharmonic partials ever desirable in musical sounds?

6. Outline a method by which a musical tone may be artificially created.

7. Explain why identical notes plucked on a guitar and a banjo have distinctly different sounds.

8. The conventional spectra in Table 11.5 represent six waves, some simple and some complex. The number preceding each "sin" is the amplitude or the amount of that particular sinusoid present. What is the fundamental frequency of each wave? What are the frequencies and amplitudes of the higher partials for each wave? Are the higher partials harmonic or inharmonic relative to the fundamental?

9. Wave 4 in Exercise 8 is the sum of two sinusoids differing in phase by 0.5 cycle. What is the net result?

10. Wave 5 in Exercise 8 is an approximately triangular wave. Plot one cycle of it on graph paper by plotting each component separately and then graphically adding the three components to form the complex wave.

11. Waves 5 and 6 in Exercise 8 have the same spectra. However, they differ in phase. If both of them are plotted and the wave shapes compared, they look different. However, they will sound the same. Plot the two waves. Then plot their spectra as bar graphs.

12. The relative amplitudes for the first few partials of a sawtooth wave are 1, 1/2, 1/3, 1/4, 1/5, and 1/6. The relative phases for all partials are zero. Plot one cycle of the sinusoid representing the first partial of the sawtooth wave on a sheet of graph paper. Next plot

Table 11.5. Various waves and their components.

Wave 1: $1.0 \sin (f_1 t)$	$f_1 = 100$ Hz
Wave 2: $1.0 \sin (f_1 t) + 1.0 \sin (f_2 t)$	$f_1 = 100$ Hz, $f_2 = 200$ Hz
Wave 3: $1.0 \sin (f_1 t) + 1.0 \sin (f_2 t)$	$f_1 = 100$ Hz, $f_2 = 205$ Hz
Wave 4: $1.0 \sin (f_1 t) + 1.0 \sin (f_1 t + 0.5)$	$f_1 = 100$ Hz
Wave 5: $1.0 \sin (f_1 t) + 0.11 \sin (f_3 t + 0.5) + 0.04 \sin (f_5 t)$	$f_1 = 100$ Hz, $f_3 = 300$ Hz, $f_5 = 500$ Hz
Wave 6: $1.0 \sin (f_1 t) + 0.11 \sin f_3 t + 0.04 \sin f_5 t$	$f_1 = 100$ Hz, $f_2 = 300$ Hz, $f_3 = 500$ Hz

68

two cycles of the second partial with the proper relative amplitude. Repeat this for the remaining partials. Now graphically add all of these sinusoids together to produce a fairly good approximation to a sawtooth wave.

13. Repeat Exercise 12 but with arbitrary phases. What happens to wave shape?

14. In what physical characteristics do loud, high violin tones differ from soft, low flute tones?

Demonstrations

1. Complex waves: analysis—A function generator is connected to a band- pass filter and its output observed on an oscilloscope. When the function generator is set to produce sinusoids at a frequency of 200 Hz, the voltages in the "sine" row of Table 11.6 are measured on the oscilloscope at the frequencies shown. Similar measurements are made when the function generator produces square waves, triangular waves, and sawtooth waves, as shown in the table. What can you say about the spectrum for each of the four wave types? Why were some voltages only slightly different from zero measured?

2. Analyze various complex waves by running them into an oscilloscope to see the waveform, and a spectrum analyzer to see the spectrum. Use a function generator, tape recordings, and microphone signals as inputs.

3. Synthesize various complex periodic waves

Table 11.6   Voltages measured at three different frequencies for four different functions.

Function	200 Hz	400 Hz	600 Hz
Sine	10	0.01	0.02
Square	10	0.03	3.3
Triangular	10	0.02	1.1
Sawtooth	10	5	3.3

with a Fourier synthesizer. Run the output to an oscilloscope, a spectrum analyzer, and a loudspeaker. Does the "output" of the spectrum analyzer agree with the "input" of the synthesizer? Change relative phases and observe the waveform and the sound. Which change with phase? Why?

Further Reading

Backus, Chapter 4
Benade, Chapter 3
Denes and Pinson, Chapter 3
Hall, Chapter 10
Olson, Chapter 1
Roederer, Chapter 4
Rossing, Chapter 7

Audiovisual

1. **Superposition** (Loop 80-3858, EFL)
2. "Fourier Analysis" (REB)
3. "Fourier Synthesis" (REB)

12. Sound Radiation and the dB Scale

The existence of sound implies a source of vibration and a medium to transmit the energy. If we consider vibrating solids or air columns as the source and air as the medium, a sound wave is, under certain limitations, the vibration of the air. When a source sets the molecules of the air into vibration, a varying pressure having the same frequency as the source is produced. This varying pressure wave can be considered a sound if it is within the frequency range audible to the human ear. A sound wave, then, is any vibration of the air between 20 Hz and 20,000 Hz. Air vibrations below 20 Hz are called "infrasound," while those above 20,000 Hz are called "ultrasound."

Almost any sort of rapid motion through the air generates sound waves. Part of the energy of motion radiates as sound waves which travel out from the source in all directions, eventually striking walls or objects in a room. Yet most sounds do not cause the walls or objects in a room to vibrate noticeably, even though we might perceive those sounds as being quite loud. Furthermore, even the feeble power of a buzzing fly can seem quite loud on a quiet night when we are trying to sleep. We are able to hear very weak sounds indeed, but even loud sounds are not very powerful compared to other energy sources such as heat and light.

In this chapter a brief overview of the radiation characteristics of sound sources is given and the dB scale for describing sounds is presented

Sound Sources

All sounds we hear ultimately derive from motion in the air around us. The air is set into motion by vibrating objects which are many and varied in nature. They include the wings of insects, the communication mechanisms of animals, wind-blown leaves on plants, and so on. We, however, are mainly interested in sound sources associated with music, speech, and sound reproduction.

Sound sources may be **free vibrators** in which the vibrator is abruptly set into oscillation, or **forced vibrators** which are kept vibrating by a continuously applied force. Free vibrators, which give rise to transient sounds, include the percussion instruments of the orchestra which are struck or plucked. The actual vibrator may be a string, a membrane, a bar, a plate, or a bell. A large vibrator, such as a drum, may set the air into motion directly. On the other hand, a small vibrator, such as a string, passes its energy on to a larger surface so that energy can be coupled to the air more efficiently. String instruments use either a soundboard or the instrument body to accomplish this.

Forced vibrators give rise to fairly steady sounds which are approximately constant in amplitude and frequency. Included in this category are the woodwind, brass, and bowed string instruments of the orchestra as well as the human speech mechanism. The efficiency with which the motion of a vibrating source is converted into air motion depends in part on the size of the vibrator. Most of the sound coming from bowed strings is due to motion of the instrument body. The wind instruments typically have a reed that interrupts an air stream to set the air column into vibratory motion. The air column is the vibrating source in these instruments. Openings in the air column determine the coupling of air vibration to the outside air. The brass instruments employ a single large opening—the bell. The woodwinds employ many small openings—the tone holes.

Electromechanical instruments, such as the electric guitar, electronically amplify and modify an acoustical signal. Electronic instruments, such as the synthesizer, use electronic circuitry to produce the desired oscillations. In both cases, a varying electrical voltage is used to drive a loudspeaker. The loudspeaker sets the air into motion.

Radiation of Sound

Radiation is the emission or "giving off" of energy. A loudspeaker provides a familiar example and illustrates two important features of acoustic radiators: (1) a relatively large radiating surface is required to radiate long wavelengths (low frequencies) efficiently; and (2) radiators do not radiate high-frequency sound uniformly in all directions. The sound power output of a radiator depends on its surface area and on the square of the volume velocity of the air it moves. Volume velocity in turn depends on the area of the radiator, the amplitude of its vibration, and the frequency at which it vibrates. Since the amplitude of vibration is limited for most vibrators, an efficient low frequency radiator must have a large area.

The low frequencies (long wavelengths) are radiated fairly uniformly around a loudspeaker because of diffraction. Figure 8.9 could represent a loudspeaker rather than a hole in a wall. The long wavelength sound waves are greatly diffracted by "passing through" the loudspeaker, which is of much smaller dimensions than the sound wave. High-frequency sounds are radiated more strongly in the direction in which the speaker is "aimed" and less strongly to the sides because they have short wavelengths compared to the size of the speaker. Figure 12.1 shows approximate distributions of sound energy emerging from a loudspeaker mounted in a wall for low, medium, and high frequencies. The low-frequency waves can be regarded as having wavelengths several times the size of the speaker. Wavelengths of the high frequency waves are only a fraction of the size of the speaker. The medium-frequency waves have wavelengths of about the same size as the speaker.

The radiators of the human voice are the mouth and nose. The radiators of wind instruments are the tone holes or the bell opening. The string instruments' radiators are their bodies. Whether an instrument radiates energy uniformly in all directions or beams most of it in a particular direction depends on its radiator size relative to the wavelength of the sound being radiated. Many radiators, such as the strings on a piano, do not efficiently transfer their vibrations to the air without some assistance. By coupling the piano strings to a sound board, the vibrations are transferred to the air much more efficiently. Many musical instruments provide examples of this type of coupled vibration in that they consist of two vibrating systems. The first determines the frequency of vibrations and forces the second to vibrate at this same frequency and the second system radiates the sound to the air.

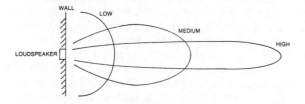

Figure 12.1. Simplified radiation patterns from a loudspeaker for low, medium, and high frequencies. Low frequency sounds have wavelengths longer than the dimensions of the loudspeaker and so radiate uniformly in all directions. High frequency sounds have wavelengths shorter than the dimensions of the loudspeaker and so radiate mostly to the front.

Decibel Scale

The human ear can respond to sound intensities that vary by a factor of one trillion between the lowest intensity to which the ear responds and an intensity that produces a sensation of feeling. This tremendous range of sensitivity is rather hard to imagine. A scale sensitive enough to weigh a single human hair with the same range of response as the human ear could also be used to weigh a 30-story apartment building! It has been known for quite some time that our ears deal with this fantastic range of sound intensities in a particular way. When the intensity of a sound is increased about ten times, the sensation of loudness is approximately doubled, regardless of whether the sound is of low or high intensity. In general, we judge the relative loudness of sound in terms of ratios of sound intensities—not in terms of differences of sound intensities.

If we listen to a buzzing fly and then the sound intensity is suddenly doubled by another buzzing fly entering the room, our ears perceive a certain increase in loudness. If we listen to a loud buzz saw and then hear two buzz saws we may find the sound intensity again doubled. Although there is a much greater intensity difference between the sounds of the two buzz saws than between the two flies, the increase in loudness in each case seems about the same. In other words, the increase in loudness is very nearly independent of the sound intensity of the original sound; a particular difference in sound intensity that is negligible in comparing two loud sounds can be a tremendous factor in comparing two faint sounds.

The above discussion suggests that it may be convenient to use a scale of measurement for sound based on sound intensity ratios rather than absolute sound intensities. The decibel (dB) scale is just such a scale. While the dB scale did not have its origin in the above considerations—it originated in the communications industry—it is possible to say that the dB scale does approximate the way our ears perceive sound better than a

linear intensity scale would. The decibel scale expresses a ratio between two sound intensities in dimensionless units known as decibels (dB). Mathematically, the **sound intensity level**, expressed in decibels, is defined as 10 times the logarithm of the ratio of the intensity to some standard intensity as SIL = $10 \log(I/I_s)$ where I is the intensity and I_s is a standard reference intensity. By convention, the reference intensity is taken to be $I_s = 10^{-12} W/m^2$. This is approximately the smallest intensity that a human can detect under ideal conditions in the frequency range where the ear is most sensitive. When a sound wave has an intensity of $10^{-12} W/m^2$, the intensity level will be 0 dB. Note that 0 dB does not mean there is an absence of sound; rather a 0 dB intensity level corresponds to an intensity equal to the standard reference intensity.

Although the dB scale may seem like an arbitrary contrivance, there is some justification for it. A change in sound level of one dB is about the smallest change that a human can perceive, and a change of 10 dB at mid-range frequencies corresponds roughly to a doubling of perceived loudness. The dB scale is typically used between about 0 dB, which is the threshold of hearing and 140 dB, which is the threshold of pain. Furthermore, since the ear perceives pressure and intensity ratio changes rather than absolute changes, the dB scale mimics the perceptual processes of the ear to a considerable extent.

A sound pressure level can be defined in a way analogous to that of intensity level. The definition of **sound pressure level** (in dB) is SPL = $20 \log(p/p_s)$. The factor of 20 appears in the expression for sound pressure level rather than the factor of 10 for intensity level because intensity varies as the square of the pressure. When a standard pressure of 20 μPa (micro pascals or millionths of a newton per square meter) is used for p_s in the above definition, sound pressure level and intensity level may be regarded as equivalent for waves travelling in one direction. For situations in which reflected waves are present the two may differ by a few dB. Most measuring devices measure sound pressure so usually we will be referring to sound pressure level when we speak of sound level.

An interesting rule of thumb describes sound level changes as the distance from the sound source changes for sounds being radiated uniformly and with no reflections. Each doubling of distance decreases sound level by 6 dB; each halving of distance increases sound level by 6 dB. (Refer to Exercise 15 for a development of this concept.) A change in sound level should not be confused with a change in loudness, even though the two phenomena are related. Loudness, as we will see in Chapter 14, is a perceptual phenomenon, while sound level is a physical property that can be measured.

To compare the intensities of several sound sources, proceed as follows: If one sound has a pressure amplitude 10 times that of another sound the more intense sound has a sound level 20 dB greater than the first sound. (A pressure ratio of 10 would gives an intensity ratio of 100 because intensity varies as the square of pressure.) Suppose that the SL of a trombone is recorded at 80 dB. When two identical trombones play the same note, exactly in phase, the sound pressure will double, but the sound pressure level goes up to only 86 dB. If 76 trombones could all play the same note in phase, the pressure level would increase by a factor of 76, but the sound pressure level would go up only to 118 dB. To reach the threshold of pain at 140 dB would require no fewer than one thousand trombones all playing in unison. Whenever the pressure increases by a factor of 10 the sound pressure level increases by 20 dB. Whenever intensity increases by a factor of 100 the sound level increases by 20 dB. Table 12.1 shows some sound pressure ratios and some sound intensity ratios with their decibel equivalents. Note that the decibel scale is somewhat more convenient to use than a linear pressure scale because of the huge range of pressures to which the ear is sensitive.

Table 12.1. Sound pressure ratios, sound intensity ratios, and their decibel equivalents.

Sound pressure ratio	Sound intensity ratio	Decibel equivalents
1:1	1:1	0
10:1	100:1	20
100:1	10^4:1	40
1000:1	10^6:1	60
10^4:1	10^8:1	80
10^5:1	10^{10}:1	100
10^6:1	10^{12}:1	120

A **sound level meter** is a device (usually portable) used for measuring sound pressure levels. It consists of a microphone, an amplifier, various filters, and a meter for displaying the measurement results. A typical sound level meter is illustrated in Figure 12.2. The A-weighting filter modifies the incoming sound so that the meter will respond in a manner similar to the way the human ear responds at low and mid frequencies. As we will see in Chapter 14, the ear attenuates the low frequencies of weak sounds. It therefore requires more low-frequency sound energy than mid-frequency energy to cause a given deflection on the meter. Presumably, then, a decibel reading on the A-scale, expressed as dBA, is a fairly accurate gauge of how humans will perceive the sound. The B-filter attenuates low frequencies, but to a lesser extent than the A-filter. The C-filter provides very little attenuation.

In order to obtain accurate measurements, it is necessary to exercise considerable care when using the sound level meter. The operator should hold the meter some distance from the body so that the effects of body reflections are reduced. The operator should move from place to place while performing the measurement to insure a truly representative sample of the overall sound field. When the measurements are performed inside a building, the observer must be especially careful because of the many reflected sounds, all of which can influence the reading.

Table 12.2 gives some sound levels produced by typical sound sources. Note that quiet background noise corresponds to a level of about 40 dBA, while 120 dBA is the threshold of feeling, in which case the sound is so loud that we feel a buzzing sensation in our ears.

Exercises

1. If a tree falls in a forest and there is no one to hear it, is a sound produced? Explain.

Figure 12.2. Sound level meter. (Courtesy of GenRad, Inc.)

2. What is the sound level at a point two meters from a source if the level is 80 dB at a distance of one meter? (Assume uniform spherical radiation. The surface area of a sphere is $4\pi r^2$, where r is the radius.) What is the sound level at a distance of 5m?

3. Draw a simplified clarinet diagram and label points of energy input and energy loss.

4. Several of the transducers that we called acoustic generators in Chapter 6 radiate sound energy. What are some of these?

5. Name several musical instruments and give an estimate of the size of their radiators.

6. Which of the instruments named in Exercise 5 radiate low frequencies the most efficiently?

7. Estimate the size of the two most significant radiators associated with the vocal system. Which is generally the most efficient of these vocal radiators?

8. What is done in hi-fi systems to radiate low frequencies efficiently?

Table 12.2. Typical sound levels produced by various sources. (Distance from source shown in parentheses.)

Source	Sound Level (dBA)
Atmospheric pressure	190
Threshold of pain	140
Threshold of feeling	120
Jet takeoff (100m)	90-120
Heavy truck (15m)	85-90
Gas lawnmower (1m)	80-100
Apartment near freeway	75-90
Food blender	70-85
Auto interior (55 mph)	65-80
Automobile (15m)	65-75
Normal conversation (1m)	60-70
Quiet residential-day	50-55
Quiet residential-night	45-50
Soft whisper	40
Wilderness	20
Threshold of hearing	0-10

9. Do we expect brass instrument tones to be more "brilliant" sounding when we sit in front of the instrument or to the side of the instrument? Why?

10. Plot the spectrum of Tone 6 from Table 11.5 in terms of sound pressure level.

11. What is the difference in sound levels between pressures of 1000 and 200 μPa?

12. The C-network of a sound level meter weights all frequencies about equally. For a complex sound composed primarily of low frequencies, will the A-scale reading be larger than, smaller than, or about equal to the C-scale reading? Answer the same question for a complex sound composed mainly of high frequencies.

13. Suppose you walk at a constant speed away from an outdoor sound source and away from all obstructions. Does the level fall off more quickly when you are near to or far from the source?

14. If you were listening to an outdoor concert from a position behind an obstruction, how would you perceive the low and high frequencies relative to listening with no obstruction? What role does diffraction play?

15. A source of sound produces an intensity of I_1 at distance d_1 and I_2 at distance $d_2 = d_1/2$. The sound levels are $SL_1 = 10 \log (I_1/10^{-12})$ and $SL_2 = 10 \log (I_2/10^{-12})$. $SL_1 = 120 + 10 \log I_1$ and $SL_2 = 120 + 10 \log I_2$ so that $SL_2 - SL_1 = 10 \log I_2 - 10 \log I_1$. But a halving of distance results in four times the original intensity because the sound power is spread over one quarter the area. Thus, $SL_2 - SL_1 = 10 \log (4I_1) - 10 \log I_1 = 10 \log 4 + 10 \log I_1 - 10 \log I_1 = 10 \log 4 = 6$ so that a halving of distance results in a 6 dB increase in sound level. Go through similar steps for a doubling of the distance where $d_2 = 2d_1$.

Demonstrations

1. Use a column loudspeaker or a linear array of four or five identical loudspeakers each driven by the same source. Illustrate directivity with the long dimension of the array and lack of directivity with the short dimension.

2. Use a sound level meter to measure some sounds in a classroom. Measure a background level with everyone quiet, a level when everyone is clapping and shouting, and so on.

Table 12.3. Dependence of measured sound level (in dB) on frequency (Hz), distance (m), and direction.

Frequency	Distance	Position	Sound level
100	2	front	81
100	4	front	74
100	8	front	69
100	8	side	65
1,000	8	front	71
1,000	8	side	65

3. Intensity level variation with distance and direction—A single diaphragm, enclosed-cabinet, direct-radiator loudspeaker is placed out of doors away from all reflecting objects. It is driven with sinusoids of 100 Hz and 1,000 Hz, as indicated in Table 12.3. Sound levels are measured with a sound level meter using the C-scale, which gives approximatey equal weightings at all frequencies. Measurements are made in front and to the side of the loudspeaker as shown in the "distance" and "position" columns, and the sound levels are recorded in the table. What approximate rule can you state relating intensity level variation to distance from the source? What can be said about the directional effects of the loudspeaker?

Further Reading

Backus, Chapters 3-5
Denes and Pinson, Chapter 3
Hall, Chapters 3, 5
Kinsler and Frey, Chapters 7, 10
Olson, Chapters 4, 6
Rossing, Chapters 6, 24

Audiovisual

1. **The Science of Sound** (Album FX6007, 1959, FRS)

2. **Sound, Energy, and Hearing** (30 min, color, 1957, EBE)

III. The Ear and Hearing

Courtesy of Instructional Science, Brigham Young University.

13. Response of the Ear to Sound

The hearing mechanism is one of the most intricate and delicate structures found in the human body. Hearing takes place in three stages: the outer-ear stage, the middle-ear stage, and the inner-ear stage. In the outer-ear stage, the incoming sound is converted to mechanical motion of the eardrum. In the middle-ear stage, this mechanical motion is transmitted to the inner ear. In the inner-ear stage, the motion is converted into a series of nerve impulses which are then sent to the brain. We will consider the anatomy and function of the hearing mechanisms as they apply to the various stages of the hearing process. Encoding of pressure and frequency by the ear will also be discussed.

The range of sound pressures over which the ear can respond is truly phenomenal. Not only can the ear withstand extremely intense sounds, but it is also capable of responding to pressures which are so small that displacement of the eardrum is less than the diameter of the air molecules striking it. The frequency range of audible sounds to which the ear can respond is most commonly specified as 20 Hz to 20 kHz (20,000 Hz) even though the limits are not precise. (The lower limit may vary from 15-20 Hz and the upper limit from 16-25 kHz.) But the ear is much more than a wide-range microphone; it also acts as a sophisticated time and frequency analyzer.

Some sounds have frequencies outside the frequency range of the ear. Inaudible low frequency (below about 20 Hz) sounds are called infrasonic. Inaudible high frequency (above about 20 kHz) sounds are called ultrasonic sounds. Ultrasonic and infrasonic sounds can influence living organisms even though they cannot be heard by humans. They will be considered at the end of this chapter.

Anatomy and Function

As shown in Figure 13.1, the ear consists of three main parts: the outer, middle, and inner ear. The **outer ear** consists of the visible portion of the ear, the ear canal, and the eardrum. The **middle ear** consists of a small chamber containing the three tiny bones which transmit vibrations to the inner ear. The **inner ear** is a complex cavity filled with a fluid and the structures necessary to convert mechanical vibrations into nerve pulses.

The outer ear serves to protect the sensitive middle and inner ear mechanisms from harsh external environments and to maintain internal temperature and humidity, independent of external conditions. Since the ear canal is an acoustic resonator open at one end and closed at the other, it enhances sounds whose frequencies are close to its resonance frequency of about 3000 Hz.

Consider the schematic representation of the ear

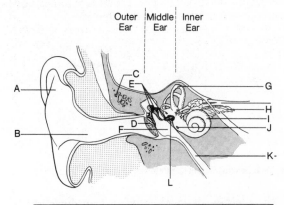

A. Pinna	G. Semicircular canals
B. External auditory meatus (canal)	H. Cochlear nerve
C. Temporal bone	I. Cochlea
D. Tympanic membrane	J. Round window
E. Ear ossicles	K. Eustachian canal
F. Tympanic cavity	L. Oval window

Figure 13.1. Overall cutaway view of the ear. (After Chaffee and Greisheimer, 1974.)

shown in Figure 13.2. The middle ear cavity is connected to the throat by means of the eustachian tube. The eustachian tube is normally closed, thus sealing the middle ear cavity, but may open during swallowing or yawning. The opening of the tube equalizes the pressure between the middle ear chamber and the external air pressure. Most people who have flown in an airplane or who have driven a car in the mountains have noticed the discomfort which often accompanies rapid changes in altitude. This discomfort is caused by an increase or decrease in the external air pressure relative to the pressure in the middle ear. Swallowing or yawning opens the eustachian tube, allowing the middle ear pressure to equalize with that of the surrounding atmosphere—thus relieving the discomfort.

The three small bones of the middle ear, called the **auditory ossicles**, are the **malleus** (hammer), the **incus** (anvil), and the **stapes** (stirrup). The ossicles transmit the vibrations of the eardrum to the fluid of the inner ear. Not only do they transmit sound vibrations to the inner ear, they also amplify the pressure from the outer ear. This is necessary to cause the fluid of the inner ear to vibrate appreciably. If the incident vibrations of the air impinged directly on the opening to the fluid-filled inner ear, most of the sound energy would be reflected and very little would actually enter the inner ear.

The combination of the eardrum and ossicles amplifies the sound pressure in two ways. First, the bones act as a lever, converting the force exerted by the eardrum on the hammer into a somewhat larger force on the stirrup. This accounts for a small increase in the force. Secondly, the eardrum is about 25 times as large as the membrane covered entrance to the cochlea, called the oval window. A small pressure over a large area (as at the eardrum) produces a large pressure over a small area (as at the oval window) when the total force in each case is the same. These two effects result in a pressure at the oval window about 30-40 times greater than that at the eardrum. Thus, we can detect sounds about one one-thousandth as intense as we could otherwise.

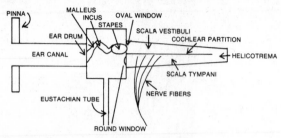

Figure 13.2. Schematic representation of the ear.

The **cochlea**, or inner ear, is a transducer shaped something like a snail's shell (see Figure 13.1), which transforms mechanical vibrations into nerve impulses. A simplified cochlea is shown uncoiled in Figure 13.2. The cochlea can be viewed as a fluid-filled, rigid-walled tube divided along its length into an upper compartment (the **scala vestibuli**) and a lower compartment (the **scala tympani**) by the slightly flexible cochlear partition. The cochlear partition extends along almost the entire length of the cochlea, ending only near its apical end. This opening, which provides a path for fluid flow between the upper and lower compartments, is known as the **helicotrema**. At the junction between the scala vestibuli and the middle ear is a flexible membrane, the **oval window**, to which the stapes is attached. Where the scala tympani joins the middle ear there is another flexible membrane called the **round window**. When the oval window is pushed inward by the stirrup, the round window is pushed outward by the fluid of the inner ear. Similarly, an outward motion of the oval window causes an inward motion of the round window. Thus, through the mechanical linkage of the middle ear, air vibrations are transformed into vibrations of the inner ear fluid.

Encoding in the Cochlea

The cochlear partition is actually a fluid-filled duct, as shown in Figure 13.3. Sound waves striking the ear result in vibrations in the cochlear fluid. The lower part of the cochlear partition, the **basilar membrane**, is somewhat flexible and can be set into vibration by the cochlear fluid. The conversion of the mechanical vibrations of the basilar membrane into electrical impulses is accomplished by the **organ of Corti** which lies on the basilar membrane. The organ of Corti contains about 30,000 small cells, called **hair cells**, which are attached to the nerve transmission lines going to the brain. The "hairy" ends of the hair cells are embedded in the **tectorial membrane**, as shown in the electron micrograph of Figure 13.4 and the drawing of Figure 13.5. When the basilar membrane vibrates, the "hairs" of the hair cells are bent, producing electrochemical impulses. These nerve impulses travel to the brain where they are interpreted as sound.

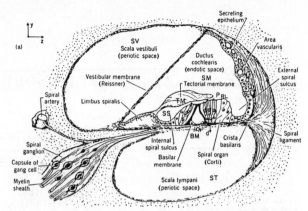

Figure 13.3. Cross-section of the cochlea showing the scala vestibuli and scala tympani separated by the cochlear partition which contains the transducing elements of the ear. (From Allen, 1977.)

Figure 13.4. Electron micrograph showing some of the hair cells in the organ of Corti. (From **Tissues and Organs: A Text Atlas of Scanning Electron Microscopy** by R.G. Kessel and R.H. Kardon. Copyright 1979 by W.H. Freeman and Co. All rights reserved.)

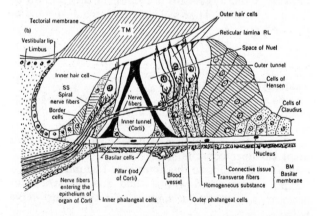

Figure 13.5. Detail of the subtectorial region showing hair cells, their nerve pathways, and the contact of the "hairs" with the tectorial membrane. (From Allen, 1977.)

Frequency information is encoded in part by the region along the basilar membrane that moves the most when vibrations occur in the cochlear fluid. Waves of different frequencies cause different regions of the basilar membrane to respond. If a high-frequency sinusoid strikes the ear, the area of the basilar membrane adjacent to the oval window responds most strongly. The far end of the basilar membrane near the helicotrema responds most vigorously to low-frequency sinusoids. Georg von Bekesy won a Nobel prize for his experimental research with actual mammalian cochlea demonstrating the manner in which the basilar membrane responds to different frequencies. Bekesy was able to observe actual traveling waves on the basilar membrane. He also noted that the place of the wave's maximum amplitude was dependent on the frequency of the excitation. Figure 13.6 shows simplified computer simulations of the response of the cochlear partition to sinusoids of different frequencies. The diagrams show "time exposures" of several cycles of vibration, so we do not see the wave at each instant, but only an overall wave envelope which shows the region of maximum amplitude. Frequency information is also encoded in the repetition rate at which groups of hair cells fire.

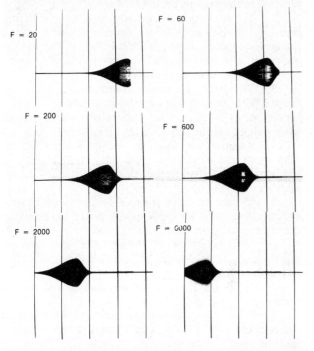

Figure 13.6. Simplified computer simulations of the response of the cochlear partition to sinusoids of 20, 60, 200, 600, 2000, and 6000 Hz. The grid is marked off in 1 cm increments; the whole "uncoiled" cochlea is about 3.5 cm long. The oval window is on the left in each diagram.

The encoding of a sound wave's intensity is determined by the number of nerve impulses produced each second. A hair cell must be stimulated by some minimum vibration amplitude or it does not produce a nerve impulse. When stimulated by large-amplitude vibrations (resulting from intense sounds), a hair cell "fires" more often than with smaller-amplitude vibrations. Each hair cell is limited to a maximum of about 1000 firings each second. More intense sounds also cause a greater portion of the basilar membrane to vibrate, resulting in the firing of still more hair cells. Simplified representations of the encoding of frequency and intensity information by the ear are shown in Figure 13.7A and Figure 13.7B.

When a complex sound consisting of a variety of frequencies strikes the ear, many different regions of the basilar membrane are excited. Figure 13.6C is a simplified representation of the encoding of the frequency, amplitude, and spectrum of a complex tone. The many frequencies of a complex tone not only stimulate different regions of the basilar membrane, but for each region of the membrane stimulated, the same overall nerve impulse repetition rate results. That is, each region of the basilar membrane which is stimulated sends out nerve impulses in a sort of "Morse code," with different codes being used for each partial, even though the codes have the same repetition rate. The brain then receives the codes as illustrated in Figure 13.7C and interprets them as a complex tone.

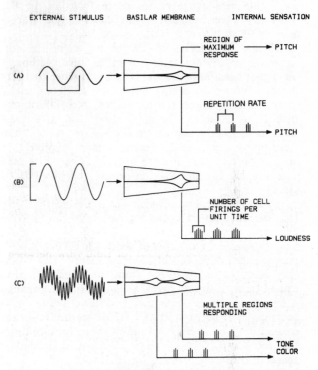

Figure 13.7. Simplified representation of the encoding of frequency, amplitude, and spectrum by the ear.

Ultrasound and Infrasound

Unlike the audible frequencies we have been discussing, the human ear does not respond to **ultrasound**, having frequencies above 20,000 Hz and **infrasound**, with frequencies below 20 Hz. This is not to say that ultrasound and infrasound do not affect living organisms. Ultrasonic sound waves can be heard by many animals. By sweeping through a range of frequencies from 22 kHz to 60 kHz, many different insects and rodents as well as some household pets, are repelled. Intense ultrasonic generating devices are sometimes used as rodent repellents in food processing plants.

Ultrasonics has found many useful applications in medicine. An image of the interior of the body, an ultrasonagraph, can be obtained by passing ultrasound through body tissues and by detecting the reflections from different body structures. This ultrasound technique can be used to prepare an image of the heart, or the liver, or even blood vessels. Since sound waves are not electromagnetic waves as are X-rays, ultrasonography can be used during pregnancy, where X-rays would be too dangerous. Ultrasound can be used to detect birth defects and other abnormalities as well as to diagnose multiple births. Other applications in biology and medicine include localized internal healing, destruction of bacteria, and cancer therapy. In spite of all the useful applications, however, we know far too little about the possible harmful effects of ultrasound on humans.

The applications of ultrasonics are wide ranging. Many areas of science and technology benefit from this tool. High-frequency sound is significant because it can be focused into narrow beams, and because it can be used to generate very high sound intensities. Industrial uses of ultrasonic radiations include cleaning, drilling, and welding. Ultrasonics can also be used as a sheet metal thickness gauge or to locate flaws in metals. Ultrasonic waves have also been used as burglar alarms and as a substitute for chemical pesticides.

Below frequencies of 20 Hz we enter the still somewhat mysterious region of infrasound. Although we cannot hear these sounds we can still sense them by feeling. Infrasonic energy comes to us from many sources, such as thunder, sonic booms, or the vibrations from heavy vehicles. The applications for infrasound seem to be mostly destructive and sinister rather than constructive and useful. Recent studies suggest that because of the power of these waves they could be used as a weapon, literally shaking people and property to pieces.

Military weapons research has indicated that very low frequency sound having high intensity can produce profound physical disturbances inside the human body. Vibrations of 8 Hz at a level of about 130 dB can cause a person's chest walls to vibrate by resonance. This vibration produces internal pain from the intense friction caused by the stomach, heart, and lungs rubbing against each other. Other experiments dealing with sound pressure levels between 95 and 115 dB in the frequency range from 2 to 16 Hz have shown that exposure times between 5 and 15 minutes are sufficient to produce a sense of ill-feeling, including dizziness and loss of balance. At a frequency of 7 Hz violent and sudden nausea was generally produced at somewhat lower pressure levels. All of these symptoms indicate a disturbance of the balance organs, located in the inner ear. In fact, sounds in the range of 2 to 20 Hz are within the natural resonance frequency range of the semicircular canals, and may cause large vibrations in the canal fluid.

The hazard of infrasound is that it is produced, often at a considerable intensity, in a wide range of man-made environments. Reports show, for example, that a car traveling at 80 km/hr will generate fairly intense infrasonic waves at frequencies below 16 Hz. The infrasound, which the driver may not notice, could produce an effect similar to drunkenness and may be the explanation for some otherwise inexplicable car accidents.

Perhaps even more ominous are the possible effects of naturally produced infrasound in the atmosphere. Thunderstorms are known to produce significant levels of infrasound; perhaps this accounts for the bad tempers that thunderstorms arouse in some people. It is also known that before an earthquake some people and many animals experience feelings of unease. Can these psychological disturbances be attributed to the infrasonic waves generated as a precursor to the earth-quake? Although we cannot answer such questions with any degree of certainty, the limited number of experiments performed to date indicate that infrasound, in addition to the physical effects mentioned previously, also has very profound psychological effects. One study did find a statistically significant relationship between the presence of strong infrasonic waves generated by natural phenomena and increases in absenteeism and car accidents. Perhaps then, many behavioral problems and unpleasant feelings of malaise previously blamed on the climate may actually be due to infrasound. Since our mechanized society is continuously producing infrasound, which more and more people are being exposed to, these preliminary findings warrant more detailed study.

Exercises

1. Describe and discuss the construction and function of the various parts of the ear.

2. The ear senses and helps to interpret oscillating pressures. It is somewhat insensitive to slow, small changes in steady pressure. However, if the steady pressure changes appreciably in a short time, the ear experiences discomfort or pain. Why?

3. The steady atmospheric pressure decreases by about 3% for each 300 m increase in altitude. Does this explain why discomfort or pain is often experienced when flying or mountain climbing? Explain.

4. Assume the outer ear canal is a cylindrical tube, open at one end and closed at the other, with a length of 3 cm. Calculate the resonant frequency of this tube.

5. If you consider the ear to be a mechanical system, what is the probable nature of its response to a transient acoustic impulse?

6. Discuss ways in which the ear and a microphone are analogous in terms of their respective inputs and outputs.

7. How are nerve impulses transmitted to the brain?

8. When sine waves strike the ear, what parts of the cochlear partition respond most strongly to high, medium, and low frequencies, respectively? Show this on a sketch of an "uncoiled" cochlea.

9. How do the encoding of frequency and intensity information differ? Discuss the mechanisms.

10. Does the maximum rate of hair cell firing place any limits on the sending of frequency or intensity information to the brain? Discuss.

11. Prior to the "place model" of hearing discussed in this chapter one accepted model was the "resonance model" of hearing. According to this model the cochlear partition acted like a group of many small piano strings, each string having a different resonance frequency. An incoming sinusoid would cause the corresponding string to resonate. Complex waves would excite several strings; thus the frequency components of the wave would be resolved. Criticize this model. (Use the references at the end of the Chapter to obtain more information.)

12. Do the "resonance" and "place" models of hearing (see Exercise 11) deal primarily with frequency information or with intensity information? Discuss.

13. Qualitative mechanical behavior of the inner

ear—By referring to Figure 13.6, determine the distance from the oval window at which maximum displacement occurs for each frequency. How well do the values you have determined agree with those in the published literature? Are the positions of maximum displacement distributed linearly or logarithmically relative to frequency?

14. Explain how the focusing of ultrasonic waves might be used to produce localized heating within a person's body.

15. How can infrasound make you sick?

16. How does infrasound warn animals of an impending earthquake?

17. If a tube open at both ends were to be used for producing infrasonic waves in air at a frequency of 10 Hz, how long would it need to be?

Demonstrations

1. Use a mechanical model of the ear to illustrate anatomy and demonstrate functions of various parts of the ear.

2. Determine how well you can hear while swimming underwater. Why is sound in air not transmitted very effectively into water? What clues does this provide for the important functions of the middle ear?

3. Put your forefinger in your ear and place a coat hanger on it. Tap the hanger and notice the rather different hearing sensation.

Further Reading

Davis and Silverman, Chapter 3

Denes and Pinson, Chapter 5

Flanagan, Chapter 4

Fletcher, Chapters 7, 14

Hall, Chapter 6

Roederer, Chapter 2

Rossing, Chapter 5

Stevens and Warshofsky, Chapters 2, 3

Allen, J.B. 1977. "Cochlear Micromechanics—A Mechanism for Transforming Mechanical to Neural Tuning within the Cochlea," J. Acoust. Soc. Am. **62**, 930-939.

Anastassiades etal. 1973. "Infrasonic Resonances Observed in Small Passenger Cars Travelling on Motorways," J. Sound and Vibration **296**, 257- 259.

Brown, R., J. Harlen, and J. Gable. 1973. "New Worries About Unheard Sound," New Scientist **60**, 414-416.

Chaffee, E.E., and E.M. Greisheimer. 1974. **Basic Physiology and Anatomy**, 3rd ed. (J.B. Lippincott Co.)

Evans, M., and W. Tempest. 1972. "Some Effects of Infrasonic Noise in Transportation," J. Sound and Vibration **22**, 19-24.

Hudspeth, A.J. 1982. "The Hair Cells of the Inner Ear," Sc. Am. **248** (Jan), 54-64.

Kessel, R.G., and R.H. Kardon. 1979. **Tissues and Organs: A Text Atlas of Scanning Electron Microscopy** (W.H. Freeman and Co.)

Schuman, W. 1980. "Ultrasound, Exploring the Womb," Parents Magazine (Apr), 56-61.

Zwislocki, I.I. 1981. "Sound Analysis in the Ear: A History of Discoveries," Am. Scientist **69**, 184-192.

Audiovisual

1. **Ears and Hearing**, 2nd edition (22 min, color, 1969, EBE)

2. **Simulated Basilar Membrane Motion** (11 min, 1966, GSF)

14. Sensory Perceptions and Their Physical Bases

In the previous chapter we considered the hearing mechanism. We will now attempt to relate perceptions of sound to actual physical attributes of the sound wave. It has often been noted that the eye is a remarkably precise instrument. In many ways however, auditory processes are even more remarkable and more sensitive than those involved in vision. Consider first that the ear can detect frequencies ranging between 20 and 20,000 Hz—a frequency range of approximately 10 octaves. The eye, on the other hand, responds to light having a frequency range of barely one octave. The range of intensities to which the ear can respond is equally phenomenal (see Chapter 12), and once again the eye must be relegated to second place. Finally, the actual pressure variations to which the ear can respond are almost indescribably minute. For example, if you stand one meter away from a workman hammering on a steel plate, you will be subjected to a very intense sound field that is almost at the threshold of feeling. If this energy were converted to electrical power, however, it would barely be enough to run a one-watt light bulb. When the ear responds to the very softest sound on the threshold of hearing, the displacement of the eardrum is less than the diameter of the air molecules striking it. The displacement of the basilar membrane is only about one-tenth of that of the eardrum. As a matter of fact, if our ears were any more sensitive than they are—that is, if they responded to even softer sound—we would hear a permanent, soft buzzing sound due to random thermal motions of the air molecules.

In Part II we considered physical characteristics of sound waves, such as sound level, frequency, and spectrum. We are now considering corresponding perceptual quantities which result when sound waves enter a listener's ear. The three perceptual quantities which correspond most closely with the three physical characteristics given above are loudness, pitch, and tone color. We do not mean to imply that pitch, for instance, depends only on frequency, or that loudness depends only on sound level. They do not. Nor do we mean to imply that there is a linear relationship between the physical and perceptual quantities—i.e., that doubling the sound level doubles the loudness. That is also not the case. The actual relationships, as we will discover, are actually much more complicated and subtle.

Loudness and Sound Level

We begin by considering the relationship between loudness and sound level. As was pointed out in Part II, the sound level of a sound wave is conveniently expressed in decibels; 0 dB being the sound level of a pressure of 20 μPa. A 0 dB sound level corresponds to just audible sound (under quiet conditions) for an "average person" when listening to a frequency of about 3000 Hz. The threshold of hearing can be determined at different frequencies by presenting tones having different sound levels to listeners. If the results are averaged for many listeners having "normal" hearing the dashed curve shown in Figure 14.1 is obtained. This "threshold curve" shows the sound level which is barely perceptible at any given frequency. For example, a sound with a frequency of 100 Hz must have a sound level of about 40 dB to be audible.

In order to differentiate loudness level (a perceptual characteristic) from sound level (a physical characteristic), a new unit, the **phon** (rhymes with John), is used to specify loudness level. Though sounds may have different frequencies, if they sound equally loud they are assigned the same loudness level. A loudness level of 20 phons is defined as the loudness level produced by a 1000-Hz tone at a sound level of 20 dB. Likewise, the loudness level in phons of any 1000-Hz tone is defined as being numerically equal to its sound level in dB.

By further psychoacoustic experimentation, tones of various frequencies and sound levels can be matched in loudness level. For example, the 40 phon "equal loudness contour" can be obtained by adjusting the sound level of sinusoids at each of many different frequencies until they are perceived as having the same loudness as a 1000 Hz sinusoid at a sound level of 40 dB. The resulting equal loudness level contours are shown in Figure 14.1. The sound level is plotted along the vertical axis with frequency along the horizontal axis. The contours shown are the lines of equal loudness level as expressed in phons. The equal loudness level contours are not straight lines. This is because of two factors: (1) loudness level depends on frequency as well as sound level and (2) all equal sound levels do not sound equally loud. The shape of the threshold (dashed) curve results from a number of conditions. The resonance of the ear canal (around 3000 Hz) makes the ear more sensitive at frequencies between 2000 and 4000 Hz. A sparser density of hair cells in the low-frequency area of the cochlear partition makes the ear less sensitive at low frequencies. The combined filtering effects of the middle ear and cochlea make the ear less sensitive at high frequencies. These effects all combine to make the ear most sensitive at frequencies around 3000 Hz.

You may notice from Figure 14.1 that as higher loudness levels are approached the contours flatten out to some extent. The contour which corresponds to 120 phons is called the threshold of feeling, because these sound levels produce a "tickling" sensation in the ear. At still higher levels (135 phons), the sensation is one of pain.

We can predict from Figure 14.1 which of two sounds having different frequencies and levels will be louder. But loudness level does not specify how much louder one sound is than another. In order to describe how many times louder one sound is than another, a new unit, the **sone**, has been developed. When perceived

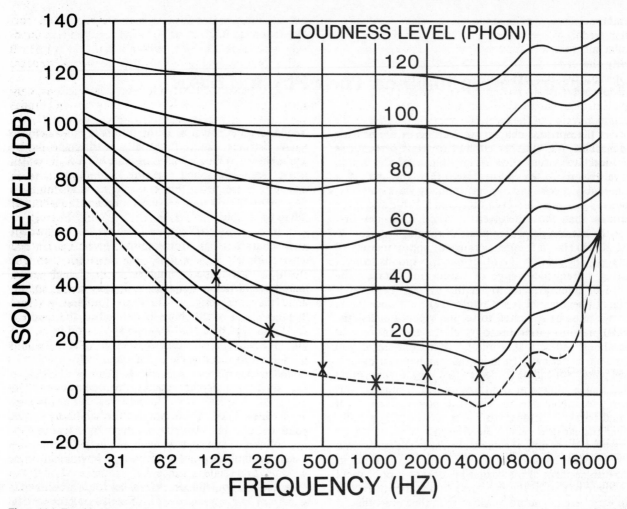

Figure 14.1. Equal loudness level contours for sinusoids. Sound level is expressed in decibels relative to 20 μPa. (After Robinson and Dadson, 1957 and Davis and Silverman, 1978.)

loudness doubles, the **loudness** (expressed in sones) also doubles. The relationship between loudness level expressed in phons and loudness expressed in sones is given in Table 14.1.

Table 14.1. Relationship between loudness level and corresponding loudness. A loudness level of 40 phons is arbitrarily assigned a loudness of 1 sone.

Loudness level (phons)	Loudness (sones)
20	0.08
30	0.3
40	1
50	2.8
60	6
70	12.5
80	25
90	45
100	80
110	150
120	250

The threshold of hearing is the minimum level at which a sound stimulus creates the sensation of sound for a listener. Another threshold of interest is the differential threshold or **just noticeable difference** (JND). The JND is the smallest change in stimulus a listener can detect. For convenience, we will limit our discussion to pure tones and the JND in sound level when frequency is constant. The JND is not constant but depends upon both the frequency and the sound level of the tone for

which it is determined. Suppose two tones having the same frequencies but slightly different sound levels are sounded one after the other. If the difference in sound levels is large enough, the tones can be distinguished. Otherwise the tones will be perceived as identical. JNDs for various sound levels are shown in Table 14.2. Note that the JNDs range from as little as 0.2 dB for tones of 2000 Hz at 80 dB to 4-6 dB for low-frequency tones presented at low levels. In speech, JNDs vary from 0.4 to 1.5 dB for overall sound levels. The JND for formant amplitude is about three percent.

Table 14.2. JNDs for intensity level (dB) of sinusoids as a function of frequency (Hz) and level above threshold (dB).

Frequency	Level (dB)			
	10	20	40	80
40	5.5	3.8	1.7	—
100	4.0	2.5	1.0	—
200	3.4	1.8	0.7	0.3
400	2.8	1.4	0.6	0.2
1000	2.4	1.3	0.5	0.2
2000	1.9	1.2	0.5	0.2
4000	1.8	1.2	0.6	0.3
10,000	3.5	1.8	0.9	0.5

Critical Band

In the preceding chapter it was noted that any given frequency stimulates a specific portion of the basilar

84

membrane. Two simultaneous signals having very different frequencies will stimulate different regions of the basilar membrane (Figure 14.2A). Two signals having nearly the same frequency will stimulate a common region of the basilar membrane (Figure 14.2B). A **critical band** (CB) is a frequency range within which two signals strongly interact; this interaction occurs because a common region of the basilar membrane is stimulated. Sensory perceptions such as loudness, pitch, tone color and masking may be expected to change rather abruptly at critical band boundaries. The "frequency width" of a CB varies with frequency as can be seen in Table 14.3.

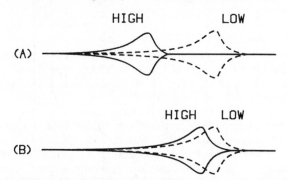

Figure 14.2. Qualitative response of basilar membrane when (A) tones are widely separated in frequency and (B) tones are close together in frequency.

Sound intensity is coded by the ear in terms of the number of nerve impulses produced per unit of time. Because two signals lying within a CB stimulate the same portion of the basilar membrane they will be using many of the same hair cells to encode intensity. For this reason, the loudness of a tonal complex, all of whose components lie within one CB, depends only on sound level. This is true for a narrow band of noise as well as a tonal complex of sinusoidal components. When a band of noise is wider than a single CB or when a tonal complex has components extending over a wider range than a single CB, the loudness increases with the width of the band. This is because a larger portion of the basilar membrane is stimulated and more hair cells are excited. The loudness of wide band signals can be considered as the sum of the loudnesses associated with each CB. The loudness of combined, widely separated sinusoids can be considered as the sum of their respective loudnesses.

As an example of determining the loudness of complex tones consider a tone consisting of two sinsoids, with respective frequencies of 500 Hz and 530 Hz and each with a sound level of 60 dB. Heard separately, each has a loudness level of 58 phons and a loudness of 5.4 sones (estimated from Figure 14.1 and Table 14.1). They lie within a CB and thus, when they are heard simultaneously we must add their intensities and com-

pute loudness on that basis. A sound level of 60 dB corresponds to an intensity of 10^{-6} W/m². Adding the intensities gives 2×10^{-6} W/m² and a sound level of 63 dB with a loudness level of about 61 phons and a loudness of 6.7 sones.

As a second example of loudness calculations, consider a sawtooth wave whose component amplitudes vary as 1/n and whose intensities thus vary as $(1/n)^2$. The loudness of the first ten components of a sawtooth having a fundamental of 500 Hz can be calculated by adding the loudnesses of the first six components which lie in separate CB to the loudness of components seven and eight (which are within a single CB) to the loudness of components nine and ten (also within a single CB). If the sound level of the first partial is 60 dB, the levels of partials two through six are 54, 50, 48, 46, and 44 dB respectively. The level of combined partials seven and eight is 49 dB and that of nine and ten is 46 dB. Estimates of the loudness levels of these eight partial combinations (from Figure 14.1) are 58, 54, 51, 50, 48, 46, 46, and 42 phons, respectively. Loudnesses (from Table 14.1) are 5.4, 4.1, 3.1, 2.8, 2.4, 2.1, 2.1, and 1.4 sones, respectively. The loudness is 23.4 sones. A 500 Hz sinusoid at an equivalent sound level of 65 db would have a loudness of 8 sones.

Everyone has experienced the difficulty of trying to hear someone speak when in a noisy environment. The reason for the decreased ability to "hear" in a noisy environment is due to **masking**—the drowning out or covering up of one sound by another. Experiments conducted on the masking effect of noise as applied to pure tones show an approximately linear relationship between masking and noise level. In other words, the greater the noise, the greater the masking. Figure 14.3 shows amount of masking in dB vs. noise level in dB. Masking occurs over the entire range of audible frequencies, but since noise can be considered as many unrelated sinusoidal components sounding simultaneously, masking due to noise can be considered as being caused by many simultaneous pure tones. As might be expected from critical band data, the most effective "maskers" are the noise components closest in frequency to the tones being masked.

Three of the most important results of masking experiments are: (1) tones close together in frequency mask each other more than do tones widely separated in frequency; (2) low-frequency tones mask high-frequency tones more effectively than high-frequency tones mask low-frequency tones; and (3) the greater the intensity of the masking tone, the broader the band of frequencies for which masking is evident. These results can be explained by considering the CB concept and the manner in which the basilar membrane responds at dif-

Table 14.3. Critical band, semitone, whole tone, and third octave bandwidths at different frequencies. All values are in hertz.

Frequency	Critical band	Semi-tone	Whole tone	Third octave
100	90	6	12	26
200	90	12	24	52
500	110	30	60	130
1000	150	60	120	260
2000	280	120	240	520
5000	750	300	600	1300
10,000	1200	600	1200	2600

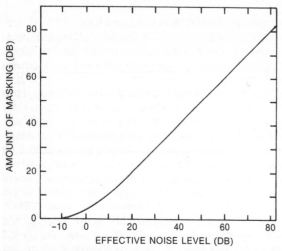

Figure 14.3. Relation between noise level and amount of masking. (After Hawkins and Stevens, 1950.)

ferent frequencies as shown in Figure 14.2. Tones widely separated in frequency do not stimulate the same part of the basilar membrane and thus do not interact and mask each other appreciably. Tones close together in frequency stimulate the same part of the basilar membrane, thus interacting and masking each other. Note in Figure 14.2 that a low frequency stimulates the part of the membrane that also responds maximally at high frequency, whereas the high-frequency pattern does not extend into the region of largest response for the low frequency. This is because the stimulation pattern of the membrane decreases more rapidly on the low-frequency side than on the high-frequency side. Because the low-frequency stimulation region extends into the high-frequency region, a low-frequency tone masks nearby high-frequency tones more effectively than vice versa. Intense masking sounds cause a larger region of the basilar membrane to be stimulated than do less intense sounds. This stimulation by more intense sounds of a larger region on the basilar membrane results in the masking of a broader range of frequencies.

The phenomenon of masking is often observed in musical performances. For example, one instrument can mask a second instrument playing the same note if the first instrument is played more loudly. On the other hand, when the fundamental frequencies are substantially different (e.g., a piccolo and a tuba playing simultaneously) we can easily distinguish each sound.

Pitch and Frequency

Frequency is measured in hertz. Our perception of frequency is **pitch**. Since there is not a one-to-one relationship between frequency and pitch, a new unit, the **mel** is used for pitch. The mel is to pitch what the sone is to loudness—that is, the mel is defined so that doubling the number of mels represents a doubling of the perceived pitch, regardless of the change in frequency. The relationship between frequency and pitch for sinusoids is shown in Table 14.4.

A pitch of 1000 mels is produced by a 1000-Hz tone at a level of 40 dB. Note that when the frequency doubles to 2000 Hz the pitch has not doubled, but has increased to 1500 mels. The pitch of pure tones depends to some extent on their sound level. Low frequency

Table 14.4. Relationship between frequency and pitch for sinusoids. A frequency of 1000 Hz at a sound level of 40 db is arbitrarily assigned a pitch of 1000 mels.

Frequency (Hz)	Pitch (mets)
10	0
20	20
50	75
100	150
200	300
500	600
1000	1000
2000	1500
5000	2300
10000	3000

sinusoids show a decreasing pitch of about two percent from low to high sound levels; high frequency sinusoids show the opposite effect; 3000 Hz sinusoids show little change.

JNDs for frequency of pure tones are shown in Table 14.5. Note that the JNDs for frequency range from as little as 0.2% for tones of 2000 Hz at 60 dB to about 4% at low frequencies and levels. JNDs for frequency of complex tones are generally smaller than those for pure tones. JNDs related to speech are 0.3-5% for fundamental frequency, 3-5% for formant frequency, and 20-40% for formant bandwidth.

Table 14.5. JNDs for frequency (percent) of sinusoids as a function of frequency (Hz) and level above threshold (dB).

Frequency (Hz)	Level (dB)			
	10	20	40	60
100	—	3.4	2.9	2.7
200	2.7	1.8	1.5	1.4
400	1.4	0.9	0.8	0.7
1000	0.7	0.4	0.4	0.3
2000	0.5	0.3	0.3	0.2
4000	0.5	0.3	0.3	0.2
10,000	0.6	0.5	0.4	—

Even though JNDs for frequency are of the order of one percent, a listener is unable to "hear out" each of two simultaneously occurring tones unless they are separated by about 30 JNDs. The ability to hear out two components of a complex tone that lie close to each other in frequency seems to be more nearly related to critical band than to JND. It is possible to estimate the number of pure tones a normal listener can distinguish from the tables of JNDs. If loudness is kept at 40 phons, the average person, under ideal conditions, can distinguish approximately 1400 different frequencies. If the frequency is kept constant at 1000 Hz, about 280 different sound levels (between 0 and 120 dB) can be distinguished. If we multiply these two numbers, to account for all frequency and intensity changes, we find that the total number of distinguishable tones (under ideal conditions) is almost 400,000! The ear indeed has amazing powers of discrimination.

Pitches of sinusoids related to the mel scale are of minor use in a world of sound consisting mostly of complex sounds. The natural unit for pitch in our "real" world is the octave. Change of pitch by an octave corresponds very closely to a doubling of frequency.

Pitches of complex tones have been the subject of much study over many years and several models have been developed for the analysis of these pitches. In one model, the perceived pitch is associated with the fundamental frequency of the complex tone. This model fails to account for the pitch of a complex periodic tone in which the fundamental frequency is missing unless it is recreated by nonlinearities in the ear at sufficiently high sound levels (as discussed in Chapter 15). For example, consider speech with a fundamental frequency of 200 Hz and higher partials of 400, 600, 800 Hz and so on. In face-to-face conversation the perceived pitch will be associated with the 200 Hz fundamental. What happens to pitch perception when the same speech is heard via a telephone where frequencies below 300 Hz are filtered out? Will the pitch be an octave higher and associated with the 400 Hz partial? No, the pitch will be associated with the missing 200 Hz partial. This is true even when the sound level is so low that nonlinearities in the ear cannot recreate the fundamental.

A more recent model of pitch perception claims that we perceive the periodicity or repetition rate of a complex signal and that this gives rise to the perceived pitch even when the fundamental is missing. Periodicity is a measure of how often a wave form repeats each second. (See Figure 13.7 for a simplified view of the periodicity of the nerve impulses coming from the cochlea.) In our example for telephone speech, the complex tone has frequencies of 400, 600, and 800 Hz and the complex tone will repeat itself 200 times per second, even though there is no frequency of 200 Hz present. Hence, if the ear perceives pitch on the basis of periodicity it will perceive a pitch associated with 200 Hz as the fundamental, even though a 200-Hz component is not present. **Periodicity detection of pitch** depends on a measurement, probably by the brain, of the period of a complex signal.

The periodicity model also accounts for the pitch of inharmonic tonal complexes. A tone with partials of 800, 1000, 1200, 1400, and 1600 Hz will have a pitch corresponding to 200 Hz. A tone with partials of 850, 1050, 1250, 1450, and 1650 Hz has a pitch corresponding to 210 Hz. The time periodicities in the latter case are somewhat shorter than in the former and hence the higher pitch. The periodicity pitch model has a deficiency in that it assigns greater importance to unresolved partials lying within a critical band than to lower partials. The experimental evidence is just the opposite— partials, especially the first five, resolved by the ear in separate critical bands are most important. Partials with frequencies below 2000 Hz are most important in determining the pitch of complex tones.

Recent models of pitch perception incorporate pattern recognition in one form or another. One model assumes pattern transformation in which a limited-resolution spectrum is operated on at higher perceptual levels to extract pitch. Another model looks for a best match to an assumed harmonic partial complex. A third model assumes a learning matrix that is filled with patterns during childhood; pitch recognition depends on matching an incoming pattern with a stored pattern. All three models include an initial limited frequency analysis in the auditory periphery followed by a central-nerve based pitch perception stage.

Tone Quality, Tone Color, and Spectrum

When considering tones from two different musical instruments, each having the same pitch and the same loudness as the other, we are able to distinguish between them on the basis of tone quality, tone color, or timbre. The foregoing terms are often used interchangeably, but we will be more restrictive in our definitions (see references).

Tone color is that attribute of a steady tone which permits it to be distinguished from other tones having the same loudness and pitch. Loudness and pitch are one dimensional quantities; tone color is multidimensional. Loudness depends on total neural activity and can be described on a scale from weak to strong. Pitch depends on local regions of maxima from which neural activity originates and on repetition rates associated with a complex tone. Pitch can be described on a scale ranging from low to high. Tone color cannot be described on a one dimensional scale, but typically requires from three to five dimensions.

Tone color depends most strongly on spectrum. However, there is some dependence of tone color on the relative phases of the partials composing the complex tone. The effect is most obvious in extreme cases such as when a tone whose relative phases are all zero is compared to a tone having the same spectrum but whose alternate partials have phases of zero and one quarter cycle. Phase effects become less apparent as the fundamental frequency is increased. Sensitivity to phase differs widely from listener to listener.

A spectrum is multidimensional because a specification must be made as to the sound level in each of several frequency bands. The ear resolves spectral components only when they are separated by at least one critical band. Thus division of a frequency range into critical bands seems plausible. A similar procedure is to divide the spectrum into third octave bands, about 15 of which are required to represent the frequency range of interest for music and speech. Each of the third octave bands is a "dimension" and the sound level in that band is the "distance" along that dimension in the resulting 15 dimensional space. The distance between spectrum one and spectrum two in this 15 dimensional space can be written as

$$D_{12} = \sum_{k=1}^{15} (L_{1k} - L_{2k})^2$$

where L_{1k} is the sound level in the kth third octave of spectrum one and so on.

Fewer than the original number of dimensions are often adequate to represent the spectra. Computational techniques are available that extract the most important dimensions from a multidimensional space. For example, consider tones composed of a fundamental and a second partial having a frequency twice that of the fundamental and a sound level 6 dB less than the fundamental. The sound level of the fundamental can be measured by a third octave filter within whose frequency band it lies. The sound level of the second partial can be measured by a second third octave filter within whose frequency band it lies. Suppose the level of the fundamental is adjusted to values of 50, 60, 70 dB and so on. When the sound levels of the second partial are plotted versus the sound levels of the fundamental, the graph shown in Figure 14.4 is obtained. We note that the original two dimensional representation can be represented with only one dimension (the line connecting the points in Figure 14.4) by reorienting our dimensions. This is because anytime the level of the fundamental is known, the level of the second partial is also known. The same thing can be done with 15 dimensions although the procedure is less obvious than in our two dimensional example.

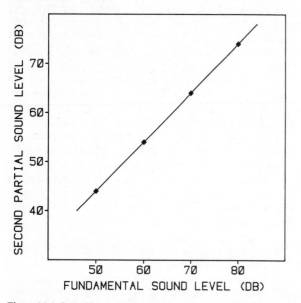

Figure 14.4. Sound level of the second partial versus the sound level of the fundamental for tones whose second partial amplitude is always 6 dB less than that of the fundamental.

The above procedure provides a means for mapping given spectra into a reduced dimension spectral space. The next step must be to provide a means for mapping perceptual judgments about the spectra into a multidimensional tone color space. The final step must be to compare mappings in the spectral space to those in the tone color space. If the spectra lie on top of each other after the two spaces have been "warped" together, then spectral distances can be used to predict tone color distances.

A method of triadic comparisons can be used to determine distances or dissimilarities in tone color space. Three complex tones are evaluated in pairs. The pair judged most dissimilar is assigned a distance of two, the least dissimilar pair is assigned a distance of zero, and the remaining pair is assigned a distance of one. Comparisons among many tones defines a multidimensional tone color space which can be reduced to three or four dimensions.

Tone color space was matched with spectral space for nine musical instrument tones. The results are shown in Figure 14.5. The triangles (tone color space) and the circles (spectral space) lie very close to each other in the three dimensions shown, indicating that tone color dissimilarites can be predicted quite well from spectral distances. A similar result is seen in Figure 14.6 for vowel sounds as plotted in three dimensional space.

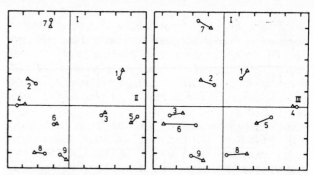

Figure 14.5. Result of matching tone color space (triangles) with a spectral space (circles) for nine complex tones. Dimension I is in the vertical; dimensions II and III are in the horizontal in the two separate frames. (From Plomp, 1976.)

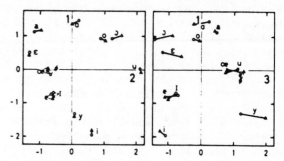

Figure 14.6. Result of matching perceptual space (circles) with spectral space (triangles) for 12 vowels. (From Klein etal, 1970.)

Studies have been done on verbal attributes of tone color. One such study included 35 different spectral configurations (Figure 14.7) which were rated with 26 adjective pairs. Figure 14.8 illustrates where the various spectra were judged to lie in a two dimensional adjective space (dull-sharp and compact-scattered).

Spectra can be characterized in two basic ways: (1) by amplitudes of higher partials relative to that of the fundamental or by (2) absolute frequency positions of spectral maxima. Harmonic spectra in which the partial frequencies are integral multiples of the fundamental frequency exist for tones that are periodic. Harmonic spectra can be characterized by specifying the frequency and sound level of each partial. This would be an especially convenient characterization if the harmonic structure were approximately the same for all notes of an instrument (i.e., if the harmonic partials have the same intensities relative to each other). However, in examining spectra of the voice and of musical instruments at many different fundamental frequencies, a spectral description of fixed relative intensities among the partials does not hold.

A "spectral envelope model" is an alternative to the "harmonic model" for describing musical tone spectra. (A spectral envelope is a curve drawn over the top of and enclosing a spectrum.) In the spectral envelope model the spectrum for a particular instrument is characterized by one or more spectral envelope peaks at appropriate frequency positions for that instrument. In this model the partial levels do not maintain the same relative values for different notes; the partials that take on the largest values are those that lie within a spectral envelope peak. For instance, if an oboe has a spectral peak at 1000 Hz its fourth partial would be most intense when the oboe is sounded at a frequency of 250 Hz, but

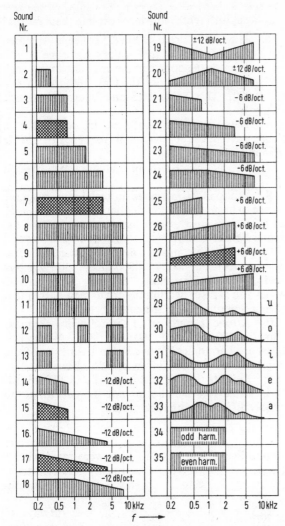

Figure 14.7. Spectral envelopes of 35 sounds rated with adjective pairs. Sound 1 is a sinusoid; sounds 4, 7, 15, 17 and 27 are noises; others are complex sounds composed of harmonic partials. (From von Bismarck, 1974.)

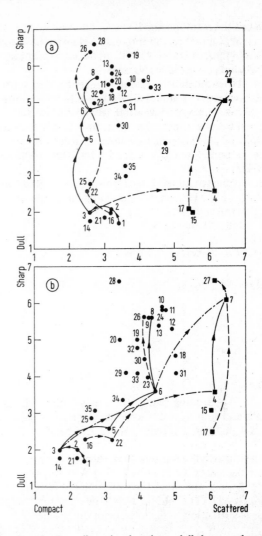

Figure 14.8. Two dimensional ratings—dull-sharp and compact-scattered— for the 35 sounds specified in Figure 14.7. (From von Bismarck, 1974.)

when sounding at 500 Hz the second partial would have the highest level. The spectral envelope model appears to be more appropriate than the harmonic model. The spectral envelope model has found considerable application in the description of musical instrument spectra (see references in Chapter 31). It provides a very good description of the vowel spectra in speech where the spectral envelope peaks are the formants produced by vocal tract resonances (see Chapter 25).

Tone quality encompasses aspects of successive changes and fusions of tone color, pitch, and loudness of a tone. Tone color, pitch, and loudness may be thought of as properties of an instantaneous snapshot of a tone, while tone quality results from the progression and changes of these instantaneous pictures—much like a motion picture. Transients—non steady parts of musical tones—play a significant role in determining tone quality. Attack transients at the beginning of a musical tone play a particularly important role in the identification of nonpercussive musical instrument tones (see Chapter 31). For percussion tones the decay transient at the conclusion of the tone is more important. Uncertainty in musical tones due to vibrato, tremolo, and choral effects are also important parts of

tone quality. Other aspects of quality might well include tone duration, rate, and rhythm. Even so, the area of quality is still defined only in rather nebulous terms. All we can do at present is to speak vaguely of a sound being oboe-like, or "ee"-like, etc. An attempt to quantify a particular tone quality by specifying the appropriate transients and all the time-varying spectra of this musical tone, would still be doomed to failure as it would still be an oversimplification. Additional aspects of this complicated subject will be discussed in Chapter 31.

Exercises

1. What sound level is required to produce just audible sound for frequencies of 50, 100, 500, 1000, 5000, and 10,000 Hz?

2. Why is a lower sound level required to produce just audible sound at 3000 Hz than at other frequencies? What does the mechanism of the ear, discussed in Chapter 13, have to do with this effect?

3. What is the meaning of "threshold of feeling"? "Threshold of pain"?

4. What is the lowest audible frequency for an "average" person? What is the highest audible frequen-

cy for an "average" person?

5. Will an 80-phon loudness level be twice as loud as a 40-phon loudness level? Will an 80-sone loudness be twice as loud as a 40-sone loudness?

6. How much greater sound level (in dB) must a 100-Hz tone have to sound as loud as a 1000-Hz tone if the 1000-Hz tone is at a level of 0 dB? 40 dB? 80 dB?

7. What is the range of sound levels between threshold of hearing and threshold of feeling to which the ear responds at frequencies of 100 Hz? 1000 Hz? 10,000 Hz?

8. If the sound level rises from 40 dB to 80 dB at a frequency of 200 Hz, what is the change in loudness (expressed in sones)?

9. Fill in Table 14.6.

Table.14.6. For completion with Exercise 9.

Frquency (Hz)	dB	Phons	Sones
100		20	
100		40	
100		60	
1000	40		
10,000	60	60	
10,000			
10,000			60

10. A marching band produces sound at the frequencies and sound levels shown in Table 14.7 when the band is 4 m from a listener. When the band is 128 m from the listener, what sound levels are produced at the listener's position? (Assume that each doubling of the distance decreases the level by 6 dB.) Which sounds are audible if we assume that they become inaudible at the 20-phon level? (This shows how "band color" might change as a function of distance from the listener.)

Table.14.7. For completion with Exercise 10.

Frequency (Hz)	Level (in dB at 4 m)	Level (in dB at 128 m)	Audible?
100	80		
200	70		
500	70		
1000	60		
5000	60		
10,000	50		

11. The text discusses loudness as being primarily dependent on sound level. However, loudness depends also on frequency and spectrum. Discuss the dependence of loudness on frequency using the equal loudness contours of Figure 14.1. How might loudness depend on spectrum?

12. In Chapter 12, we discussed sound levels as a function of the distance from the source. An approximate rule of thumb is that each doubling of the distance decreases the intensity level by 6 dB, or halving the distance increases the level by 6 dB. Suppose at a distance of 10 meters you measure a level of 60 dB for a 100-Hz tone. What is its loudness in sones? If you now move to a distance of 200 meters, what sound level might you measure? What would the loudness be at a distance of 200 meters?

13. What are two important results of masking experiments? How does the behavior of the basilar membrane at different frequencies explain these results?

14. How is the "critical band" concept explained in terms of the behavior of the basilar membrane?

15. What is the approximate critical band at each of the following frequencies (in Hz): 100; 200; 1000; 3000; 10,000? (Use Table 14.3 and interpolate where necessary.)

16. Determine the loudness in sones for a sawtooth wave whose fundamental frequency of 150 Hz has a sound level of 60 dB.

17. Repeat Exercise 16 for a square wave. Explain the difference in loudness.

18. What pitch (measured in mels) must a simple tone have to have a pitch double that of a 100-mel tone? What frequency must a tone have to have double the pitch of a 100-mel tone?

19. Suppose a complex tone having harmonic frequencies of 300, 450, 600, 750, 900, and 1050 Hz enters the ear. What fundamental frequency is perceived? How can you explain hearing the fundamental frequency when it is not present in the complex tone?

20. A tone is made up of the following frequencies (in Hz): 200, 300, 400, 500, 600, 900. Which frequencies are harmonics? What is the frequency of the fundamental? What "pitch frequency" is heard?

21. Indicate in the column labeled "Perceptible?" in Table 14.8 whether or not the two tones are noticeably different from each other.

Table 14.8. For completion with Exercise 21.

Tone 1		Tone 2		
Frequency	Level	Frequency	Level	Perceptible?
50	60	55	60	
50	60	51	60	
2000	60	2002	60	
2000	60	2020	60	
40	20	40	43	
2000	80	2000	80.5	
2000	80	2000	84	

22. What minimum change of a second formant frequency of 1500 Hz for speech would be just perceptible? Of a fundamental frequency of 130 Hz for a male talker? Of a fundamental frequency of 260 Hz for a female talker?

Demonstrations

1. Display waveforms and spectra of speech sounds, musical instrument tones, and standard waveforms.

2. Two sinusoidal generators producing approximately equal outputs are connected to an amplifier-loudspeaker system. Explain the following listener judgments of the various complex tones: When sinusoids of 260 Hz and 520 Hz are added, the resulting tone is judged to be in tune with C4 (about 260 Hz) on a piano. Why? When sinusoids of 260 Hz and 390 Hz are added, the resulting tone is judged to be in tune with C3 (about 130 Hz) on a piano. Why? When sinusoids of 130 Hz and 195 Hz are added, the resulting tone is in tune with C2 (65 Hz). Why?

3. Comparison of subjective and objective measures for hearing—Produce sinusoids of 1000 Hz

through a loudspeaker at a 60 dB level. What is the loudness level? What is the loudness of the tone? Increase the sound level to 90 dB. What is the loudness level? What is the loudness? Is the second sound half again as loud as the first? Adjust the amplifier gain until a tone is produced that sounds twice as loud as the first one. What is its loudness? What is its loudness level? What is its sound level? Use function generators to produce a sawtooth wave and a square wave of 300 Hz. Adjust their sound levels so that they sound equally loud and observe that they have the same pitch. Do the two waves sound the same? Why might this be so? Use your knowledge of the spectra of the two waves in your explanation. (Refer to the spectra in Chapter 6.)

4. Have someone with a reasonably low-pitched voice talk to you face to face and then via the telephone. Do you perceive a different pitch in the two cases? Why?

5. Run the corner of a stiff card along a comb. Move it slowly so the sound comes in separate clicks; then speed up until a definite pitch is heard. Try to match the comb pitch to your voice. Compute the frequency of your comb pitch by dividing the number of comb teeth by the time spent moving the card across them. Does it agree with the frequency of a piano note of the same pitch?

Further Reading

Backus, Chapters 5-7
Benade, Chapters 5, 13, 14
Chedd, Chapters 5, 6
Davis and Silverman, Chapters 2, 7
Denes and Pinson, Chapters 5, 6
Fletcher, Chapters 8, 10-12, 14
Hall, Chapters 6, 17
Plomp, Chapters 1, 5-7
Roederer, Chapters 2-4
Rossing, Chapters 6, 7
Seashore, Chapters 2, 5-9
Stevens and Warshofsky, Chapters 4, 5
Winckel, Chapter 2

Klein, W., R. Plomp, and L. C. W. Pols. 1970. "Vowel Spectra, Vowel Spaces, and Vowel Identification," J. Acoust. Soc. Am. **48**, 999-1009.

Robinson, D.W., and R.S. Dadson. 1957. "Threshold of Hearing and Equal- Loudness Relations for Pure Tones, and Loudness Function," J. Acoust. Soc. Am. **57**, 1284-1288.

von Bismarck, G. 1974. "Timbre of Steady Sounds: A Factorial Investigation of Its Verbal Attribute," Acustica **30**, 146-159.

Wightman, F. L. and D. M. Green. 1974. "The Perception of Pitch," Am. Scientist **62**, 208-215.

Audiovisual

1. **The Science of Sound** (Album FX6007, 1959, FRS)

2. **Sound Energy and Hearing** (30 min, 1957, EBEC)

3. "Frequency, Amplitude, and Tone Quality" (REB)

15. Additional Aspects of Sound Perception

As we have seen, the ear can detect and interpret a considerable variety of sounds over substantial ranges of sound level and frequency. Although physicists are primarily interested in describing the physical characteristics of sound, physiological effects and psychological sensations are also extremely important in understanding the perception of sound. In the previous chapter, we considered the correlation between the three main physical characteristics of sound waves (frequency, sound level, and spectrum) and their perceptual counterparts (pitch, loudness, and tone color). In this chapter we will consider additional aspects of sound perception including tonal roughness due to beats and combination tones due to nonlinear effects. We will also discuss temporal effects which are important for perception of loudness, pitch, and other phenomena. Binaural effects important for localization of a sound source will also be discussed, and finally, auditory illusions and cerebral specialization will be considered.

Perception of Simultaneous Tones

Consider two sinusoids presented simultaneously, one with a constant frequency FC and the other with a frequency F which varies from FC to 1.5 FC. Figure 15.1 illustrates how the sensation will change as F is varied. When F is equal to FC a single tone with a pitch corresponding to frequency FC will be heard. As F is made slightly larger than FC a single tone with a pitch between FC and F will be heard. The amplitude of this tone will fluctuate due to the "beating" of the sinusoids, and audible beats at a frequency of F - FC will be heard. Beyond a certain frequency difference (10-15 Hz) the beating sensation gives way to a sensation of "roughness". As F is further increased, the sensation is one of hearing two separate tones—though the roughness persists. When F is sufficiently increased, two separate tones are perceived and the roughness disappears.

Audible beating, roughness, and single tone sensation occur when two sinusoids lie approximately within a single critical band. The degree of roughness at a particular average frequency of two sinusoids depends on the frequency difference between them. A specific frequency difference at a specific average frequency produces maximum roughness. Either a smaller or larger frequency difference will decrease the perceived roughness. Table 15.1 gives representative values for frequency differences that cause maximum roughness. If two tones are more than a critical band apart, they are approximately "noninteracting" and little roughness results. As the two tones come closer in frequency than a critical band, roughness is produced, with the maximum roughness occurring when the two tones differ by a frequency approximately equal to one-fourth the critical band. Then, as the frequency difference grows

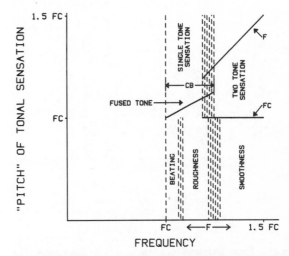

Figure 15.1. Sensation of two simultaneously presented sinsoids. One sinusoid has a constant frequency FC and the other has a frequency F which varies between FC and 1.5 FC. A single "fused" tone is sensed when F and FC are separated by less than a critical band. Two tones (labelled F and FC) are sensed when F and FC are separated by more than a critical band. Sensations in the striated regions are ambiguous. Refer to the text for further detail. (After Roederer, 1975.)

still smaller, the roughness decreases.

Figure 15.2 illustrates tonal dissonance (roughness) for sinusoids and complex tones where f_1 = 440 Hz and $440 \leq f_2 \leq 880$ Hz. Note that tonal roughness increases toward the bottom of the graph and tonal smoothness, (i.e., lack of roughness) increases toward the top. For the sinusoids, the smoothness is at a maximum when f_2 = f_1, decreasing to a minimum at about a semitone interval and then increasing beyond that up to the octave. For the complex tones there are other points of smoothness maxima and minima which depend on the number of components in the complex tone. More will be said about this in the following chapter.

Table 15.1. Frequency difference (in Hz) producing maximum roughness as related to average frequency (in Hz) for pairs of sinusoids. (After Plomp and Levelt, 1965.)

Average frequency	Frequency difference
125	5
250	15
500	28
1000	66
2000	76

Smoothness and roughness depend not only on the notes written but also on the combination of instruments producing the tones. Instruments having many higher partials are more likely to produce rough sounding combinations than instruments having few partials, due primarily to the fact that there is more opportunity for partials to interact with each other. Pro-

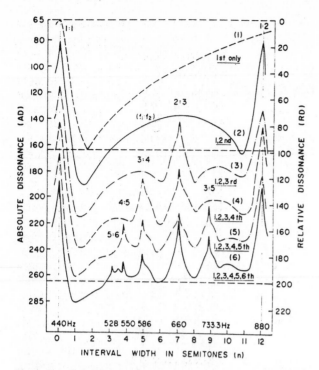

Figure 15.2. Tonal dissonance for a tone with a constant fundamental frequency of 440 Hz heard simultaneously with a tone whose fundamental frequency varies from 440 to 880 Hz. The labels on the curves indicate how many harmonics are present. (From Kameoka and Kuriyagawa, 1969.)

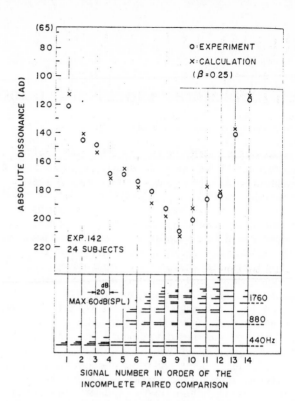

Figure 15.3. Comparisons of dissonances determined from listeners (O's) and calculations (X's) for the spectra shown at the bottom of the figure. (From Kameoka and Kuriyagawa, 1969.)

cedures have been developed for calculating the relative dissonance or roughness of arbitrary complex tone combinations when the resulting spectra are known. These predictions agree well with perceptual experiments. Figure 15.3 illustrates both experimental and numerical results for various complex tones. Figure 15.4 illustrates some interesting phenomena for complex tones with even partials only, odd partials only, and "octave partials" only. Note that the "odd partials" tones don't produce maxima at a frequency ratio of 3/2 and that the "octave partials" tones don't produce well-defined smoothness maxima.

A second order beating effect can occur when f_2 is slightly different than two times f_1—as well as at other intervals. This beating may be related to tone color and depends on the ability of the hearing mechanism to sense slow changes in waveform. If f_2 is two times f_1 the waveform does not change. If f_2 is slightly larger than two times f_1, the waveform changes slowly as shown in Figure 15.5.

Nonlinear Systems and Combination Tones

In linear vibrating systems the response of the system is directly proportional to the input (stimulus). For example, if we were to graph the amount of displacement (response) versus the force applied (input) to a mass and spring system we would get a straight line. This would show us that the relation between displacement and force is linear—one of direct proportionality. In other words, doubling the force doubles the displacement of the mass. Many vibrating systems in music and speech are nonlinear. Imagine, for example, that we find our mass and spring system getting progressively "stiffer" when displaced by larger amounts in the up-

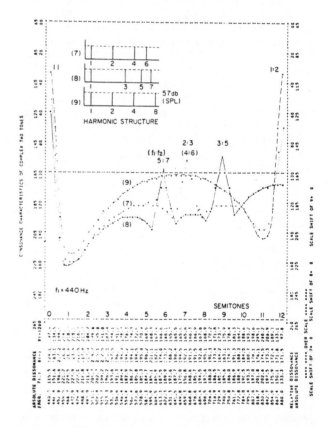

Figure 15.4. Dissonance characteristics for the different harmonic structures shown in the upper left. The graphs show dissonance for two complex tones presented simultaneously, one with a constant fundamental frequency of 440 Hz and the other with a frequency that varies from 440 to 880 Hz. (From Kameoka and Kuriyagawa, 1969.)

Figure 15.5. Second order beating when the second partial has a frequency somewhat greater than twice that of the fundamental.

ward direction—that is, more force is required to produce an additional unit of displacement when the displacement is already large. The displacement is no longer directly proportional to the force; this mass and spring system is a **nonlinear system**. If we were to plot the displacement of our nonlinear mass and spring system we might get a sinusoid with distorted peaks as shown in Figure 15.6A. The motion is still periodic, but it is not sinusoidal. The vibration recipe of the motion includes the fundamental and higher harmonics of the fundamental.

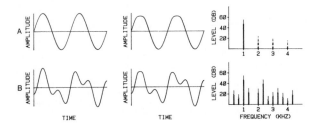

Figure 15.6. (A) Sinusoidal input and the corresponding output and its spectrum for a nonlinear system. (B) Complex wave input and the corresponding output and its spectrum for a nonlinear system.

When linear systems are driven by sinusoidal forces, only those frequencies associated with the applied sinusoids appear in the motion of the system. When a nonlinear system is driven, frequencies other than the applied frequencies may appear in the motion of the system. Typically, the greater the driving force applied to a nonlinear system, the more apparent the additional frequencies become. This has some interesting implications with regards to the ear. If the ear were a perfectly linear system we would hear only those frequencies supplied to it. However, because the ear is somewhat nonlinear, we perceive frequencies in addition to those supplied. When a sinusoid of frequency f_1 is presented to the ear we may perceive frequencies of two times f_1, three times f_1, and so on, the so called **aural harmonics** generated by nonlinear response of the ear.

When two sinusoids having different frequencies are presented to the ear, new tones, called **combination tones**, may be heard in conjunction with the original tones. Combination tones are due to nonlinear action of some mechanism as illustrated in Figure 15.6B. For example, when two frequencies f_1 and f_2 are supplied to the ear, combination tones such as $f_2 - f_1$ and $2f_1 - f_2$ (referred to as difference tones) and $f_1 + f_2$ and $2f_1 + f_2$ (referred to as summation tones) may be perceived.

What nonlinearity causes combination tones? Formerly it was believed that the middle ear was nonlinear, but it has been shown that this is not the case. Thus, any nonlinearity must occur in the inner ear. A study of direct neural pulse measurements has established that there are regions on the basilar membrane which are activated at positions corresponding to the frequencies of combination tones. The combination tones are thought to be caused by a distortion of the waveform in the cochlea itself. Although summation tones have been assumed to exist for many years, very few researchers have been able to detect them. It may be that they get masked more severely than do the difference tones. Difference tones like $2f_1 - f_2$ are not only heard, but are heard at levels in excess of those expected based on nonlinear models of the ear.

Difference tones have been used to explain perception of a pitch associated with a missing fundamental frequency. For example, suppose we have a moderately intense complex tone having components of 400, 600, and 800 Hz. If we consider only difference tones produced by nearest neighbor pairs, we get a difference tone of 200 Hz from two pairs: 600 - 400 = 200 Hz and 800 - 600 = 200 Hz. But 200 Hz is the fundamental frequency of which the 400, 600, and 800 Hz components are higher partials. In fact, the perceived pitch of this complex tone corresponds to a frequency of 200 Hz, even though there is no 200 Hz component in the original tone. In some cases this may be a contributing factor in pitch perception—but the mechanisms discussed in Chapter 14 are still the fundamental factors.

Temporal Aspects of Hearing

In many respects, the ear may be considered a time-sensing device much as the eye is considered a space-sensing device. The ear is sensitive to the time duration of a tone and to the time lapse between successive tones. The duration of a tone can affect both its perceived loudness and pitch. Increasing the duration of a tone increases its perceived loudness up to a point in time of about 0.5 seconds. Beyond a duration of 0.5 s, loudness decreases slightly due to a reduced firing rate of the hair cells because of their increased recovery time under continuous stimulation. It is much like placing one's hand on a hot stove: the extent of the burn is determined by the temperature of the stove and length of time the hand is in contact with it, or in other words the total heat absorbed by the hand.

There is an interesting uncertainty relationship that appears in the description of many physical systems. This relationship is stated as $\Delta f \times \Delta t \geq K$—meaning that the uncertainty in frequency times the uncertainty in time is greater than or equal to some constant K. It suggests that if we have a signal that is too short in duration (Δt small) the frequency (pitch) will be poorly defined (Δf large). If we listen to a sinusoid of a few milliseconds duration, we will not be able to ascribe a pitch to it; as the duration is increased to 10 milliseconds or so, we will be able to ascribe a "click pitch" to it. The "tone pitch" we would ascribe to a tone of even longer duration would be more definite and would be higher than the click pitch. Table 15.2 illustrates some time intervals required to ascribe a definite pitch to a sinusoid. Note that to some extent a constant number of cycles is required at low frequencies and that a constant duration is required at high frequencies to make the judgment.

Another question of interest is related to the perception of successive tones. When two tones

Table 15.2. Number of cycles and duration required for a listener to ascribe a definite pitch to a sinusoid.

Frequency (Hz)	Number of cycles	Duration (ms)
80	3.5	44
100	4.0	40
200	5.5	28
400	7	18
1000	12	12
2000	23	11
4000	56	14

presented in time succession are separated in time by less than 2 ms they are perceived as being simultaneous; when their time separation exceeds 2 ms they are perceived as successive, although their order of presentation is not perceived until the time separation is about 20 ms. A listener is able to "count" the number of events when they occur at rates of 10 or fewer per second.

Echoes occur when a reflected or delayed sound is heard by a listener as separate and distinct from the direct sound. This typically occurs when the time of arrival of the reflected sound is about 50 ms later than that of the direct sound. If the difference in the distances traveled by the direct and reflected sounds is about 20 meters or more, a delay in time of 50 ms or more will exist between the direct and reflected sound and echoes may be heard.

Another temporal phenomenon of interest is the **precedence effect** in which the apparent direction of a sound source is that from which the first sound arrives. When listening to a sound, in a room for example, the direct sound identifies the source direction because it arrives earlier than reflected sounds. In some cases, where electronic reinforcement is used, the source direction may be made ambiguous.

The ear has an "integration time" of about 10 ms and a "recognition time" of about 50 ms. This suggests that transients of musical instruments and of speech production may play a substantial role in the perception of such signals. Transient times in speech are of the order of 5-15 ms (considering plosive bursts of consonants or "turn-on" times for vowels). Musical instrument transients vary all the way from a few milliseconds for percussives, to 20-40 ms for an oboe to 300-400 ms for bowed strings. The consonants of speech, which are basically transients, and the attack transients of musical instruments play an important role in the identification and recognition of their respective sounds. (See Chapter 31 for further details on musical transients.) This may be due in part to the spectral changes that occur in the transients. The perception of one part of a changing sound is influenced by other parts of the sound—particularly those adjacent in time. This again offers us a partial explanation as to why the transients are so important in the identification of musical instruments and the intelligibility of speech.

Binaural Effects

Because normal hearing is binaural ("two-ear"), we have an ability to locate a sound source in a manner similar to our visual depth perception resulting from having two eyes. This **localization of sound** means that we are able to tell the horizontal direction from which a sound originates. In the vertical direction, however, our ability to localize sound is less accurate. In normal hearing situations the sound reaching one ear differs slightly from the sound reaching the other ear because the ears are separated by the head, a sphere with a diameter of about 20 cm. Consequently, when we listen to pure tones, each ear perceives a slightly different loudness. The sound waves also reach each ear at a slightly different time. Figure 15.7 shows how sound localization for pure tones depends on frequency. The vertical axis represents the average localization error in degrees. Notice that high and low frequencies can be localized rather well, but there is a curious hump (between 2000 and 4000 Hz) where larger errors are made.

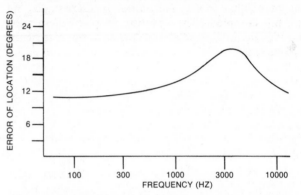

Figure 15.7. Angular location of the source of pure tones by humans. (After Stevens and Newman, 1934.)

Low-frequency tones, having a long wavelength in comparison to the size of the head, diffract around the head, while wavelengths substantially smaller than the head do not diffract appreciably. When the sound wave does not diffract, the head creates an appreciable "sound shadow" which causes the sound to be less intense at the ear within the shadow. We are able to localize the sound source because the magnitude of this intensity difference varies with the direction of the source. The shorter the wavelength relative to the head, the more distinct the sound shadow created. For low-frequency tones there is no appreciable sound shadow, and thus intensity differences are negligible. However, the slight time delay between the reception of a particular portion of a long wave by one ear and its reception by the other ear becomes important. Because of the spatial separation of the two ears, a long wave will have a slightly different phase at one ear than at the other. For each frequency, the magnitude of this phase difference varies according to the position of the sound source.

Differences in arrival time for transient sounds can play a role similar to that of phase differences for low frequencies. When high-frequency sounds seem equally intense at each ear, or when low-frequency or transient sounds arrive at each ear simultaneouly, the apparent location of the sound is directly in front of or behind the listener. This suggests that since intensity comparisons help localize high frequencies and phase comparisons help localize low frequencies, there will be a frequency range in which neither method will be very effective; thus the "hump" in Figure 15.7.

Sound localization is of considerable importance when a signal is detected in the presence of other interfering sounds. Almost everyone has had the experience of excluding irrelevant background noises at a noisy party while listening to a conversation. That this effect (sometimes called the "cocktail party effect") is related to our ability to localize sound can be demonstrated by a simple experiment. If you record a monaural tape of a friend talking to you at a loud party, you most likely won't even be able to pick out your friends voice much less understand his words when the tape is played back. Our ability to localize sounds enables us to "focus" our hearing on a particular conversation coming from a specific direction and to discriminate against other sounds. In order to achieve this effect, however, the listener must decide which sound he wishes to hear and then concentrate upon it. This introduces a new psychological concept, "attention," as a hearing parameter. In listening to music, we use our ability to localize sound in order to listen selectively to certain instruments. We may decide to concentrate on the sound of a solo violin, even though it is much quieter than the concurrent sound of the brass section. Stereophonic sound systems, by reproducing a binaural effect, restore some of a listener's sense of "presence" at a musical performance by enabling him to listen selectively.

The pinna, or outer ear, aids in sound localization—particularly between front and back. The pinna receives sound more efficiently from the front than from the back—especially at higher frequencies. There is some evidence that filtering effects caused by interference of high frequency sounds on reflection within the pinna provide both front and back directional information.

Auditory Illusions

In some listening situations, such as spoken communication, the perception of sound may be very context-dependent. Our perception may be conditioned by constraints imposed on the situation. If unknown vocabulary or topical items are suddenly introduced in a conversation we may hear them as something we expected rather than as what was actually said. In extreme cases it is a matter of perceiving what we anticipate even in the absence of what might be considered adequate stimuli.

An interesting auditory illusion can be constructed from an especially contrived set of complex tones. The theory for this effect is based on constructing very unnatural tones with many sinusoidal components, each consecutive partial having twice the frequency of the previous partial. Musical tones produced by natural means have harmonics which are integer multiples of the fundamental (e.g., 100, 200, 300, 400, 500 Hz, etc.), while these unnatural tones would have partials something like 100, 200, 400, 800 Hz, etc. Furthermore, the amplitudes are large only for components of intermediate frequency. The amplitudes of the highest and lowest extremes taper off to subthreshold levels. The spectrum for such a tone is shown in Figure 15.8. The lowest-frequency component is about 8 Hz, well below the lower limit of hearing. Note that several of the low-frequency components are also below the

threshold of hearing. Imagine now that an envelope is drawn over the frequency components, as represented by the upper curve in the figure. We now generate a second tone, in which all components are shifted upward in frequency by a constant percentage but where the amplitude of each component is still shaped by the same spectral envelope. The spectral composition of this second tone will differ from that of the original tone in the manner in which the dashed vertical lines of the figure differ from the original solid vertical lines. Note that the upward frequency shift of the components has been offset, to some extent at least, by the increased contribution of the lower components and the decreased contribution of the higher components. Indeed, if the second tone is shifted up by an entire octave (i.e., the fundamental frequency is doubled), it becomes identical to the original tone, for at the octave the highest component which has already faded below threshold is introduced an octave below the previous lowest component. By generating a set of 12 tones in this manner, each tone representing one note of an octave, we can generate a chromatic scale.

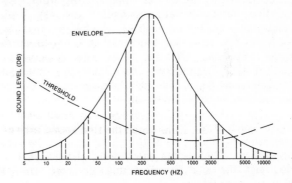

Figure 15.8. Spectrum for producing auditory illusion of constantly increasing pitch. Sinusoidal components are spaced at octave intervals. (After Shephard, 1964.)

These tones are very unnatural, and are constructed by means of a digital computer. By having the 12 tones follow each other consecutively and then repeating, an illusion of constantly increasing pitch is produced. Each consecutive tone is heard as being higher in pitch than the preceding, but after several seconds of listening the tones do not seem to be getting any "higher." In a similar way, going through the tone sequence in reverse gives the illusion of a constantly decreasing pitch. These auditory illusions are analogous to the visual illusion in Figure 15.9.

Other auditory illusions can be created by presenting sinusoidal signals dichotically to the two ears. Earphones must be used for dichotic presentation so that the signal to each ear can be controlled explicitly. When high, low, high, low sequences are presented to one ear and low, high, low, high sequences are simultaneously presented to the other ear, listeners typically report hearing only a single tone high, low sequence that shifts from ear to ear in such a way that the right ear hears the high tone and the left ear the low tone. Other illusions are produced with more complex sequences of tones and musical sequences of tones.

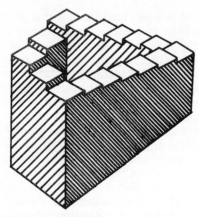

Figure 15.9. "Circular" staircase illusion.

Cerebral Specialization

Humans posses "intuitive" perceptual and cognitive abilities that seem to function in different channels than their "rational" cognition. Modern research provides evidence that "rational" functions are controlled by the left hemisphere of the brain, while intuitive functions are controlled by the right hemisphere. The left hemisphere, which is dominant for most humans, carries out sequential and analytic operations. Speech, writing, language processing, mathematical calculations, and logic are thus left hemisphere activities. The right hemisphere can process several simultaneous inputs and is used for spatial construction and pattern recognition. Many musical activities, such as the recognition of melodies and the appreciation of music, seem to be centered in the right hemisphere. The reading and writing of musical notation and the analysis of a musical score are left hemisphere activities.

Because of the way in which the auditory senses connect the inner ear to the brain, signals from the right ear are processed primarily by the left cerebral hemisphere and vice-versa. This implies a right ear dominance for speech related activities and a left ear dominance for music which have been observed to some extent. However, even though the left hemisphere is generally dominant for speech activity, any right ear dominance for speech is much less pronounced because processing occurs in the left hemisphere when sound enters either ear.

Exercises

1. Suppose that F in Figure 15.1 varies from FC to FC/2. What sensations will result? Sketch the "other half" of Figure 15.1.

2. Indicate in Table 15.3 on a scale from 1 to 5 (where 1 = very smooth and 5 = very rough) the degree of roughness of the tone pairs in the table.

3. Complete Table 15.4 by supplying a tone 2 frequency that will produce approximately maximum roughness with tone 1 in each case.

4. Indicate on a scale from 1 to 10 (10 = maximally rough) the roughness for the following combinations: (a) Tone 1 = 100, 200, 300, 400 Hz; tone 2 = 200, 600 Hz. (b) Tone 1 = 100, 200, 300, 400 Hz; tone 2 = 150, 300, 450 Hz. (c) Tone 1 = 100, 200, 500, 1000 Hz; tone 2 = 110, 220, 520, 1020 Hz. (d) Tone 1 = 100,

Table 15.3. For completion with Exercise 2.

f1	f2	Smoothness/Roughness
100	100	
100	102	
100	104	
100	106	
100	110	
100	120	

Table 15.4. For completion with Exercise 3.

f (tone 1)	f (tone 2)	f (beat)
100		
225		
490		
1,010		
1,950		

220, 490 Hz; tone 2 = 105, 240, 520 Hz.

5. Which instruments are most likely to produce dissonant combinations? Why?

6. What frequencies may emerge from a nonlinear loudspeaker when frequencies of 700 and 1000 Hz are input to the speaker?

7. What frequencies may be heard if we put two sinusoids of frequencies 700 and 1000 Hz into the ear at fairly high levels?

8. Suppose you build a room with hard walls. What are the largest dimensions that you might use so that no echoes would be produced?

9. What two mechanisms are at work in sound localization? Which is best for low frequencies? Which is best for high frequencies? Explain.

10. Can a person listening with only one ear tell much difference between mono and stereo sound? To get a more realistic stereo effect, a person should listen over earphones rather than through loudspeakers. Why? Even with earphone listening to stereo there is a lack of realism. Why?

11. Offer some explanations as to why we can localize sounds much more accurately in the horizontal plane than in the vertical.

Demonstrations

1. "Binaural Hearing," Freir and Anderson, page S-26.

2. A group of five listeners are asked to listen to two sinusoids presented simultaneously at the same level but at slightly different frequencies. A sine wave generator is set to the frequencies shown in Table 15.5. The listeners are asked to adjust the second generator until a maximally rough sound is produced. The resulting setting for each of the five listeners is shown in the table. Calculate the difference between the frequency of generator 1 and the average frequency of generator 2 for each case and record in the table. How do these differences compare with those in Table 15.1?

Further Reading

Backus, Chapter 7
Benade, Chapters 12, 14
Denes and Pinson, Chapters 5, 6
Flanagan, Chapters 4, 7

Table 15.5. For completion with Demonstration 2.

Generator 1 (Hz)	Generator 2 (Hz)	Average of 2 (Hz)	Difference (Hz)
125	129, 130, 131 132,130		
250	262, 266, 269 268, 265		
500	527, 522, 532 529, 524		
1000	1060, 1074, 1072, 1070, 1066		
2000	2081, 2083, 2071 2077, 2085		

Fletcher, Chapter 9

Hall, Chapter 17

Olson, Chapter 7

Plomp, Chapters 2-4

Roederer, Chapters 1, 2, 5

Rossing, Chapter 8

Von Bekesy

Winckel, Chapters 3, 5, 6

Deutsch, D. 1975. "Musical Illusions," Sc. Am. **233** (Oct), 92-104.

Goldstein, J. L. 1966. "Auditory Nonlinearity," J. Acoust. Soc. Am. **41**, 676-689.

Goldstein, J. L. 1970. "Aural Combination Tones," in **Frequency Analysis and Periodicity Detection in Hearing**, R. Plomp and G. F. Smoorenburg, eds. (A. W. Suithoff).

Hebrank, J., and D. Wright. 1974. "Are Two Ears Necessary for Localization of Sound Sources on the Median Plane?" J. Acoust. Soc. Am. **56**, 935-938.

Kameoka, A., and M. Kuriyagawa. 1969. "Consonance Theory Part II: Consonance of Complex Tones and Its Calculation Method," J. Acoust. Soc. Am. **45**, 1460-1469.

Plenge, G. 1974. "On the Differences between Localization and Lateralization," J. Acoust. Soc. Am. **56**, 944-951.

Rodgers, C. A. P. 1981. "Pinna Transformations and Sound Reproduction," J. Audio Eng. Soc. **29**, 226-234.

Shepard, R. 1964. "Circularity in Judgments of Relative Pitch," J. Acoust. Soc. Am. **36**, 346-353.

Stevens, S., and E. Newman. 1934. "The Localization of Pure Tones," Proc. Nat. Acad. Sci. **20**, 593.

Audiovisual

1. **The Science of Sound** (Album FX6007, 1959, FRS)

2. **Auditory Illusions and Experiments** (Cassette 72232, 1977, ES)

3. **Auditory Demonstration Tapes - Harvard University** (Laboratory of Psychophysics, Harvard University)

4. **"'Staircase' Acoustical Illustion"** (GRP)

16. Musical Scales and Harmony

In discussion of JNDs we discovered that, at a constant intensity level, the ear can discriminate several thousand frequencies. However, this represents a far too numerous set of frequencies to be useful in music where a few hundred frequencies are found to be sufficient. The actual set of frequencies used in creating music has evolved over many centuries. Melody (homophony), or tones played in sequence, probably exerted some influence on the notes (or frequencies) chosen. Multiple concurrent melodies (polyphony) may have exerted further influences on the tones selected. However, the appearance of harmony, in which two or more tones are sounded concurrently to form chords or other harmonic musical structures, has probably had the most significant influence on the selection of notes used in the scales of Western music. In fact, in parts of the world where harmony has played a less important role in music, the tones selected differ from those of Western music. In this chapter, after some basic definitions, we will consider some of the systems of tuning that have been given serious consideration over the years. We then briefly consider a physical basis for the evolution of Western harmony.

Intervals and Scales

No one knows when humans created the first musical melody, but we can be reasonably certain that the human voice was the first musical instrument. The word melody itself derives from two Greek words, melos (song) and aoidos (singer), indicating the important role of our vocal apparatus in producing melodies. Although we usually associate the word "melody" with a pleasing succession of tones, it is almost impossible to obtain agreement as to which succession of tones are pleasing and which are not. We have a natural bias toward the succession of tones which our particular culture has dictated as "pleasant," while the tones used in other cultures may seem strange, or "out- of-tune." In an attempt to avoid cultural bias, we define **melody** as a succession of tones arranged in a particular order. Melodies in classical music of the Western World are arranged as a series of definite pitches. Melodies of the epic singing of Eastern European countries may slip and slide in a seemingly random way around a tonal center. Between these two extremes there are other sytems, such as the music of India and American "soul" music which combine a set of basic pitches with fluctuating tones (called microtones) having a much smaller frequency variation.

The particular set of basic tones used to construct melodies, when arranged in order of ascending and descending pitches, is called a **scale** (from the Latin, scala, "a ladder"). An individual frequency or pitch of a scale is called a **note**, or tone, of the scale. (The word "note" is also used to connote a symbol on a sheet of music paper of the name of a musical tone.) Different scales are characterized by the number of notes per oc-

tave; a chromatic scale has 12 notes per octave, a diatonic scale has seven, and a pentatonic scale has five.

The "pitch spacing" between two notes is called an **interval** and may be given special names such as octave, fifth, third, etc. Intervals may be expressed in terms of frequency ratios such as 2/1, 3/2, 5/4, etc. An **octave** is an interval between two tones such that the ratio of their fundamental frequencies is 2/1. Since the octave appears in virtually every system, we will limit our discussion of scales to a description of the way in which the notes within an octave are defined. The notation used to specify which octave is being referred to will be that of the USA Standards Association. In this notation, the octave numbering starts on C, with C4 being middle C. A4 is the A above C4; C5 is the C an octave above middle C; C3 is an octave below middle C; and so on.

Musical intervals need not be played successively as melodic intervals. When tones are played concurrently the interval is called a **harmonic interval**. Harmonic intervals can basically be divided into two groups: consonant and dissonant. Consonant intervals result when the two tones sounded together give a "smooth" or pleasant sensation. Dissonant intervals are judged to be "harsh" or "rough" sounding.

In the sixth century B.C., Pythagoras of Samos used a monochord (a one- stringed instrument in which a movable bridge allows the string to vibrate in two adjustable segments) to demonstrate consonance. He found a consonant sound when the two vibrating segments had lengths in the ratios of 1:1, 1:2, 2:3, and 3:4. We learned in Chapter 10 that the frequency is inversely proportional to the length of a vibrating string. The frequency ratios for the strings are then 1:1 (unison), 2:1 (an octave), and two new intervals having ratios of 3:2 and 4:3. A musician listening to these new intervals would identify them as the "fifth" and the "fourth." Since this terminology has reference to a piano keyboard, let us digress momentarily and briefly consider the evolution of the piano keyboard.

The piano keyboard was borrowed from a much older instrument, the organ. In 1361 in the Saxon city of Halberstadt an organ builder, Nicholas Faber, completed a three-manual instrument which was destined to exert a substantial influence on all future organs. The upper two manuals of the Halberstadt organ had a series of 9 front keys and 5 raised rear keys in groups of two and three, as shown in Figure 16.1. Even though

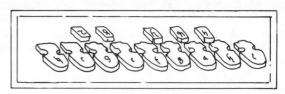

Figure 16.1. Third manual of the Halberstadt organ. (From Praetorius' **Syntagma Musicum**, 1619.)

these keys were made to be struck by the fists, this was the prototype of the now well-known 7 white and 5 black keys per octave. ("Prototype" does not imply color, as black and white keys were not used until 1475, and even at that time the convention was reversed, a practice that prevailed for the next 300 years.) The modern piano keyboard is shown in Figure 16.2 along with an example of musical notation. If we number the white keys starting with C as one, we find that F is the fourth note or "fourth," and G is the "fifth." The octave ends on the eighth note, which is C again. Consecutive white keys, however, actually contain two different types of interval: whole tones and the semitones. The interval between any two consecutive keys, white or black, is a semitone, regardless of whether we go from white to black, white to white, or black to white. A whole tone is arrived at by passing through two semitones. The interval from C to D is a whole tone, while the interval from E to F is a semitone.

for example, is tuned slightly flat from the exact fifths of Pythagorean and just tunings.

Now let us repeat Pythagoras's experiment, but this time using complex tones, each having six partials. The first tone is a constant C4 while the other tone is gradually increased in pitch from C4 to C5. The "degree of dissonance" perceived as the pitch increases is represented in Figure 16.3. At several intervals there is relatively little dissonance. These intervals correspond to the third (E), the fourth (F), the fifth (G), the sixth (A) and the octave (C). If we add two somewhat dissonant intervals, the second (D) and the seventh (B), to the collection we already have, we have one example of a scale. This particular scale (which utilizes only the white keys of a piano) is known as a C major diatonic scale. An examination of Figure 16.2 shows that the interval sequence for the scale of C major is whole, whole, semi, whole, whole, whole, semi. This sequence of intervals defines any major scale, the name of the scale being the

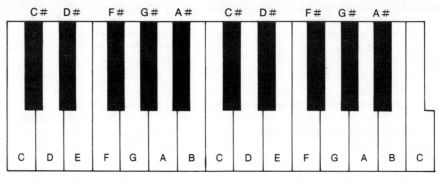

Figure 16.2. Section of a modern piano keyboard with note names (upper) and corresponding treble cleff and bass cleff musical notation for the white keys (lower). (Courtesy of C. J. Papachristou.)

Musical intervals can be defined from any starting note. Figure 16.2 shows that the interval from C to G (a fifth) consists of 7 semitones, while a third (C to E) consists of 4 semitones. The third of G is thus seen to be B, while the fifth of G is the D above. But what is the fifth of B? If we go up by 7 semitones we end on a black key, F#, which is the fifth of B.

Musical intervals can exist in the three varieties of major, diminished (or minor), and augmented. A diminished interval has been lowered one semitone, while the augmented interval has been raised one semitone. A major interval refers to the interval which has neither been diminished nor augmented. For historical reasons, whenever the major fourth or major fifth is referred to, it is called the perfect fourth or the perfect fifth. The use of the word "perfect" has nothing to do with the manner in which the interval is tuned. We will find later that a perfect fifth in equal temperament,

starting note. A minor scale (harmonic form) is defined by the following sequence of intervals: whole, semi, whole, whole, semi, whole + semi, semi. Again, the starting note determines the name of the scale, and is known as the **key** in reference to the actual white or black key on which the scale begins.

Scales and Tuning

Before describing tuning systems we digress briefly to develop a very useful alternative for expressing musical intervals. We have seen that intervals can be expressed as frequency ratios. An octave is a frequency ratio of 2/1, a fifth is a ratio of 3/2, a fourth is a ratio of 4/3, and so on. When two intervals are combined, the frequency ratio of the combined interval is given by the product of the original two ratios. For example, a fifth and a fourth give an octave whose frequency ratio is $(3/2) \times (4/3) = (2/1)$.

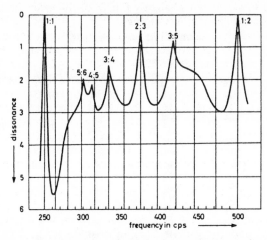

Figure 16.3. Consonances of two simultaneously presented complex tones each consisting of six harmonics. The fundamental of one tone is held constant at 250 Hz and the other is varied over an octave from 250 to 500 Hz. Note that consonance increases at certain musical intervals. (From Plomp and Levelt, 1965.)

It has been found useful to have an additive unit for expressing intervals, in addition to the multiplicative frequency ratios. For this purpose the octave is divided into 1200 equal ratios (the twelve hundredth root of 2) equal to 1.0005778. If we think of the octave as a ladder, and the notes of the scale as the rungs, there are many different ways to arrange the steps. The simplest way is to divide the octave into equal steps. On a piano keyboard the octave is divided into 12 equal semitones. One **cent** in this system is one one-hundredth of a semitone. If we multiply the frequency ratio corresponding to one cent by itself 1200 times we obtain a ratio of 2, the ratio for an octave. If we add one cent to itself 1200 times we get 1200 cents, corresponding to an octave. [When two or more numbers are multiplied together the logarithm of their product is equal to the sum of the logarithms of the numbers. The cent as an additive unit for expressing intervals is defined as 1200 times the logarithm of $2^{1/1200}$ to the base 2 or 3986 times the logarithm of $2^{1/1200}$ to the base 10 or 3986 log (1.0005778). The unusual factor 3986 appears because of the difference in logarithms between base 2 and base 10 and need not concern us here. The number of cents in any interval is given by 3986 times the logarithm of the frequency ratio of the interval.]

In different parts of the world, and in different tuning systems, intervals which are not exact multiples of 100 cents abound. Microtonal intervals of less than 100 cents are common in Asia and India. The neutral third, between the major and minor third, is prevalent all over the world. In the United States it is used in conjunction with several other microtones to give the special sound associated with the "blues" style. Some of the different ways in which the octave is divided and tuned in various cultures is shown in Figure 16.4. Each of these scales has its own unique character, which in association with the microtones often present, produces the sounds peculiar to a particular culture. The 12 tone chromatic scale (1) and 7 tone diatonic scale (4) are shown for comparison. The 12 tone chromatic scale (1), the 7 tone scale from Thailand (2), and the 4 tone scale from Uganda (3) all use equal intervals between tones. The diatonic scale (4), the 5 tone scale from India (5), the five tone scale from

Japan (6), and the five tone scale from China (7) each employ different unequal intervals between tones.

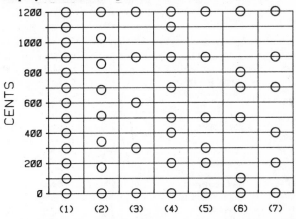

Figure 16.4. Various ways of dividing the octave in equal intervals: (1) 12 tone chromatic scale, (2) 7 tone Thai scale, and (3) 4 tone Ugandan scale. Various ways of dividing the octave in unequal intervals: (4) diatonic scale, (5) five tone scale from India, (6) five tone scale from Japan, and (7) five tone scale from China.

After a particular scale has been selected, there are still many ways in which its notes can be tuned. We next consider three different ways that a keyboard instrument (such as a piano or an organ) can be tuned—just tuning, Pythagorean tuning, and equal tempered tuning. Many other ways have been and still are being used, but these three systems will provide adequate illustrations for our purposes. The Pythagorean and just systems have many consonant intervals having no beats between the partials of the tones. The equal tempered system has "tempered" or adjusted the intervals so that many different musical keys can be performed without excessive dissonance on a keyboard with twelve keys per octave. Table 16.1 gives intervals in terms of notes, names, frequency ratios, and cents for just, Pythagorean, and equal tempered tunings. Note that the table includes twelve semitone intervals so that both white and black keys on a piano are represented.

Just Tuning

Suppose we wish to create a scale in which complex tone pairs sounded together produce a maximally consonant result. Looking at Figure 16.3 we see that a unison produces maximum consonance. This is because the partials of the two tones coincide and no rough interaction results. As a general rule, we can state that as more harmonics of two complex tones coincide, the result becomes smoother. **Just tuning** is interesting from this point of view and gives intervals defined in terms of ratios of small whole numbers.

The octave is another interval which gives maximum smoothness. This is because harmonics 1, 2, 3, etc. of the higher tone coincide with harmonics 2, 4, 6, etc. of the lower tone since the frequency ratio of their fundamentals is 2/1. The next smoothest interval is the fifth with a ratio of 3/2. Harmonics 2, 4, 6, etc. of the higher tone coincide with harmonics 3, 6, 9, etc. of the lower tone and other harmonics are quite widely separated so little roughness occurs. Figure 16.5 illustrates the harmonics of an octave and a fifth in musical notation. Other "smooth" intervals of impor-

103

Table 16.1. Just, Pythagorean, and equal-tempered tunings and their frequency ratios and intervals. (Bracketed notes are identical in equal-tempered tuning.)

Note	Musical interval	Frequency ratio to first note			Interval from first note (cents)		
		Just	Pythagorean	Equal	Just	Pythagorean	Equal
C	unison	1.000	1.000	1.000	0	0	0
C#	semitone (chromatic)	1.042	1.068	1.059	71	114	100
D♭	minor second	1.067	1.053	1.059	112	90	100
D	major second	1.125	1.125	1.122	204	204	200
D#	augmented second	1.172	1.201	1.189	275	317	300
E♭	minor third	1.200	1.185	1.189	316	294	300
E	major third	1.250	1.266	1.260	386	408	400
F♭	diminished third	1.280	1.249	1.260	427	385	400
E#	augmented third	1.302	1.352	1.335	457	522	500
F	perfect fourth	1.333	1.333	1.335	498	498	500
F#	augmented forth	1.406	1.424	1.414	490	612	600
G♭	diminished fifth	1.440	1.405	1.414	631	589	600
G	perfect fifth	1.500	1.500	1.498	702	702	700
G#	augmented fifth	1.563	1.602	1.587	773	816	800
A♭	minor sixth	1.600	1.580	1.587	814	792	800
A	major sixth	1.667	1.688	1.682	884	906	900
A#	augmented sixth	1.758	1.802	1.782	977	1020	1000
B♭	minor seventh	1.800	1.778	1.782	1018	996	1000
B	major seventh	1.875	1.898	1.888	1088	1110	1100
C♭	diminished octave	1.920	1.873	1.888	1129	1086	1100
B#	augmented seventh	1.953	2.027	2.000	1159	1223	1200
C	octave	2.000	2.000	2.000	1200	1200	1200

tance are the sixth (5/3), the fourth (4/3), and the third (5/4). There are three different interval sizes between notes in just tuning. The interval for C-D, F-G, and A-B is 204 cents; the interval for D-E and G-A is 182 cents; the interval for E-F and B-C is 112 cents. Thus there are two sizes of whole tones (204 and 182 cents) and neither is equal to twice a semitone (112 cents).

16.2 that the third partial of C3 (390 Hz) corresponds to the fundamental frequency of G4. The sixth partial of C3 (780 Hz) corresponds to the fundamental frequency of G5. We can follow this procedure further and find that the fifth partial of C3 (650 Hz) corresponds to E5 and the eighth harmonic of C3 (1040 Hz) corresponds to C6. The seventh harmonic of C3 (910 Hz) presents some

Figure 16.5. Harmonics of an octave and a fifth in musical notation. (Courtesy of C.J. Papachristou.)

If a periodic musical tone has a fundamental frequency of 130 Hz, (C3), its higher harmonics will be 260 Hz, 390 Hz, 520 Hz, etc. The second harmonic of C3 (260 Hz) has the same frequency as the fundamental of C4, an octave above C3, while the fourth harmonic (520 Hz) has the same frequency as the fundamental of C5. (Observe that the tuning to C3 = 130 Hz rather than to A3 = 220 Hz is for convenience and to reduce the use of decimal numbers in this example.) We see from Table

difficulties. Since 910 Hz lies between the second partial of A4 (866.7 Hz) and that of B4 (970 Hz), we consider B flat and A sharp, notes between A and B. However, calculations show that neither B flat nor A sharp corresponds exactly to the desired frequency of 910 Hz. The best we can do is to say that the seventh partial of 130 Hz approximates B5♭ (which is used rather that A5# for historical reasons). Figure 16.6 shows the relationship between the partials of C3 and the notes of the scale

Table 16.2. Just tuning showing frequency relationships among notes in a C major scale and higher partials.

Musical interval	Note	Ratio to C3	Partials					
			1	2	3	4	5	6
Unison	C3	1/1	130	260	390	520	650	780
Second	D3	9/8	146.3	292.5	438.8	585	731.3	877.5
Third	E3	5/4	162.5	325	487.5	650	812.5	975
Fourth	F3	4/3	173.3	346.7	520	693.3	866.7	1040
Fifth	G3	3/2	195	390	585	780	975	1170
Sixth	A3	5/3	216.7	433.3	650	866.7	1083.3	1300
Seventh	B3	15/8	243.8	487.5	731.3	975	1219	1463
Octave	C4	2/1	260	520	780	1040	1300	1560

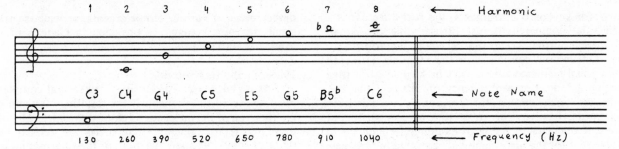

Figure 16.6. Relationship between the partials of C3 and notes of the scale in musical notation. (Courtesy of C. J. Papachristou.)

in musical notation.

Even though it can be shown (see exercises) that the just system produces the smoothest tone pairs, it suffers from a major drawback. When scales are constructed in the just system (i.e., each scale starts on a different key), the notes have different frequencies in different keys. For example, if we start a scale on C with a frequency of 260 Hz the frequencies for E, F, G, A, and B will be 325, 346.7, 390, 433.3, and 485 Hz. If we now start a scale on G with a frequency of 390 Hz its third will be on B and should have a frequency of 487.5 Hz which we see is slightly different from B defined as the seventh of C. We would need two or more B keys to play scales starting on all possible notes in the octave. The just scale has never been used extensively because at least 53 keys per octave should be available to obtain reasonable accuracy in all keys.

Pythagorean Tuning

Pythagorean tuning is based on fourths and fifths having small whole number frequency ratios. Suppose we construct a C major scale where the frequency of each note is an exact fifth or fourth from one of the other notes. In other words, the frequency of G is exactly 1.5 times the frequency of C, and the frequency of F is exactly 1.333...(4/3) times the frequency of C. A frequency 1.5 times that of G4 gives an exact fifth, D5, but an octave too high. Taking half the frequency of D5, then yields the frequency of D4, which is back in the octave where we started. The frequency 1.5 times that of D4 gives A4, and so on.

Although the idea of having a system of tuning with exact fourths and fifths looks appealing, there is one severe discrepancy. If we assume an octave has twelve notes, then by advancing frequencies by exact fifths we eventually arrive back at the same note, but several octaves higher. This concept is illustrated in Figure 16.7 as the "circle of fifths." In other words, if we take the frequency ratio of a fifth (3:2) and multiply it by itself 12 times we should get an exact multiple of the original frequency. In actual fact, the frequency we obtain is about 24 cents too high. With Pythagorean tuning the only way around this problem is to have two different sizes of semitones: the diatonic semitones of 90 cents and the chromatic semitones of 114 cents. (Note that these semitones differ by the Pythagorean discrepancy of 24 cents.) From the information given in Table 16.1 we can see that there are five 204 cent intervals (C-D, D-E, F-G, G-A, and A-B) and two 90 cent intervals (E-F and B-C) in a C major scale with Pythagorean tuning. Notice that a whole tone (204 cents) is not equal to twice the diatonic (90 cent) or twice

the chromatic (114 cent) semitone, but is equal to the sum of a chromatic and diatonic semitone.

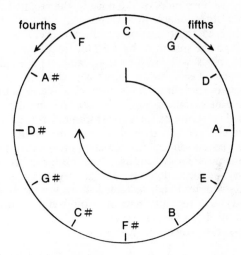

Figure 16.7. Circle of fifths.

With the coming into use of thirds as consonant intervals, Pythagorean tuning was somewhat deficient because its major third (408 cents) and minor third (294 cents) sounded sharp and flat respectively. Mean tone tuning attempted to compensate for these and similar difficulties by making upward or downward adjustments of certain intervals. However, mean tone tuning proved too restrictive as modulations from one key to another developed, and ultimately equal temperament prevailed.

Equal Tempered Tuning

Equal temperament is the system which is currently most widely adopted. It is based on compromises of the exact intervals used in other systems. Although the mathematics of equal temperament had been solved some fifty years earlier, it was not until 1688, three years after J. S. Bach's birth, that the first organ was tuned in accordance with this system. Bach himself favored equal temperament, and his many compositions (written in almost all keys) for harpsichord and organ undoubtedly had tremendous influence in moving the musical world to the almost universal adoption of this system. In **equal-tempered tuning** the octave is divided into twelve equal intervals, or semitones, any two consecutive notes having the same frequency ratios. The frequency ratio of two consecutive semitones is given by the twelfth root of two, which is 1.059463. In an equal-tempered octave, a whole tone is equal to exactly two semitones. Although equal temperament is very convenient for keyboard instruments, the fifths are about 2 cents flat compared to just fifths, while the equal-tempered fourth is 2 cents sharper than a just fourth.

Let us now consider how the intervals work out with equal-tempered tuning. Let us begin with a tone having a fundamental frequency of 130 Hz. The frequencies which comprise the C major scale are listed in the partial 1 column of Table 16.3. A comparison of these frequencies to the corresponding frequencies in Table 16.2 for the just system indicates that the two sets are, in most cases, fairly close. For instance, the just fifth has a frequency of 195 Hz, while the equal tempered fifth of 194.7 Hz is flat by only 0.3 Hz. Although the harmonics of an equal-tempered fifth will beat with the harmonics of the unison, the result is not considered objectionable in most musical contexts. The equal-tempered third fares less well, however, as it is sharper than the just third by 1.3 Hz. Equal-tempered harmonic thirds will produce noticeable, and usually objectionable, beats in long sustained chords. Organ music from the Seventeenth Century often featured prominent use of sustained thirds because the instruments were tuned to eliminate (or at least minimize) the beats in the two or three keys then in common use. When this music is played on an instrument tuned for equal temperament the harsh beating of the thirds may sound dissonant and objectionable. In the organ music of Bach, on the other hand, sustained thirds are rarely encountered, and the dissonances produced by equal-tempered thirds, in the context of the changing fabric of the composition, are not noticeable.

development of various temperaments, as well as proposals for new systems, is provided in **Genesis of a Music**, by Harry Partch.

Harmony and Its Evolution

As we have seen previously, when several complex tones are heard simultaneously the resulting dissonance is a function of the interactions among partials. Pairs of tones whose fundamental frequencies are related by ratios of small integers have more of their harmonics in unison and fewer harmonics that can interact to create dissonance. It is generally recognized that Western harmony is based on the major triad, which is a chord composed of the following intervals: the unison, a major third, and a perfect fifth. Why was this particular combination of intervals chosen, rather than some other? While we cannot answer this question with certainty, there are several interesting hypotheses which are worth considering. First, consider a major triad consisting of A3, C4#, and E4, with respective frequencies (in the just system) of 220 Hz, 275 Hz, and 330 Hz. Table 16.4 gives the first six harmonics of each of these tones. Notice that a number of frequencies (indicated by an *) are common to two of the tones. Because of the duplication of a number of partials, and because there are no instances of any two partials being separated by the frequency which would produce maximum roughness, this triad is perceived as having a "smooth" sound. Perhaps, then, because the major triad sounded so consonant it was first adopted as the basis of musical harmony.

Table 16.3. Equal-tempered tuning showing frequency relationships among notes in a C major scale and higher partials.

Musical interval	Note	Ratio to C3	1	2	Partials 3	4	5	6
Unison	C4	1.00	130	260	390	520	650	780
Second	D4	1.122	145.9	291.7	437.6	583.4	729.3	875.2
Third	E4	1.260	163.8	327.6	491.4	655.2	819	982.8
Fourth	F4	1.335	173.6	347.1	520.7	694.2	867.8	1041
Fifth	G4	1.498	194.7	389.5	584.2	779.0	973.7	1168
Sixth	A4	1.682	218.7	437.3	656.0	874.6	1093	1312
Seventh	B4	1.888	245.4	490.9	736.3	981.8	1227	1473
Octave	C5	2.000	260	520	780	1040	1300	1560

Other methods for dividing the octave have from time to time been devised. A quarter tone scale has been proposed that would involve 24 notes per octave, each differing by 50 cents from its neighbor. This system would include all of the equal-tempered notes, but would have an additional note between each pair of notes of the tempered system. The division of the octave into nineteen equal parts has been suggested, and a nineteen-tone harmonium was actually constructed over 100 years ago. This system gives very good thirds, but the fifths are 5 cents lower than the fifths in a twelve-toned system. A division of the octave into 53 equal parts has been proposed, and a harmonium having 53 keys for each octave has even been constructed. Of all the various proposed octave divisions, this one is the most nearly "perfect" in that most of the important intervals are within 2 cents of being exact. It is thus possible to play with nearly just tuning in any key, provided one is willing to invest the time and effort that would be required to master the keyboard. A detailed discussion (from a musician's viewpoint) of the history and

Some people have gone beyond merely considering the foundations of Western harmony and argued that the entire superstructure evolved in accordance with the physical nature of the harmonic series. In spite of the fact that their case is usually presented in an erudite manner, there are some serious problems with such a viewpoint. Consider first the following information in support of this hypothesis. If a harmonic series is constructed based on C4 (such as shown in Table 16.5), using just intervals we find the following notes: C4, C5,

Table 16.4. Harmonics of tones of major triad A3, C4#, and E4. Harmonics marked with * are common to two tones.

Harmonic	A3	Note C4#	E4
1	220	275	330
2	440	550	660*
3	660*	825	990
4	880	1100*	1320*
5	1100*	1375	1650*
6	1320*	1650*	1980

Table 16.5. Harmonic series starting on C. Intervals between neighboring harmonics are shown at the bottom.

Harmonic	1	2	3	4	5	6	7	8	9	10	11	12
Note	C	C	G	C	E	G	B♭	C	D	E	F#	G
		oct	5th	4th	maj 3rd	min 3rd			maj 2nd		min 2nd	

G5, C6, E6, G6, B6♭, C7, D7, E7, F7#, G7, A7♭, B7♭, B7#, and C8. The exact intervals (the octave, the fifth, and the fourth) occur quite early in the series. That is, the interval between the first two harmonics is an octave, the interval between the second and third harmonics is a fifth, and the interval between the third and fourth harmonics is a perfect fourth. (Some notes, namely B6♭, F7#, A7♭, and B7♭, would be out of tune, even in the just system.) Harmonics 1, 2, 3, and 4 formed the basis of medieval harmony (which consisted of chords in octaves, fifths, and fourths). After the year 1200, harmonic 5 (the major third) was added, giving the major triad, which was the basis of music for the next three hundred years. Around the year 1600, the seventh harmonic (B6♭) was added, thus contributing the dominant seventh chord. In the eighteenth century harmonic nine provided the ninth chord, and in the late nineteenth century the eleventh harmonic gave the eleventh chord. In the twentieth century music has been characterized by much dissonance, and the high harmonics give rise to dissonant sounds such as are found in the minor second.

Although the above argument presents a convincing view of the evolution of harmony from the harmonic series, this theory contains several serious flaws. If nature is the basic source of harmony, why are certain notes (such as the seventh harmonic) out of tune, even in just tuning? Why are the fourth and sixth notes of the scale (F and A) missing from the series, while they have played an important part in harmony? Furthermore, why are the sharp fourth and flatted sevenths present in the harmonic series but not in the scale? Also, if the harmonic series is the basis of music, why did modal scales precede the major scales by many centuries, and why does the minor mode not appear in the harmonic series? Finally, the interval of a fourth was considered for many centuries to be a dissonant interval, and yet it appears quite early in the harmonic series. Obviously then, the viewpoint that all of Western harmony developed and evolved on the basis of the harmonic series is greatly oversimplified, if not inaccurate. We can probably say, however, that the natural structure of the harmonic series exerted some considerable influence on the development of harmony, but other nonscientific factors were probably equally important.

There has been an evolution of what is considered consonant and dissonant. What was called dissonant some years ago could very well be considered consonant today, even though the amount of roughness is the same. Consonance in this sense is a function of conditioning and is learned from the culture. Consonance may still depend on the relative roughness, but the relationship is neither simple nor fixed. If we equate musical dissonance with roughness and consonance with smoothness, we have a scheme for determining how dissonant or consonant particular tone pairs or compositions will be. This approach can be applied across compositions by considering beat combinations between all possible pairs of partials. Such an analysis predicts that J.S. Bach's Trio Sonata for Organ No. 3 in C Major should sound less rough than Dvorak's String Quartet Op. 51 as shown in Figure 16.8. The solid curves in the left part of the figure show intervals between partials of the Bach Sonata; those on the right show partial intervals for the Dvorak Quartet. The Dvorak quartet intervals lie closer to one quarter of a critical band (lower dashed curve) and therefore produce greater roughness.

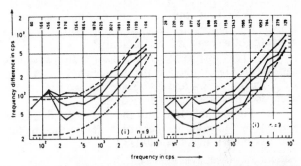

Figure 16.8. "Consonance" analysis of J.S. Bach's Trio Sonata (left) and Dvorak's String Quartet (right) as related to critical band. See the text for further explanation. (From Plomp and Levelt, 1965.)

Intonation and Standard Pitch

Intonation is the extent to which the frequency of an instrumental tone corresponds to the expected value in the desired scale. In practice the accuracy of intonation of bowed strings is limited primarily by the skill of the performer. Instruments employing an air column are subject to temperature-caused variation in intonation as well as to intonation problems due to instrument design. Woodwind players can compensate to some extent and brass players to a much greater extent for instrumental intonation problems. The human voice, like the bowed string, has intonation precision limited primarily by the skill of the performer.

Many of the more subtle differences among different scales are probably not realized in musical performance. The inability of a performer to play "perfectly in tune," as well as aesthetic considerations and ornamentations like vibrato, help to obscure perceptual differences. Sustained tones would be most likely to show such differences, but sustained tones are not common in musical performance. The just scale has often been considered to be the most "natural." However, recent experiments have shown that musicians not limited to equal temperament (such as violinists and vocalists) make more use of the Pythagorean intervals than the just intervals.

It is necessary to have some standard pitch to tune against and then to tune all other notes relative to this standard within the intonation accuracies of an instrument and performer. It is important that the standard pitch be retained at a constant value over a long period of time because many instruments (both string and wind) are designed to perform optimally at a particular tuning and may perform less well at higher or lower tunings. Over the years the tuning of A4 has varied from as low as 415 Hz to as high as 460 Hz, an interval of almost two semitones. The currently accepted **standard pitch** is the tuning of A4 to 440 Hz, although not all practicing musicians adhere to this value.

Exercises

1. Tables 16.2 and 16.3 give the ratio of fundamental frequencies between the first tone and each additional tone for just and equal-tempered tunings. The frequencies of the first six partials are also given for each tone. Assume that each tone is played with C3 and determine the partials that will produce roughness; order the pairs of tones in terms of relative roughness and smoothness. Work with each table separately. Note that the tuning is to C3 = 130 Hz.

2. Which of the two tunings, just or equal-tempered, produces the smoothest tone pairs? Why do we use the other scale more commonly?

3. Try constructing a major scale with just tuning starting on G4. What are the frequencies of A4 for scales starting on C4 and G4? Since the two are not the same, does this present any difficulty?

4. Fill in Table 16.6 with the correct fundamental frequencies for equal temperament. (Note that in this case the tuning is to A4 = 440 Hz.)

Table 16.6. For completion with Exercise 4.

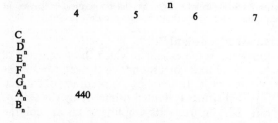

5. Consider the discussion of the A major triad given in the section entitled "Harmony and Its Evolution." Assume that each note of the chord has the first six harmonics present. Calculate the frequency of every difference tone that will be present.

Demonstrations

1. "Harmonics of a String," Freier and Anderson, page S-22.

2. "Harmonious Notes," Freier and Anderson, page S-22.

3. "Chords," Freier and Anderson, page S-23.

4. Use a monochord to recreate some of the Pythagoras consonance experiments.

5. Set up two function generators to operate into a mixer, scope, and speaker system. Check perceived roughness as a function of waveform and frequency difference.

6. Explore many two and three note combinations on the piano in the low, mid, and high octaves.

Further Reading

Bakus, Chapter 8
Benade, Chapters 15, 16
Bartholomew, Chapter 4
Hall, Chapters 7, 18, 19
Helmholtz, Chapters 8-19
Jeans, Chapter 5
Olson, Chapter 3
Roederer, Chapters 4,5
Rossing, Chapter 9
Seashore, Chapter 10
Winckel, Chapter 8
Wood, Chapters 10, 11

Kameoka, A., and M. Kuriyagawa. 1969. "Consonance Theory Part I: Consonance of Dyads" and "Consonance Theory Part II: Consonance of Complex Tones and its Calculation Method," J. Acoust. Soc. Am. 45, 1451-1469.

Partch, H. 1974. **Genesis of a Music** 2nd ed. (Da. Caps Press).

Plomp, R., and W.J.M. Levelt. 1965. "Tonal Consonance and Critical Bandwidth," J. Acoust. Soc. Am. 38, 548-560.

Young, R.W. 1939. "Terminology for Logarthmic Frequency Units," J. Acoust. Soc. Am. 11, 134-139.

Audiovisual

1. **Temperaments: Equal, Meantone, Pythagorean** (Tape available from R. C. Nicklin, Physics Dept., Apalachian State Univ., Boone, N.C. 280708)

2. **The Science of Sound** (Album #FX6007, 1959, FRS)

3. **Auditory Illusions and Experiments** (Cassette #72232, 1977, ES)

4. **Science of the Musical Scale** (30 min, color, 1957, EBE)

5. "Relative Consonance of Tone Pairs and Temperaments" (GRP)

6. "The Overtone Series: (REB)

17. Hearing Impairments and Hazards

The ear is a delicate instrument which, even though well protected against normal environmental exposure, is subject to impairment. Impairments can occur in any of the three parts of the ear and usually result in some hearing loss. A loss occurring because of problems in the outer or middle ear is called a **conductive loss** because it results in a reduced conduction (or transmission) of sound to the sensing elements of the ear. A loss occurring because of problems of the inner ear is called a **sensorineural loss**. Such a loss is caused by a partial or complete breakdown of the nerve-sensing (or mechanical-to-nerve-pulse-converting) elements of the ear. Conductive losses are usually amenable to correction. Sensorineural losses are not amenable to correction but may be compensated for with various aids.

Hearing losses typically result in an inability to perceive some of the more subtle aspects of sounds. When the loss is more than a mild one, enjoyment of music may be reduced. A more critical situation occurs when the loss progresses to the point where speech reception is impaired. At that point an individual's ability to function effectively with everyday speech communication is reduced. For this reason, most laws dealing with hazards to hearing are concerned with the hearing loss that will impair speech communication. The first sounds to be lost in sensorineural deafness are the high-frequency sounds of speech. The most common complaint of people suffering from noise-induced deafness is "I can hear you, but I don't understand what you say." These people confuse words, because some similar speech sounds are distinguished by their upper frequencies.

In this chapter we will consider several common ear impairments, the assessment of hearing loss, hearing aids, the hazards of intense sound, regulation of hazardous sounds, and protection against hazardous sounds.

Ear Impairments

It has been estimated that more than 16 million Americans are suffering from impairments that affect their hearing and that many of these people do not receive proper treatment. We will discuss some common ear impairments, their related hearing losses, and their probable treatments.

Blockage of the ear canal can be caused by a buildup of waxy secretions or by an infection in the canal. Waxy secretions found in the ear canal serve a useful purpose by trapping foreign particles, but if too much wax accumulates the ear canal will become clogged, and less sound will be conducted to the eardrum. Removal of the waxy material by a physician (so that care will be taken not to damage any part of the ear structure) is the most usual solution to this problem. Inflammation can be caused by contaminated water which has not been completely drained from the ear canal. Various ointments are effective in controlling these in-fections, but preventative medicine, such as draining the ears after swimming, is a better practice.

A perforated eardrum may result from sticking objects into the ear, a sharp blow to the side of the head, or an infection. Small perforations usually heal themselves, but larger ones may require an operation where a piece of skin from the ear canal is grafted over the perforation.

Inflammation of the middle ear called **otitis media** is the most common cause of conductive hearing loss. Infections are generally transmitted through the eustachian tube into the middle ear. Children are more susceptible to such infections because of their shorter eustachian tubes. With otitis media the ear drum is initially displaced inward, but then as fluids build up it is displaced outward. In either case the ossicular chain is displaced to near its limit, causing a tensing of the ossicular ligaments, reducing conduction. Treatment involves the use of antibiotics and small incisions in the eardrum if needed.

Disablement of the mechanical link in the middle ear can occur because of a break in the bone chain. Disablement can also be caused by a condition known as **otosclerosis** which is a calcifying, or "freezing," of the stapes. In either case the mechanical link between the eardrum and the cochlea is partly or wholly impaired, and the vibrations arising from a sound are not passed on to the cochlea with their normal stength. In some instances the bone chain will repair itself. Otherwise the stapes must be freed using surgical techniques. In other instances it is necessary to remove the damaged parts and complete the mechanical link with a prosthetic appliance.

Tinnitus, or "ringing in the ears," is characterized by ringing, rushing, or roaring noises in the ear that come and go spontaneously without any sound stimulus. Only about 10% of the cases of tinnitus can be treated medically because the causes of specific cases are not easily determined. One "treatment" employs a small noise generator, similar in appearance to a hearing aid which produces a constant sound at the same frequency as the tinnitus sounds in order to mask them.

Meniere's disease is caused by an excess pressure in the cochlear fluid of the inner ear. Its symptoms are sensorineural hearing loss, tinnitus, nausea, and episodes of vertigo. Timely medical attention is necessary in its treatment.

Loss of or damage to the hair cells in the cochlea is the most serious of the various types of hearing impairments. Some hair cells may be missing; some may increase spontaneous firings; some may require unusually strong stimuli to excite them; some may suffer a loss in sharpness of tuning. At present there is no known means for repairing damaged hair cells or replacing lost ones. The most effective thing that has been done to date is to compensate through the use of hearing aids.

A certain kind of hearing loss, **presbycusis**, seems to affect all people as they grow older. The ear gradually becomes insensitive to high-frequency sounds, presumably because of deterioration of the hair cells that respond to high frequencies, loss of neurons in the central nervous system, and middle-ear impairment. In industrial nations, most of the population 30 years and older cannot hear above 15,000 Hz. At 50 years the limit is 12,000 Hz; at age 60 it is 10,000 Hz; and at age 70 it is 6000 Hz on the average. (We will see later that there is evidence that loss of hearing with age may be due to a basically noisy environment.)

Permanent hearing loss occurs most frequently in people who work in noisy industrial environments. Single exposures to very intense sounds or blows to the head can also produce hearing loss. In addition, fevers, tumors, infections, drugs, and poisons can produce hearing loss. As a matter of fact, before World War II ear infections were the leading cause of hearing loss.

Hearing Assessment

Various methods of measurement must be used in order to determine the nature of hearing loss. For example, if a person can hear sounds when the bony structure behind the ear is caused to vibrate, then sensorineural loss is less likely. If a person hears bone-conducted sound, but is unable to hear air conducted sound, a middle ear impairment may be the cause. Middle ear impairments may also be assessed by measuring the response of the ear drum to different stimuli.

There are several crude hearing tests using whispered speech and conversational speech, that permit easy and approximate testing. However, more precise measurements of hearing acuity or loss are important in clinical situations involving the fitting of hearing aids or the prescription of therapy, as well as in situations where fitness for service is at issue. A pure tone **audiometer** is a device used to test hearing acuity. It generates sinusoids at different frequencies and permits variation of their intensity levels. A person listening to a tone produced by the audiometer responds when he can barely perceive the tone. In this manner his individual threshold of hearing can be established and compared with a standard threshold curve. If the subject has sustained a hearing loss, the degree of loss, expressed in dB above the normal threshold, is determined at each of the measured frequencies.

Figure 14.1 shows a common reference level for audiometry as represented by the X's at certain frequencies. You will note that the reference level differs from the threshold curve. Clinical audiometers typically perform the subtraction between the measured threshold and the reference level. A reading is given directly in terms of hearing level relative to the reference with 0 dB corresponding to the reference level.

A graph showing hearing level at different frequencies is called an **audiogram**. The O's represent right ear values and the X's left ear values in the sample audiogram shown in Figure 17.1. The audiogram shows hearing levels ranging from 10 - 30 dB "above" the reference level at frequencies of 250, 500, 1000, and 2000 Hz for the left ear. The hearing levels at 4000 and 8000 Hz show a substantial loss in the left ear. Hearing levels for the right ear are somewhat unusual in that they are

normal at 2000 and 4000 Hz, but show losses at lower frequencies that may give rise to problems with speech perception. A patient whose audiogram shows a 20 dB average loss at frequencies of 500, 1000, and 2000 Hz may have difficulty hearing distinctly in a theater. An average loss of 40-55 dB will result in frequent difficulty with normal conversation.

Pure tone audiometry depends on cooperation of the subject and cannot tell the full story when hearing impairments are involved. Other methods such as bone conduction audiometry, speech audiometry, and evoked response audiometry are used to gain additional information.

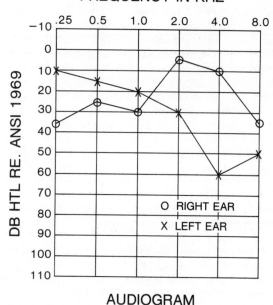

Figure 17.1. Audiogram showing hearing threshold level in decibels relative to some standard reference values (shown by the X's in Figure 14.1). A level of 0 dB in the audiogram corresponds to the reference value. Positive values shown below 0 dB in the audiogram represent hearing threshold levels greater than the reference and thus a "loss" of hearing. (Courtesy of R. H. Brey.)

Hearing Aids

When hearing is impaired, higher sound levels must be produced. A **hearing aid** is any instrument that makes sound more accessible to a listener. The first objective of any hearing aid is usually to make speech more intelligible—even if speech quality must be sacrificed.

Acoustic hearing aids were the first to be used. "Ear trumpets" were designed to "funnel" sound into the ear. These aids were often concealed so their use was not obvious to others and were thus more aesthetically acceptable. Cupping the hand behind the ear provides a rudimentary acoustic hearing aid.

With the miniaturization of electronics and transducers, electroacoustic hearing aids have become widely used. An **electronic hearing aid** consists of a microphone which converts sound into electrical energy, an amplifier, and an earphone or receiver to convert the electrical signal back into sound. Hearing aids are useful for people with a hearing loss that is not too severe. A hearing aid is, in a sense, a "private public address system," which attempts to compensate for the decreased sensitivity of the ear by increasing the intensity of the signal sent to it.

Hearing aids are of several types, including body-worn aids, ear-level aids, and in-the-ear aids. The latter two types are usually cosmetically more acceptable to the user, but body-worn aids can supply greater amplification when it is needed. A hearing aid should be prescribed by a qualified clinician on the basis of the patient's audiogram and other pertinent information. The clinician can also make an assessment as to whether a monaural or binaural aid should be used.

Components to be used in hearing aid systems must meet demanding requirements. The microphones must be compact, sensitive, rugged, and inexpensive. The industry has produced magnetic, ceramic, and electret microphones packaged in volumes of about 100 mm^3. Each type of microphone has certain desirable features relative to acoustical, mechanical, and environmental sensitivities. Directional properties are also sometimes incorporated. Small solid state amplifiers have volume and tone controls while consuming small amounts of power. Magnetic earphones are most commonly used. The earphone may be worn in the ear, or a plastic tube may carry the sound from the earphone to the ear.

The sound sent to the ear is often electronically and acoustically tailored to eliminate frequencies below about 200 Hz—the range where environmental noise is a major problem and there is little speech information. Some aids supply compression amplification to accommodate the reduced dynamic range of impaired hearing. (Discomfort thresholds are typically the same for normal hearing and hearing impaired subjects, but the threshold of audibility is much higher for the hearing impaired. Thus they have a smaller dynamic range.) Hearing aids are subject to the deficiencies common to electroacoustic systems. Feedback between earphone and microphone may produce "squealing." Nonlinear response produces distortion, especially at high sounds levels.

For persons who have lost essentially all usefulness of the cochlea as a sound transducer there still exists a possibility that the auditory nerve is intact and functional at the cochlea. There has been some experimentation with "auditory implants" in which electrodes are surgically implanted on or in the cochlea. Electrical signals are sent via the electrodes in an attempt to directly stimulate the auditory nerve. Patients in whom the electrodes have been implanted experience a sensation of hearing, but frequency discrimination is poor and speech is not intelligible with only the simple signals currently available.

Hazards of Intense Sounds

It has long been assumed that substantial hearing loss is an inevitable consequence of growing older. However, there is evidence from "primitive" civilizations that this is not necessarily the case. In a remote part of the Egyptian Sudan live the Mabaans, a tribe whose life is essentially free of the intense sounds of modern civilization. Although the Mabaans are subject to presbycusis like everyone else, even Mabaans 70 years old have hearing acuity similar to that of young boys.

Furthermore, there is no noticeable difference in hearing acuity between men and women. This is in sharp contrast to the industrialized nations, where men show greater hearing losses than women of the same age. Presumably this result is due to the higher sound levels encountered by men working in industry. This effect is summarized in a striking manner in Figure 17.2. This research suggests rather strongly that a substantial hearing loss with age is not a necessary part of growing older. Rather, it seems to be due to our total lifetime exposure to unnecessarily intense sounds.

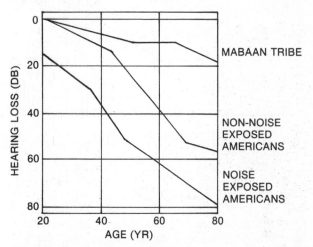

Figure 17.2. Hearing loss with age for Mabaans, non-noise and noise exposed Americans. (After Rosen, 1962.)

Short-term hearing loss is an experience that almost everyone has had. When exposed to intense sounds we become "partially deaf" for a period of time until the ears recover. This temporary hearing loss means there is a temporary upward shift of our threshold of hearing. Since the shift is not permanent it is called a **temporary threshold shift** (TTS). A loud hand clap close to the ear will produce a short-term hearing loss, as will several hours of exposure to very intense sounds. Recovery from a TTS of 30-40 dB usually requires several hours. For a shift of 50-60 dB even several days may be insufficient for full recovery. These temporary losses may not be too critical if they are not incurred often. However, studies show a positive correlation between TTS acquired on a regular basis and a **permanent threshold shift** (PTS). Figure 17.3 illustrates temporary and permanent losses of band members and dancers exposed to very intense rock music. The vertical axis gives the threshold shift in dB, 0 dB being the reference threshold. Note that the dancers suffer a smaller loss than the band members, but that the nature of the permanent loss (solid curve) is similar to the temporary loss (dashed curve) for both dancers and band members.

Despite the data just presented, it should not be assumed that a permanent threshold shift depends only upon the overall noise level. Whether or not a TTS will become permanent depends upon a person's total exposure to sound. This includes the overall sound level (average dB level), the time distribution of the sound, whether it is continuous or intermittent, and the person's total lifetime exposure.

Exposure to a short duration of an extremely intense sound, such as an explosion, can produce a sudden

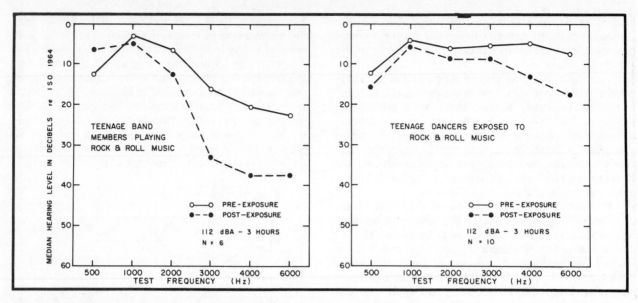

Figure 17.3. Hearing levels of teen-age rock-and-roll musicians and dancers, measured just before and 5-11 minutes after a three-hour "rock session" with average sound levels of 112 dBA. (From Cohen etal, 1970. Data from PHS sample observations.)

and often permanent, hearing loss. Hearing loss can be an occupational hazard for people who work in very noisy environments, such as factories or around jet aircraft. The hearing losses in these cases are not as sudden as for explosions, but they can be appreciable over comparatively short periods of time. In either case, exposure to adverse environmental conditions can give rise to premature deterioration of the hair cells and thus to early hearing impairment. Figure 17.4 illustrates hearing loss due to exposure to a firecracker explosion. The loss appears to be permanent since little recovery is noted after five months. Figure 17.5 illustrates hearing loss due to regular use of a lawnmower without adequate ear protection. Figure 17.6 illustrates hearing loss due to regular use of firearms without adequate ear protection. Note that the left ear, which has greater exposure to sounds from the muzzle, shows an appreciably greater loss than the right ear.

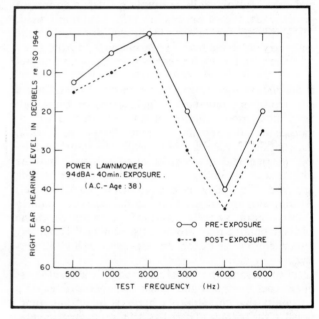

Figure 17.5. Differences in pre-exposure and 2-8 minutes post exposure hearing levels of a lawnmower operator. (From Cohen etal, 1970. Data from PHS observations.)

As a means of preventing noise-induced hearing loss, a set of curves called the "damage risk criteria" have been developed. These curves, shown in Figure 17.7, show the duration of a given combination of sound pressure level and frequency range that the average adult can be exposed to with limited risk. An examination of the graph yields the following information: (1) in any frequency region the tolerable exposure period decreases as the sound level increases; (2) at a constant sound pressure level the exposure period decreases as frequency increases up to the 2400-4800 Hz band; and (3) for a constant exposure time, the sound pressure level decreases as frequency increases. Briefly, then, the damage-risk criteria specify the maximum period of time a person can be exposed to sound of various sound levels and frequencies with minimal risk of hearing loss.

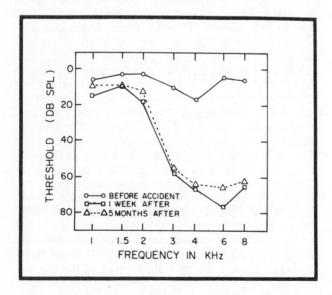

Figure 17.4. Traumatic unilateral hearing loss caused by accidental exposure to a firecracker. (From Cohen etal, 1970. Data from Ward and Glorig.)

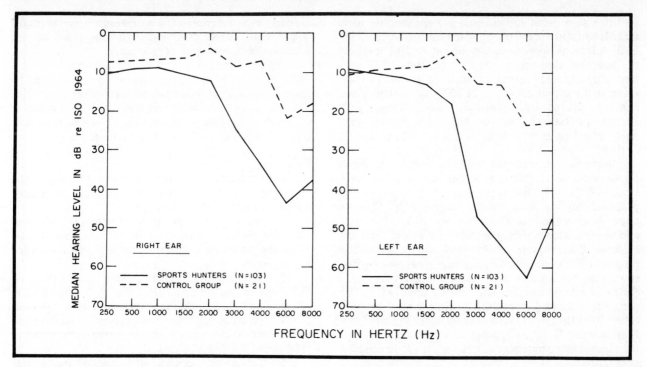

Figure 17.6. Comparison of hearing levels of sports hunters with those of a control group. (From Cohen etal, 1970. Data from Taylor and Williams.)

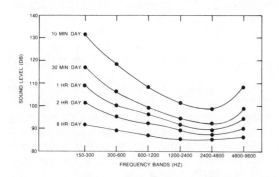

Figure 17.7. Damage risk for a single daily exposure. (After Kryter etal, 1966.)

Although occupational noise exposure is still the principal cause of hearing loss, many people are being subjected to undue hazard due to loud rock music. In one study of sound levels produced by rock bands it was found that sound levels ranged from 105 to 120 dB—about the same level of noise one would encounter in a boiler shop! A study dealing with the effects of loud rock music on the ears of guinea pigs was conducted. After ninety hours of intermittent exposure, the cells of the guinea pigs' cochleas were examined. It was discovered that they shriveled and collapsed. In a study of the effects of rock music on teenagers, the music was recorded in a discotheque and played back to subjects at 100- and 110-dB levels for various time periods. Plans to play the music at a 120-dB level were dropped because the TTS was so great for those exposed to the 110-dB level. The report concluded that it is not likely the young can be restricted to reasonable sound levels for reasonable exposure times, and so society will have to provide social rehabilitative and medical support to the most susceptible youth who suffer large losses.

Regulation and Protection

The potential hazard posed by intense sounds has been recognized by the government. Legislation has been enacted in an attempt to define noise hazards and methods for their measurement. The Occupational Safety and Health Act (OSHAct) of 1970 is a federal law which, in part, attempts to limit worker exposure to intense sounds. The law applies specifically to industry, but the sound levels recommended by the law are probably liberal guidelines for exposure to any intense sound. Table 17.1 shows maximum permissible exposure times and levels for occupational noise measured with the A-network of a sound level meter. Table 17.1 also shows some recommended maximum allowed exposure times for nonoccupational noise.

Table 17.1. Maximum allowed exposure times for occupational noise and recommended maximum exposure times for nonoccupational noise per 24-hour day at the sound levels shown. (After Cohen etal, 1970.)

Sound level (dBA)	Occupational exposure	Nonoccupational exposure
80		4 hours
85		2 hours
90	8 hours	1 hour
95	4 hours	30 minutes
100	2 hours	15 minutes
105	1 hour	8 minutes or less
110	30 minutes	4 minutes or less
115	15 minutes or less	2 minutes or less

The 1970 OSHAct is based on the assumption that no more than 21-29 percent of a population exposed to the equivalent of 90 dBA sound levels for 8 hours per day will suffer significant hearing impairment. Since enactment of the 1970 OSHAct, there has been much discussion in industry, government, audiology, and medicine as to what sound levels are appropriate with regards to future regulations. Many have advocated an 85 dBA level as opposed to the current 90 dBA level on the basis that then supposedly no more than 10-15 percent of an exposed population would be at risk. In 1981 an amendment was added to the 1970 Act that requires hearing protectors, exposure monitoring, and audiometric testing for workers exposed to occupational noise exceeding an eight hour time-weighted average of 85 dBA.

Some have argued that the 1970 OSHAct is based on incorrect assumptions and that lowering the level from 90 dBA to 85 dBA is the wrong way to go. In one proposed hearing conservation program, the allowed upper level would remain set at 90 dBA. However, audiometric testing, noise control at the source, and use of personal hearing protectors would also be part of the program. With this type of program it may be possible to prevent hearing loss in the people who are most affected by intense sounds. The premise upon which the program is based is that each worker should be protected from the intense sounds that impair the worker's hearing individually, as opposed to supposing that there is an acceptable percentage of risk for the entire exposed population.

Protection from excessive exposure to intense sounds can be accomplished in a number of ways. We will discuss some ways as they relate to the work place, though similar methods could be used in other situations involving exposure to intense sounds. Efforts can be made to reduce the amount of sound produced by a given source such as a noisy machine. However, it is not practical to do this with all machines. Noisy machines can sometimes be placed in tight enclosures to reduce the amount of noise being radiated, but the enclosure must not interfere with the function of the machine or with required accessibility to the machine.

Assembly line workers are often exposed to the intense sounds of the their coworkers' machines in addition to the sound produced by their own machine. When sound barriers can be placed between adjacent workers it is often possible to reduce excessive sound exposure.

Use of personal hearing protectors may complement other methods which alone provide incomplete solutions to the problem of noise exposure. There are two basic types of hearing protectors—muff protectors worn over the ear and insert protectors worn in the ear. One study found that one insert ear protector provided as much as 20-30 dB attenuation under laboratory conditions where the protector was properly fitted and where nothing was done to hinder the functioning of the ear protector. A similar study found that the same ear protector provided only 8-20 dB attenuation in typical work place situations. Apparently workers had improperly fitted protectors or, because of discomfort, wore them too loosely. If one is depending on ear protectors to provide attenuation of intense sounds, the ear protectors must be properly fitted and then worn as intended so there are no noise "leaks."

One final means of controlling noise exposure is referred to as "dose control." In this case, workers are allowed only so much noise exposure time during the working day. They are then required to perform tasks in quieter surroundings during the remainder of the day.

Dosimeters are devices used to measure time averaged sound levels. Small dosimeters worn by a worker are often useful in dose control programs.

Whether one is involved in occupational or recreational activity, exposure to intense sounds should be limited and proper personal ear protectors should be worn as needed. Audiometric tests should be taken at reasonable intervals to detect hearing losses, especially if one is often exposed to a noisy environment.

Exercises

1. Of hearing impairments due to problems in the outer, middle, or inner ear, which is likely to be most serious? Least serious? Why?

2. Indicate what type of hearing loss is associated with the outer, middle, and inner ear, respectively. What are the possible causes of the loss and means for correcting it?

3. Which frequencies are affected most in hearing loss due to aging (presbycusis)? Which frequencies are affected most by exposure to intense sounds?

4. Hearing loss due to exposure to adverse environmental conditions might be termed "premature presbycusis." In what sense is this so?

5. Determine the amount of hearing loss at frequencies of 100, 200, 1000, 2000, 5000, and 10,000 Hz if the measured thresholds are 50, 35, 15, 15, 30, and 60 dB respectively. (Use the threshold values in Figure 14.1 as the reference).

6. After attending weekly rock sessions for a period of three years a person is observed to have a hearing threshold of 20 dB at 2000 Hz. What is the hearing loss (in dB)?

7. After a day at the rifle range, a person's right ear is found to have a threshold of 20 dB at 2000 Hz. What is the temporary hearing loss (in dB), assuming that the hearing was normal at the beginning of the day?

8. What occupations are most likely to cause hearing impairment? In these occupations, what precautions might be taken to reduce hearing loss?

9. Farmers who spend a great deal of time operating power equipment are observed to have greater hearing loss on the average than people who live in urban areas. Why?

10. What noise hazards exist in a typical house or yard?

11. Which feature, noise hazard or annoyance, is addressed in the 1970 OSHAct?

12. In a noisy listening environment we might assume that background noise level is constant everywhere. Under this condition will a hearing aid user be able to perceive speech best when close to or far from another person? Why?

13. Suppose a subject has a hearing level half way between the threshold level and a 100 dB discomfort level. What amplitude compression factor should a hearing aid use to compensate?

Demonstrations

1. Schools that provide training in audiology are often willing to conduct hearing tests and provide audiograms free of charge (or for a very nominal fee) in order to provide experience for their students. Check your locale and make arrangements to get a personal

audiogram. Get a copy of the audiogram and account for any losses it may show. What is the reference level relative to which losses are shown in an audiogram? Is it the same at all frequencies? Interpret the reference line in a typical audiogram in terms of an "average" threshold of hearing curve.

2. Make an ear trumpet by rolling heavy paper into a cone. Insert the small end into your ear. Perform various experiments to check the effectiveness of the ear trumpet.

3. Use a sound level meter to measure A-scale readings in different residential areas. Describe the areas and record the readings. Repeat for business areas. Repeat for industrial areas. Do any of the measured levels represent a hazard to hearing?

4. Set up a sound system capable of producing high sound levels in a classroom (or living room). Use a sound level meter to measure various sound levels.

Further Reading

Baron, Chapters 1, 2

Chedd, Chapter 7

Davis and Silverman, Chapters 4-8, 10-12

Fletcher, Chapters 19, 20

Levitt, Pickett, and Houde, Parts II, III, VIII

Rossing, Chapters 30 - 32

Stevens and Warshofsky, Chapters 7, 8

Botsford, J.H. 1972. "Ear Protectors—Their Characteristics and Uses," Sound and Vibration 6 (Nov), 24-29.

Cohen, A., J. Anticaglia, and H. H. Jones. 1970. "Sociocusis—Hearing Loss from Non-occupational Noise Exposure," Sound and Vibration 4 (Nov), 12-20.

Dear, T.A., and B.W. Karrh. 1979. "An Effective Hearing Conservation Program—Federal Regulation or Practical Achievement?" Sound and Vibration 13 (Sep), 12-19.

Dey, F. L. 1972. "Auditory Fatigue and Predicted Permanent Hearing Defects from Rock and Roll Music," New England J. of Medicine 282, 467.

Edwards, R.G., W.P. Hauser, N.A. Moiseer, A.B. Broderson, and W.W. Green. 1978. "Effectiveness of Earplugs as Worn in the Workplace," Sound and Vibration 12 (Jan), 12-22.

Fletcher, D. H., and C. W. Gross. 1977. "Effects on Hearing of Sports-Related Noise or Trauma," Sound and Vibration 11 (Jan), 26-27.

Heggie, A.S. 1978. "The OSHA Noise Standard—How to Live With It," Sound and Vibration 12 (Sep), 20-25.

Kryter, K.D., W.D. Ward, J.D.Miller, and D.H. Eldridge. 1966. "Hazardous Exposure to Intermittent and Steady State Noise," J. Acoust. Soc. Am. 39, 451-464.

Lipscomb, D. M. 1969. "High-Intensity Sounds in the Recreational Environment," Clinical Pediatrics 8, 63.

Miller, J. D. 1974. "Effects of Noise on People." J. Acoust. Soc. Am. 56, 729-764.

Rosen, S., etal. 1962. "Presbycusis Study of Relatively Noise-Free Population in Sudan," Annals of Otolaryngology, Rhinology and Laryngology 71, 727.

Sound and Vibration News. 1972. "Summary of Noise Control Act of 1972." Sound and Vibration 6, (Nov).

Yerges, L.F. 1977. "Control the Noise—or the Exposure?" Sound and Vibration 11 (Sep), 12-14.

Audiovisual

1. **Death Be Not Loud** (26 min, 1970, MGHT)

2. **Ears and Hearing**, 2nd edition (22 min, color, 1969, EBE)

3. **Noise and Its Effects on Health** (20 min, color, 1973, FLMFR)

4. **Quiet Please** (21 min, color, 1971, JACBMC)

5. **Who Stole the Quiet Day** (16 min, color, 1973, AHP)

18. Noise Pollution

By and large, noise pollution is a result of our modern industrial society. Modern technology has provided more powerful sources of noise, and it has made them available in greater numbers. Our industrial machines, our transportation vehicles, our labor-saving devices, and our entertainment devices generally involve the production of mechanical power. Some of the mechanical power is turned into acoustical power which in turn becomes part of our listening environment.

Noise pollution, much like chemical pollution of air and water, can prove hazardous to health—specifically to hearing. In Chapter 17 we discussed hazardous noise pollution that results from sound levels in excess of 80 dBA. In this chapter we will discuss noise pollution that results from sound levels of about 40 to 80 dBA. While these levels do not generally represent a hazard to hearing, they do result in annoyances which can interfere with daily routines as well as body physiology. We will consider some physical measurements of noise and how they relate to human responses. We will also consider various sources of noise, effects of noise on people, and regulation of noise.

Individual Response to Noise

Noise can be defined either physically, in terms of frequencies and sound levels, or in terms of its effect upon humans. We will begin with the physical definition. Noise is considered to be any vibration lacking the "regularity" which characterizes musical sounds. In other words, noise is an erratic, intermittent, or statistically random oscillation. Most musical and speech sounds, even though complex, display a regular periodic wave pattern. Keep in mind we have already seen that a periodic complex wave can always be broken down into sinusoidal components whose frequencies are harmonically related. If we plot the time varying pressure for a noise, as illustrated in Figure 18.1, the irregular pattern is apparent. Waves which are not periodic, such as noise, can also be broken down into sinusoids, but their frequencies are not harmonically related.

Figure 18.1. Pressure wave of noise.

Any unwanted or annoying sound can also be considered noise independent of its physical regularity or irregularity. People are often annoyed by the unwanted sounds produced by other people, events, and things not under their control. Sounds produced by an individual are rarely perceived as annoying by that individual. For many people, producing loud noises is emotionally satisfying because the noise represents the accomplishment of something. To those listening, however, the effect may be unpleasant. In one survey where almost all respondents thought noise pollution was a problem over ninety percent of those responding felt they did not contribute to the problem.

Many noises, while not constituting a hearing hazard, may still constitute a nuisance, leading to frustration on the part of the recipient. Even mild noise can have a strong psychological effect on humans. Thousands of years ago the Chinese realized that to a man in quiet isolation the sound of slowly but steadily dripping water assumes the proportions of a loudly beaten drum. Such noise was used to drive a man mad or to break his will. Today, extremely quiet conditions are very difficult for city people to tolerate. Noise has become such an intimate part of their lives that silence is a condition to be avoided.

Surveys have shown that for small cities without a jet airport the following sources of noise are most annoying: motorcycles, cars, trucks, dogs, buses, sirens, neighbors, construction, aircraft, and trains. For larger cities one might expect sources of noise such as aircraft, trains, construction, and freeways to appear higher up on the list.

Individual responses to noise can depend on a number of factors in addition to the sound itself. Even so, any noise must be perceptible above background sounds if it is to be annoying. Sounds with unique spectra, periodicities, and impulsive natures are often heard above the background even though their overall sound levels may be lower than that of the background. If a sound has special qualities, such as being musical, having a rhythm, or being speech-like, it may be particularly noticeable.

Past experiences with a sound may result in the conditioning of a person to a particular sound so that there is an increased or decreased sensitivity. A person who has been "emotionally involved" with a sound will be more sensitive to it, either positively or negatively. Different cultures respond differently to the same sound. In addition, the manner in which a recipient responds is often determined by his mood or other emotional factors. The same sound at different times of the day may evoke different responses, depending upon whether the recipient is tired, angry, busy, or relaxed.

Another factor which can influence our reaction to sound is whether or not the noise is essential. We may accept the hum and buzz of an air conditioner or a refrigerator since we deem them necessary, but the sound of hard rock music coming from the stereo may be perceived as annoying because we believe it to be nonessential. Also, sounds accompanied by visual information are often more acceptable to a listener than sounds without an accompanying visual stimulus. The sounds of a T.V. gun battle could be quite noisy and bothersome unless accompanied by the visual part of the program. Likewise the sound of loud laughter from the apartment next door could be very annoying; but in conjunction with the funny antics of two puppies at play the same sounds could be quite enjoyable.

Other factors which influence our reaction to sounds include their frequency of occurrence and their predictability. An annoying sound becomes increasingly bothersome the more often it is repeated. We adjust to some steady sounds and become almost unaware of them. On the other hand, very loud, unpredictable sounds, such as sirens, horns, bells, and sonic booms can be extremely annoying. Such sounds may cause startle reactions which may result in increased adrenaline flow, increased heart rate, tensing of the body, and a general fear reaction.

Persons with some sensorineural hearing loss may be very insensitive to low level noises and yet be very sensitive to high level noises. This is due to the phenomenon called **recruitment** which results from a raised threshold of hearing at low levels and a normal threshold of hearing at high levels. Low level sounds may be imperceptible while high-level sounds are perceived normally.

Measures of Noise

We have considered sound levels measured with the A-scale of a sound level meter to be roughly indicative of the loudness of a sound. However, it might be desirable to have a sound level measure for noise that would relate more directly to annoyance. One such measure is the **equivalent level** which is made using the A-scale of a sound level meter; the reading is averaged over time to give an equivalent level of steady sound that would produce the same total sound energy in the same time.

Another measure of sound level is based on the observation that people are generally less sensitive to noise during the day than during the night. The **day-night average level** is measured in a way similar to the equivalent level. Average levels are measured for two time periods: 7:00 a.m. to 10:00 p.m. (day) and 10:00 p.m. to 7:00 a.m. (night). An additional 10 dB is added to the night levels to compensate for greater annoyance potential at night.

The percentage of people who are "highly annoyed" by a given sound is highly correlated with the day-night level of the sound. This is illustrated in Figure 18.2 where the open and closed circles represent aircraft noise and the square represents motor-vehicle noise. There is no community action taken for the condition labeled A in the figure, but there is vigorous community action taken for the condition labeled E. From condition A we might conclude that day-night levels in a range of 50 to 60 dB are desirable for quiet residential neighborhoods.

Sources of Noise

The annoyance level of a noise source depends on the "amount" and "quality" of noise as perceived by the listener. The "amount" of noise will depend on how much sound power a source produces, how far the listener is from the source, and what barriers to sound exist between source and listener. Several ways exist for controlling the sound reaching a listener from a source. In many ways the most desirable method is to reduce the

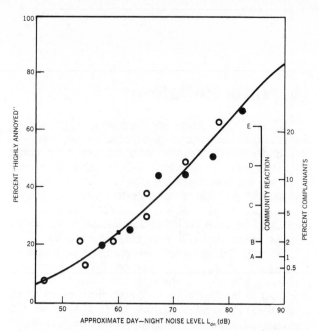

Figure 18.2. Percent of people "highly annoyed," percent of complaints, and community reaction as functions of day-night noise level. Community reactions vary from none (A) to vigorous (E). See text for further explanation. (From Shaw, 1975.)

sound power produced by the source. If the source is distant from the listener (or vice versa), the sound power is spread out and partially absorbed as it travels through the atmosphere. This will result in lower sound levels for the listener. Barriers can be used to surround the source, to surround the listener, or to deflect and absorb noise as it travels. These will be considered further in Chapter 19.

Machines that produce large amounts of mechanical power typically also produce large amounts of sound power. Figure 18.3 illustrates the relationship between acoustical and mechanical power for some common machines. We see from the figure that aircraft, trucks, autos, motorcycles, and lawnmowers might be expected to be sources of noise annoyance. Fortunately, the most powerful sound sources are usually at greater distances from listeners.

It is sometimes held that the sound power produced by a machine is in direct proportion to its mechanical power. Figure 18.3 provides some examples to the contrary. We see that some aircraft—the DC8 and the DC10 for example—having comparable mechanical powers produce very different sound powers. As illustrated in the figure, substantial steps have been taken to reduce engine noise. Further steps promise further reductions in engine noise—probably the major noise problem with aircraft.

Noise levels recorded at several outdoor locations are shown in Figure 18.4. (Notice the key in the lower right corner of the figure. The white diamond indicates that a given sound level was exceeded 50% of the time. The left and right ends of the black bar indicate sound levels that were exceeded 90% and 10% of the time, respectively. The left and right ends of the total bar indicate levels that were exceeded 99% and 1% of the time, respectively.) We see that aircraft and traffic noises are major contributors to the noise problem. There are special noise problems associated with aircraft

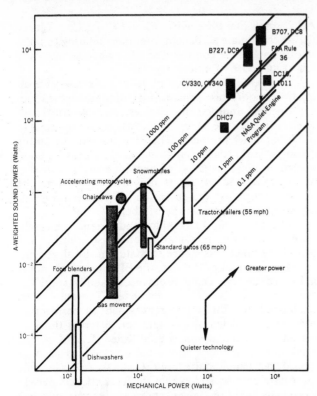

Figure 18.3. Estimated A-weighted sound power versus mechanical power for various machines. The diagonal lines represent constant sound power to mechanical power ratios. The lower lines represent less noisy machines for a given mechanical power. (From Shaw, 1975.)

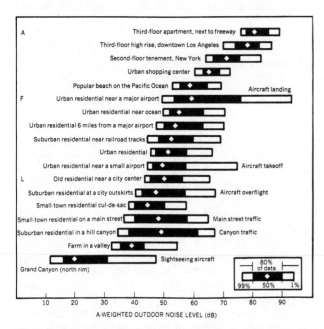

Figure 18.4. Outdoor noise levels recorded at various locations, See text for further explanation. (From Shaw, 1975.)

take-off and landing. In order to minimize these effects, special standards and limits are often imposed on aircraft during take-off and landing.

Special noise problems are associated with aircraft operating at speeds greater than the speed of sound. The resulting **sonic boom** is a short duration "N-shaped" pressure transient as illustrated in Figure 18.5. The initial rise in pressure to as much as 1000 Pa above at-

mospheric pressure occurs in about 25 ms. This is followed by a "smooth" decrease (in about 250 ms) in pressure to as much as 1000 Pa below atmospheric pressure. The final transient which returns the pressure to atmospheric occurs in about 20 ms. The origin of these very powerful transients can be seen by referring to Figures 8.9 and 18.6. Circular waves produced by a sound source moving at less than the wave speed in a medium do not overlap each other (Figure 8.9b). Circular waves produced by a source moving at greater than the wave speed in a medium do overlap each other (Figure 18.6). The combination of overlapping waves produces the powerful pressure transient of a sonic boom which can be experienced at distances up to 100 km from the source. Sonic booms can be destructive to property as well as producing severe startle reactions in people.

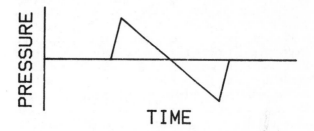

Figure 18.5. N-shaped pressure transient of a sonic boom.

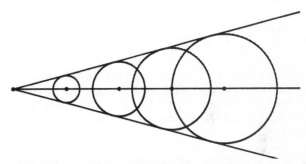

Figure 18.6. Overlapping circular waves produced by a source moving at speeds greater than the wave speed in the medium.

Medium- and heavy-duty trucks contribute the most energy to traffic noise even though there are many more cars in use. Engines, transmissions, and cooling systems are the major sources of noise in trucks; tire noise is also a prime concern at higher speeds. Improved engine and transmission enclosures and larger cooling systems could reduce truck noise levels to near those of automobiles. Truck tire noise could be reduced to a similar extent only with new concepts in tire design and highway surfacing.

Industrial plants and most types of construction equipment generate large amounts of sound power. Motorcycles and off-road recreational vehicles are another source of noise pollution.

One can only wonder about the possible effects that our noisy modern appliances are having upon homemakers. Their ears are bombarded by the sounds of mixers, blenders, garbage disposals, dishwashers, clothes washers, dryers, and vacuum cleaners—not to mention radio, television, stereo, and crying babies!

Effects of Noise

Hazardous effects of noise were considered in Chapter 17. Other auditory and some nonauditory effects of noise will now be discussed. Unwanted sounds can interfere with speech communication. We are able to distinguish speech sounds from a background of unwanted noise if the noise level is not too high relative to the speech level. The higher the noise level the greater the masking of the speech and the more difficult the communication process becomes (see Chapter 28). At sufficiently high noise levels, speech communication becomes impossible. The interference effect of noise with regards to speech also depends on other factors such as the predictability of the message and the opportunity for lip reading. Noise can also interfere with other auditory signals including those of pleasurable listening activities such as radio, TV, plays, and concerts.

Noise can interfere with the various stages of sleep. It can prevent a person from going to sleep, awaken a person from sleep, or cause a person to change from a deeper stage of sleep to a lighter stage. Women and older people are generally more susceptible to noise-related sleep disturbance. Different noise types can have markedly different effects on different individuals. Although conventional wisdom holds that one can adapt and sleep through almost any noise, research indicates that adaptation is minimal at best. However, some people can successfuly use a constant noise source to mask out other noise sounds; they learn to sleep with the known, steady noise.

Other activities disturbed by noise include talking on the telephone, reading, eating, and relaxing. Noise can impair performance of demanding tasks. Various forms of anxiety and stress are also associated with noise exposure. Chronic exposure to noise can produce other behavioral and physiological effects such as narrowed focus of attention, increased blood pressure, and increased hormone excretion.

Some members of society, particularly the young, the ill, and the elderly, are especially susceptible to noise annoyance. Recuperation time from medical treatment may be lengthened. Children growing up in noisy environments learn to "tune out" much of the noise, but later in school must be trained to listen and concentrate.

The effects of noise pollution have their concommitant costs to our health, our hearing, and our environment. There are also noise-related economic costs incurred from lost time on the job, hearing-loss lawsuits, and excess medical expenses.

Regulation of Noise

Noise, like other types of pollution, has tended to increase with the increase in applied technology and population. In the future, considerable effort will be necessary just to prevent a further increase in community noise levels. Lack of progress toward noise abatement in the past has been due to widespread apathy on the part of the public and government agencies. The Federal Government, however, did begin to address the problem with the enactment of the Noise Control Act of 1972. This law deals with noise emission standards for products distributed commercially as well as noise standards for aircraft, railroads, and motor carriers. This Act requires that the Environmental Protection Agency (EPA) coordinate all Federal programs relating to noise research and noise control. The EPA was also made the "clearing-house" for noise regulations.

The Noise Control Act of 1972 was aimed at dealing with major noise sources (including transportation, construction, motors and engines, and electrical and electronic equipment) by requiring manufacturers to produce quieter products. The EPA was also given the authority to require environmental impact studies (including noise impacts) of new highways, industrial installations, and so on. If it is determined that the negative effects on the environment will be too great, approval of a project may be withheld.

The EPA was to conduct and finance research, disseminate public information on noise control, and provide technical assistance to state and local governments. As a result of the Act, the EPA prepared model noise ordinances. These ordinances usually specify a sound level that is not to be exceeded. These are often established for different areas—such as residential, commercial and industrial—and for different times of the day. The actual values established depend on the particular community and usually take into account typical background noise levels. (Even so, levels in some areas have been set so low that the local crickets were in violation.) Table 18.1 gives representative values. Intermittent noises (10% of any hour) are often permitted to be 5 dBA and impulsive noises (short bursts) 10 dBA higher than the values in the table.

Table 18.1. Representative values of allowed sound levels for continuous noise in a typical local noise ordinance for annoyance.

Area class	Day (dBA)	Night (dBA)
Residential	55	50
Commercial	65	55
Light Industrial	70	60
Heavy Industrial	80	70

Currently allowed noise emission levels for cars are generally set at around 74 dBA or less at a distance of 15 meters. Allowed noise emission levels for trucks and other vehicles over 10,000 pounds GVW is typically set at around 82 dBA at a distance of 15 meters. Allowed levels for motorcycles is typically set around 80 dBA at a distance of 15 meters.

A final matter of considerable importance with regards to noise regulation is that of land use planning. Residential areas should be kept separated from industrial areas, airports, railroads, and freeways. Planning commissions should resist the temptation to let residential zones encroach on buffer areas near airports and freeways. Commercial zones can be used as buffer zones between residential and industrial zones.

One difficulty in writing noise ordinances to control noise annoyance is that the reaction of different individuals to particular noises varies so widely. Most ordinances are written to satisfy some "average" person and, therefore, may fail completely to satisfy some individuals or groups.

Exercises

1. We have defined noise in two different ways. If

120

you are enjoying the experience of very loud rock music, does it qualify as noise by either of these definitions? How could you interpret the word "undesirable" so that very loud, but enjoyable musical sounds are included in one of these definitions?

2. List several sources of noise that you find annoying when you are engaged in activities such as sleeping, reading, listening to music, and cleaning the house. Are the same noise sources most annoying for all activities, or are some sources more annoying for some activities and less annoying for others?

3. List some sounds that you find especially annoying because of their musical or speech-like qualities. Under what circumstances are they most annoying?

4. An older couple was quite annoyed by sounds from an amusement park across a four-lane highway from their home. They seemed quite insensitive to traffic noise on the highway even though it produced sound levels at their home some 15 dBA higher than sound levels produced by the amusement park. Why?

5. Suppose that you try to achieve a day-night level of 55 dBA in your neighborhood. If you measure an average 55 dBA level during the day, what must the average level be at night?

6. DC8 and DC10 aircraft generate approximately equal mechanical power. However, their acoustical powers differ greatly. If a DC8 produced a sound level of 80 dBA at a location near an airport, what sound level would a DC10 produce under similar circumstances?

7. Assuming that sound powers add directly, how many "standard autos" are required to generate as much sound power as one tractor trailer? (See Figure 18.3.)

8. Normal conversation occurs at a sound level of about 65 dBA. By how much would you need to raise your speech level if you were conversing in a third floor apartment next to a freeway? (See Figure 18.4 and assume that your speech level must be equal to the background noise level.)

9. Which aspect of noise pollution, hazard or annoyance, is most important in typical community noise ordinances?

10. What criteria would you use if you were responsible for writing a noise ordinance for your community? How would the allowed levels compare with those in Table 18.1?

11. If noise levels of 55 dBA for day and 50 dBA for night are allowed as shown in Table 18.1, what will the overall day-night level be?

12. Compare the sound intensities for sound levels of 60 dB and 80 dB.

13. How long could you be exposed to a sound having a level of 60 dB before receiving the same amount of sound energy as produced by a sound with a level of 80 dB sounded for one hour?

14. Explain how constructive wave intarference plays a role in the production of a sonic boom? Can an aircraft traveling at less than the speed of sound produce a sonic boom?

15. List a number of societal "costs" of noise.

16. A friend tells you that the various forms of noise pollutions are an inevitable part of technological progress. How would you respond?

Demonstrations

1. Sit quietly in the living room of your apartment. Use a sound level meter to measure various annoying sounds. Rate the annoyance of each different sound on a scale from one to five. How effective is sound level in determining the rated annoyance of each sound?

2. Try conversing with a friend when a "noise" source such as a TV or stereo is turned to low, medium, and high levels. Does your "conversational effort" increase?

3. Try performing a demanding task such as dividing a series of multidigit numbers under conditions of low, medium, and high sound levels.

4. Visit a neighborhood in which you might think of buying a house. Make evaluations both inside and outside the house for potential noise problems.

5. Have someone startle you with a loud sound. Estimate changes in your respiration rate, heart rate, adrenalin production, and so on. Compare to normal conditions.

Further Reading

Berendt etal, Chapters 2, 10

Baron, Chapters 1, 2

Chedd, Chapter 7

Rossing, Chapters 30 - 32

Stevens and Warshofsky, Chapter 8

Angevine, O.L. 1975. "Individual Differences in the Annoyance of Noise," Sound and Vibration 9 (Nov), 40-42.

Bragdon, C.R. 1974. "Quiet Product Emphasis in Consumer Advertising," Sound and Vibration 8 (Sep), 33-36.

Bragdon, C.R., J. Burton, M.A. Mock, C.A. Head, and D. Taylor. 1974. "Establishing Georgia's Statewide Noise Control Program," Sound and Vibration 8 (Dec), 12-15.

Bragdon, C.R., 1974. "Municipal Noise Ordinances: 1974," Sound and Vibration 8 (Dec), 28-30.

Bragdon, C.R. 1977. "Environmental Noise Control Programs in the United States," Sound and Vibration 11 (Dec), 12-16.

Bragdon, C.R., and R.K. Miller. 1978. "The Regulation and Control of Animal Noise in the Community," Sound and Vibration 12 (Dec), 8-11.

Broderson, A.B., R.G. Edwards, D.F. McCoy, and W.S. Coakley. 1981. "Proposed State Noise Regulations—An Urban Attitude Survey," Sound and Vibration 15 (Dec), 8-13.

Cohen, S., D.S. Krantz, G.W. Evans, and D. Stokols. 1981. "Cardiovascular and Behavioral Effects of Community Noise," Am. Scientist 69, 528-535.

Dwiggins, D. 1969. The SST: Here It Comes, Ready or Not (Doubleday).

Haag, F.G. 1979. "State Noise Restrictions: 1979," Sound and Vibration 13 (Dec), 16-17.

Jackson, G.M. and H.G. Leventhall. 1975. "Household Appliance Noise," Applied Acoustics 8, 101-111.

Kryter, K.D. 1970. Noise and Man (Academic Press).

Miller, J.D. 1974. "Effects of Noise on People," J. Acoust. Soc. Am. 56, 729-764.

Rossing, T.D. 1979. **Environmental Noise Control** (AAPT).

Schultz T.J. 1978. "Synthesis of Social Surveys on Noise Annoyance," J. Acoust. Soc. Am. **64**, 377-405.

Schwartz, J.M., W.A. Yost, and A.E.S. Green. 1974. "Community Noise Effort in Gainsville, Florida," Sound and Vibration **8** (Dec), 24-27.

Shaw, E.A.G. 1975. "Noise Pollution—What can be Done?" Physics Today **28** (Jan), 46-58.

Simmons, R.A., and R.C. Chanaud. 1974. "The 'Soft Fuzz' Approach to Noise Ordinance Enforcement," Sound and Vibration **8** (Sep), 24-32.

Sound and Vibration News. 1972. "Summary of Noise Control Act of 1972," Sound and Vibration **6** (Nov).

Sound and Vibration News. 1974. "Identification of Products as Major Sources of Noise," Sound and Vibration **8** (Sep).

Stempler, S., H. Sanders, H. Watkens, and E. Boronow. 1977. "Development of Environmental Noise Codes for the City of New York," Sound and Vibration **11** (Dec), 18-22.

Welch and Welch, ed. 1970. **Physiological Effects of Noise** (Plenum).

Audiovisual
1. **Death Be Not Loud** (26 min, 1970, MGHT)
2. **Noise and Its Effects on Health** (20 min, color, 1973, FLMFR)
3. **Quiet Please** (21 min, color, 1971, JACBMC)
4. **Noise Boom** (26 min, color, 1969, HAR)
5. **Who Stole The Quiet Day** (16 min, color, 1973, AHP)

IV. Listening Environments

Courtesy of the Boston Symphony.

19. Controlling the Environment

When listening to musical performances or engaged in similar activities, isolation from unwanted sounds is important. Noise may arise from sources outside of the listening enclosure such as aircraft and traffic. Noise can also arise from sources inside the listening enclosure such as air conditioning and the audience. These unwanted sounds must be reduced to acceptable levels. Desirable sounds should be enhanced and controlled in such a way as to make them most effective for their intended purpose. In this chapter we will discuss means of isolation and insulation from unwanted sounds. Methods of controlling sounds "in the open" will be considered. Control of sound in rooms such as class rooms and music practice rooms will be discussed at the conclusion of the chapter.

Isolation from External Noise

One way to deal with external noise is to reduce the amount of noise reaching the location of a listening environment. This can be accomplished in part by limiting the amount of sound emitted by any particular device, be it an aircraft, an automobile, or a truck. However, there are practical limits below which sound emission cannot be reduced, especially near an airport or a heavily traveled highway. However, even in these circumstances zoning regulations can be imposed so that certain types of buildings (e.g., homes, schools, hospitals) are not allowed within particularly noisy environs. However, zoning must be consistent to produce satisfactory results.

Under even the best circumstances, with zoning and control of noise emissions, significant amounts of noise still remain. These can be prevented from entering a building by surrounding the building with a sound barrier. This barrier should satisfy two basic conditions: (1) it should be massive, and (2) it should be airtight. Sound waves are transmitted effectively from one medium to another when the two media have similar densities and sound speeds. However, when sound waves in air strike a massive barrier, they are mostly reflected because of the mismatch in sound speed and density between the air and the barrier. Even though a barrier is massive, it will not be effective unless it is made fairly airtight. For example, suppose a window does not fit well, so that the cracks around the edge are equal in area to 1% of the area of the window. Of the total sound energy striking the window, about 1-4% (depending on frequency) will leak into the building. This limits the attenuation produced by the window to 14 dB for 4% leakage or 20 dB for 1% leakage, while an airtight window may provide a loss of 30 dB. Thus, as poor-quality construction can negate the value of an otherwise good sound barrier, care in construction to provide tight fits is equally important to the production of proper sound barriers.

As an example of a structure where the external environment has been a particular problem, consider the J.F. Kennedy Center for the Performing Arts in Washington, D.C. The Center is near the Potomac River, close to the National Airport. Aircraft often fly as low as 100 meters above the roof, and low-flying helicopters are often seen in the immediate vicinity of the Center. In addition, the structure is surrounded by the usual noises of automotive traffic. The design used to suppress external noise is the "box-within-a-box" concept. The three auditoriums are completely enclosed with an exterior shell. Furthermore, the columns supporting each auditorium have been designed to isolate both airborne noise and mechanical vibrations from the interior surfaces. The exterior shell is a double-walled construction with enclosed air spaces. At all outside entrances special doors are used, and there is a "sound lock" region between the foyer and the interior of each auditorium. Because of this special construction it is possible to enjoy performances without undue interference from external noise.

When selecting materials to be used in sound barriers, their sound transmission loss characteristics must be considered. Typical transmission loss characteristics are shown in Table 19.1 for various materials and structures. Generally, the greater the mass of the material or structure, the greater its transmission loss and the more effective it will be as a barrier. Note that layering various materials in a structure increases transmission loss.

Most materials and structures have greater transmission loss at higher frequencies as seen in Table 19.1. The **sound transmission class** (STC) is a single number specification for the sound transmission loss properties of a material or a structure. A standard curve is adjusted to fit transmission losses at different frequencies as shown in Figure 19.1. The standard curve shows less loss at lower frequencies. This is typical of most materials. The standard curve is adjusted so that no measured value is more than 8 dB below the curve and so that the sum of measured values below the curve is no more than 32 dB. The STC value of the material is the transmission loss of the standard curve at 500 Hz. Typical STC values for various constructions are shown in Table 19.2.

A common misconception about sound insulation is that materials which are effective as heat insulators are also useful for sound insulation. This is untrue in almost all cases. Porous materials which are good heat insulators are often good sound absorbers, but they are usually poor sound insulators. Acoustical absorbers can reduce the noise level within a room, but absorbing material will have little effect on noises transmitted into a room from outside.

Table 19.1. Typical sound transmission losses for various acoustical barriers. The losses expressed in dB are for the frequencies shown and assume no leakage around the barrier.

Barrier	Frequency (Hz)		
	125	500	2000
Solid wall			
Density of 0.25 kg/m²		23	average
Density of 1.0 kg/m²		29	for
Density of 5.0 kg/m²		38	100 Hz
Density of 25 kg/m²		50	to 3200 Hz
Double wall with air core - increase in loss over solid wall with same mass.			
4 cm air core	1	2	8
15 cm air core	4	11	18
Double wall with filled core			
17 cm foam core	28	51	61
6 cm foam and 1 cm sound board	27	45	58
10 cm mineral wool and 1 cm sound board	28	46	60
Hollow core door (14 kg)	11	16	22
Solid door (27 kg)	15	14	25
Solid core door (42 kg)	20	14	26
Solid glass window (3 mm)	12	20	25
Insulating glass window (10 mm)	17	19	27
Insulating window with storm sash	16	27	35

Table 19.2. Typical sound transmission class (STC) values for various barriers.

Barrier	STC value
Wall materials	
9mm gypsum wall board	25
16mm gypsum wall board	30
Solid walls	
10 cm brick, 12mm plaster both sides	40
60 cm stone, 12mm plaster both sides	55
Walls on 2 × 4 studs	
16mm wall board both sides	35
16mm wall board & 12mm plaster both sides	45
16mm wall board, 6 cm isolation blanket, resilient mounting	50
16mm wallboard, 6 cm isolation blanket, double row of studs	55
Doors	
hollow core (10kg/m²)	20
solid core (20kg/m²)	25
Windows	
2mm single pane	25
3mm double pane, 10 cm air space	35

Isolation from Internal Noise

Noise problems arise from noises produced inside a structure as well as from noises outside the structure. Some of the ways in which noise can travel from a source to a listener are illustrated in Figure 19.2. External noises from the airplane, the auto, and the paper boy can be insulated against most effectively by closing the windows to provide sound barriers as discussed in the last section. However, some internal noises in the house may be more difficult to control. These internal noises can travel through the air as airborne noise or through the solid structure as structure-borne noise. The

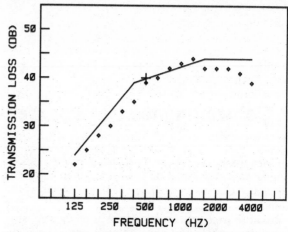

Figure 19.1. Sound transmission class (STC) as determined from transmission loss measurements. A standard curve is fitted to transmission losses measured at different frequencies. The transmission loss for the curve at 500 Hz gives an STC value of 40 in this example.

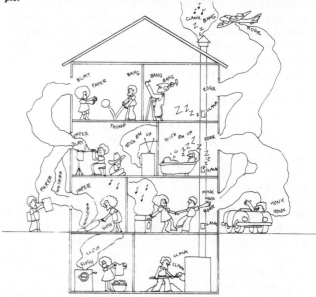

Figure 19.2. Some of the many sources of sound and some of the paths by which sound is transmitted.

means used to reduce the two problems may be quite different.

The same two barrier rules—massiveness and airtightness—discussed in the last section are applicable to providing room-to-room isolation from noise within a house. Open windows and doors are the most common culprits in allowing sound from one room to reach another room, but even the crack under a closed door can transmit a substantial amount of incident noise. Ducts used in central heating and air conditioning systems provide openings into each room and a common path connecting these openings. A noise produced in one room (such as the "clank-clank" of the furnace in the figure) becomes the common property of all rooms connected by the ductwork. To reduce this transmitted noise, the ducts can be lined with sound-absorbing materials. Lining of ducts with sound absorbing materials is often regarded as too expensive for residential structures. However, it is used almost without exception in many large public buildings such as concert halls where there is a strict demand to reduce

126

noise from heating and air conditioning sources to very low levels. Air gaps around water pipes and electrical fixtures are additional sources of airborne noise transmission. Careful construction and proper caulking can reduce these problems.

Airborne sound can be transmitted through barriers as a forced vibration when a noise in one room is transmitted through the wall to an adjacent room. The wall in the adjacent room is caused to vibrate, thus transmitting the sound. A listener in the adjacent room hears a "reproduction" of the original sound provided by the partition. Reradiated airborne noise can best be attenuated by a more massive partition which offers greater resistance to vibration. Layered structures are often used to provide walls with high transmission loss. Layered structures can give high losses because of sound speed and density mismatches at layer boundaries.

Structure-borne noises originate where a vibration is set up in the solid structure of a building as illustrated in Figure 19.2 by the thump of a ball, the banging of a hammer, the vibration of a vacuum cleaner, etc. Massive solid walls and floors are not usually adequate to prevent transmission of structure-borne sounds even though they work well for airborne sounds. Structure-borne noise and vibration should be suppressed at the source. If the noise cannot be adequately suppressed at the source, the various structures should be isolated from each other. Flexible mountings or antivibration pads may be necessary to isolate machines from building structures. A floor can be acoustically isolated from the ceiling below if the ceiling supports are connected to the floor joists by resilient mounts. Impact noises such as footsteps can be greatly reduced by resilient flooring, such as carpeting or foam-backed tile. An effective layered construction for a common floor-ceiling boundary might be a concrete barrier overlaid with carpet on top and with a resiliently mounted ceiling below.

Sounds in the Open

Open structures for listening have certain advantages over enclosed structures. An open structure is less expensive to construct, it can easily accommodate very large audiences, and it usually lends an aura of informality. The main problems associated with open structures are sound absorption of the audience, spreading of the sound, and interfering noises from external sources. To overcome these difficulties, open structures should be designed so that the listening areas are as far from noise sources as possible. Placement should also utilize any natural sound barriers (e.g., hills, vegatation) between noise sources and the performers and listeners. Not much can be done beyond this to control external noises. Sound in the open "spreads out," becoming less intense as it travels (see Chapter 3), thus creating listening levels that are too low to be acceptable.

The auditorium (which literally means "hearing place") developed from the Greek open-air theater (which can be loosely translated as "seeing place"). Although the Greeks did not deliberately consider acoustical principles when constructing the open-air theaters, it is interesting to see how some of these theaters dealt with acoustical problems. The best of the Greek theaters were located where background noise

and wind noise were at a minimum. The arrangement of the seats relative to the areas used by performers provided listeners with many early reflected sounds. Early reflected sounds are those which reach a listener after being scattered from empty seats and other listeners. The average angle of the seating area is steep which helps to avoid too much absorption as the sound travels over an audience. Some of the Greek theaters provided a reflecting surface behind the performers to reinforce direct sound. The use of large masks may have helped to make the actors' voices more directional. The best of the Greek theaters also provided good acoustics for speech reception and solo musical performance. The Greek theaters were superior in siting and design to similar Roman theaters.

A site for an open theater should be chosen to have minimum background noise. Desirable acoustics for an outdoor arena can be achieved by having a large reflecting surface, called the "band shell," behind the stage. The band shell directs the sound toward the audience as well as helping the members of a musical group to stay in tune and in time with one another. The slope of the seating area should be fairly steep to reduce sound absorption by the audience. Hard-surfaced seats should generally be used so that sound will be scattered and not absorbed.

Electronic reinforcement of sound is also an important means of overcoming the problem of inadequate sound power. However, even with sound reinforcement, the listeners close to the performers will hear much louder sound than those farther away. This situation does not exist in a well-designed concert hall.

Sound in Enclosures

An enclosed structure has considerable advantages over an open structure in keeping out noises and in retaining desirable sound. In an enclosed space, very little sound is lost by transmission through the walls so the energy remains within the room. Figure 19.3 illustrates how sound waves, in striking various surfaces, can be reflected, absorbed, dispersed, diffracted, or transmit-

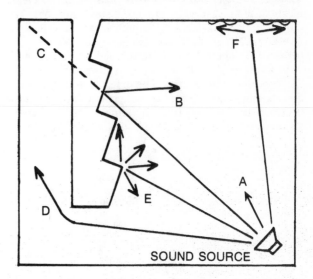

Figure 19.3. Paths of sound in an enclosure: (A) direct from a source; (B) reflected from a wall; (C) transmitted through a wall; (D) diffracted around a corner; (E) diffused from an irregular object; (F) absorbed at a surface.

ted elsewhere. Let us now consider each of these processes in more detail.

The laws governing sound reflection were discussed in Chapter 8. Hard, rigid, flat surfaces (concrete, plaster, etc.) reflect most of the sound energy which strikes them. Flat surfaces produce specular reflections—sound originally travelling in one direction is reflected in a single different direction. Concave surfaces tend to focus (or concentrate) reflected sound waves, while convex surfaces tend to disperse the waves. Rough surfaces produce diffuse reflections—sound originally travelling in one direction is reflected in many directions. Diffuse reflections occur when the wavelength is small compared to the size of details on a reflecting surface.

The change of sound energy into another form of energy is known as **absorption**. When sound energy strikes on an absorptive material, some of the sound energy is changed into heat, leaving less energy to be reflected. Sound can be absorbed by appropriate materials on walls, floors, and ceilings, or by objects such as drapes, furniture, and people in a room.

Sound diffusion occurs when there is an approximately equal distribution of sound energy in a room. Adequate diffusion is highly desirable in a concert hall, as it helps accentuate the natural qualities of music. Diffusion of sound is usually produced by irregularly shaped objects which scatter the sound. Diffraction, which was discussed in Chapter 8, helps to spread the low-frequency sounds (below 200 Hz) beneath balconies and into other places where there may be no direct sound path.

As pointed out previously, an enclosure serves the important function of blocking out external noise. Once a location has been selected for an enclosure, typical noise conditions at the site should be measured and then construction materials should be selected to provide acceptable internal background noise levels. The "acceptability" of noise levels depends on the nature of the sound, as discussed in Chapter 18. However, for noises without distinct tone colors it is possible to specify a meaningful noise level in terms of a single number. The **noise criteria** (NC) is a single number used to rate noise levels. An NC is given approximately by NC = 1.25 (dBA - 13), where dBA is the A-weighted sound level. Typical acceptable NC values are shown in Table 19.3 for various environments. The external sound level at the site minus the acceptable background level gives the transmission loss required. Building materials can be selected on the basis of transmission loss required.

Table 19.3. Typical acceptable noise levels in enclosures.

Enclosure	NC Value
Studio	15-20
Concert Hall	15-20
Theater	20-25
Auditorium	25-30
Bedroom	25-30
Living Room	30-35
Business Offices	30-35
Restaurant	35-45

The **direct sound** in a room is the sound going directly from a sound source to a listener with no reflec-

tions. **Early sound** is all reflected sound arriving at a listener within about 35 ms of the direct sound. Early sound may undergo one or more reflections. **Reverberant sound** is all reflected sound arriving at a listener later than 35 ms after the arrival of the direct sound. The purpose for which an enclosure is designed largely determines how the early and reverberant sounds should be managed. Rooms designed for speech generally require a reduction in the reverberant sound so that spoken sounds at a given instant are not blended with and masked by reverberation of previously-spoken sounds. The object is to make the speech intelligible so a listener can understand what is being said. Small musical ensembles, opera, orchestra, and organ, in that order, typically require increasingly more reverberant sound.

Certain acoustical defects such as dead spots and echoes should also be avoided. However, these problems are generally associated with larger rooms and will thus be considered in Chapter 21 on auditorium design. Two categories of small and medium-sized rooms—classrooms and music practice rooms—will be discussed in the remainder of this chapter.

Classrooms

Classrooms should provide adequate insulation from outside noises. Reverberant sound should be reduced in a classroom so that speech will be intelligible. Reverberant sound is controlled by absorption at walls, ceilings, and floors in such rooms. Reverberant sound is generally weak in a small room because the sound undergoes many absorptive reflections in a short time. In medium-sized and large classrooms it may be necessary to use absorptive material on some of the room surfaces to keep the reverberant sound at acceptable levels. It is sometimes necessary to place sound absorbing materials on rear walls and ceilings to eliminate undesirable echoes.

Open-plan classrooms, which hark back to the one room school in many ways, have become popular in recent years. In the open-plan scheme, a large room is divided into smaller rooms with barriers that do not extend all the way to the ceiling (or floors or walls in many cases). Several acoustical factors need to be considered in the design of open-plan rooms to successfully reduce excess noise.

Since sound is attenuated by spreading out, open-plan rooms should be spaced as widely apart as possible. Since background noise in one room depends on activities in adjoining rooms, sound barriers should be placed between rooms. The speaking level of a teacher determines not only how well students in the teacher's room hear, but also how well students in adjoining rooms hear the same teacher. A talker's orientation can affect the amount of sound going from one room to the next and thus teachers should orient themselves to face away from adjoining rooms wherever possible.

Highly effective sound absorbing materials should be used on ceilings, floors, walls, and partitions. Partial barriers having considerable density can help to reduce sound transmission from room to room. However, these surfaces need to be highly absorbing since sound will tend to flank the barrier wherever it does not make contact with a floor, a wall, or a ceiling.

Open-plan schemes typically require more floor space than schools with complete walls and should be chosen over conventional designs only where there is a clear educational advantage—not in an attempt to save on construction costs. If the decision is made to use an open-plan, it should be designed, implemented, and used with great care.

Music Practice Rooms

Music practic rooms are often clustered together in some larger structure. The most annoying sounds typically result from other people practicing—especially if the other person is practicing the same instrument. In general, a noise criteria value of NC-40 is acceptable for a practice room's background noise. An NC-30 value is recommended when the noise is another instrument and NC-25 when background noise is caused by the same kind of instrument. The last requirement can be met to some extent by arranging practice schedules so that similar instruments are not practiced in adjacent rooms. Ventilation noise might be used to "mask" some of the noise from similar instruments.

Students usually prefer to practice in rooms with a considerable amount of reverberant sound. However, they do tend to feel that practicing in a room with less reverberant sound is probably more effective for as long as you can stand it.

Exercises

1. List several external sources of noise that may have a significant effect in a classroom, a living room, a bedroom, or a concert hall. Describe the features of the source and the path the sound travels. Describe steps that might be taken to eliminate or reduce problems.

2. Repeat Exercise 1 for internal sources of noise.

3. Why are massive partitions needed to eliminate external sound? Why are airtight partitions needed to eliminate external sound?

4. How successful will a massive partition be in eliminating external sound if it is not airtight? How successful will an airtight partition be if it is not massive? Give examples.

5. What sound barriers (e.g., doors, windows, walls) in a typical house are most likely to permit external sounds to enter? What practical steps can be taken to correct the problem?

6. A person attempting to sleep in a hot hotel room opens the window to get some cool air. The window provides a transmission loss of 30 dB when closed. When opened it allows 25% of outside energy to get in. What is its transmission loss when open? How effective is this transmmission loss for providing quiet sleeping conditions?

7. What role does zoning play in dealing with external noise sources?

8. The level of reveberant sound necessary for optimum results varies between speech and music. Why should the optimal level of reverberant sound generally be less for speech than for music?

9. What type of music requires the greatest reverberant sound for optimum results? What type of music requires the least reverberant sound for optimum results?

10. Suppose people in the front row receive strong reflections from the back wall of a room 20 m away. Will it be perceived as an echo?

11. Suppose a time delay of 200 ms produces maximum interference for a lecturer. How far away from the lecturer is a rear wall that will produce reflected waves giving rise to maximum interference? How does this distance compare with the distance between the stage and back wall in typical theaters and concert halls?

12. A sound level of 90 dB is measured through an open door. When the open door is draped, the measured level is 88 dB. When the door is closed, the level is 76 dB; and when the door is closed and the cracks around the opening stuffed, it is 72 dB. What is the transmission loss in each case? How well do these values agree with those in the text? How can you account for any discrepancies?

13. A sound level of 100 dB is measured through an open door. When a 60-cm layer of dense mineral wool fills the opening, the level is reduced to 70 dB. When a 15-cm-thick solid-core door fills the opening, the level is reduced to 65 dB. When both are used, the level is 55 dB. How do these values agree with those in the text? How do you account for any discrepancies?

Demonstrations

1. Place a sound source with reasonable power on one side of an open door. Measure the sound level on the other side of the door opening with a sound level meter. Place a light drape over the opening and measure the sound level. What is the transmissin loss? Close the door and repeat measurements and calculations. Stuff any cracks or openings around the door and repeat.

2. Place a battery operated sound source on a hard surface. Measure the sound level it produces. Place a box built from heavy plywood over the source. Measure the sound level. Line the box with sound absorbing material and repeat. Repeat with the absorbing material only. Place the source on a resilient pad and repeat. Carefully seal all openings between box, pad, and table and repeat.

Further Reading

Backus, Chapters 9, 15

Beranek, Chapters 1-12

Berendt etal, Chapters 2-9

Doelle, Chapters 4, 13, 16

Kinsler etal, Chapter 12

Olson, Chapters 8, 9

Rossing, Chapter 32

Stevens and Warshofsky, Chapter 8

Winckel, Chapter 4

Bishop, D.E., and P.W. Hirtle. 1968. "Notes on the Sound-Transmission Loss of Residential-Type Windows and Doors," J. Acoust. Soc. Am. **43**, 880- 882.

Ellwood, E. 1972. "The Anatomy of a Wall," Sound and Vibration **6** (Jun), 14-18.

Emme, J.H. 1970. "Composite Materials for Noise and Vibration Control," Sound and Vibration **4** (Jul), 17-21.

Harris, C.M. 1972. "Acoustical Design of the John F. Kennedy Center for the Performing Arts," J. Acoust. Soc. Am. **51**, 1113-1126.

Heebink, T.B. 1970. "Effectiveness of Sound Absorptive Material in Drywalls," Sound and Vibration **4** (May), 16-18.

Heebink, T.B. 1975. "Sound Reduction of Windows in Exterior Wood-Framed Walls," Sound and Vibration **9** (Jun), 14-18.

Keast, D.N. 1979. "Energy Conservation and Noise Control in Residences," Sound and Vibration **13** (Jul), 18-22.

Lamberty, D.C. 1980. "Music Practice Rooms," J. Sound and Vibration **69**, 149-155.

Philbrick, R.L. 1978. "The Acoustics of Open Plan Spaces," The Physics Teacher **16**, 144-149.

Purcell, W.E. 1981. "Systems for Noise and Vibration Control," Sound and Vibration **15** (Aug), 4-28.

Purcell, W.E. 1982. "Materials for Noise and Vibration Control," Sound and Vibration **16** (Jul), 6-31.

Rossing, T.D. ed. 1979. **Environmental Noise Control** (American Association of Physics Teachers).

Shankland, R.S. 1968. "Rooms for Music and Speech," The Physics Teacher **6**, 443-449.

Shankland, R.S. 1973. "Acoustics of Greek Theatres," Physics Today **26** (Oct), 30-35

Strumpf, F.M. 1969. "Water Piping Systems and Noise Control," Sound and Vibration **3** (May), 23-25.

Wetherill, E.A. 1975. "Noise Control in Buildings," Sound and Vibration **9** (Jul), 20-26.

Yerges, L.F. 1971. "Windows—The Weak Link?" Sound and Vibration **5** (Jan), 19-21.

Yerges, J.F., and J.G. Bollinger. 1972. "Development and Utilization of Open-Plan Educational Facilities," Sound and Vibration **6** (Jun), 19-24.

Audiovisual

1. **The Science of Sound** (33 1/3 rpm, 4 sides, FRSC)

2. **Acoustics of the Classroom** (19 min, color, USNAC)

20. Reverberation Time

As noted in the previous chapter, listening to sound in an enclosed space has several advantages over listening to sound in an open space. One advantage is that external noises can be eliminated from an enclosed space. Perhaps of equal importance is the degree of control that can be exercised over sounds produced in enclosed spaces. Sounds are confined to the enclosure and are continually bouncing from reflecting surfaces of the enclosure. In this chapter we consider the relationships among direct, early, and reverberant sounds as determined by (1) the locations of the sound source and the listener relative to reflecting surfaces, and (2) the nature of the reflecting surfaces. In particular, we consider how reverberant sound depends on the absorption characteristics of the room. We discuss some of the properties of a room that determine reverberation time and what values of reverberation time are optimum for various auditoriums.

Early and Reverberant Sound

The three types of sound—direct, early, and reverberant—discussed in Chapter 19 can be visualized as in Figures 20.1 and 8.4. Direct sound travels from the source to the listener and undergoes no reflection. All other sound reaching the listener is reflected sound. Suppose we place a microphone in a two dimensional room with reflecting walls as illustrated in Figure 20.1. At another place in the room a toy balloon is popped to provide a source of sound. The first sound picked up by the microphone travels directly from the source. After a short time the first reflected sound, followed by more and more reflected sounds, will be picked up by the microphone. After a while the original wave will be traveling in all directions, and so it will be spread around the room in a fairly uniform manner. Figure 20.2 illustrates what the microphone would pick up in the above situation. After the direct sound and the first reflection, later reflections would be so close in time as to produce a nearly continuous, diffuse mixture of sound.

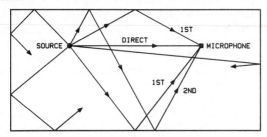

Figure 20.1. Paths of direct sound and some reflected sounds from source to "listener."

Any reflected sound arriving at a listener (microphone) within about 35 ms of the direct sound plays a very special role in how a sound is perceived. The direct sound establishes the direction of a sound source for a listener. Because of the precedence effect

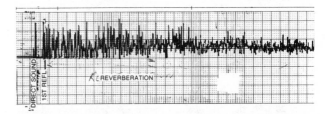

Figure 20.2. Simulated microphone response for sound in a two-dimensional room such as shown in Figure 20.1. (Courtesy of S.E. Stewart.)

(Chapter 15), early sound tends to reinforce the direct sound and be part of the same "sound image." In order for early sound to play this role it must have a spectrum not too different from that of the direct sound and it must not be too much louder than the direct sound.

The arrival time of early sound is controlled by the distance of reflecting surfaces from source and listener. The first reflection from the nearest wall (shown at the top of Figure 20.1) will be the first early sound to arrive at the listener. First reflections from more distant walls, along with multiple reflections, will arrive later, but still soon enough to be part of the early sound.

Early sound arriving within 20 ms of the direct sound plays a special role in determining the "intimacy" of a hall. This early sound should come from the walls rather than the ceiling because people prefer sounds lying in the plane of the ears and the source. Halls designed to provide early lateral reflections should be rectangular in shape and should not be too wide. The time delay is too long for halls that are too wide. In fan-shaped halls, the earliest reflections to reach the listener do not come from the walls.

Reverberant sound—reflected sound arriving later than 35 ms after the direct sound—plays an important role in determining the "liveness" and the sense of being immersed in sound. The longer the reverberant sound lingers the greater the sense of immersion. The reverberant sound is controlled by the absorbent properties of surfaces and objects in a room.

Absorption and Reverberation

When a sound wave strikes a wall (including ceiling, floor, and other surfaces) in a room, part of the energy is reflected, but part of the energy is absorbed by the wall and converted into heat or other nonsound energy, as illustrated in Figure 20.3. The proportions of sound reflected and absorbed each time a sound wave strikes a wall is determined by the characteristics of the wall. At one extreme is a **reverberation room,** which has hard, smooth walls that reflect most and absorb little of a sound wave each time it strikes the wall. Sound in a hard walled room takes a long time to be absorbed because the sound must strike the walls many times. (Losses of sound energy to the air through which the sound travels may be important when losses to the walls are small.) At the other extreme is an **anechoic room**

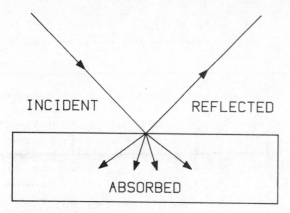

Figure 20.3. Partial reflection and partial absorption of sound incident on an absorbing surface.

(Figure 20.4) which has highly absorbent walls, usually covered with sound absorbing wedges a meter or so in length. The word "anechoic" means "without echo." An anechoice room simulates the condition of being outdoors and away from all sound reflecting objects. The small amount of sound reflected from a wedge gets trapped in the space between wedges where it is absorbed on successive reflections. In an anechoic room almost all of the sound energy is absorbed the first time a wave strikes the wall—thus, the sound dies out almost immediately. Reverberation rooms and anechoic rooms are used for performing various acoustical experiments and tests.

Figure 20.4. Interior view of an ancheoic room showing subjects in the process of a listening test. The large wedges are highly sound absorbent.

"Reverberation time" is the time required for a sound to "die out." The less absorptive the room, the greater the reverberation time. As an example of this relationship, consider a hard-walled tube 34 m in length, as shown in Figure 20.5. Imagine that a pulse of sound is somehow introduced into the tube so that it bounces back and forth between opposite ends of the tube. Assume that no sound energy is lost to the air in the tube or to the sides of the tube, but that all losses are at the ends of the tube. (In many cases, such as a wind instrument, this assumption does not hold true. As a wave travels in a tube, part of the sound energy is absorbed by the walls of the tube so that the pulse weakens as it travels.) If the ends of the tube were completely nonabsorbing, the pulse would bounce back and forth forever.

If the ends of the tube were moderately absorbing, the pulse would bounce back and forth for a long time, but would eventually die out. If the ends of the tube were highly absorbing, the pulse would die out quickly. We could place a microphone at some point in the tube (see Figure 20.5) and measure how quickly the pulse died out. The microphone would detect the pulse once every trip along the tube, or, since t = ℓ/v, once every t = 0.1 s. Figure 20.6 shows the results of microphone measurements for the cases of small absorption and large absorption. The energy level of each reflected pulse is plotted in dB for each of the two cases. The energy level decrease (in dB) after each reflection can be determined by taking 10 times the log of the ratio of one to the fraction of energy reflected. Thus, energy level loss = 10 log (1.0/fraction of energy reflected). Case 1 assumes that 27% of the energy is absorbed and that 63% is reflected, or a 2 dB loss per reflection. Case 2 assumes 75% absorption and 25% reflection, or a 6 dB loss per reflection.

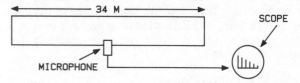

Figure 20.5. Hard-walled tube in which a simulated pulse of sound bounces back and forth between the partially absorbing ends. The "display" on the scope is the output of the microphone.

In actual rooms, the microphone response will be much more complicated than that shown in Figure 20.5 for a "one dimensional room," or even than that shown in Figure 20.2 for a "two dimensional room." However, the same governing principles apply in all three cases. The length of time reverberant sound lingers in a room depends on how often it is reflected and on what fraction of its energy is lost on each reflection.

Reverberation Time

On examining Figure 20.2 we see that reverberant sound gradually decreases in amplitude. The **reverberation time** (RT) is specifically defined as the time required for the sound level to decrease by 60 dB. Note that for the case of 63% reflection in Figure 20.6 RT is 3 seconds and for 25% reflection RT is 1 second. The larger the reflection (or the smaller the absorption), the longer the RT—thus, RT may be expressed as being inversely proportional to absorption. For the sound pulse in a tube the RT also depends on the length of the tube and on the speed of sound in the tube. If the pulse moved faster or if the tube were shorter, the RT would be shorter because the pulse would strike the ends of the tube more times per second.

The RT in three-dimensional structures, such as rooms, is of primary interest in building design. As might be expected on the basis of the one-dimensional example, RT varies inversely with absorption—but it depends on the volume of the enclosure rather than on length as in the one-dimensional example. An expression for reverberation time called the Sabine equation is RT = 0.16 V/TA—where V is the room volume in cubic meters and TA is total absorbtion associated with the interior surfaces of the room. If all the surfaces of

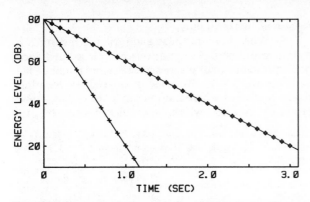

Figure 20.6. Energy levels of succesively reflected sound pulses in the 34 m tube of Figure 20.5. No energy is lost to the walls. Two conditions at the ends of the tube are considered: one in which 63% of the energy is reflected and the other in which 25% of the energy is reflected.

the room were completely absorbing, TA would be equal to the surface area of the room in square meters. (The only perfect absorber, however, is an opening.) If instead of a perfect absorber, we have twice the area of some material which absorbs half of the incident sound energy, the resulting TA will be the same. Let us define the **absorption coefficient** (AC) of any material as the fraction of energy absorbed on each reflection of a sound wave. Then we can see that for any surface area, S, TA = S × AC. When many different surfaces are present the total absorption is given by TA = S_1 × AC_1 + S_2 × AC_2 + S_3 × AC_3 . . .—where the subscripts represent different surfaces. It will be noted that TA has the dimensions of an area and that it is smaller in magnitude than the total surface area of the room. Typical absorption coefficients are shown in Table 20.1. Note that the values vary with frequency for any given material and that most materials are more highly absorbing at higher frequencies.

The Sabine equation will provide an estimate of reverberation time when the sound is diffuse so that it dies away in a fairly smooth manner. However, several conditions can occur in a room which give rise to non-diffuse sound including: (1) concentration of absorbing materials on one surface; (2) covering of many surfaces with highly absorbing materials; (3) use of curved surfaces which focus the sound; (4) having one room dimension substantially different from the other two; and (5) having very large enclosed air volumes. Since most listening environments do not have a totally diffuse sound field there is often a considerable discrepancy between the measured and the calculated RT. Other RT formulas have been developed to account for some of these discrepancies. Computer modeling methods have been used to trace the paths of sound "rays" as they are reflected from room surfaces and thus numerically calculate RT in better agreement with measured RT.

We now consider as an example a room 8 × 10 × 6 m with heavy carpet on the floor, acoustical plaster on the ceiling, and 0.30 cm plywood paneling on the walls. Assume that 50 people are present in the room. We calculate the RT at 500 Hz in the following manner. The room volume is given by V = 8 × 10 × 6 = 480 m³. The total absorption is given by TA = 8 × 10 × 0.60 (floor) + 8 × 10 × 0.50 (ceiling) + 2 × 8 × 6 × 0.10 (end walls) + 2 × 10 × 6 × 0.10 (side walls) +

50 × 0.45 (people) = 132.1 m². We then obtain the RT from the Sabine equation as RT = 0.16 × 480/132.1 = 0.58 s.

Ambient Sound Levels

So far we have discussed the relationship of the RT of a room to the room volume and the room absorption. The buildup time of the sound and the final sound level produced by a constant-power sound source in a room are two related aspects of interest. If a sound source having constant power output is placed in a room, the sound level in the room will increase until it reaches some final value, at which time the sound being absorbed is just equal to the sound being supplied by the source.

The relationships between absorption, power input, RT, buildup time, and final sound level can be illustrated with the following example. Imagine a bucket with small holes (absorption) punched in its side from bottom to top. Water from a garden hose is turned on to some steady flow rate (source of constant power). Water from the hose runs into the bucket and the time required (buildup time) for the water level to reach its highest possible point (final sound level) is observed. The hose is removed and the time required for the water to drain from the bucket (RT) is observed. The buildup time and the RT both tend to be of about the same length. If the holes in the bucket are small (small absorption), both times are relatively long and the final water level (sound level) is high. For large holes both times are relatively short and the final level is low.

Consider a room with a volume of 1000 m³ and a total absorption of 320 m² so that it has RT = 0.5 s. A one-watt sound source is placed at one point in the room and a microphone at another point (similar to Figure 20.1). After the sound source is turned on, the sound level (measured by the microphone) initially increases as illustrated in Figure 20.7. When the sound level reaches its final value, it remains constant as long as the sound source is on. When the source is turned off, the sound level decreases until some background level is reached. Consider a room having the same volume but only one quarter the total absorption as the first room so that its RT = 2.0 s. When the same sound source is used in this room the curve labelled "RT = 2.0 s" in Figure 20.7 will result. Again, the greater the absorption, the shorter the RT and buildup time, and the lower the final sound level.

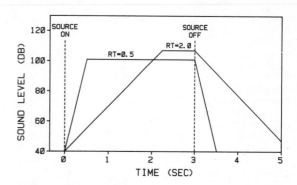

Figure 20.7. Curves showing sound level increase, final sound level, and sound level decrease for two rooms. Both rooms have the same volume but differ in acoustical absorption and RT.

An expression that gives the final sound level in a room in terms of the sound power of the source and the total absorption of the room is SL = 10 log(P/10^{-12}) + 10 log(4/TA), where P is the source power in watts and TA is the total absorption in square meters. Referring to the first example in Figure 20.7, we calculate the final sound level as SL = 10 log(1/10^{-12}) + 10 log(4/320). This gives a final sound level of 101 dB as shown in the figure. A similar calculation for the second example in Figure 20.7 is SL = 10 log(1/10^{-12}) + 10 log(4/80). This gives a final sound level of 107 dB.

Optimum Reverberation Times

Speech generally requires the shortest RT because too much reverberant sound tends to mask speech making it less intelligible. Different kinds of music require different RT for optimum results. Optimum RT increases for opera, chamber music, orchestral music, and organ in that order. Sufficient sound levels are required to produce optimum results for both speech and music. In the previous section we found that sound level is reduced when TA is increased. From this it can be seen that the RT should be made longer for large rooms so that a sufficient sound level can be achieved. As the volume increases the RT should be increased to maintain a roughly constant sound level in an enclosure.

Optimum values of RT are listed in Table 20.2 for a frequency of about 500 Hz. From Table 20.1 we see that absorption coefficients for different materials vary with frequency which means that RT will be different at different frequencies. The value of RT should be about 20-40% greater at 125 Hz than at 500 Hz for optimum results with music. An increased RT at lower frequencies is less desirable for speech because speech energy at these frequencies does not contribute much to intelligibility.

Table 20.1. Typical absorption coefficients for some building materials. Actual values depend on backing and mounting of material. TA values for adult person and upholstered chair are expressed in square meters.

Material	Frequency (Hz)		
	125	500	2000
Acoustical plaster	0.15	0.50	0.70
Acoustical tile	0.20	0.65	0.65
Brick	0.02	0.03	0.05
Carpeted floor			
heavy, on heavy pad	0.10	0.60	0.65
light, without pad	0.08	0.20	0.60
Concrete	0.01	0.01	0.02
Draperies			
heavy	0.15	0.55	0.70
light	0.03	0.15	0.40
Fiberglass blanket			
2.5 cm thick	0.3	0.70	0.80
7.5 cm thick	0.6	0.95	0.80
Glazed tile	0.01	0.01	0.02
Paneling—plywood supported at 1 m intervals and backed with 5-cm air space			
.15 cm thick	0.10	0.20	0.06
.30 cm thick	0.30	0.10	0.08
Plaster	0.04	0.05	0.05
Vinyl floor on concrete	0.02	0.03	0.04
Wood floor	0.06	0.06	0.06
Adult person	0.30	0.45	0.55
Upholstered chair	0.20	0.35	0.45

Measurements in Rooms

Reverberation time in an auditorium is measured in a number of ways. A sound source is placed at a location representative of where actual sources might be in a performance. A microphone is placed at locations representative of where a listener might be. The source produces sounds which are picked up by the

Table 20.2. Optimum reverberation times (in seconds) at 500 Hz for different room sizes and uses. (Data from Knudsen, Olson, and U.S. Gypsum.)

Use	Room size—cubic meters			
	30	300	3000	30,000
Office—speech	0.4	0.6	—	—
Classroom—speech	0.6	0.9	1.0	—
Workroom—speech	0.8	1.2	1.5	—
Rehearsal room—music	0.8	0.9	1.0	—
Studio—music	0.4	0.6	1.0	—
Chamber music	—	1.0	1.2	—
Classical music	—	—	1.5	1.5
Modern music	—	—	1.5	1.5
Opera	—	—	1.4	1.7
Organ music	—	1.3	1.8	2.2
Romantic music	—	—	2.1	2.1
Room in home—speech	0.5	0.8	—	—
Room in home—music	0.7	1.2	—	—

microphone and which are used to determine RT and other characteristics of the sound for the particular source and microphone locations. Uniform RT and sound characteristics are generally considered marks of a good listening hall. However, not all listening locations are equivalent even in the best of halls, and may be far from uniform in poorly designed halls. For this reason, the microphone is typically moved to other listening locations (or multiple microphones are used) so that the RT is known at different points in the auditorium. Sometimes the source is moved to determine how source location affects RT and other characteristics.

The RT and sound characteristics of a hall depend on frequency as well as location. For this reason the source must produce sounds that allow RT to be tested at different frequencies. An impulsive sound, such as a pistol shot, is often used as a source because it is rich in frequency content. A quasi- steady "white noise" can be used as a source to produce some final sound level. When the source is turned off, the RT of the decaying sound can be measured. Octave band filters are used to separate the sound into its various frequencies. Warbled sinusoids can be used in a way similar to the white noise. (Sinusoids with constant frequency should not be used since complications can occur because of the natural mode frequencies of the room.) When either white noise or warbled sinusoids are used, RT at frequencies of 125, 250, 500, 1000, 2000, and 4000 Hz are usually measured.

Modifications to Rooms

Often one encounters an auditorium or a classroom with undesirable acoustical properties, one of the most common being excesssive reverberation. As a hypothetical example of ineffective treatment, suppose that this problem confronts the congregation of a newly constructed church. A committee appointed to solve the problem seeks advice from a vendor of acoustical products whose solution is to cover all surfaces with acoustical tiles or panels. The "solution" is worse than the orginal problem, because the congregation is left with an expensive and excessively dead room in which it is difficult to hear the spoken word.

An example of effective sound absorbing treatment for a room is to be found on the campus of Frostburg State College in Frostburg, Maryland. One large lecture hall with a seating capacity of 212 students was originally constructed with the ceiling and floor of concrete and walls of brick. Since the volume of the hall was also large with a 6 m high ceiling in the front, the reverberation time was excessive. Corrective measures were taken because the room as constructed was unsuitable for classroom use. Acoustical consultants were hired, recommendations were made, and remedial action was taken. The recommendations, which were followed, specified that (1) no absorbing material was to be used on the large front wall; (2) a small amount of absorption was to be placed on the ceiling in the form of widely separated individual panels; (3) large panels of absorptive material were to be hung on the side walls, alternating with equally large untreated areas; and (4) the rear wall was to be almost entirely covered with absorbing panels. The final results of the modifications were the creation of an optimum acoustical environment for a lecture hall. Teachers can now lecture without having their words garbled by the room and no supplementary sound system is necessary.

From the two foregoing examples we may draw some useful conclusions. (1) It is possible to reduce excessive reverberation without making a room excessively dead. (2) No acoustically well designed hall or medium-sized listening space has need of a sound system. (3) An acoustical consultant should be hired to achieve optimal results. The expense of hiring a consultant may well be offset by savings in acoustical treatment and sound reinforcing equipment.

Exercises

1. Describe qualitatively the absorption characteristics and reverberation time you might expect to find in each of the following rooms: concrete-walled room, living room, bathroom, anechoic room, concert hall, recording studio, and wood-paneled room.

2. Consider an auditorium 30 × 20 × 15 m high. Assume that the floor is concrete, with an average of one upholstered chair every square meter. Assume that the ceiling and walls are plastered. Calculate the RT. What is the sound level in the hall if the orchestra produces a sound power of 2 W?

3. A listener is seated 5 m from the rear wall, 7 m from one sidewall, and 13 m from the other sidewall in the auditorium of Exercise 2. Does the first reflected sound come from a wall or the ceiling? Determine where the first several reflections come from and their time delays relative to the direct sound. Will the precedence effect play an important role in this auditorium?

4. Discuss the acoustical reason for each of the modifications made in the Frostburg State College lecture hall.

Demonstrations

1. Produce a sharp clap or pop a balloon to provide an impulsive sound source in various rooms, large and small. Comment on the RT you experience.

2. Choose an auditorium known to have good acoustics. Perform various qualitative tests to estimate its RT and other properties.

3. Choose a poorly designed auditorium and perform the tests of Demonstration 2 in it.

4. Perform experiments using cans or pipes with various size holes punched in them as described in the text to demonstrate the effect of total absorption and source power on final sound level and RT.

Further Reading

Backus, Chapter 9
Benade, Chapters 11, 12
Beranek
Doelle, Chapter 5
Hall, Chapter 15
Kinsler etal, Chapter 13
Knudsen and Harris, Chapters 6-8
Olson, Chapter 8
Rossing, Chapter 23
Knudsen, V.O. 1963. "Architectural Acoustics," Scientific Am. 210 (Nov). (In Rossing, 1977).
Sabine, W.C. 1915. "Architectural Acoustics," J. Franklin Int. 179. (In Rossing, 1977.)
Schroeder, M.R. 1980. "Toward Better Acoustics for Concert Halls," Physics Today 33 (Oct), 24-30.
Shankland, R.S. 1968. "Quality of Reverberation," J. Acoust. Soc. Am. 43, 426-430.
Shankland, R.S. 1979. "Acoustical Designing for Performers," J. Acoust. Soc. Am. 65, 140-144.

21. Auditorium Design

Auditoriums as we know them, evolved from the open-air theater referred to previously. The first enclosed theater was the odeion. The odeion was basically a moderately-sized open-air theater enclosed with a wooden roof. The odeion seated between 200 and 2000 people. The enclosed theater not only provided shielding from extraneous noise, but the additional reflecting surfaces made it considerably easier to hear the actors. By the seventeenth century theaters had become completely enclosed and the seating area had evolved into a U-shaped arrangement with multiple balconies. Because of the relatively small size of the theater and the high absorptivity of the audience, reverberation times were short. Since members of the audience were also located in close proximity to the stage, many of these theaters provided almost ideal listening conditions. During the eighteenth century auditoriums became large, multiple balconies disappeared, and hard plaster began to be used for the walls. All of this had a deleterious effect on the acoustics. As the halls grew larger during this period, the instrumental ensembles grew larger so as to be able to produce the greater sound volume required. This is the origin of the modern symphony orchestra.

During the nineteenth century scientists began to recognize and understand some of the acoustical problems with auditoriums, but it was not until the present century that any systematic acoustical research was performed. The principles of acoustics are now fairly well established, and it is possible to utilize them to achieve desirable results when new auditoriums are designed. In this chapter we will consider basic acoustic requirements as well as defects to be avoided in the design of various auditorium types. We will then consider listener preference and design criteria for a "good" concert hall in some detail. Finally, we will present several examples of some well known halls (see Figure 21.1) and will comment on both their good and bad points.

Basic Acoustic Requirements

The hearing conditions in any auditorium are determined primarily by purely architectural considerations. Practically every detail within the enclosed space contributes in some manner to the overall acoustical character of the room. Nevertheless, by considering several basic acoustical rules and by making judicious use of the variety of materials available, almost any structure can be designed so as to have adequate acoustics. A primary requisite for a good acoustical environment is that the structure be free of exterior and interior noises (see Chapter 19). The second requisite is that the room have optimum reverberation (see Chapter 20) for its volume and for its intended purpose. If an auditorium is to provide good listening conditions, the

Figure 21.1. Interior view of Boston Symphony Hall as seen from the stage, looking toward the rear balcony. (Courtesy of the Boston Symphony.)

137

sound should be adequately loud in all parts of the room and the sound energy should be diffused (uniformly distributed) throughout the auditorium.

Adequate loudness is achieved in small or medium-sized auditoriums by directing as much of the sound energy as possible toward the audience and by preventing excessive absorption of the sound. Sound can be directed toward the audience by several means as illustrated in Figure 21.2. First, it is important that the room be shaped so that the audience is as close as possible to the sound source. Generally this means that long, narrow rooms should be avoided and balconies should be used whenever possible. Second, the floor of the auditorium should be "raked" (constructed on a ramp). Third, the stage should be raised to elevate the sound source. Fourth, large sound-reflective surfaces (e.g., plaster or heavy plywood) should be located as close to the sound source as is feasible. Finally, the ceilings and walls need to be designed for favorable sound reflection, especially for those seats farthest from the stage. These measures, while extremely important, will not perform miracles. The first provision for adequate loudness must come from the performer. For instance, a talker must speak slowly and clearly and with volume sufficient for the room. In very large auditoriums even the above conditions may prove inadequate and the installation of a sound reinforcement system (see Chapter 22) may be in order.

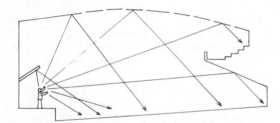

Figure 21.2. Some direct and reflected sounds in an auditorium.

Raking of the floor permits sound to travel more directly to the listeners and helps avoid the attenuation that would occur if the sound were to graze the heads of the highly absorptive audience. Raking of the floor also helps to bring the audience closer to the performers.

The proper diffusion of sound is important for the achievement of uniform sound distribution in an auditorium. Surface irregularities can be implemented in the construction of the room to diffuse sound. Irregularities such as exposed beams, coffered ceilings, sculptured balcony railings, and protruding boxes are effective diffusers. Small, medium, and large surface irregularities should be used so as to provide diffusion for various sound wavelengths. For smaller rooms, diffusion can be achieved by alternating areas of sound reflective and sound absorptive materials. Finally, irregular or random arrangement of materials having different absorption coefficients will also produce diffusion. Listening conditions in rooms with excessive reverberation improve considerably when properly sized diffusers are installed.

In addition to the usual requirement of low ambient noise level, auditoriums that are used primarily for speech-related activities have other requirements. They should have a directional sound source to provide more direct sound to the listeners and allow less sound to be reflected. They should generally have a short RT for good speech intelligibility. Even so, RT should be long enough to provide adequate sound levels in large halls where sound reinforcement is not used. In such halls, it may be necessary for a talker to speak more succinctly and slowly so that speech is intelligible in spite of longer RT.

Defects to Avoid

Although the acoustical attributes described above are very important positive aspects of auditorium design, it is equally important that all potential acoustical defects be minimized, if not entirely eliminated. The most common of these defects are echoes, flutter, focusing of sound, distortion, room resonances, and sound shadows. Each of these defects will be described briefly.

Echoes are probably the most serious of the acoustic defects listed above. While reverberation is a desirable acoustical property of an auditorium, echoes are to be avoided at all costs. Echoes, like early sound and reverberant sound, are caused by reflections. Reflections producing early sound typically have spectral properties similar to those of the related direct sound. These reflections arriving soon after (within 35 ms) the direct sound are perceived as part of the original sound image. Multiple reflections which typically produce reverberant sound are weaker than reflections producing early sound and are less similar (spectrally-speaking) to the direct sound. These sounds are not perceived as part of the original sound image. When such a reverberant sound is too loud it will be perceived as an echo. **Echoes** result from loud reflected sounds arriving later than about 50 ms after the direct sound. With this time interval between direct and reflected sound, the ear can resolve the reflected sound separately and it no longer sounds like a continuation of the original sound.

A sound-reflective rear wall in an auditorium is often the source of an annoying echo on the stage as illustrated in Figure 21.3. In this case, the echo can be eliminated by treating the rear wall with sound absorptive material. A second way of eliminating echo from a rear wall is to slope the rear wall forward as illustrated in Figure 21.4 for a concert hall. A sloped rear wall reflects sound into the back rows of seats, but not to the front seats and stage. This method has the further adavantage of not wasting sound energy that would be lost to absorption in the first method.

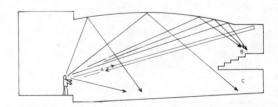

Figure 21.3. Some acoustical defects: (A) reflection from rear wall resulting in echo; (B) focusing of sound by a concave surface; (C) sound shadow under a deep balcony.

Figure 21.4. Interior view from the stage toward the back of the De-Jong Concert Hall, Brigham Young University. The rear walls under and above the balcony are designed with a slope to avoid echoes. The balcony is designed so that its height is greater than half its depth to avoid sound shadows. (Courtesy of Brigham Young University Archives.)

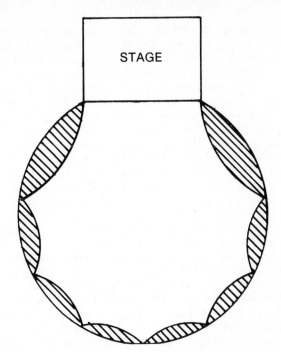

Figure 21.5. A circular auditorium with convex diffusers.

Flutter is a rapid succession of small, but noticeable, echoes. It occurs most commonly between two parallel but highly reflective surfaces. The easiest way to avoid this defect in auditoriums is to avoid parallel reflecting surfaces. If parallel surfaces cannnot be avoided, one solution is to treat them with sound-absorbing materials. If such acoustical treatment is not feasible, another solution is to incorporate sound diffusing elements on the parallel walls or to slightly angle sections of the walls.

Sound focusing is the product of concave surfaces. The effect of the focused sound is to produce areas where the sound level is unnaturally high while the sound level is excessively low in other areas (see Figure 21.3). Since we are attempting to achieve uniform distribution of sound in the auditorium, focusing effects are highly undesirable. If concave surfaces are unavoidable, they should either be treated with sound-absorbing materials or covered with convex surfaces (which diffuse sound) as illustrated in Figure 21.5.

Distortion, as it applies to auditorium acoustics, is any undesirable change in the quality of a musical sound due to the uneven or excessive absorption of sound at certain frequencies. For instance, if an auditorium has a RT of 2.0 s at 2000 Hz, but the RT is only 1.0 sec at 200 Hz, low-frequency sounds will attenuate much more rapidly than the mid-frequency sounds. The listener will perceive a sound which seems deficient in bass, even though the original sound produced by the orchestra may have been balanced. This defect can be avoided if the interior surfaces and the acoustical materials applied to them have absorption characteristics which are fairly uniform over the audio frequency range.

Room resonances exist when some of the natural frequencies of the room are excited by a sound source. Room resonance problems are most acute in small rooms with hard interior surfaces, especially when several natural modes occur near the same frequency. The natural modes of a room are usually not uniformly distributed across all frequencies. Some frequency ranges have more modes, or in other words, greater response than other ranges. Irregularities in the shape of a room or an irregular distribution of absorptive materials throughout a room will contribute to uniform spacing of natural modes. In addition, increasing the amount of absorptive material in a room will broaden and lower the natural mode peaks, thus giving more uniform room response.

A **sound shadow** occurs in a region of an auditorium where there is direct sound energy but little reflected sound, as illustrated in Figure 21.3. This defect is most often noticed under balconies, especially low and deep balconies. To avoid a noticeable under-balcony sound shadow, the depth of the balcony should not exceed twice the height of the opening (see Figure 21.4). It is also desirable to utilize the underside of the balcony as much as possible to reflect sound to the seats below.

Listener Preference

Criteria for concert hall designs are different and more complex than criteria for halls designed for speech-related activities. The designer of the hall must achieve appropriate perceptual characteristics through control of the hall's physical characteristics. Some major halls designed to accepted design criteria in the 1950's were not well received by musicians and the listening public. This led to extensive efforts in the 1960's and 1970's to determine what perceptual characteristics listeners prefer in a concert hall, and how these are related to the physical characteristics of a hall.

In one series of studies, recordings of an orchestra performing a symphony were made in an anechoic room. The "anechoic recordings" were played through loudspeakers located on the stage of each hall to be tested. Stereo recordings were made at different seat locations in the hall by means of a dummy with microphones for ears. These stereo recordings were played in an anechoic room through loudspeakers located in front and to the left and right of listeners. (Special electronic circuitry was used to convert stereo signals appropriate for headphone listening to signals for loudspeaker listening.) In this way any given recording heard by listeners in the specially-equipped anechoic room had the characteristics of a specific hall.

Listeners were asked to compare two halls (or two seats in the same hall) by listening to the recordings for each of the two halls. They were then asked which of the two halls they preferred. The halls were plotted in a two-dimensional listener preference space as illustrated in Figure 21.6. The horizontal dimension represents average listener preference. Halls labelled H1 and H2 were more preferred than those labelled H3 and H4 and thus appear farther to the right in the figure along the preference axis. The vertical dimension represents disparities among individual judgments.

In order to relate perceptual and physical properties of a hall, various physical properties of the halls were plotted in the preference space. Reverberation time (RT) shows up on the right of the graph indicating that longer RT (up to about 2.2 s) is preferred. Hall width (HW) and ear-to-ear coherence (EC) are on the left side of the graph indicating that greater coherence and greater hall width are less preferred. Ear-to-ear coherence is a measure of the similarity of sounds heard at a listener's two ears. Listeners prefer the sounds to be dissimilar, which explains why ceiling reflections are less desirable than wall reflections. Sounds coming from the ceiling will reach both ears at about the same time and with the same phase so they will be similar. Sounds coming from a wall will be dissimilar because they reach the ears at slightly different times and with different phases. This also explains why wide halls are less preferred. In wide halls, the first reflections come from the ceiling for many listeners and produce coherence.

Concert Hall Criteria

An extensive study was made of many of the world's best concert halls in the late 1950's. The study included acoustical measurements in each hall along

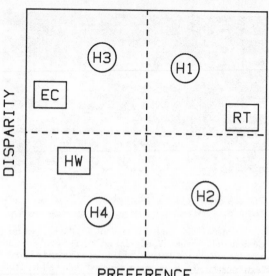

Figure 21.6. Two-dimensional perceptual space. Horizontal dimension is listener preference. Vertical dimension is disparity among listener judgments. H1, H2, H3, and H4 represent four halls. RT, EC, and HW represent reverberation time, ear-to-ear correlation, and hall width, respectively. (After Schroeder, 1980.)

with comments of musicians and critics who were familiar with the halls. The study identified important concert halls attributes and correlated acoustical properties with the musicians' perceptions. Table 21.1 summarizes the most important perceptual attributes, their physical correlates, and optimum values of the physical correlates.

The two most important criteria for a concert hall are liveness and intimacy. **Liveness** is dependent on adequate reverberation time. During each major musical era, the style of music has been influenced to some extent by the type of hall in which the music has been performed. There is therefore no such thing as an ideal con-

Table 21.1. Perceptual attributes, physical correlates and optimum values of physical correlates for concert halls. (After Beranek, 1962.)

Perceptual attribute	Physical correlate	Optimum value
Liveness	Reverberation time	Depends on music (See Table 21.2)
Intimacy	Time delay between direct and first reflected sound.	Less than 20 ms
Warmth	RT longer for low frequencies	Low frequency RT 20-40% longer than high frequency RT.
Definition	Ratio of early sound level to reverberant sound level	Level of direct plus early sound somewhat greater than reverberant sound level.
Uniformity	Diffuse sound field. Absence of focusing and shadows.	As diffuse a sound field a possible
Envelopment	Space and time pattern of reflections reaching a listener	Several relfections from different directions during first 60 ms.
Dynamic range	Sound level of fortissimo relative to background noise	Fortissimo passages should not exceed 120 dBA. Background noise level should be less than NC 20.
Ensemble	Stage enclosure so musicians can hear each other	Stage with ample reflecting surfaces. Stage less than 20 m wide.
Blend	Stage enclosure to provide mixing of various sounds.	Wider stages should be less high. Higher stages should be less wide.

140

Table 21.2. Optimum "values" of attributes of concert halls for different periods.

Period	RT(s)	Liveness	Intimacy	Definition
Baroque (secular)	under 1.5	low	high	high
Baroque (sacred)	2.0	high	low	low
Classical (1750-1820)	1.5-1.7	medium	medium	high
Romantic (1800-1900)	1.9-2.2	high	low	low
Modern (1900-)	varies	varies	varies	varies
European opera	under 1.5	low	high	high
Wagnerian opera	1.6-1.8	high	low	low

cert hall, but rather halls in which RT and other parameters have been optimized for music of a particular period (as shown in Table 21.2). Note that although there is considerable variation in these characteristics, the trend has been toward longer RT as time progressed, with the noticeable exception that the modern period is characterized by a complete lack of uniformity in any of these properties. (Perhaps this is because modern music seems to stress non-uniformity in general.)

An additional aspect of reverberation not obvious from the above discussion is that reverberant sound should decay uniformly. An ideal reverberation curve is shown in Figure 21.7a in which the reverberant sound level dies away uniformly with time in almost a straight line manner. This kind of uniform reverberation response is highly desirable because the sound is getting "soft" at an even rate. Figure 21.7b shows a highly irregular response in which the sound level decreases, then increases, then decreases again in a type of erratic warble. Generally, the more diffuse the sound field, the less likely the chance of finding this type of irregular curve. Finally, Figure 21.7c shows another situation which is very undesirable even though the reverberation curve is smooth. In this illustration, the reverberant sound falls off rapidly during the first half second, thus telling the ear that the RT will be short. In actuality, the RT is 2.0 s as with the other two curves, but the result is a type of auditory confusion. Even though the RT is 2.0 s, the perceived RT corresponds to the rapid initial decay of sound and the room will seem somewhat dead. The best way to realize a uniform reverberation curve is to fill the concert hall with diffusing elements of all sizes so that the sound will bounce evenly throughout the room.

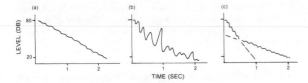

Figure 21.7. Curves showing different reverberation characteristics: (a) ideal; (b) irregular; (c) double sloped. (After Bliven, 1976.)

The **intimacy** of a hall depends on the time delay between the direct sound and the first reflected sound. When the initial time delay gap is too long, the hall seems cold and impersonal, even though it may have optimum reverberation characteristics. Placing reflecting screens near a performer as illustrated in Figure 21.8 will provide early reflections to the audience and may improve the acoustical intimacy of a hall. However, as

discussed in the previous section, the early reflections should come from the sides rather than from overhead to produce results most preferred by listeners.

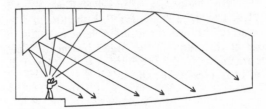

Figure 21.8. The use of ceiling reflectors to shorten the initial time delay gap.

Warmth is related to fullness of bass tone and is achieved when low frequency RT is about 20-40% longer than high frequency RT.

Definition (or clarity) depends on the level of direct and early sound relative to the level of reverberant sound. The sum of the direct and early sounds should be somewhat greater at all locations than the level of the reverberant sound.

Uniformity results from having good sound diffusion. Diffusion is achieved, as previously discussed, through the use of irregular shapes, non-parallel surfaces, and surfaces with convex shapes. The defects discussed previously should be avoided—especially focusing and shadows.

Envelopment is the feeling of being surrounded by sound. Envelopment depends on having several reflected sounds arrive from different directions within about 60 ms after the direct sound. Reflected sounds should come from the side walls and ceiling as well as from front and rear walls.

Dynamic range depends on the loudness of "fortissimo" passages relative to "pianissimo" passages. The loudnesses of the passages depend on their sound levels. The background noise level must be kept sufficiently low (NC-20 or lower) so as not to obscure pianissimo passages. In a poorly designed hall with too high a background noise level, the dynamic range will be limited to the difference in levels between fortissimo passages and background noise.

Ensemble is that attribute of a hall—more particularly the stage enclosure—that enables musicians to hear each other, allowing them to play in time, in tune, and in balance with one another. There should be ample reflecting surfaces on the sides of, behind, and above the stage to provide early reflections to the performers. These surfaces should have enough structural detail to provide diffuse reflections to the performers. The

141

preferred time delay for reflections is about 20 ms; delays of less than about 15 ms or greater than about 35 ms are not desirable. Higher frequency (above 500 Hz) reflected sound is most important for ensemble.

Blend is that attribute of a hall—more particularly the stage enclosure—that produces blending and mixing of sound on stage so that a reasonable representation of all performers is radiated to the audience. Some of the stage design criteria for ensemble are also appropriate for blend. There should be an inverse relationship between stage width and height. A stage wider than about 15 m shoud be irregular in shape and no higher than 10 m. A stage higher than about 10 m should have a width no greater than 15 m and a depth no greater than 10 m.

An oft cited example of an excellent stage design is that in the Troy, New York, Concert Hall as shown in Figure 21.9. The stage "enclosure" is shallow and has many sound reflecting and scattering elements including the boxes at the side and the organ overhead. These features are important for producing excellent ensemble and blend attributes in the hall.

Figure 21.9. Troy, New York, Concert Hall stage. (From Shankland, 1979.)

Too little consideration has been given to designing of halls so that performers can achieve proper ensemble. Halls in which a performing group is surrounded by the audience violate the ensemble requirement explicitly and will generally not be highly regarded. In other halls, the ensemble requirement is violated by making the stage enclosure too efficient as a radiator of sound to the audience, thus "starving" the performers of much needed reflected sound.

Examples of Concert Halls

We will now consider several well known concert halls which illustrate some of the problems, pitfalls, and potentials in working toward excellent concert hall acoustics. Two of the halls, the Royal Albert Hall in London and the Mormon Tabernacle in Salt Lake City

were designed before acoustical knowledge of acoustics had been applied to architecture. Boston Symphony Hall was the first hall ever designed using the expertise of an acoustical consultant. The recently constructed Avery Fisher Hall in New York City was carefully designed to have the best possible acoustics.

The Royal Albert Hall, opened in 1871, has long been considered an acoustical disaster since it exhibits nearly all the defects which should be avoided in concert hall design. The cavernous size of the Hall, which seats 5080 people with room for another 1000 to stand, is the source of most of the problems. The hall is approximately elliptical in shape (see Figure 21.10), 70 meters long and 60 meters wide, enclosing a total volume of 85 thousand cubic meters. The principal acoustical problems which plagued this hall during its first 100 years were the result of three main causes. First, the reverberation time (3.7 s when unoccupied) was excessive, even for a structure of its size. Second, there were very pronounced and annoying echoes, intensified in some areas by the dome-shaped ceiling. Finally, the loudness of the direct sound decreased as it traveled such great distances in this hall. The echoes were recognized as being the most serious defect, since, in certain locations, the reflected sound from the ceiling would reach a listener nearly one fifth of a second after the direct sound and be nearly as loud because of focusing effects caused by the curved ceiling (see Figure 21.10). A convex cloth velarium was suspended below the dome which reduced echoes and the reverberation time.

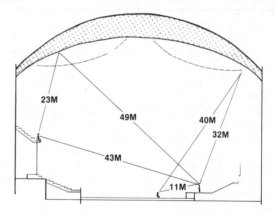

Figure 21.10. Simplified sectional view of Royal Albert Hall. The lines show paths of travel for different sound waves. A listener at the front would receive reflected sound more than a sixth of a second after the direct sound. (After Knudsen, 1963.)

In 1961 substantial changes were made in an effort to eliminate the echo and to further reduce the mid-frequency reverberation time. Some 109 fiberglass saucers (2 to 4 meters in diameter) were hung 25 meters above the floor. The disks were made so that mid-range frequencies would be absorbed while lower and upper frequencies would largely be reflected. These saucers were hung together so as to form a convex dome at the base of the true dome. A large 20-meter reflector was also installed behind the orchestra to help bounce the sound toward the rear of the hall. These two changes have greatly improved the hall; the worst of the notorious echoes are no longer present, the mid-frequency reverberation has been reduced substantially, and the amount of direct sound has increased slightly.

The Mormon Tabernacle in Salt Lake City, completed in 1867, was not designed as a concert hall, but is often used for that purpose. The Tabernacle, shown in plan view in Figure 21.11, has several features that often cause acoustical problems. However, by a fortuitous set of circumstances the acoustics are not so bad as might be expected. Focusing effects due to the elliptical shape cause serious disturbances at relatively few locations. Sound focusing has also been reduced somewhat by the installation of an elaborate sound-reinforcing system. Since the interior of the tabernacle is mostly plaster, one would expect the reverberation time to be excessive. However, the absorptive properties of the plaster are exceptionally good because at the time the building was constructed large quantities of cattle hair were mixed with the plaster. Even so, the empty tabernacle has a mid-frequency RT in excess of 4.0 sec, which is about double the optimum time. When the room is about one-third full, the RT is just about right, but when completely filled (8,000 people) the room becomes somewhat dead (RT = 1.0 s).

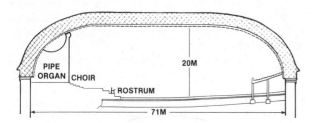

Figure 21.11. Simplified sectional view of Mormon Tabernacle in Salt Lake City, Utah. (After Knudsen, 1963.)

Boston Symphony Hall, contructed in 1900, is recognized as the first concert hall designed according to principles of acoustics. Wallace Sabine, the originator of the classical reverberation equation and the father of architectural acoustics, was the acoustical consultant. His intuition, combined with his understanding of acoustical principles, resulted in a concert hall regarded as one of the world's finest. The sound in Symphony Hall has been described as clear, live, and warm, with good ensemble. Very few of the 2631 seats are considered acoustically deficient. Many excellent acoustical features were incorporated in the Hall, some of which can be seen by examining Figure 21.1. The hall is rectangular in shape, with a high, horizontal, coffered ceiling of plaster. The walls above the upper balcony are a series of large recessed niches, in front of which stand replicas of Greek and Roman statues. The walls and ceiling are constructed of plaster. The main floor is a concrete slab covered with finished wood. Most of the stage enclosure, including the ceiling, is constructed of thick wooden panels. The basic rectangular design with high ceiling and hard interior surfaces gives the Hall an optimum reverberation time of 1.8 seconds at mid-frequencies when fully occupied. The coffered ceiling, the wall niches, the statues, and the various other Victorian details all help to provide a diffuse sound field—one of the most important acoustical characteristics of an excellent concert hall. The initial

time delay gap is 15 ms at the center of the main floor and about 7 ms in the upper balcony, which provides good intimacy. The shallow balconies minimize sound shadows. The stage enclosure contains sufficient architectural detail, including exposed organ pipes, to provide good ensemble and blend. The stage enclosure projects sound to the audience and the wooden stage and main floor help to transmit vibrations directly from large instruments, such as the double-bass, to the audience.

The final concert hall which we will consider is the Philharmonic Hall at Lincoln Center, New York City. This hall, which opened in the fall of 1962, was to be designed so that acoustic considerations were paramount. The acoustical consultant spent several years studying 54 of the world's best concert halls. As a result of this study he discovered and applied the primary requisites of good concert halls to the design criteria of Philharmonic Hall. It was anticipated that Philharmonic Hall would rank with the world's greatest halls, verifying that good concert hall acoustics can be achieved by the application of scientific principles.

Unfortunately, when the hall opened many musicians and music critics commented very unfavorably on the hall and its acoustics. Apparently the hall was so plagued with acoustical problems that major changes were made to the interior three times over the next dozen years in an effort to correct the problem. After each change the hall was improved slightly, but the problems remained. What went wrong? While no one can be completely certain, we do know what the main acoustical problems have been. We also realize that many of these defects were of the type which the consultants recognized and sought to avoid. Perhaps the unconventional design of the hall introduced more unknown variables than could be satisfactorily reckoned with. Apparently the "final" architectural plans the consultant had worked on and had expected to see constructed were modified and expanded into a larger, more modern design without his consent. Also, plans for an adjustable ceiling and many sound-diffusing features were discarded as too expensive.

The main complaints about the hall were that: (1) The orchestra sounded dry and lifeless. There was no effect of being surrounded by sound; rather, the sound seemed to come primarily from the stage area. (2) The hall was rated as being deficient in bass. (3) There were several annoying echoes. (4) The musicians on the stage could not hear each other well. In more scientific terms, there was not enough sound diffusion, which resulted in irregular fluctuations in the reverberation decay curve. Also, the initial reverberation decay was too fast, giving a double-sloped reverberation curve (see Figure 21.7c). Although the RT was about right for the size of the room, the room seemed dead because of the rapid initial decay. Also, the sound of the orchestra was not well distributed, so that it did not seem well balanced in all parts of the hall.

In 1974 an acoustical consultant was engaged who concluded that the only feasible way to make it a first-rate concert hall was to demolish it and build it over! He felt that it would be hopeless, as well as being extremely expensive, to make any further renovations to the structure. Early in 1975 money was found to build an entirely new concert hall within the concrete shell of the original hall. The acoustical consultant insisted that everything be ripped out down to the girders, acoustics would have priority over visual aesthetics, and the acoustical consultant's views would have priority over those of the architect. The design philosophy was to use only the "tried and true" methods of acoustical design. The consultant attempted to construct an acoustical setting which would be as much like that of Boston's Symphony Hall as possible. The new Philharmonic Hall was to have the same interior dimensions as the Boston Hall, with a wooden floor and stage area. The diffusion of sound was to be achieved by a clever ceiling construction of a montage of inverted plaster pyramids of varying sizes and angles. In 1976 the rebuilding commenced. When the hall, renamed Avery Fisher Hall after its chief benefactor, opened during the fall of 1976 musicians and critics were lavish in their praise. Since then, however, some musicians have apparently waned in their enthusiasm for the Hall. The main criticism seems to be that the Hall projects sound too efficiently to the audience and fails to provide the reflections necessary for good ensemble and blend. Consequently, although soloists and small ensembles sound marvelous, the sound of the full orchestra is somewhat lacking.

Some important acoustical principles and their relationships to perceptual attributes of concert halls have been elucidated by the research of the 1960's and the 1970's. Experience in the design and construction of concert halls has greatly expanded during the twentieth century and it now appears possible to use acoustical principles to design concert halls with excellent acoustics. However, well established principles and practices must be adhered to and not compromised to other architectural demands if the excellence of a concert hall is to be realized.

Exercises

1. What are the basic requirements for a well-designed auditorium? Which of these is most important? Least important? Explain.

2. What things can be done to provide adequate loudness in a hall?

3. What things can be done to provide adequate diffusion of sound in a hall?

4. What features should be included in an auditorium designed primarily for speech-related activities?

5. List several acoustical defects that should be avoided. What design details need special attention to avoid each of these defects?

6. What physical property correlates positively with listener preference for concert halls? Should it be increased or decreased? Is there a limit?

7. What two physical properties correlate negatively with listener preference for concert halls? Should they be increased or decreased to increase preference?

8. How far can a rear wall be located from a stage before an echo is perceived on the stage?

9. How would you design a balcony for an auditorium in order to get the maximum number of seats, but without any acoustic defects under the balcony? Make sketches and give explanations of your design.

10. Which of the perceptual attributes of Table 21.1 are most important for a good concert hall? Which are least important? Explain why.

11. Do the data of Table 21.1 imply that there might be a maximum size to a concert hall, which, if exceeded, would render the hall less desirable for orchestral concerts? Explain.

12. Is it desirable to use the same hall for orchestral performances and for opera? Why or why not?

13. If you want the RT of an auditorium to be approximately independent of audience size, would you use cushioned or hard surfaced seats? Explain.

14. Is it better to design a concert hall with too high an RT or too low an RT? Which situation is easier to correct? Why?

15. A municipal auditorium is observed to sound cold and impersonal during orchestral performances. Also, the RT for low frequencies is substantially longer than the high-frequency RT. What recommendations would you make to help rectify these situations?

16. Suppose you are an acoustical consultant. An architect asks you for advice on the design of a high school auditorium to seat 1200. What specific advice would you give him? (Your recommendations should include rough dimensions and a top view and side view of your proposed design).

17. If you are designing a multipurpose auditorium, to be used for drama and musical concerts, which will seat 1000 people, what RT would you recommend? What volume should the room have in order to have this RT be optimum (see Chapter 20)? What recommendations would you make concerning the ceiling of the room? The stage enclosure? The rear wall?

18. Suppose you are called upon to act as an acoustical consultant for a small church. Would you recommend hard interior surfaces (and no carpeting) or acoustically treated surfaces and the introduction of a sound system? Explain which design is preferable and why.

19. Can you imagine any type of music where the undesirable acoustical properties of the Royal Albert Hall could be used to advantage? Explain.

20. Which of the attributes of Table 21.1 were probably present in the original Philharmonic Hall at Lincoln Center? Which were probably deficient? Why?

21. What is the most likely deficiency of Avery Fisher Hall today? Does it affect performers or listeners or both?

Demonstrations

1. Visit several large auditoriums such as concert halls, drama theaters, large classrooms, and sports arenas. Make qualitative tests for acoustical defects. Specify design details to overcome potential defects.

2. Make qualitative tests for desirable acoustical features in several large auditoriums.

Further Reading

Benade, Chapters 11, 12
Beranek, Chapters 3, 4, 9-12, 15
Doelle, Chapter 6
Hall, Chapter 15
Knudsen, Chapter 9

Olson, Chapter 8

Rossing, Chapter 23

Allen, W.A. 1980. "Music Stage Design," J. Sound and Vibration **69**, 143-147.

Barron, M. 1971. "The Subjective Effects of First Reflections in Concert Halls—The Need for Lateral Reflections," J. Sound & Vibration **15**, 475-494.

Benade, A.H. 1978. "Room Acoustic Requirements for the Performer," J. Acoust. Soc. Am. **63**, S36.

Beranek, L.L. 1975. "Changing Role of the 'Expert'," J. Acoust. Soc. Am. **58**, 547-555.

Beranek, L.L., et al. 1964. "Acoustics of Philharmonic Hall, New York, During Its First Season," J. Acoust. Soc. Am. **36**, 1247-1262.

Bliven, B. 1976. "Annals of Architecture (Avery Fisher Hall)," New Yorker (Nov 8), 51-135.

Clark, R.W. 1958. **The Royal Albert Hall** (Hamish Hamilton).

Fry, D.B. 1980. "The Singer and the Auditorium," J. Sound and Vibration. **69**, 139-142.

Hales, W.B. 1930. "Acoustics of the Salt Lake Tabernacle," J. Acoust. Soc. Am **1**, 280-292.

Jordan, V.L. 1975. "Auditoria Acoustics: Developments in Recent Years," Applied Acoustics **8**, 217.

Knudsen, V.O. 1963. "Architectural Acoustics," Scientific Am. **210** (Nov), (In Rossing, 1977).

Marshal, A.H., D. Gottlob, and H. Alrutz. 1978. "Acoustical Conditions Preferred for Ensemble," J. Acoust. Soc. Am. **64**, 1437-1442.

McGuinnes, W.J. 1968. "Adjusting Auditoriums Acoustically," Progressive Architecture (Mar), 166.

Sabine, W.C. 1922. **Collected Papers on Acoustics** (Harvard University Press).

Schroeder, M.R. 1980. "Toward Better Acoustics for Concert Halls," Physics Today **33** (Oct), 24-30.

Shankland, R.S. 1968. "Rooms for Music and Speech," The Physics Teacher **6**, 443-449.

Shankland, R.S. 1972. "The Development of Architectural Acoustics," Amer. Sci. **60**, 201-209.

Shankland, R.S. 1977. "Architectural Acoustics in America to 1930," J. Acoust. Soc. Am. **61**, 250-254.

Shankland, R.S. 1979. "Acoustical Designing for Performers," J. Acoust. Soc. Am. **65**, 140-144.

Vaughan, D. 1980. "Warm String Tone in Acoustics," J. Sound and Vibration **69**, 119-138.

Audiovisual

1. **The Science of Sound** (33.3 rpm, 4 sides, FRSC)

22. Electronic Reinforcement

In a well-designed auditorium of medium size (300 seats), a lecturer or a performing group can produce adequate sound levels in all parts of the room without the aid of electronic amplification equipment. However, outdoors or in very large halls, a sound amplification system is required to insure adequate loudness. In poorly-designed halls, a sound system can be used to improve the directional properties of radiated sound and thus enhance sound distribution. Sound-reinforcement systems may also be used to provide desired signal delays, to modify spectral balance, and to provide special sonic environments.

In this chapter we will consider the basic principles of sound reinforcment and characteristics of system components. We will then discuss design of a simple sound reinforcement system using an outdoor reinforcement system as an example. Certain additional design and installation features important for indoor sound reinforcement will then be considered, including equalization procedures. Several examples of indoor sound reinforcement systems used in different environments to accomplish different ends will be discussed. These examples will include systems used in a large indoor arena, a concert hall, rooms for worship, and rock music environments.

Principles of Reinforcement

A well-designed sound system should be unobtrusive to listeners, giving the illusion that the amplified sound comes from the actual sound source. The three basic components of any sound reinforcement system are microphones, amplifiers, and loudspeakers. It is important that each component of the system be carefully chosen and have fidelity adequate to meet the requirements of a particular application. Public address paging systems which, in the strictest sense, are not sound reinforcement systems require the least fidelity. They provide sound in places where there is little direct sound, and do not attempt to provide a "source illusion." Systems for reinforcing music require the highest fidelity; they should transmit a wide range of frequencies (about 30 Hz to 15 kHz) and be capable of a wide dynamic range without distortion.

A simple sound reinforcement system is shown in Figure 22.1. Sound from a talker (or other source) is picked up and converted into an electrical signal by the microphone. One of the primary advantages of a sound system is that the electrical signal representing the sound can be modified in many ways. One obvious modification, for which the system is usually built, is that of amplification. Another modification, called **equalization**, is the amplification or attenuation of certain frequency bands to produce a more uniform response at all frequencies. Other alterations of the signal, such as artificial reverberation, may also be used to produce various effects. The electrical signal may be delayed before being reconverted into sound. Ultimately, a loudspeaker converts the modified electrical signal back into sound.

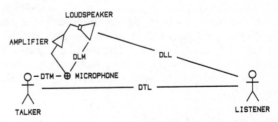

Figure 22.1. Elements of a sound reinforcement system.

Sound intensity decreases as sound travels from a loudspeaker (or other source) to a microphone or a listener as shown in Figure 22.1. The power of the corresponding electrical signal increases as it "travels" from the microphone through the amplifier to the loudspeaker as shown in Figure 22.1. Sound traveling from the loudspeaker back to the microphone is called **feedback**. When sound from the loudspeaker at a particular frequency is of sufficient amplitude and in phase with the original sound at the microphone a condition of "positive" or "self-enhancing" feedback may exist. (The term "feedback" is often used when describing self enhancing or oscillatory feedback in sound reinforcement systems. We will use the term "feedback" as defined above and will use terms such as "oscillatory feedback" or "self enhancing feedback" when such is specificially intended.) When self-enhancing feedback exists and the amplifier gain more than offsets the room loss, the whole system goes into self-sustained oscillation resulting in a loud "squealing" sound. When the amplifier gain is only slightly less than that required to offset room loss, the reinforced sound becomes "hollow" or is accompanied by a "chiming" or "ringing." A sound system operated with the amplifier gain high enough to produce ringing or squealing is of no use for sound reinforcement. In fact, depending on circumstances, the amplifier gain must be set from 2 to 6 dB below the point at which ringing would result to provide a safety margin if the system is to be truly useful.

The talker will produce a given sound level at the listener position in Figure 22.1 when the sound reinforcement system is not in use. A sound reinforcement system can be used to provide higher sound levels at a listener position than would be produced by the talker with no reinforcement. **Acoustical gain** is the increase in sound level at a listener position when the system is on, as compared with the level when the system is off. The required acoustical gain is just the difference between the desired sound level at a given location and the level produced by the sound source without reinforcement. However, the amount of acoustical gain that a system can produce before it starts ringing is limited in a number of ways. The sound level required of the loudspeaker is affected by the loudspeaker-to-listener

147

distance (DLL), as shown in Figure 22.1. Assuming that we are in the open and that no reflecting surfaces are present, each doubling of the distance requires an additional 6 dB from the loudspeaker. The sound level produced at the microphone by the loudspeaker depends on the distance DLM. The shorter this distance, the more likely the system is to ring for a given amplifier gain. The sound level produced at the microphone by the talker depends on the distance DTM. Acoustical gain is increased as the sound level produced by the talker is increased over that produced by the loudspeaker at the microphone. This means that DTM should be made short and DLM should be long. The use of a directional loudspeaker and a directional microphone can also help to reduce loudspeaker sound picked up by the microphone.

System Component Characteristics

Of the basic components of a sound reinforcement system, the loudspeakers are the most critical for achieving a desired result. Three different types of loudspeaker arrays are found in commercial sound systems: (1) cone-type loudspeakers, (2) column arrays of cone loudspeakers, and (3) horn radiators. Cone loudspeakers produce a uniform distribution of energy over the mid and low frequency range, but are fairly directional at high frequencies. A column array is a set of stacked cone loudspeakers. If the speakers are stacked vertically, the effective vertical dimension of the speaker is increased, while the horizontal dimension remains unchanged. Because of the greater "vertical size," diffraction is reduced and the column becomes quite directional along the vertical axis, while still distributing sound over a large horizontal area.

Horn radiator loudspeakers utilize a diaphragm driver at the small end of a long flaring tube. Horns are usually high-efficiency radiators which can be made quite directional. They are available in a wide variety of flares, shapes, and sizes. When space is a problem, a "folded horn" may be used. Multicellular horns consisting of several horns attached together on one driver have been used to provide wide angle coverage at high frequencies, but they are being phased out of current design in favor of constant directivity horns. Several horns may be stacked together to provide either wider or narrower coverage than is possible from a single horn. "Stack and splay" techniques can be employed in order to minimize interference effects and maintain a uniform distribution of sound.

Loudspeakers have been designed and built with a wide range of directional characteristics which are often expressed in terms of angular coverage in each of two directions. Figure 22.2 illustrates a biradial horn loudspeaker flared in two directions—tailoring the angular coverage to a nominal 90° by 40° pattern. The angular coverage is often specified as the **beamwidth** (in both the horizontal and vertical directions) within which the SPL is no lower than 6 dB from its on-axis value. The beamwidths of the biradial horn are illustrated in Figure 22.3 as they vary with frequency. We see that the beamwidth is nearly constant over a wide frequency range (about 500-10,000 Hz). **Isobar plots** of the type shown in Figure 22.4 can be useful for purposes of planning loudspeaker installations. If one imagines that the

Figure 22.2. JBL model 2360 biradial horn atop a ported low-frequency unit. (Courtesy of JBL, Inc.)

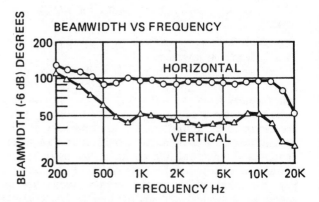

Figure 22.3. Beamwidth plot for model 2360 biradial horn. The plot shows the horizontal and vertical angular coverages within which the sound level is no lower than 6 dB below the on-axis level. (Courtesy of JBL, Inc.)

horn is aimed at the center of the figure, the curves show the angular regions within which the sound level is no more than 3dB, 6dB, and 9 dB lower than it is on axis. Curves of this kind can be superimposed on a seating area to determine whether a given loudspeaker will provide adequate coverage. However, isobar curves become distorted when the horn axis is at an angle relative to the plane of coverage. This must be taken into account when the curves are used for design purposes.

Two other loudspeaker specifications important in system design are axial pressure sensitivity and power handling capacity. **Axial pressure sensitivity** is the SPL that would be produced directly in front of the loudspeaker at a distance of one meter when one electrical watt of power is supplied to the loudspeaker. The

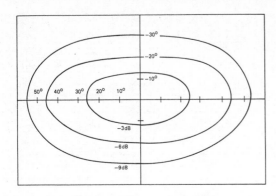

Figure 22.4. Isobar plot for model 2360 biradial horn. See text for further detail. (Courtesy of JBL, Inc.)

power handling capacity of a loudspeaker is the maximum amount of electrical power the loudspeaker is capable of handling without being damaged.

Design of a Simple System

In many respects designing of a sound reinforcement system for outdoor applications is simpler than for indoor use. In outdoor applications one must deal only with the direct sound because there are no reflecting surfaces such as are encountered in a room. However, even in this case a sound system engineer should be consulted in the design and installation of actual systems. The engineer's knowledge of available equipment and experience with equipment in actual installations will be invaluable.

The engineer should consider the needs and desires of the user and reach satisfactory compromises when the user's desires are not possible. The engineer should first determine whether sound reinforcement is necessary or desirable in a given situation. Then a determination should be made of the area to be covered by the radiated sound. Power requirements should then be determined on the basis of the required sound level at some listener position. Meeting the coverage and power requirements depends primarily on loudspeaker specifications, including directional characteristics and axial pressure sensitivity.

As a simple example of system design, suppose that outdoor sound reinforcement is to be provided having an angular coverage of 90° × 40° within which the sound level is no more than 6 dB lower than it is on-axis. The horn illustrated in Figures 22.2-22.4 should provide adequate coverage. The sound level on-axis is to be 100 dB at a distance of 32 m. The horn when equipped with a particular driver produces a sound level of 113 dB at one meter on axis when driven by one electrical watt. Distance is doubled five times in going from 1 m to 32 m with a 6 dB drop in sound level for each doubling of distance (see Chapter 12). Thus, one electrical watt of power input would result in a sound level of 83 dB (113 dB - 30 dB) at a distance of 32 m on axis. We require a 17 dB increase in sound level (100 dB - 83 dB), and since each doubling of electrical power provides a 3 dB increase the power must be doubled about 6 times. Thus, the electrical power input to the horn loudspeaker must be about 64 watts.

Reinforcement for Outdoor Concerts

Each summer, the Metropolitan Opera and New York Philharmonic Orchestra present outdoor concerts to audiences of up to 100,000 people. A stage enclosure projects sound toward the audience, but adequate sound levels are not produced beyond about 15 m. The primary requirement of the sound system is to provide high fidelity sound with sufficient power for the audience beyond 15 m of the enclosure.

The system was designed to cover a trapezoidal area starting 15 m in front of the stage enclosure as shown in Figure 22.5. The area is bounded by a 125 m side nearest the enclosure and a 300 m side farthest from the enclosure as shown in the figure. Two 12 m high loudspeaker towers flank the stage on either side as shown in the figure. The upper 6 m of each tower contains loudspeaker clusters; the lower portion of each tower contains amplifiers and other electronics. The loudspeaker clusters are designed for directional radiation of sound so that the specified area receives adequate sound without "wasting" sound to other areas. The system is designed to have flat frequency response up to 1000 Hz and to decrease 2 dB per octave beyond 1000 Hz. Artificial reverberation can be added to the electrical signals before they are radiated. The system is capable of producing sound levels of 100 dB at a distance of 100 m in front of the stage.

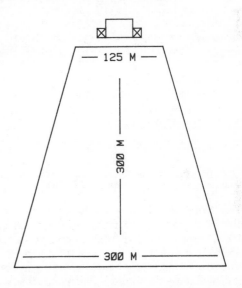

Figure 22.5. Diagram showing coverage of outdoor sound reinforcement system. (After Rosner and King, 1977.)

Design for Indoor Reinforcement

When designing outdoor sound systems one need be concerned only with direct sound from a loudspeaker or some other source. However, both direct sound and reverberant sound must be dealt with when designing systems for enclosed spaces. We will need to draw on background provided in preceding chapters in order to understand the complex interactions among source, listener, room acoustics, and sound system. Figure 22.6 illustrates the relative amounts of direct (solid dots) and reverberant sound (open circles) for a source radiating with a semicircular pattern into a room. We see in the figure that the direct sound always predominates close to the source, but we know it decreases by 6 dB for each doubling of distance from the source. Beyond a certain distance from the source the reverberant sound, which

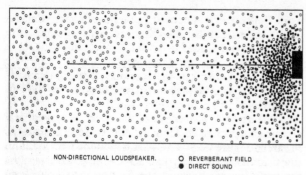

NON-DIRECTIONAL LOUDSPEAKER.   ○ REVERBERANT FIELD
                        ● DIRECT SOUND

Figure 22.6. Direct and reverberant sound fields produced by a non-directional loudspeaker in a room. (Courtesy of JBL, Inc.)

we assume is uniform throughout the room, predominates. **Critical distance** is the distance from a sound source at which the direct and reverberant sounds are equal. Critical distance is affected by the directivity of the source and the listener location in the room as seen in Figure 22.7 for a directional sound source. Critical distance also depends on the absorption of a room; the less the absorption, the shorter the critical distance.

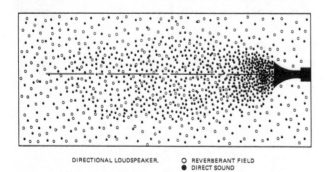

DIRECTIONAL LOUDSPEAKER.   ○ REVERBERANT FIELD
                        ● DIRECT SOUND

Figure 22.7. Direct and reverberant sound fields produced by a directional loudspeaker in a room. (Courtesy of JBL, Inc.)

The "total sound" at some location in a room is the sum of the direct and reverberant sounds at that location. Figure 22.8 shows total sound as it varies with distance from a source. At less than the critical distance total sound level decreases by about 6 dB per doubling of distance because the sound is mostly direct sound. At distances several times the critical distance the total sound level is nearly constant because the sound is mostly reverberant sound.

Most sound reinforcement systems are designed to provide adequate speech intelligibility, which is dependent upon four primary factors. The first is that of talker characteristics, including enunciation, speech rate, and any accents. The second factor is that of listener characteristics, including age, interest, and condition of hearing. The third factor is the sound level of the speech relative to background noise. The final factor is the reverberation time of the room which affects speech intelligibility in two ways. A decrease of intelligibility results when one spoken syllable overlaps the next due to a RT that is too long. At an average speech rate of 3 syllables per second a RT of less than 1.5 second does not seriously degrade intelligibility.

The ratio of direct sound to reverberant sound defines a type of signal-to-noise ratio for determining

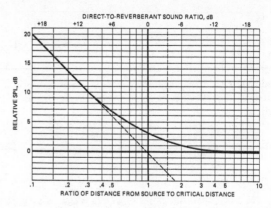

Figure 22.8. Relative sound level (solid curve), direct sound level (dashed line), and reverberant sound level (solid line at 0 dB) versus distance from source expressed in units of critical distance. (Courtesy of JBL, Inc.)

the intelligibility of speech. A moderate amount of reverberation is advantageous because it increases sound levels somewhat, but too long a RT results in too much reverberant sound thus decreasing the signal-to-noise ratio and lowering intelligibility. Practical experience has shown that in rooms having moderate RT (less than 2.0 s), good speech intelligibility is maintained if the direct sound level is no lower than 12 dB below that of the reverberant sound. This ratio of direct to reverberant sound occurs at a distance of four times the critical distance. A specific goal for a sound reinforcement system designed for speech is to provide intelligibility of 85% or more for consonants. Since language is not random, but consists of familiar words and phrases with contextual constraints, a consonant intelligibility of 85% implies that only three words out of a hundred will not be understood.

Perhaps the single most important decision that a sound system engineer must make is where to place loudspeakers to best achieve the desired result. If adequate coverage of the audience can be achieved with a single source, then the loudspeaker cluster should be on the centerline of the room above the lecturer. Since the microphone is far off the main axis of the loudspeaker, it is generally beyond the critical distance and thus in the reverberant field. This is considered desirable because the microphone can be moved about without disturbing the system, and the loudspeaker is kept close to the lecturer so he is not bothered by the delayed sound of his own voice. Although the loudspeaker array is above the lecturer, listeners perceive the amplified sound as an enhanced direct sound because sound in the vertical plane is not localized very well. A sound emitted from a point 10 meters above a lecturer's head still seems to be emanating from his mouth.

The references at the end of this chapter are a source of much additional information on sound system design. Topics covered include multiple arrays of loudspeakers, directional properties of loudspeakers, and isobar methods for planning loudspeaker arrays.

System Installation and Regulation

In sound system installation, microphones should be placed far enough from loudspeakers so that they are in the reverberant sound region. They should also be placed so that direct sound from the loudspeakers does not produce ringing. Several things can be done to improve the system's useful acoustical gain—which is limited by the strength of the reverberant sound at the microphone position. A directional microphone can be used to discriminate against the reverberant field. Directional loudspeakers can be used to direct sound to the audience where it is absorbed and thus contributes less to reverberant sound. Negative feedback can be provided electronically to offset positive feeback in the system.

For a talker at a podium, placement of the microphone some 30-50 cm from the talker should be adequate. Use of a directional microphone is desirable to reduce pickup of audience noise. Several microphones may be required for use with large performing groups. They should be placed several meters from the nearest performer (except in circumstances where one performer is to be emphasized above the group, as in the case of a soloist), so that a blended response is obtained over the whole performing group. Microphones used with theatrical productions are usually placed in special mountings on the floor at the front edge of the stage, although other placements to achieve special effects may be employed. As few microphones as possible should be used since the gain must be lowered 3 dB with each doubling of the number of microphones.

As noted above, loudspeakers should be placed so as not to produce much sound that is picked up by the microphones. Loudspeakers placed at the front of a hall and close to the source of sound tend to produce a better illusion of the sound coming from the original source. Loudspeakers placed farther from the source of sound and closer to the listeners may produce sound that will arrive at the listener's position before the sounds coming directly from the source. This is because electrical signals travel at near the speed of light which is much faster than the speed of sound in air. This may well create the effect of an apparent source of sound at the loudspeakers rather than at the actual source. Even worse, if the difference in arrival times between sound from the loudspeaker and sound from the actual source is great enough, echo will be perceived. The problem can be reduced by delaying the sound that goes to the loudspeakers, by means of digital delays.

It is possible to create loudspeaker arrays that will beam the sound into selected parts of the hall—which may be useful in concentrating the sound on the audience and avoiding undesirable reflections from back walls or ceilings in the hall. In general, a loudspeaker array with a large width will produce a beam of sound with a narrow width (see Chapter 12).

A sound system should be designed and installed to respond uniformly at all frequencies. Otherwise, the system will start ringing at those frequencies with the greatest response and the useful gain will be limited by these frequencies. After the system has been installed, the final important step is to equalize the sound system to provide uniform frequency response. To equalize a sound system is to adjust its frequency response to compensate for the non-uniform combined system and room response. When a sound system has been properly equalized, the resulting sound will be more natural and the system will yield greater acoustical gain without feedback. Equalization is accomplished by attenuating or amplifying the signal as it is passed through a set of band-pass-filters. An example of equalization will be given in the next section.

Reinforcement in a Large Enclosure

The Brigham Young University Marriott Center is a multipurpose arena with a seating capacity of 23,000. A high quality sound system was designed to accommodate athletic events, student assemblies, variety shows, and pop music concerts. The interior volume of the arena is 180,000 m³; consequently a primary requirement of the sound system is to provide adequate acoustical power. The system is adequate from that perspective, as it can produce 23 W of acoustical output. This output results in sound levels of 100 dBA at any seat in the arena. A single loudspeaker cluster is suspended above the floor at the position indicated in Figure 22.9. The off-center placement of the cluster helps to avoid sound focusing that usually occurs in an arena of this shape.

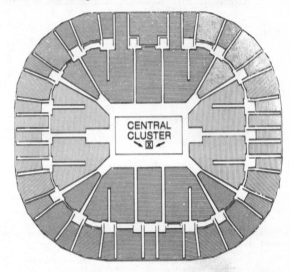

Figure 22.9. Floor plan of Marriott Center showing location of loudspeaker cluster relative to playing floor and spectator area.

The response curve shown in Figure 22.10 was obtained by exciting the Marriott Center sound system with third octave bands of noise and averaging the sound levels from several microphone locations. The combined arena and sound system responded too strongly at third octave center frequencies of 125, 315, 800, and 1600 Hz. A collection of variable-gain filters was incorporated into the sound system to equalize this nonuniform response so as to correspond to the preferred curve shown in the figure. Two common methods of equalization—narrow band and broad band—were used. Ringing modes are single frequency modes. Narrow band filters several hertz in bandwidth were tuned to the combined room and sound system responses below 800 Hz. Third octave filters were adjusted to provide broad band equalization above 800 Hz. The final response curve, shown in Figure 22.11, is very much like the empirically determined preferred curve. It has a measured flat response within a few dB up to 1250 Hz and then falls off smoothly. The Marriott Center sound equalization system is stable and effective over a wide range of audience sizes, temperature variations, microphone placements, and microphone types. It is tolerant of a

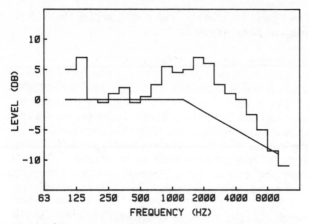

Figure 22.10. Measured Marriott Center response curve before equalization. The preferred curve is flat up to 1250 Hz and decreases by 2 dB per octave above 1250 Hz. (After Boner and Jones, 1975.)

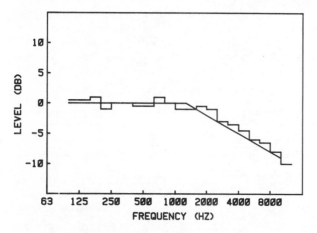

Figure 22.11. Measured Marriott Center responsese curve after equalization. (After Boner and Jones, 1975.)

limited use of tone controls and artificial reverberation devices.

Assisted Reverberation in a Concert Hall

Sound reinforcement must provide sufficient power in halls seating more than about 3000 people. Sound reinforcement is employed rather uniquely in the Royal Festival Hall in London which seats 3000 people. The RT in the finished hall was found to be much too short at low frequencies. A sound reinforcement system was designed to lengthen the RT in a range from about 60 Hz to 700 Hz.

We will now see how sound reinforcement can affect RT by referring to Figure 22.1. When the amplifier is turned off, the RT of the room is determined by the rate of energy loss in the room. The amplifier could be turned on with a gain to offset room losses. Were this to be done, a sound started in the room would continue "forever" because the electrical gain of energy would just offset the loss of acoustical energy. The RT would be infinite because the sound would never die away. If the amplifier gain were turned down slightly, the RT would still be very long because the electrical gain of energy would almost offset the acoustical loss. The RT could be made to have any value between that of the room alone and an infinite value by adjusting the amplifier gain.

The assisted reverberation design for the Royal Festival Hall uses many "channels," each consisting of a microphone, amplifier, and loudspeaker. Both acoustical and electronic filters are used with each channel so that each channel handles only a narrow frequency bandwidth. The frequency spacing between channels varies from about 2 Hz at low frequencies up to about 5 Hz at the upper frequencies. About 170 channels are required to cover the frequency range from 60 to 700 Hz. The microphone and loudspeaker for each channel are placed at antinodal positions for their particular room mode, strongly coupling them into the mode and to each other. The RT for a particular frequency can be increased to any desired value by controlling the amplifier gain of the appropriate channel. Measured RT, both with the system on and with the system off, are shown in Figure 22.12. It is clear from this figure that low frequency RT has been increased from unacceptable values to near optimum values. After assisted reverberation was added, complaints from the press about the "dry" acoustics of the Royal Festival Hall almost stopped. One conductor remarked that the acoustics of the hall had become among the finest in the world.

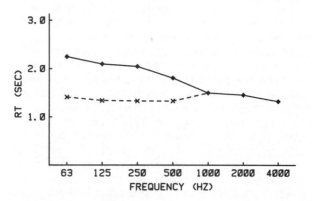

Figure 22.12. Reverberation times of Royal Festival Hall alone (dashed) and Hall with "assisted resonance" (solid) for full hall. (After Parkin and Morgan, 1970.)

Reinforcement in Rooms for Worship

In addition to providing adequate sound power, sound systems are often used in auditoriums and halls for worship to overcome acoustical deficiencies. Some auditoriums and cathederals have such long RT that speech intelligibility is severely reduced. In such cases, sound systems may be used to increase the ratio of direct sound to reverberant sound. Two basic systems—central clusters of speakers and distributed speakers—are commonly used.

A central cluster of speakers, as shown in Figure 22.13a, is the most cost effective system. The cluster is usually mounted above the speaking position to provide the illusion that the sound is coming from the person speaking. The cluster should not be mounted too far from the microphone or an echo may be produced. The cluster should direct the sound into the audience where it is absorbed, thus avoiding sound reflections from walls and ceilings. Such reflections produce unwanted echoes. This approach is especially useful for churches and multi-use auditoriums where there may be a conflict between the optimum RT for music and for speech. A large room RT may be tolerated, thus rendering the acoustics for the unamplified music reasonably live. Speech is made intelligible through amplification and directional radiation. With this approach many churches in America, designed with short RT in deference to the spoken word, could improve both the sound of the organ and the congregational singing without reducing speech intelligibility.

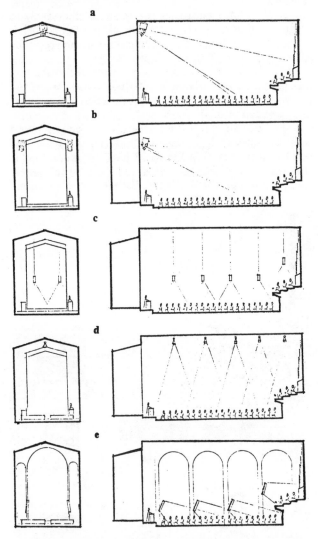

Figure 12.13. Sound systems for reverberant rooms: (a) central cluster; (b) two-cluster system; (c) distributed cone loudspeakers; (d) distributed horns; (e) distributed column loudspeakers. (From Klepper, 1970.)

A variation of the cluster system is the two cluster system shown in Figure 22.13b. This system is useful in rooms where a single cluster cannot feasibly be installed. For example, in a room, where speech originates from two locations such as a lectern and a pulpit, with each microphone having its own loudspeaker array, a single-cluster system would not be appropriate. In two cluster systems both clusters should not be used at the same time when the clusters are spaced more than 15 m apart.

Distributed loudspeaker systems are used in long rooms with low ceilings and under balcony areas for which cluster systems cannot be designed to provide adequate direct-to-reverberant sound. In a conventional distributed system, shown in Figure 22.13c, loudspeakers are typically mounted in the ceiling and no more than 5 m above the floor. Other distributed systems, such as the directional horn loudspeaker in Figure 22.13d, can be designed to meet special requirements. Mounting high above the floor is one such requirement. The horns should be designed and installed to radiate into very restricted areas and should be controlled by separate switches so that they can be turned off when there is no audience in their area. In distributed systems, the loudspeakers must be spaced close enough to provide adequate coverage of the listening area. The loudspeakers should provide good high frequency coverage off axis so that those listeners not directly under a loudspeaker will still receive adequate high frequency sound.

A basic difficulty with distributed systems is that the live sound arrives after the amplified sound. This can cause problems with localization of the source (which may appear to be the loudspeaker) and can reduce speech intelligibility. The problem can be alleviated and a more natural sound produced by electronically delaying the amplified sound with an audio delay system. The delay system should provide successively longer delays for the loudspeakers located successively farther from the source.

Although they are generally less preferred, distributed columnar loudspeakers (see Figure 22.13e) can be used in long narrow spaces where there are side columns to which the loudspeaker arrays may be attached. The columnar arrays should be mounted no farther than 12 meters apart and tilted to adequately cover the proximate seating. The use of a delay system is essential with a distributed columnar system.

Finally, a pew-back distribution system may be used where the other systems cannot be installed due to architectural or other limitations. This system consists of a large number of small cone loudspeakers mounted approximately every meter and high on the back of the pew. For high intelligibility and directional realism, a delay unit is mandatory. A major drawback of the pew-back system is its cost.

Rock Music Systems

Although sound systems designed for rock musicians may be elaborate and expensive, they differ from most other systems in that they must be portable. In one design philosophy for rock musicians the mixing of sound is controlled by a person in the audience area who adjusts the total sound for the desired effect. The individual artist does not control the outcome of the final product, but the resulting sound can be maintained uniformly from performance to performance. An example of one such design for a soft rock group is shown in Figure 22.14. The group consists of four musicians (lead guitar, bass, drum, and keyboard) and a mixer who controls the final blend from the audience or dance floor. The lead guitar is miked from a cabinet speaker since the cabinet produces part of the desired sound effect. The bass and keyboards feed directly to the mixer, while four mikes are used for the drums (bass, snare, and hi-hat), cymbals, and tom-tom. In addition, there are four mikes for vocals, as each musician also sings. All mikes and lines from the instruments are patched into a multi-conductor cable called a snake which connects to a multi-input mixing console. The mixing console can vary the frequency response or can add echo, reverbation, and exotic effects such as phasing distortion and pitch alteration. It is used principally to blend each source into a composite signal which is fed to the power amplifier and then to the loudspeakers. Often separate "mixes" and blends are sent to stage monitor loudspeakers so the performers can hear each other.

Another design philosophy is that the sound system not only provides high-quality amplified sound, but becomes an extension of the performer and his instrument. Each performer is immersed in the total acoustical output from the whole group through extensive use of stage monitors. In this way the performing artist alone controls the balance and blend of the sound without a mixer.

In one such system, traditional acoustical instruments with microphone pickup are avoided to reduce the chance of oscillatory feedback. Instead, competely non-acoustic guitars with magnetic pickups are

utilized and vocals are miked using highly directional microphones with variable sensitivity. (Refer to Chapter 40 for additional detail.) Intermodulation problems between instruments are eliminated by using separate sound systems for each instrument. Each artist performs and blends as a coordinated part of the ensemble on the basis of his immersion in the total sound. The audience hears the combination of performers plus sound system as a musical instrument in its own right.

Exercises

1. At what frequencies is a sound system likely to "squeal" due to acoustical feedback if such squeal is primarily determined by a loudspeaker to microphone distance of 10 m? Refer to Figure 22.1.

2. The outdoor sound system described in the text can produce a sound level of 100 dB at a distance of 100 m. What sound level would be produced at 200 m under the same conditions?

3. Estimate the surface area into which the sound will be spread at a distance of 100 m from the loudspeaker clusters in Figure 22.5. If the sound level is 100 dB at this distance, what total acoustical power passes through the surface? Estimate the electrical

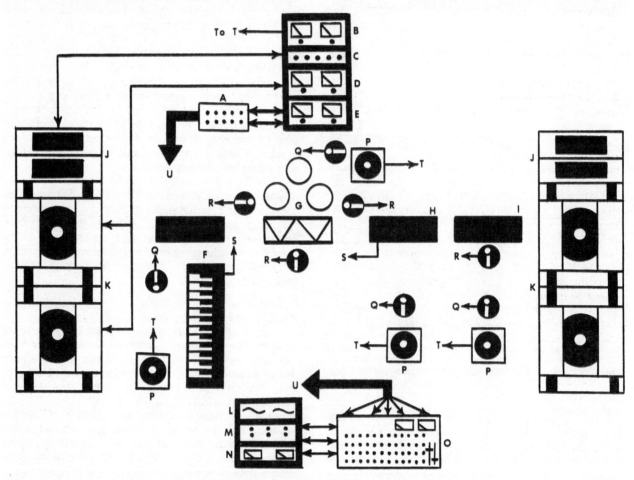

Figure 21.14. Sound reinforcement system layout for the soft rock group "Windfall" (Maryland, West Virginia, Pennsylvania). All instruments and vocals are mixed through the main power amplifier with final volume and tonal qualities controlled by the sound technician situated at some distance in front of the band. A. Multi-cable connector ("snake"). B. Monitor amplifier (200 watts). C. Electronic crossover. D. Hi-frequency amplifier (100 watts). E. Lo-frequency amplifier (800 watts). F. Keyboards and amplification system. G. Drums. H. Bass guitar amplifier. I. Lead guitar amplifier. J. Hi-frequency horns. K. Lo-frequency speakers. L. Graphic equalization. M. Delay unit. N. Compressor/limiter. O. Mixing board. P. Monitor speakers. Q. Vocal microphones. R. Instrument microphones. S. Direct instrument outputs to "snake". T. Speaker connections from amplifier. U. Main cable connections between "snake" and mixing board.

signal power supplied to the loudspeakers if they are 10% efficient in converting electrical power into acoutical power.

4. The RT in the Marriott Center is 4 s. What is the average AC of its surfaces if one assumes its shape is rectangular with a height of 14 m?

5. By how many dB does the third octave at 1600 Hz exceed the preferred response in Figure 22.10?

6. About 3900 W of electrical power are required to produce 23 W of acoustical power in the Marriott Center. What is the efficiency of the sound system?

7. How do the assisted RT in Figure 22.12 compare with optimum RT discussed in Chapter 21 for a concert hall with a volume of 25,000 m^3?

8. Suppose two loudspeaker clusters are mounted next to the walls in a church with floor dimensions of 40 m by 20 m wide. Will a person sitting next to a wall and 10 m back from one cluster hear "echo" if both clusters radiate the same signal?

9. If distributed loudspeakers are mounted 7 m apart from front to rear in a hall, how much should the signal be delayed in time between neighboring loudspeakers?

10. Suppose a digital delay line is used to provide the delay of Exercise 9. In the digital delay the signal is sampled and the samples are stored in memory. There must be enough memory to store a number of samples equivalent to the time delay. How much memory is required if the sampling rate is 20,000 samples per second?

11. If a tape system (now obsolete) is used to provide the time delay of Exercise 9, what must the tape speed be if the record and playback heads are 1 cm apart?

12. A listener is situated a distance of 100 m from a sound source where the critical distance is 25 m. How many dB below the reverberant sound will the direct sound be?

13. A room has an excessively long reverberation time so that speech intelligibility is poor. Would it be better to treat the room with absorbing panels to reduce the RT, or to install a directional sound system. Explain the advantages and disadvantages of each approach.

14. A room has a volume of 2600 m^3 and total absorption of 400 m^2. Approximately what acoustical power is necesary to achieve a sound level of 80 dB? (Refer to Chapter 20.) What electrical input power is required if the speakers are 10% efficient?

15. For a distributed sound system in a long, low ceilinged room, explain why the loudspeakers should not be placed on opposite side walls.

16. Explain the advantages and disadvantages of using horn loudspeakers in auditorium sound systems. Why are they seldom used in home hifi systems?

17. Which rock music system would be less expensive, and therefore preferable, for a small rock group? Explain.

18. Which of the two rock music systems discussed is more likely to achieve its design philosophy?

Demonstrations

1. In a room, set up a sound reinforcement system consisting of microphone, amplifier, and loudspeaker. Experiment with amplifier gain to see how it affects feedback resulting in squeal.

2. Move the microphone and loudspeaker of Demonstration 1 to different locations to see how feedback depends on location.

3. Gain access to halls in which sound systems have been installed to meet different needs. Experiment with different adjustments of the sound system if possible to see how they affect the sound. Explore the effects of signal delay and equalization if possible.

Further Reading

Rossing, Chapter 24

Altec. 1977. **Altec Training Manual** (Altec Corp.).

Boner, C.R., and E.S. Jones. 1975. "The Equalization and Subsequent Observation of the Sound System in the Marriott Center at Brigham Young University," J. Audio. Eng. Soc. **23**, 386-389.

Davis, D., and C. Davis. 1975. **Sound System Engineering** (Howard Sams).

Eargle, J., and G. Augspurger. 1982. **JBL Sound System Design Reference Manual** (JBL, Inc.)

Klein, W. 1971. "Articulation Loss of Consonants as a Basis for the Design and Judgment of Sound Reinforcement Systems," J. Audio Eng. Soc. **19**.

Klepper, D.L. 1970. "Sound Systems in Reverberant Rooms for Worship," J. Audio Eng. Soc. **18**, 391-401.

Klepper, D.L. 1978. **Sound Reinforcement: An Anthology** (Audio Engineering Society).

Mathews, W.C. 1982. "Church Sound at 105 dB SPL," db Magazine.

Parkin, P.H., and K. Morgan. 1970. "'Assisted Resonance' in the Royal Festival Hall, London: 1965-1969," J. Acoust. Soc. Am. **48**, 1025-1035.

Peutz, V.M.A. 1971. "Articulation Loss of Consonants as Criterion for Speech Transmission in a Room," J. Audio Eng. Soc. **19**, 915.

Rosner, A., and L.S. King. 1977. "New Mobile Sound Reinforcement System for the Metropolitan Opera/New York Philharmonic Orchestra Park Concerts," J. Audio Eng. Soc. **25**, 566-571.

Sands, L. 1977. **Sound System Installers Handbook** (Howard Sams).

V. The Human Voice and Speech

Courtesy of Public Communications, Brigham Young University.

23. Energy Types in Speech Production

As shown in Figure 23.1, the human vocal mechanism can be thought of as consisting of an energy source, a mechanism for interrupting the air stream, and a variable resonator for creating different sound spectra. The lungs' reservoir of air as driven by surrounding musculature is the energy source. The air can be expelled to provide an air stream which flows through the rest of the vocal system. The air flow from the lungs may be interrupted by the vocal folds or by constrictions in the vocal tract. The variable resonator is the vocal tract—the effects of which will be considered in the next chapter.

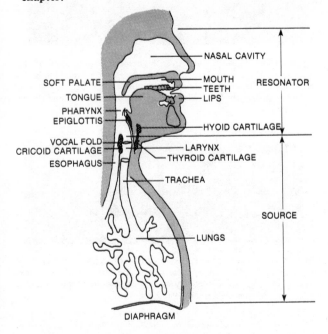

Figure 23.1. The speech production mechanism.

Anatomy and Function

The following definitions are necessary for further discussion of speech production. The **diaphragm** is a muscular structure at the bottom of the chest cavity used to control pressure in the chest cavity surrounding the lungs. The **lungs** form a passive air reservoir. The **trachea** is a cartilagenous tube that carries air between the lungs and the pharynx. The **larynx** is a structure of cartilage at the upper end of the trachea containing the vocal folds. Several cartilages—thyroid, cricoid, arytenoids, and hyoid—as well as associated musculature are incorporated in the larynx. The **esophagus** is a muscular tube behind the trachea and is used for the transport of materials from the pharynx to the stomach. The **epiglottis** is a cartilage flap used for covering the tracheal opening. The **pharynx** forms the

lower part of the vocal tract, connecting the mouth and nasal cavity to the trachea. The **hard palate** is the roof of the mouth. The **soft palate** (or velum) is a muscle used as a valve to open and close the nasal cavity.

Several parts of the speech production mechanism serve dual functions. For example, when you eat, food must pass through the mouth and pharynx into the esophagus. It is important that the epiglottis be lowered to cover the larynx so that food does not enter the air pathway to the lungs. When you breathe, the epiglottis is raised and the vocal folds must be wide open to permit free flow of air to and from the lungs.

When a person speaks, air is forced from the lungs and flows through the trachea, then through the opening between the vocal folds, then through the pharynx and mouth (and optionally, the nasal cavity) to the outside. The energy required to produce speech sounds is supplied by exhaling air from the lungs. However, exhaling alone is not adequate to produce useful speech sounds. To be audible, the otherwise steady flow of air must be converted into a pulsating flow. The interruption of the air flow for speaking purposes is accomplished in one of three ways: (1) by using the vibrating vocal folds to interrupt the airstream in a periodic manner; (2) by forming a constriction in the vocal tract, causing the air stream to become turbulent and noisy as it flows through the constriction; or (3) by completely closing off the vocal tract (thus stopping the air flow) and then releasing the air pressure suddenly. The energy types produced by the above means are labeled **voice energy**, **noise energy**, and **burst energy**, respectively. It is also possible to get a mixture of voice and noise energy by vibrating the vocal folds while constricting the vocal tract.

The **vocal folds** can be thought of as small "laminated" sheets of muscle, ligament, and mucosa. The two vocal folds meet each other at the front of the larynx and are attached to the thyroid cartilage as shown in Figure 23.2. (The thyroid cartilage produces the protrusion in the neck, most apparent in adult males, commonly called the Adam's apple.) Each fold is attached to an arytenoid cartilage at the back of the larynx. The arytenoids are normally positioned widely separated from each other to permit breathing, as can be seen in the left part of Figure 23.3. They can, however, be brought into contact with each other by the action of the interarytenoid muscles connecting them. This brings the vocal folds together—touching each other in many cases—for the production of voice energy. This process can be modeled by putting your hands in front of you with the palms down. Place the tips of the index fingers together (at what would be the inside of the Adam's apple). Each hand with the fingers close together represents a vocal fold. Each thumb represents an arytenoid cartilage. When the thumbs are extended and the thumb tips touch each other the configuration is that for breathing (left part of Figure 23.3). When the thumbs slide past each other to close the opening the configuration is that for voicing (right part of Figure 23.3).

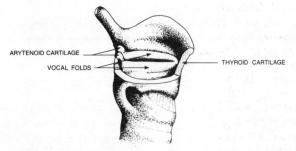

Figure 23.2. Cut-away view of the larynx showing the basic structures. (After Farnsworth, 1940.)

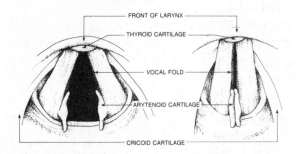

Figure 23.3. Simplified overhead view of the vocal folds shown in position for breathing on the left and in position for voicing on the right. (After Titze, 1973.)

Several muscles help to configure and control the vocal folds. Each thyroarytenoid muscle, as the name implies, is attached to the thyroid cartilage and an arytenoid cartilage. The thyroarytenoid muscle serves to tense and thus shorten the vocal folds. The inner portion of the thyroarytenoid muscle—the vocalis muscle—is the vibrating muscular portion of the vocal folds. The vocalis muscle is overlaid with a ligament layer which is in turn covered with a mucosal layer. The cricothyroid muscle tenses the vocal folds by moving the thyroid cartilage relative to the cricoid cartilage. Cricoarytenoid muscles also help to control position and tension in the vocal folds.

Vocal Fold Action

High speed photography is widely used for studying the vocal folds in action. Figure 23.4 illustrates a typical high-speed photography setup used to study vocal fold action. The high-speed photograph sequence appearing in Figure 23.5 illustrates the vocal folds as they are open for breathing and then as they close for voicing. Figure 23.6 illustrates several phases in one cycle of vocal fold vibration when voicing.

Study of many important features of vocal fold vibration can be simplified by representing each fold as a simple vibrator with associated mass, tension, and resistance. Since the vocal folds have resistance, their motion will tend to die out unless energy is supplied to them. The air "driving" them from below and rushing between them supplies the needed energy. The opening

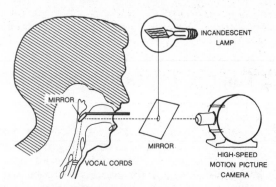

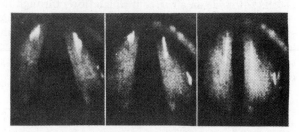

Figure 23.4. High-speed photography method of observing vocal fold motions. (After Farnsworth, 1940.)

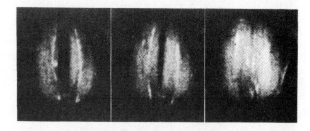

Figure 23.5. Frames obtained from high-speed photography of the vocal folds. The vocal folds are shown wide open for breathing in the left frame. In the middle and right frames the glottal opening is shown becoming progressively smaller in preparation for voicing. (From Lieberman, 1977.)

Figure 23.6. Vocal folds during three different stages of one cycle of voicing. The glottal opening is largest in the left frame. The vocal folds are in contact—completely closing the glottal opening—in the right frame. (From Lieberman, 1977.)

between the vocal folds is called the **glottis** or **glottal opening**. When the glottis is very small, the air tends to force the folds apart, but then as the glottis gets larger the rushing air creates decreased pressure between the folds (due to the Bernoulli effect) and pulls them back together. Figure 23.7 illustrates an idealization of how the Bernoulli force acts on the vocal folds. The Bernoulli force acts in an almost impulsive way twice during the vibration cycle. As the folds open, the Bernoulli force acts in the opposite direction and so takes energy from the folds. As the folds close, the Bernoulli force acts in the same direction and adds energy to the folds—enough to offset the energy taken from them on opening and the energy lost to resistance. The asymmetry we see in the Bernoulli force—larger on closing than on opening—is important to maintain the vibration. (The asymmetry in this simple model is due to the inertia of the air in the glottis. Air flows more slowly when the glottis is opening than when the glottis is closing because time is required to accelerate it.) The muscular tension in the folds also serves to pull them back toward their rest positions. The interaction of the inertia and tension of the folds, and the pushing and pulling of the air stream serves to produce and maintain an oscillatory motion of the folds. The resistance and nonlinear tension of the folds tend to prevent the amplitude of the oscillatory motion from becoming too large.

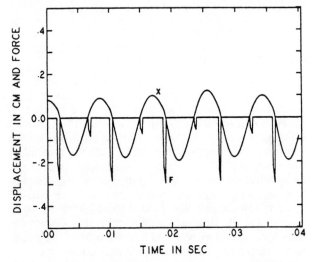

Figure 23.7. Simulated oscillation of a simple mass and spring vocal fold model. The curve labelled X represents vocal fold displacement. The curve labelled F represents an impulse-like Bernoulli force. (From Titze, 1980.)

Two primary variables serve to control the vibration frequency of the vocal folds: (1) tension in the vocal folds, and (2) mass of the vocal folds. The muscular control of the vocal folds is very complicated and not completely understood at the present time. However, researchers have developed mathematical models that are capable of modeling various aspects of observed vocal fold behavior. Simplified, the tension in the vocal folds is increased when the muscles cause the folds to be stretched. This stretching may also produce a slight reduction in effective mass by "locking out" part of the vocal fold mass from the vibratory motion. In the simplified view of the folds as mass and spring vibrators, increased tension and decreased mass will both increase the vibration frequency. In various studies it has been noted that an increase in tracheal air pressure results in increased vibration frequency of the folds. Recently it has been convincingly argued that this is a secondary effect. Larger amplitude vibrations occur when the pressure is increased; the effective tension in the vocal folds increases with amplitude. Hence, frequency increase with pressure is actually due to increased effective tension.

An interesting feature of the vocal folds that seems to contradict our general description of vibrating systems is that the folds tend to lengthen for the production of high-frequency sounds. For a normal male talker the folds are short and thick for the production of low frequencies (about 9 mm long and 5 mm thick for a frequency of 125 Hz) and comparatively long and thin for the production of high frequencies (about 18 mm long and 3 mm thick for a frequency of 250 Hz). As seen in Chapter 10, longer strings produce lower frequencies than do otherwise identical shorter strings. However, in the case of the vocal folds tension is increasing along

with vocal fold lengthening. The increased tension tends to produce higher frequencies, more than offsetting any tendency to produce lower frequencies due to increased length.

Other basic features of simple vibrators, however, do hold true when comparing the vocal folds of adult males, adult females, and children. In general, male vocal folds are both longer and more massive than female vocal folds, which in turn are longer and more massive than the vocal folds of children. Furthermore, the range of fundamental voicing vibration frequencies produced by adult males (about 80-240 Hz) is lower by about a factor of two than is the range of fundamental voicing vibration frequencies produced by adult females (about 140-500 Hz)—which is in turn lower than the corresponding frequencies for children. (Note: ranges given are appropriate for speech. Ranges for singing would be more like 80-700 Hz and 140-1100 Hz for adult males and females, respectively.) The relationship that shows more massive simple vibrators to have lower vibration frequencies than less massive ones is consistent in this case considering the different masses and frequencies of vocal folds for males, females, and children. The relationship as derived for strings between frequency and length also holds in a qualitative sense for vocal folds—the shorter female vocal folds have a higher frequency than do the longer male vocal folds.

The single mass vocal fold model represents many important features of vocal fold action, but is still deficient in several ways—it represents the glottis as rectangular, it does not represent motion of the folds along the tract, it cannot represent wavelike behavior of the vocal folds. Figure 23.8 illustrates the behavior of a many mass model of the vocal folds. Note the two views of the folds—one looking down from above and the other looking along the folds—at several different phases of a glottal cycle. In the upper set of curves the model has been adjusted to produce "normal" voicing. In the middle set the glottal opening was decreased in a way typical of "creaky" voicing. In the lower set the glottal opening was increased in a way typical of "breathy" voicing.

Glottal Air Flow

The valving action of the vocal folds results in an

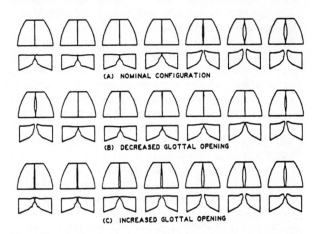

Figure 23.8. Computer generated vocal fold movement in horizontal and vertical cross sections for three adjustments of vocal fold model parameters. (From Titze and Talkin, 1980.)

approximately periodic flow of air through the glottis. The volume of air flowing through the glottis per unit of time is called the **glottal air flow**. The glottal air flow is generally not sinusoidal, but contains energy at many harmonics of the fundamental frequency in addition to the energy at the fundamental. The harmonic energy arises from the nonlinear behavior of air flow through the glottis. Figure 23.9 illustrates air flow (in cm^3/s) through the glottis for the three glottal conditions of Figure 23.8. If we were to observe the glottal opening we would find that it changes with time in a more nearly symmetrical way than does the glottal air flow. This is due in part to the inertia of the air in the glottis which takes time to accelerate when the glottis opens. The resulting glottal air flow waveforms are rather asymmetric.

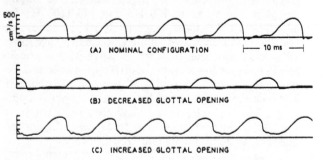

Figure 23.9. Glottal air flow waveforms for the three adjustments of the vocal fold model corresponding to Figure 23.8. (From Titze and Talkin, 1980.)

Glottal air flow does not vary sinusoidally for one or more of the following reasons: (1) The area of the glottis itself does not vary sinusoidally. This occurs because the forces on the vocal folds are nonlinear and not sinusoidal. For example, when the folds collide on closing, the forces suddenly (and nonlinearly) become much larger. Furthermore, when the vocal folds open wide more force is required to stretch them than when they are less open. When the vocal folds are driven with higher pressures, they open and close more abruptly, the motions are more violent, and the vocal folds tend to remain closed for larger portions of the whole cycle. These features give rise to a glottal area that changes in a way very different from a sinusoid. (2) The airflow does not vary in direct proportion to the glottal area. This is because the resistance of the glottis to air flow changes rather drastically as the glottal area changes. In practice, doubling the area more than doubles the air flow (especially for small areas) if all other conditions are constant. (3) The pressure across the glottis is not constant. Even if we assume the pressure in the trachea to be constant the pressure in the larynx will be fluctuating because of standing waves in the vocal tract. The fluctuating pressure across the glottis gives rise to fluctuations in the air flow.

An idealized spectrum typical of the "normal" glottal air flow of Figure 23.9A is shown in Figure 23.10. In this example the air flow is shut off for part of a cycle. The resulting flow is pulselike. The flow begins rather gradually after being shut off, but then is shut off rather abruptly after reaching some largest value. The spectrum shows a rich harmonic content due to the pulselike nature of the wave.

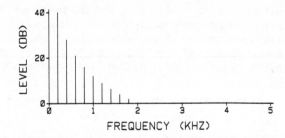

Figure 23.10. Idealized spectrum of glottal air flow.

Standing waves in the vocal tract will affect the pressure across the glottis which in turn affects the glottal air flow. The glottal air flow affects the strength of the Bernoulli force which drives the vocal folds. Strong standing waves in the vocal tract could influence the vibration frequency of the folds, although standing waves that typically occur in the vocal tract do not have much influence on the vocal folds themselves. There is, however, some influence when vocal tract resonances occur at low frequencies. Furthermore, when we come to consider the lips as a similar vibrating system for wind instruments we will find that the tube of the instrument exerts a very substantial influence on the lips.

Other Sound Sources

The voicing source is located at the vocal folds, but noise sources may occur at any one of several different locations in the vocal tract. Any time air is forced through a small or irregular constriction, the smooth flow of the air stream is disrupted and the flow becomes irregular and turbulent. The "hissy" sounds of speech are produced at a noise source. Typical points of constriction in the tract are teeth to teeth, teeth to tongue, and teeth to lips. When the vocal tract is constricted and the vocal folds are also caused to vibrate, a mixture of both voice and noise energy results. The voicing causes alternate puffs of air to pass through the constriction, thus giving rise to repetitive bursts of noise. This procedure is used in the production of voiced fricatives.

It is possible to completely block off the vocal tract while trying to force air from the lungs, increasing the pressure on the lung side of the constriction. If the constriction is suddenly removed, the sudden release produces a burst of energy which is characteristic of many of the unvoiced plosive sounds. A more gentle release of a smaller amount of pressure, typical of the voiced plosive sounds, usually results in a negligible burst of energy.

You can discover for yourself what energy types are used in the production of the various speech sounds by applying the following techniques. Test for the presence of voicing by placing the fingertips gently on the Adam's apple; a slight vibration will be felt when voice energy is present. Test for the presence of noise energy simply by listening for a "hissy" character in the sounds produced. If both of these conditions are satisfied, we say the result is a mixture of voice and noise energy. Look for burst energy by noting whether a large amount of pressure is built up in a constricted tract and, if so, whether or not it is suddenly released.

Classification of Speech Sounds

A convenient concept in dealing with speech is that

of the **phoneme**, defined as a "distinguishable speech sound." The number of phonemes to be used depends to some extent on how finely one wishes to divide the world of speech sounds. The human vocal mechanism is capable of producing an almost infinite variety of different sounds; however, most of these are not readily distinguishable from one another. In the present context we will use the phonemes listed in Table 23.1. Grouping of speech sounds in the table is done partly on the basis of energy type. In later chapters other ways of grouping and classifying phonemes will be discussed.

Table 23.1. Phoneme classification with corresponding symbols used in the text and International Phonetic Alphabet (IPA) symbols. An example of each phoneme used in a word is given in the last column.

	Text symbol	IPA symbol	Example
Vowels	EE	i	beet
	I	I	hit
	E	ε	bed
	A	ae	had
	O	a	hot
	U	U	put
	OO	u	cool
	UH	ʌ	fun
	AE	e	make
Nasals	M	m	me
	N	n	no
	NG	η	sing
Liquids	L	l	law
	R	r	red
Glides	W	w	we
	Y	j	you
Unvoiced fricatives	WH	hw	when
	H	h	he
	F	f	fin
	TH	θ	thin
	S	s	sin
	SH	ʃ	shin
	CH	tʃ	chin
Voiced fricatives	V	v	view
	DH	ठ	then
	Z	z	zoo
	ZH	ʒ	mirage
	JH	dʒ	judge
Unvoiced plosives	P	p	pea
	T	t	tea
	K	k	key
Voiced plosives	B	b	bee
	D	d	down
	G	g	go

Exercises

1. Describe the basic parts of the human vocal mechanism and the functions of each.

2. What is the primary force that causes the vocal folds to vibrate?

3. What controls the fundamental vibration rate of the vocal folds?

4. Tell why you would expect the vibration frequencies of adult female vocal folds to be higher than those for an adult male. Frame your description in terms of a simple vibrator model of the vocal folds.

5. What ranges of fundamental voicing frequency are associated with the speech of adult males, adult females, and children? Are these ranges comparable to the respective singing ranges?

6. What effect does a higher blowing pressure have on the behavior of the vocal folds? How does this affect the spectrum of the vocal fold waveform?

7. What features of vocal fold action are well-represented by the single mass model? What features are poorly represented if at all?

Demonstrations

1. Convince yourself that the act of exhaling in and of itself is not sufficient to produce disturbances that are useful in the speech communication process. To do this, exhale without using the vocal folds and without constricting the vocal tract.

2. Bernoulli force—Place one small sheet of paper on top of another small sheet. Tape them together at one end, leaving a small opening in the middle. Insert one end of a soda straw into the opening and tape around it to prevent air leakage. Blow into the straw so that air flows between the two sheets of paper. What happens? Why?

3. Produce each of the phonemes listed in Table 23.1 and determine its energy type (voice, noise, mixture, burst). Apply the tests described in the text.

Further Reading

Borden and Harris, Chapter 4

Denes and Pinson, Chapter 4

Dickson and Maue-Dickson, Chapters 3, 4

Flanagan, Chapters 2-4

Hall, Chapter 14

Ladefoged, Chapters 1-4, 6, 7, 11, 12

Pickett, Chapters 1, 4, 5

Rossing, Chapter 15

Farnsworth, D.W. 1940. "High Speed Motion Pictures of the Human Vocal Cords," Bell Lab. Record **18,** 203-208.

Flanagan, J.L., and L. Landgraf. 1968. "Self-oscillating Source for Vocal Tract Synthesizers," IEEE Trans. Audio Electroacoust. **AU-16,** 57-64.

Hollien, H., D. Dew, and P. Philips. 1971. "Phonational Frequency Ranges of Adults," J. Speech Hearing Res. **14,** 755-760.

Lieberman, P. 1977. **Speech Physiology and Acoustic Phonetics** (Macmillan Publishing Co.)

Monsen, R.B. and A.M. Engebretson, 1977. Study of Variations in the Male and Female Glottal Wave," J. Acoust. Soc. Am. **62,** 981-993.

Titze, I.R. 1973; 1974. "The Human Vocal Cords: A Mathematical Model, Parts I and II," Phonetica **28,** 129-170; **29,** 1-21.

Titze, I.R. 1980. "Comments on the Myoelastic-Aerodynamic Theory of Phonation," J. Speech & Hearing Res. **23,** 495-510.

Titze, I.R. and D.T. Talkin. 1979. "A Theoretical Study of the Effects of Various Laryngeal Configurations on the Acoustics of Phonation," J. Acoust. Soc. Am. **66,** 60-74.

Audiovisual

1. **Function of the Normal Larynx** (20 min, color, 1956, ILVAD)

2. "Vocal Formants" (REB)

24. Vocal Tract Modifications of Speech Energy

In the previous chapter we listed most of the phonemes used in English. Any number of phonemes smaller than that of those listed would make English as we know it impossible to use. The energy produced at the vocal folds or at a constriction cannot be varied in enough different ways to produce this many useful and distinguishable speech sounds. However, the speech energy must pass through at least portions of the vocal tract on its way to the outside. It is possible to modify the spectrum as the speech energy passes through the vocal tract by causing the vocal tract to assume different shapes.

Vocal Tract Features

The vocal tract is distinguished from most musical wind instruments by the manner in which it can be modified. Most wind instruments make use of a tube that maintains an approximately constant shape but whose length can be varied through the use of tone holes or valves. The vocal tract length, on the the hand, is of a fixed length (ignoring the effects of lip rounding, etc.), but can be caused to assume a large number of different shapes. This variable shape capability of the vocal tract permits the selective modification of speech energy produced at the vocal folds or at a constriction and thus makes possible the creation of enough distinguishable speech sounds to make English a viable spoken language. (You might at this point do a small experiment to convince yourself that vocal tract shaping plays an important part in helping to create a sufficient number of distinguishable speech sounds. Try opening your mouth to some comfortable position and then, keeping it in this position, see how many different and useful speech sounds you can produce by only controlling your vocal folds.)

The **vocal tract** consists of three main sections of "tube": the pharynx, the mouth, and the nasal cavity. These can be modified with the vocal organs—the tongue, the teeth, the lips, and the soft palate—as shown in Figure 24.1. The mouth can be varied more extensively in shape than either the pharynx or the nasal cavity because of the greater maneuverability of the forepart of the tongue. The nasal cavity is basically fixed in terms of internal dimensions and can be coupled to the pharynx only to a degree. The size of the pharynx is somewhat variable and is controlled by the back of the tongue.

We will restrict our attention to modification of voice energy as it passes through the vocal tract. First, let's consider the case in which the velum (or soft palate) is raised and brought into firm contact with the back of the pharynx so that air is not allowed to flow into the nasal cavity. This means that energy produced at the vocal folds must flow through the pharynx and then through the mouth to reach the outside world. (Cases involving a lowered velum—thus coupling the nasal cavity to the pharynx—will be considered later.)

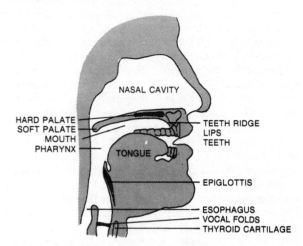

Figure 24.1. Simplified cross-section of vocal tract.

Neutral Vocal Tract

In order to study the way in which the vocal tract modifies speech energy, we begin with a greatly simplified tube and consider its effects. A uniform cylindrical tube closed at one end and open at the other end is termed a **neutral tract** because it has constant area from one end to the other. This tract is probably most closely approximated in an actual vocal tract configuration by the vowel /UH/. (All phoneme representations, such as /UH/, are listed in Table 23.1.) The neutral tract gives us a model that is simple enough to consider in some detail. Figure 24.2. shows a neutral tract where the closed end is at the vocal folds and the open end is at the mouth. In the figure the tract has been unbent to form a straight tract. (The unbending has no appreciable effect on the behavior of the waves which pass through it.)

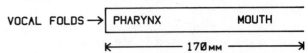

Figure 24.2. A neutral vocal tract 17 cm (170 mm) in length.

The vocal tract modifies speech energy by, figuratively, preferring some frequencies and discriminating against others. If the vocal folds produce a "preferred" frequency, this frequency is "amplified" by the vocal tract via constructive interference. However, if the vocal folds produce a "nonpreferred" frequency, this frequency is "attenuated" by the vocal tract via destructive interference. In practice, of course, the vocal folds produce both "preferred" and "nonpreferred" frequencies.

The "preferred" frequencies turn out to be the natural frequencies of the tube. The methods developed in Chapter 10 can be used to gain information as to what frequencies the vocal tract will emphasize. Our neutral vocal tract is equivalent to the closed-open tube considered in Chapter 10, whose lowest natural frequency was determined as $f_1 = v/4\mathcal{L}$, and whose higher natural frequencies were given by taking odd-integer multiples of f_1 as $f_n = (2n-1)f_1$. The first natural mode (or its frequency) of the vocal tract is called the first formant (F1), the second natural mode is the second formant (F2), the third is the third formant (F3), and so on. A **formant** in this sense is just a resonance of the vocal tract (or the corresponding frequency). If we assume that the tract in Figure 24.2. is 17 cm long and that the speed of sound in the tract is 34,000 cm/s, we calculate a value of 500 Hz for the first formant (F1) of the neutral tract. Similar calculations give values of 1500 Hz for F2 and 2500 Hz for F3. We can use the method described in Chapter 10 to measure the input impedance of the neutral tract or, for that matter, that of a tract having any shape. A quantity of greater interest than vocal tract impedance in the present situation is the **vocal tract transmission** which is a measure of air flow at the mouth relative to air flow at the vocal folds. The transmission spectrum shows us what frequencies will be emphasized—as illustrated in Figure 24.3 for the neutral tract. Note that the peaks occur at frequencies of 500, 1500, 2500 Hz, the same as for the impedance. Note too that the peaks are smaller at higher frequencies because there is greater resistance at higher frequencies.

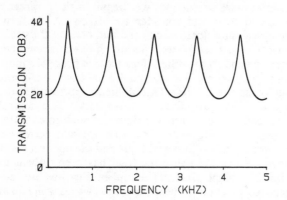

Figure 24.3. Neutral vocal tract transmission as it varies with frequency.

The speech signal we would measure with a pressure microphone placed outside the "mouth" of our neutral tract depends on three things: (1) glottal air flow, (2) vocal tract transmission, and (3) the way in which air flow at the mouth is radiated as pressure. Radiated pressure has a high frequency boost relative to air flow because the radiation of air flow as pressure is more efficient at higher frequencies. Let us take the glottal air flow spectrum of Figure 23.10 and add the radiation boost (about 6 dB per octave) to give a source pressure spectrum which incorporates items (1) and (3) above as illustrated in Figure 24.4. The spectrum of the radiated speech illustrated in Figure 24.5 is the sum of the source spectrum and the transmission spectrum. (It is possible to add the source and transmission spectra because they are expressed in dB.)

The source spectrum in Figure 24.4 is for a fundamental frequency of 200 Hz. Suppose we had a source spectrum as shown in Figure 24.6 for a fundamental fre-

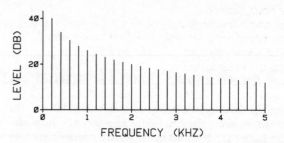

Figure 24.4. Source pressure spectrum resulting from 6 dB per octave boost of glottal air flow spectrum of Figure 23.10. Fundamental frequency is 200 Hz.

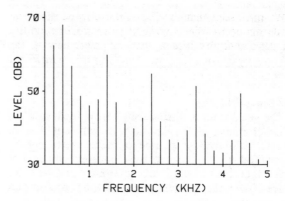

Figure 24.5. Spectrum of speech radiated from neutral tract. Fundamental frequency is 200 Hz.

quency of 140 Hz. The radiated spectrum would be as illustrated in Figure 24.7. Note that the formants remain unchanged from Figure 24.5 to Figure 24.7 because the tract is unchanged. However, now the third harmonic in the spectrum is largest rather than the second harmonic because the third harmonic is emphasized by the first formant, and so on.

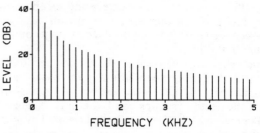

Figure 24.6. Source pressure spectrum at a fundamental frequency of 140 Hz.

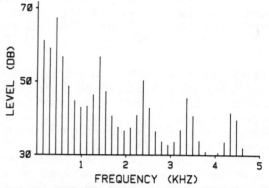

Figure 24.7. Spectrum of speech radiated from neutral tract at a fundamental frequency of 140 Hz.

Other Tract Shapes

If small changes are made in the shape of the neutral vocal tract, we might expect the formant frequencies to be changed to some small degree. If large changes are made in the vocal tract shape the changes in formant frequencies will also be large. However, for any vocal tract shape there will be an accompanying set of formant frequencies related to the tract shape. In other words, the formant frequencies (peaks) that we observe in the spectrum of a speech signal coming from the vocal tract are determined by the vocal tract shape. (Additional peaks and valleys can occur in the spectrum if the waveform produced by the vocal folds is too "unusual.")

We have seen that the formant frequencies of the neutral tract occur when standing waves are set up in the tract; the standing waves depend on the length of the tract, the speed of sound, and reflections from the ends of the tract. In an attempt to more closely approximate a real vocal tract, we consider a "two-tube tract," each tube having a different cross-sectional area. In multi-tube tracts, wave reflections occur where tubes of different areas join—as well as at the ends of the tubes. In this case, the standing waves are more complicated and occur at different frequencies than for our single-tube neutral tract. As examples, consider /EE/- and /O/-shaped tracts, their two tube approximations, and their transmissions shown in Figure 24.8 and Figure 24.9. The two tube lengths and areas for the /EE/ tract are 9 cm, 6 cm, 8 cm^2, and 1.5 cm^2, those for the /O/ tract are 8 cm, 9 cm, 1 cm^2, and 8 cm^2. Note that constricting the tract at the mouth end for the /EE/ lowers F1 and raises F2; constricting the tract in the pharynx for /O/ raises F1 and lowers F2.

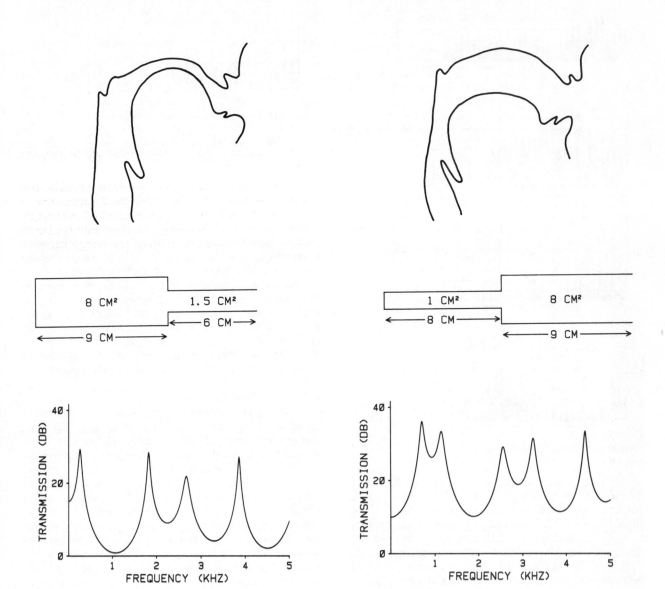

Figure 24.8. Vocal tract (upper), two tube approximation (middle), and transmission (lower) for /EE/-shaped tract. (After Ladefoged, 1975.)

Figure 24.9. Vocal tract (upper), two tube approximation (middle), and transmission (lower) for /O/-shaped tract. (After Ladefoged, 1975.)

Figures 24.10 and 24.11 illustrate spectra for several voiced sounds. Note the well defined harmonic lines in the spectra. Note also the broad groupings of high amplitude harmonic lines which indicate the presence of formants or vocal tract resonances.

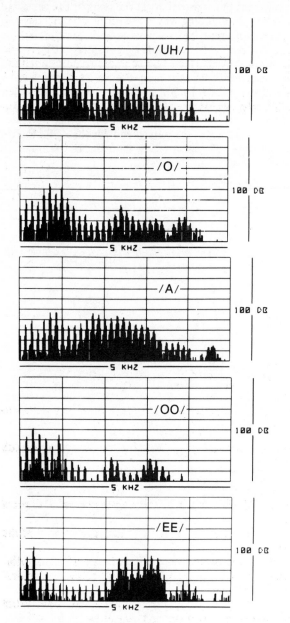

Figure 24.10. Spectra of vowel sounds.

Spectra of noise energy sounds such as /SH/ and /S/ do not have such well defined harmonic lines. However, different sounds still differ in spectral characteristics (/SH/ and /S/ in Figure 24.12) because the noise producing constriction in the vocal tract occurs at different places. Spectra of mixed energy sounds (/ZH/ and /Z/ in Figure 24.12) show spectra typical of their unvoiced counterparts at middle and high frequencies, but, because of the voicing show harmonic line spectra at low frequencies.

Different speech pressure waves result from the production of different phonemes. The differences are in part due to the energy types involved. They are also

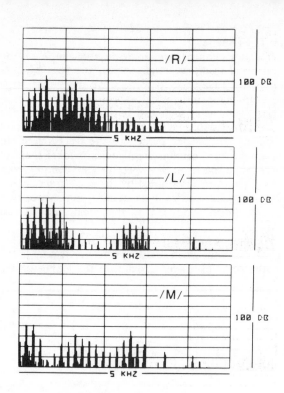

Figure 24.11. Spectra of voiced consonant sounds.

due to vocal tract shapes. When voice energy is used, the wave is approximately periodic. When noise energy is used, the wave is irregular. A pressure waveform recorded with a microphone for the sentence "Joe took father's shoe bench out," as produced by one talker, is shown in Figure 24.13. Note the relatively high-level, periodic portions of the wave associated with the vowel sounds. The periodic portions of the waveform were generated by glottal pulses of air which are much alike throughout the sentence. The differences among the various periodic portions of the waveform are due to different filtering effects of different vocal tract shapes. Other periodic, but lower-level, portions of the wave are associated with voiced sounds produced with a partially closed vocal tract, such as /DH/ and /B/. The rather low-level, non-periodic portions of the wave are associated with the /JH/, /SH/, and /CH/ sounds.

Other Tract Modifications

The diverse sounds of voiced speech can be produced by means of various vocal tract shapes. The particular spectrum produced is mostly independent of the voicing fundamental frequency; it depends primarily on the vocal tract shape and the resulting formant frequencies. So far in this discussion we have assumed an "average" male tract. The shorter tracts of adult females and children would be expected to result in higher formant frequencies. Approximate average formant frequencies for men, women, and children are shown in Table 24.1. In light of different formant frequencies for a given phoneme it may be wondered how we recognize the speech of men, women, and children. There must be some perceptual scaling to account for our ability to recognize the same speech sounds produced by different-sized talkers. Perhaps the formant frequency ratios remain roughly the same for the speech of men, women, and children.

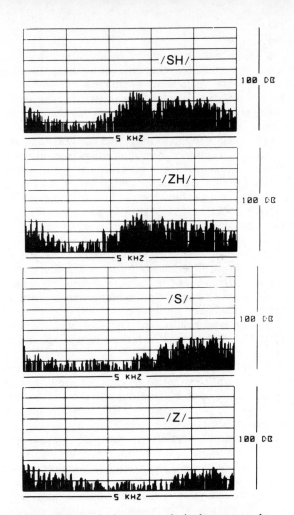

Figure 24.12. Spectra of noise energy and mixed energy sounds.

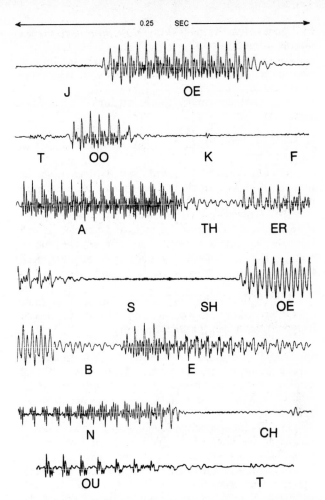

Figure 24.13. Waveform with phoneme labels for the sentence "Joe took father's shoe bench out."

Filling the tract with gas mixtures other than air would be expected to result in a different wave speed and hence different formant frequencies. Air consists of about four parts nitrogen (with a relative density of 28) to one part oxygen (with a relative density of 32) and has a relative density of about 29. If the nitrogen were replaced with helium (relative density of 4) the resulting mixture would have a relative density of about 10. The wave speed in this mixture would be about 70% higher than in air and the resulting formant frequencies would be about 70% higher. Because of this, deep sea divers using helium-oxygen mixtures have difficulty understanding one another because of the shifted formant frequencies. Likewise, if one inhales gas from a helium-filled balloon, "Donald Duck" type speech results.

Exercises

1. What are the main components of the vocal tract?

2. What is the function of each component of the vocal tract?

3. Suppose a "neutral female tract" is a cylindrical tube 14 cm long. At what frequencies will the first three formants occur?

4. A "neutral child's tract" is 11 cm long. What are the first three formant frequencies?

5. Suppose a neutral tract 17 cm in length is filled

Table 24.1. Approximate average formant frequencies for men, women, and children. (After Peterson and Barney, 1952.)

Phoneme	EE	I	E	A	O	U	OO	UH
F1								
Men	270	390	530	660	730	440	300	640
Women	310	430	610	860	850	470	370	760
Children	370	530	690	1010	1030	560	430	850
F2								
Men	2290	1990	1840	1720	1090	1020	870	1190
Women	2790	2480	2330	2050	1220	1160	950	1400
Children	3200	2730	2610	2320	1370	1410	1170	1590
F3								
Men	3010	2550	2480	2410	2440	2240	2240	2390
Women	3310	3070	2990	2850	2810	2680	2670	2780
Children	3730	3600	3570	3320	3170	3310	3260	3360

with pure helium (v = 970 m/s). At what frequencies will the formants occur?

6. What would the formant frequencies be if a neutral tract 17 cm in length were filled with krypton (v = 200 m/s)?

7. What determines the spectrum of the output speech waveform?

8. Use Figures 24.6 and 24.8 to determine the spectrum for /EE/. How does your result compare with Figure 24.10?

9. Use Figures 24.6 and 24.9 to determine the spectrum for /O/. How does your result compare with Figure 24.10?

10. How do the ratios of F2 to F1 and F3 to F1 from Table 24.1 compare for male and female speech? Does this provide any clues as to how we are able to perceive the same vowel for male and female speech even though the actual formant frequencies are different?

11. Estimate the length ratio of an average male tract to an average female tract on the basis of ratios of formant frequencies for the same vowel.

12. Use the spectra in Fig. 24.10 to determine the approximate frequencies of the first three formants for the phonemes /UH/, /O/, /A/, /OO/, and /EE/ by looking for the broad peaks in the spectrum. Record the results. How do your results compare with those in Table 24.1?

13. Estimate relative female and male vocal tract lengths on the basis of relative heights of females and males. How does your estimate compare with the text.?

14. An alternative means can be used for determining which frequencies the neutral vocal tract will emphasize. Consider, for example, the effect of the neutral tract upon air pulses produced at the rate of one pulse every 2 ms. The time required for a single pulse to travel the length of a 17 cm tract is equal to 0.5 ms. At the end of 0.5 ms the first pulse will be at the open end of the tract and will be reflected back toward the closed end as a negative pulse. After a total elapsed time of 1 ms (0.5 ms to return) the negative pulse will be back at the closed end of the tract. At the closed end of the tract the negative pulse is reflected toward the open end. After another 0.5 ms the pulse arrives at the open end of the tube, where it is reflected as a positive pulse to return to the closed end. The positive pulse arrives back at the closed end just as a new positive pulse is produced. The two positive pulses add together to produce a larger positive pulse. From this example we deduce that the tube serves to enhance the pulses produced at a rate of one every 2 ms. To what frequency does the pulse period of 2 ms correspond? How does this relate to the first formant frequency?

15. Repeat Exercise 14 for pulses produced at the rate of one every 1 ms. Are these pulses enhanced or diminished?

Demonstrations

1. Attach a sinusoidally driven loudspeaker to one end of a model vocal tract. Use a microphone hooked to an oscilloscope at the other end to pick up the pressure wave. Measure the response as the frequency is varied.

2. Construct three model vocal tracts using construction paper and masking tape (or other appropriate materials). Make them about 17 cm long, one with constant cross-sectional area and the other two with cross-sectional areas appropriate for the vowels /EE/ and /0/ (see Figures 24.8 and 24.9). Use an artificial larynx to excite the tracts. Do the sounds from the /EE/ and /0/ tracts shift perceptually from the neutral tract in a way similar to real speech? Use a spectrum analyzer to determine the formant frequencies. What effect does vocal tract shape have on the spectra? Do the formant frequencies for /EE/ and /0/ shift in the same direction from the neutral values as occurs in actual speech?

3. Analysis of energy types in speech—Identify the portions of the waveform in Figure 24.13 associated with the various phonemes in "Joe took father's shoe bench out." Determine the degree of periodicity in each part of the wave (much, some, none) for each of the 12 phonemes. Compare these "periodicity" results with the "energy type results" of Demonstration 3 from Chapter 23. What correlation is there between periodicity and voicing or noise?

4. Take short sections of tube and place them in contact with your lips so as to extend your lips while producing vowel sounds. Does the vowel color change? Why? Does the pitch change? Why? Perform spectral analyses if possible.

5. Perform spectral analyses of voiced and whispered vowels. Are the formant frequencies the same or different? Why? Do harmonic lines appear in the spectra of both? Why?

6. Estimate relative female and male vocal tract lengths by having a marshmallow stuffing contest. Determine the average number of marshmallows females and males can respectively stuff in their mouths. Divide one average by the other and take the cube root to give an estimate of relative lengths. (Taking the cube root is necessary because the stuffing measures volume which varies approximately as the cube of length.)

Further Reading

Borden and Harris, Chapter 4
Denes and Pinson, Chapter 4
Dickson and Maue-Dickson, Chapter 5
Flanagan, Chapters 3, 5
Fletcher, Chapter 3
Hall, Chapter 14
Ladefoged, Chapters 8,9
Pickett, Chapters 2-4
Rossing, Chapter 15
Singh and Singh, Chapters 3-5
Peterson, G. E., and H. L. Barney. 1952. "Control Methods Used in a Study of the Vowels." J. Acoust. Soc. Am. **24**, 175-184.

Audiovisual

1. **Your Voice** (10 min, 1947, EBE)
2. "Vocal Formants" (REB)
3. "Effect of Gas on Voice" (REB)

25. Distinguishing Characteristics of Speech Sounds

Phonemes are the individual, distinguishable sounds of speech—the building blocks of speech from which syllables, words, and sentences are constructed. In Chapters 23 and 24 we discussed how phonemes are produced by the human vocal mechanism. In this chapter we will briefly consider phoneme categorization schemes. We will then discuss the complementary methods of speech analysis and speech synthesis used to determine distinguishing characteristics of different speech sounds.

Categorization of Phonemes

Phonemes can be categorized in several different ways. In Chapter 23 we considered a categorization by energy types. Other categorizations might be on the basis of a steady-transitory dichotomy, the manner of articulation, or the place of articulation (see Table 23.1).

Some phonemes can be produced as "steady sounds." For instance, a vowel can be produced for as long as we have the breath to drive the vocal mechanism. However, there is no such thing as a "steady production" for some other sounds, such as the plosives. We will label sounds that can be produced in a steady state as **steady phonemes** and sounds that are inherently transitory as **transitory phonemes**. (In running speech, almost all of the phonemes are transitory and very seldom is a steady state achieved. However, the fact that some sounds can be produced in a steady state and others cannot is interesting and instructive. It may be that there is more dependence on transitions for perception of transitory phonemes than for steady phonemes.) Steady phonemes include the vowels, the liquids, the fricatives, and the nasals, whereas the transitory phonemes include the plosives, the semivowels, and the diphthongs. (You can determine into which category a particular phoneme falls by simply trying to produce the phoneme in a steady manner; if you are able to do so it is a steady phoneme, and if not it is a transitory phoneme.)

Steady sounds are produced when the vocal tract maintains a constant shape; transitory sounds are produced when the shape of the vocal tract varies in time. Since formant frequencies are related to vocal tract shape, we expect a constant vocal tract shape to produce formant frequencies that are constant, and a changing vocal tract shape to produce formant frequencies that change. We will find when we study spectrograms that changing formant trajectories are apparent, thus indicating the changing nature of the vocal tract shape.

There are several different manners of speech production: plosive production involves the sudden release of pressure; fricative production involves forcing the air through a constriction; nasal production involves coupling of the nasal cavity with the pharynx; production of liquids involves a fairly small constriction in the tract; and the semivowels are produced by starting from a vowel-like configuration and then moving to the following vowel configuration. Plosives and fricatives can be either voiced or unvoiced. Note that the phonemes listed in Table 23.1 are grouped according to manner of articulation, which factor is closely related in many cases to the energy types discussed in Chapter 23.

The various articulators used in speech production determine the place of articulation. Places of articulation (or constriction) are: between the lips (labial), between the teeth (dental), between lips and teeth (labiodental), between the tongue and gum ridge (alveolar), between the tongue and hard palate (palatal), between the tongue and soft palate (velar), and in the vicinity of the glottis (glottal). Different places of articulation produce different vocal tract shapes, which in turn give rise to different formant patterns.

Speech Analysis

In Chapter 24 a description was given as to how the vocal tract, depending on its shape, produces characteristic formant patterns. Suppose that spectral analyses of the kind described in Chapter 24 were performed successively on a changing speech signal. Placing such spectra contiguously along an axis representing time provides a three-dimensional representation showing the speech level at different frequencies and at different times. Figure 25.1 illustrates how a three-dimensional display of sound level, frequency, and time might appear.

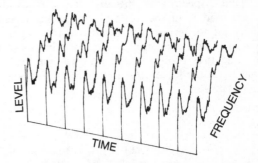

Figure 25.1. Three-dimensional display of contiguous speech spectra.

A roughly equivalent display can be produced by using relative lightness and darkness to represent the sound level dimension; in this way we can get a three-dimensional display in two dimensions. A **sound spectrograph**, shown in Figure 25.2, is a device that creates a frequency display along the vertical axis, time along the horizontal axis, and sound level in terms of relative lightness and darkness. The display produced by a sound spectrograph is called a **sound spectrogram**. The most important elements of a spectrograph are the bandpass filters that divide the incoming speech signal into many different frequency bands (from 50 to about 250 bands, depending on the application). The amount of energy that comes through each filter is used to control a "printing" mechanism. The printing mechanism of the digital sound spectrograph in Figure 25.2 is the high resolution grey scale printer shown at the right. In a conventional analog sound spectrograph the mechanism is a voltage-controlled sparking wire in contact with heat-sensitive paper. When a computer is used as a sound spectrograph the output signal from each channel is used to control the brightness of a display point on an oscilloscope screen.

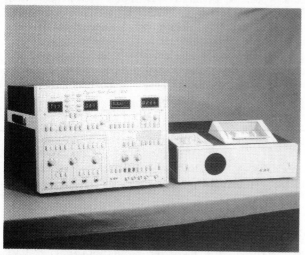

Figure 25.2. Digital sound spectrograph with the electronics for analysis shown at the left and the printer producing a spectrogram shown at the right. (Courtesy of Kay Elemetrics Corp.)

A typical analog spectrograph produces spectrograms in which sound level is indicated by relative darkness, as shown in Figure 25.3. The sentence spoken for the spectrogram of Figure 25.3 was "Joe took father's shoe bench out." This spectrogram is called a "wide band" spectrogram because wide bandwidth filters were used in the spectrograph. Individual voicing harmonics cannot be seen but the broad formants are apparent. The time resolution is very good in this spectrogram and thus transients can be seen; the vertical striations show the large pressures at the beginning of each glottal pulse.

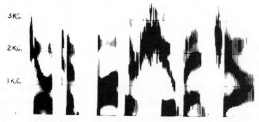

JOE TOOK FATHER'S SHOE BENCH OUT (NATURAL)

Figure 25.3. Wide-band spectrogram of the utterance "Joe took father's shoe bench out." (From Strong, 1967.)

A typical computer-generated spectrogram indicates the intensity level by relative brightness as seen in Figure 25.4; the fundamental frequency (in Hz) and sound level (in dB) are also shown. The sentence spoken for the spectrogram of Figure 25.4 was "The second planet was inhabited by a conceited man." This spectrogram is called a "narrow band" spectrogram because

narrow bandwidth filters were used in the spectrograph. Individual voicing harmonics can be seen as horizontal bars. Because the time resolution is poor short time transients cannot be seen. A comparison of the wide band and narrow band spectrograms illustrates that where frequency resolution is good, time resolution is poor, and vice versa.

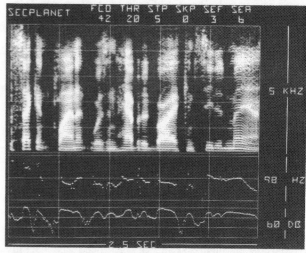

Figure 25.4. Computer-generated narrow-band spectrogram of the utterance "The second planet was inhabited by a conceited man." (From Strong and Palmer, 1975.)

Various distinguishing spectrographic features for each of the several phoneme categories can be seen in the spectrograms of these figures. Keep in mind that we are discussing idealizations here. It is not usually possible to sit down with a spectrogram and determine unambiguously what was spoken. Many subtleties occur when the phonemes blend together in running speech. A feature commonly observed for vowel sounds is the presence of three or more fairly distinct formant bands of energy, which may become approximately constant in frequency if the vowel is of fairly long duration. The diphthongs are also typically characterized by three or more distinct formant bands that change their frequency positions quite markedly from the beginning of the diphthong to its end. Nasals are typically characterized by a low-frequency formant, with much lower sound levels at high frequencies. (It should be noted that even for nasals there are bands of energy present at higher frequencies, but because they are so much weaker than the low-frequency band, in sound spectrograms, they do not appear very dark if at all.)

Voiced plosives typically start with the presence of a low-frequency formant during the vocal tract closure for the plosive production, and then higher-frequency formants appear following the vocal tract opening; these higher-frequency formants usually exhibit transitions into the speech sound following the plosive. The voiced plosives usually show very little if any "burst energy" because the release of the vocal tract closure involves the release of much smaller pressures than is usually the case with unvoiced plosives. The unvoiced plosives typically give rise to a very short burst of energy that is spread across many different frequencies. Vocal fold vibrations for unvoiced plosives usually begin later after the release of the vocal tract constriction than is

172

the case with the voiced plosives. An example can be seen in the /T/ of "took" in "Joe took father's shoe bench out," in Figure 25.3.

The unvoiced fricatives (of which the /SH/ of "shoe" in "Joe took father's shoe bench out" is a good example) typically give rise to high-frequency bands of energy in a spectrogram. These bands of energy are more ragged in appearance than the formants for voiced speech, because the vocal tract is excited with noise energy, whereas voice excitation is used for vowel sounds. Again, it should be noted that unvoiced fricatives produce energy at low frequencies—though this energy is weaker in comparison to the high-frequency energy than for voiced sounds and thus often does not appear on the spectrogram. The voiced fricatives (of which the /Z/ in "was in" of "The second planet was inhabited by a conceited man" is an example) have spectrographic characteristics similar to the unvoiced fricatives. However, spectrograms of voiced fricatives also typically exhibit a low-frequency formant, due to the presence of voice energy in their production.

Note in the spectrograms of Figures 25.3 and 25.4 that our idealizations for the spectra of phonemes spoken in isolation or for simple consonant-vowel syllables break down when phonemes are combined in "running" speech. It is not possible to determine in a spectrogram explicitly where one phoneme ends and another begins. Vowel formant frequencies very often do not achieve their steady-state values in running speech and everything seems to be in a constant state of transition. Unusual groupings of sounds also occur. For example, consider the behavior of the /CH/ sound in "bench." In the written version we think of the /CH/ as belonging to the word bench, but in the spectrogram it appears to belong to the word "out". The utterance actually produced seems more like "ben chout" than "bench out".

The same phoneme may also appear different when it occurs in different contexts. Note in the spectrogram of "Robby will like you, daddy-oh" (Figure 25.5) the much tighter closure for the first /D/ than for the second /D/ of "daddy-oh." The first formant is the only one apparent in the first instance of the /D/ sound whereas the higher formants are also apparent in the second. Further study of the spectrograms will reveal other interesting features of actual speech.

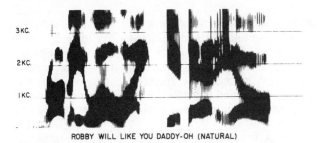

ROBBY WILL LIKE YOU DADDY-OH (NATURAL)

Figure 25.5. Wide-band spectrogram of the utterance "Robby will like you, daddy-oh." (From Strong, 1967.)

Another way of presenting spectrographic information is in terms of formant frequencies and amplitudes. The spectrogram in Figure 25.5 is for the sentence

"Robby will like you, daddy-oh." The "formant spectrogram" in Figure 25.6 is for the same sentence. In this case the formant frequencies are traced out in the upper part of the figure and the formant amplitudes are traced out in the lower part of the figure.

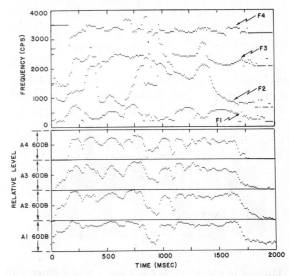

Figure 25.6. "Formant spectrogram" of the utterance "Robby will like you, daddy-oh." Formant frequencies are shown in the upper part of the figure and formant amplitudes in the lower part. (From Strong, 1967.)

Speech Synthesis

Speech synthesis is complementary to speech analysis in exploring distinguishing features of speech sounds. In speech synthesis, speech parameters obtained from analysis are used to synthesize speech and thus check the adequacy of the parameters. One device that is useful in speech synthesis is the **formant synthesizer,** the basic elements of which are shown in Figure 25.7. Each of the band-pass filters represents a different formant. The frequency and amplitude of each filter can be controlled independently, thus making it possible to simulate formant frequencies and amplitudes. The two types of excitation used as inputs to the synthesizer are "pulse trains" and noise. The pulse train input is used to simulate voicing energy, and the noise input is used to synthesize phonemes requiring noise energy. A switch controls which of the two types of energy goes into the band-pass filters. Both the pulse train and the noise have large amounts of energy at many different frequencies. The band-pass filters selectively eliminate most of the energy outside their own band and are thus able to simulate formants by permitting only a selected range of frequencies to pass through. The outputs of the filters, when added together and played through a loudspeaker or headphones, produce the synthetic speech. Various versions of formant synthesizers have been successfully used to produce synthetic speech of good quality.

The formant frequency and formant amplitude parameters shown in Figure 25.6 were used to control a formant synthesizer of the type shown in Figure 25.7. A spectrogram of the resulting synthetic speech is shown in Figure 25.8. A comparison of the spectrogram for the synthetic speech (Figure 25.8) with that for the natural

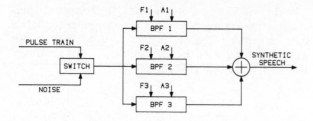

Figure 25.7. Elements of a formant speech synthesizer.

speech (Figure 25.5) shows the two to be very similar. The synthetic speech also sounds very much like the natural.

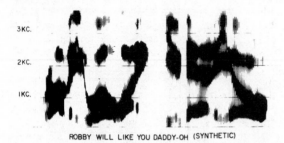

Figure 25.8. Spectrogram of the synthesized utterance "Robby will like you, daddy-oh." Compare with Figure 25.5. (From Strong, 1967.)

Stylized, simplified speech parameters are often used in speech synthesis to ascertain which parameters are most significant. This is accomplished by presenting the synthetic speech to listeners. The Pattern-Playback machine, designed and used at the Haskins Laboratories, provided much of the early information regarding significant parameters for speech synthesis. Basically, it used either spectrograph-generated spectrograms or hand-drawn spectrograms to produce the synthetic speech. By modifying an existing spectrogram, or by hand-drawing a spectrogram, it was possible to eliminate or add features, thus helping to determine which features of the spectrogram were most significant. Figure 25.9 illustrates a stylized spectrogram and the original spectrogram from which it was derived.

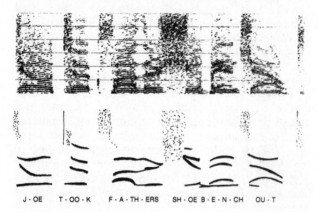

Figure 25.9. Spectrograms of the sentence "Joe took father's shoe bench out." The upper part of the figure is from a naturally produced utterance. The lower part of the figure is a stylized pattern that can be used to control a speech synthesizer. (Courtesy of J. H. Clegg and W. C. Fails.)

Stylized spectrogram patterns illustrating first and

second formant transitions for the voiced plosives in the context of seven vowels appear in Figure 25.10. The transitions of F2 for /B/ seem to have originated from the same place for all vowels; this common origin has been called the "locus." Similar conditions hold for /D/ and /G/ except that /G/ apparently requires two loci. Fig. 25.11 illustrates spectrogram patterns for consonants having common features. All consonants in a given row are produced in the same manner and all in a given column are articulated in the same place.

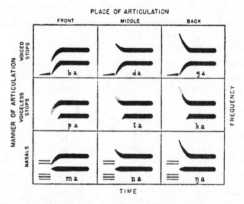

Figure 25.10. Stylized first and second formant transitions for /B/, /D/ and /G/ in the context of seven vowels. (From Liberman etal, 1959.)

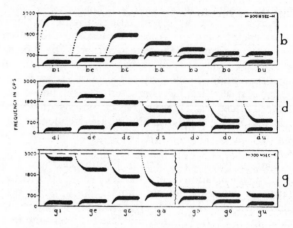

Figure 25.11. Stylized first and second formant transitions for the stop and nasal consonants. (From Liberman etal, 1959.)

Simulation of Speech Production

Speech analysis and synthesis as we have discussed them thus far have been concerned primarily with the acoustical signal. This is natural since the acoustical signal is easily accessible. However, it may be of interest to understand and simulate the motions which the vocal folds and the vocal tract go through in speech production.

Various models have been devised to simulate speech production. Some have been mechanical models; others have used analog electronic circuitry to represent the vocal tract. Most current models are digital models in which vocal fold motion, glottal air flow, and wave propagation in the vocal tract are represented mathematically. Representative of these models are single mass, two mass, and multi-mass vocal fold models. These models represent the vocal tract in terms of contiguous cylindrical sections whose areas change in time.

The parameters needed to control a formant synthesizer can be obtained from tape recordings of speech, via spectral and spectrographic analysis. The parameters needed to control a voice production simulation synthesizer are very different. Mass, tension, and configuration of the vocal folds are needed to give proper fundamental frequency and spectral content. The pressure just below the larynx is needed because it provides the driving force for the vocal folds. The shape of the vocal tract as it changes in time is needed so that the correct filtering will be specified in the model.

Many instruments and techniques have been devised for measuring the above parameters. Air flow measuring devices are used. Tracheal punctures and miniature microphones are used to permit measurement of the subglottal pressure. High speed motion pictures using both visible light and x-rays are used to view vocal fold and vocal tract motion. Tiny wire electrodes are placed in various muscles to measure their activity. Transducers are used to measure lip, jaw, and tongue movements. As these efforts continue, more accurate parameters for controlling simulation synthesizers will become available. Then simulation synthesizers will be controlled in more meaningful ways.

Exercises

1. Consider the phonemes in Table 23.1. Indicate whether each is steady or transitory. Indicate the place of articulation (where appropriate) as one of the following: labial, labio-dental, alveolar, palatal, velar, glottal. Indicate the manner of articulation as one of the following: vowel, semivowel, liquid, nasal, voiced fricative, unvoiced fricative, voiced plosive, unvoiced plosive.

2. List the distinguishing features typically seen in a spectrogram for the following categories of sounds: nasals, voiced fricatives, unvoiced fricatives, diphthongs, voiced plosives, unvoiced plosives. (For example, vowels are typically characterized by three or more fairly well defined energy bands or formants.) Describe what the vocal folds and the vocal tract are doing to produce the observed spectrographic features. Sketch basic spectrographic features for each category, as illustrated in Figure 25.12 for a vowel.

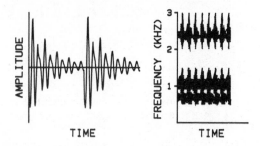

Figure 25.12. Waveform and simplified spectrogram for /O/ Note the three formant energy bands in spectrogram.

3. Compare the sound pressure wave in Figure 24.13 with the spectrogram in Figure 25.3. Both were obtained for the sentence "Joe took father's shoe bench out." Describe the corresponding features of the two.

4. Fundamental frequency can be determined by measuring the frequency of some higher harmonic (such as the 10th) and then dividing by the harmonic number (10 in this case). Use this technique to determine the fundamental frequency at various points in the narrow-band spectrogram of Figure 25.4.

5. Compare formant frequencies obtained from approximately steady portions of the spectrograms in Figures 25.3 and 25.5 with values obtained from the spectra of Figures 24.10 and 24.11. How well do they agree? List some variables that might account for the discrepancies.

6. Compare the spectrogram in Figure 25.5 to the "formant spectrogram" in Figure 25.6. Both are for the same utterance "Robby will like you, daddy- oh." Describe similarities and differences between the two spectrograms.

7. Interpret the spectrograms in Figures 25.3, 25.4, and 25.5. Label each of the phonemes in the spectrograms with an appropriate phoneme symbol and show the approximate time duration of each by bounding each with vertical lines. Note transition regions and the influences of one phoneme on another.

8. Imagine you have a sheet of spectrogram paper. Draw a stylized spectrogram for the diphthong /O/EE/. First draw a spectrogram at the left side for /O/ and then draw a spectrogram at the right for /EE/. Connect the first, second, and third formants of the /O/ to those for the /EE/. Compare your stylized spectrogram with the long "i" in "like" of Figure 25.5.

9. Suppose a two-formant speech synthesizer is to be controlled by means of a lap board. A lap board is shown in Figure 25.13 and is used by placing a pointer on the board in a position appropriate for the first two formant frequencies of the desired sound. The first formant frequency is controlled by the horizontal position of the pointer as it increases from left to right, and the second formant frequency is controlled by the vertical position of the pointer as it increases from bottom to top. Show on the diagram of the lap board where the pointer should be placed to produce the following vowel sounds: /EE/, /I/, /E/, /A/, /O/, /U/, and /OO/. Also show approximate paths that might be traced out by the pointer to produce the diphthongs /OI/ and /OW/, and to produce the syllables /D/O/, /G/O/, /B/EE/, /D/EE/, and /G/EE/. Use information from Figure 25.10.

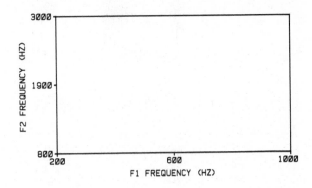

Figure 25.13. Lap board for controlling first and second formant frequencies of a two-formant synthesizer.

Demonstrations

1. Use a sound spectrograph to produce a sound spectrogram for your own voice while speaking a sentence that is about 2 seconds in length. Interpret the sound spectrogram by labeling different parts of the spectrogram with appropriate phoneme symbols. Also indicate points of high and low pitch.

2. Use a speech synthesizer to synthesize some steady vowel sounds and some consonant-vowel syllables.

3. Listen to synthetic speech produced by a "Speak 'n' Spell" or other commercial synthesizer. Comment on the intelligibility and naturalness of the synthetic speech. Suggest what the deficiencies might be.

Further Reading

Borden and Harris, Chapters 4-6
Denes and Pinson, Chapters 4, 7
Flanagan, Chapters 5, 6
Flanagan and Rabiner
Fletcher, Chapters 3-5
Ladefoged, Chapters 6-9
Painter, Chapters 2, 3, and Appendix
Pickett, Chapters 2, 4, 6-10
Rossing, Chapter 16
Schafer and Markel
Singh and Singh, Chapters 5-10
Cooper, F.S. 1980. "Acoustics in Human Communication: Evolving Ideas About the Nature of Speech," J. Acoust. Soc. Am. **68**, 18-21.
Flanagan, J.L., K. Isizaka, and K.L. Shipley. 1975. "Synthesis of Speech from a Dynamic Model of the Vocal Cords and Vocal Tract," Bell System Tech. J. **54**, 485-506.
Liberman, A.M., F. Ingemann, L. Lisker, P. Delattre, and F.S. Cooper. 1959. "Minimal Rules for Synthesizing Speech," J. Acoust. Soc. Am. **31**, 1490-1499.
Potter, R.K., G.A. Kopp, and H.C. Green. 1947. **Visible Speech** (D. Van Nostrand Co.).
Strong, W.J. 1967. "Machine-aided Formant Determination for Speech Synthesis," J. Acoust. Soc. Am. **41**, 1434-1442.
Strong, W.J., and E.P. Palmer. 1975. "Computer Based Sound Spectrograph System," J. Acoust. Soc. Am. **58**, 899-904.

Audiovisual

1. "Real and Synthetic Speech" (GRP)
2. "Audio Spectrograms" (REB)

26. Prosodic Features of Speech

Prosodic features are those features of speech that while not specifically associated with phonemes, still transmit information to the listener. Prosodic features often extend over more than one phoneme or speech segment. The more common prosodic features in spoken English are stress, intonation, and rhythm. These prosodic features play a secondary role to that of the phonemes in terms of communicating meaning. However, they are often used to clarify otherwise ambiguous sentences and emphasize selected words within a sentence. In some spoken languages, such as tone languages, prosodic features play a more significant role. In such cases prosodic features can actually change word-meanings. Prosodic features also provide information about the emotional and physical characteristics of a talker. Prosodic features may even be given more attention than the actual phonemes in activities such as reciting poetry or singing. In general, this occurs when the embellishments (or the way in which something is said) are more important than the content (or what is said).

In order to carry out experiments and measurements relating to prosodic features we must know what their physical counterparts are. Keep in mind that there is not a simple one-to-one relationship between the two. Hence, the relationships that we describe are not without some ambiguity. **Stress** is dependent on fundamental frequency, duration, sound level, and spectrum, in declining order of importance. **Intonation** is variation of pitch which is most significantlly dependent on fundmental frequency. **Rhythm** is dependent on relative durations of syllables and pauses and on the number of syllables produced per second.

Normal Spoken English

Consider first some examples showing the role of prosody in spoken English where stress is used contrastively. As a noun "digest" is pronounced with stress on the first syllable (di'gest). As a verb "digest" is spoken with stress on the second syllable (di gest'). These contrasting stresses are illustrated in Figure 26.1. Subglottal pressure is higher on the stressed syllable, resulting in a higher sound level which helps to produce a higher fundamental frequency on the stressed syllable. The stressed syllable also is longer in duration as can be seen on the fundamental frequency contours. The spectrum may be modified for stressed sounds because of a richer spectrum resulting from greater vocal effort and because of a modified vocal tract shape.

A second example of stress can be seen in the two sentences "The light housekeeper is gone" and "The lighthouse keeper is gone". Both involve essentially the same phonemes; yet their meanings are quite different. In Figures 26.2 and 26.3 fundamental frequency and sound level as functions of time are traced out for each of the two sentences. Note that in the "light

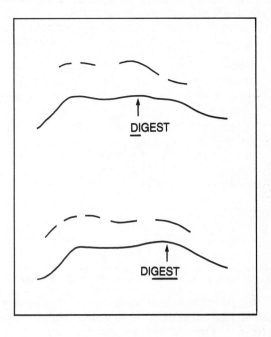

Figure 26.1. Relation between contours of voice fundamental frequency (broken curves) and subglottal pressure (solid curves) with two different patterns of word stress. The upper contours are for the utterance "That's a di'gest." The lower contours are for the utterance "He didn't digest'." The stressed syllable is underlined in each case and an arrow points to the stressed syllable in the contours. (After Ladefoged, 1963 and Pickett, 1980.)

housekeeper" version (Figure 26.2) the emphasis or stress is on "house," as shown by higher relative fundamental frequency, longer duration, and higher relative sound level. In the "lighthouse keeper" version (Figure 26.3) the emphasis is on "light." Commonly (though by no means always), the fundamental frequency, duration, and sound level change together in a particular direction. For instance, when the intensity is increased the fundamental frequency has a natural

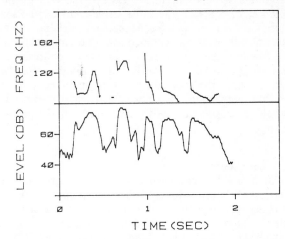

Figure 26.2. Fundamental frequency and sound level contours for "The light housekeeper is gone." The word "house" is emphasized.

tendency to increase unless other adjustments are made in the speech apparatus. It often happens that when a particular sound is intensified, its duration is increased. When sound intensity decreases, duration usually decreases also. What we expect to observe is that in many cases fundamental frequency, intensity, and duration tend to increase or decrease together.

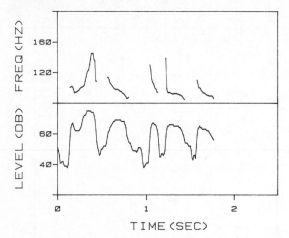

Figure 26.3. Fundamental frequency and sound level contours for "The lighthouse keeper is gone." The word "light" is emphasized.

A further example of the influence of stress is provided in four versions of the sentence "John drove to the store." One of each of the four versions were spoken, respectively, without stress, with stress on "John," with stress on "drove," and stress on "store." Fundamental frequency and sound level contours for these four versions are shown in Figure 26.4. There is a very pronounced peak in the frequency and level contours of the stressed word in each case and a noticeable increase in duration. This use of stress permits the speaker to emphasize that it was "John" and not "Mary" or someone else who acted (see Figure 26.4).

Intonation (variation of pitch) is used over phrase and sentence length units. A common use of intonation in spoken English is to distinguish a declarative sentence, such as "They are going now," from its interrogative counterpart, "They are going now?" Note in Figure 26.5 that fundamental frequency for the declarative sentence decreases toward the end of the sentence. In contrast, the same sentence as a question exhibits a fundamental frequency that is rising (or at least not falling) at the sentence end. (You might try uttering several declarative-question sentence contrasts to see how this works.) A non-falling intonation may be used to signal that the talker has not finished talking and that there is more to come. A particular intonation contour in conjunction with a pause may signal a phrase ending.

Tone Languages

In atonal languages, such as English, words can be pronounced with different pitch variations without changing their meanings. In tone languages a word can take on one of several different meanings depending on the tone with which it is spoken. **Tones** are word length pitch variations that affect word meaning. Tone languages are categorized in terms of **register tones,**

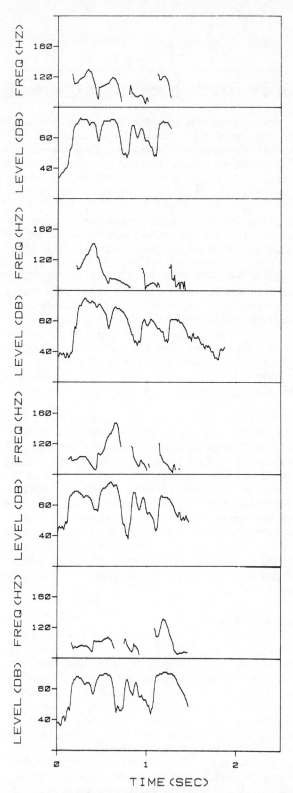

Figure 26.4. Fundamental frequency and sound level contours for four versions of "John drove to the store." The version without stress appears at the top. The version with stress on "John" is second from the top. The version with stress on "drove" is second from the bottom. The version with stress on "store" is at the bottom.

those tones in which contrasts are made among two or more constant pitch levels, or **contour tones,** those in which contrasts are made among changing pitch contours. Navajo is a register tone language in which two pitch levels are used. Yoruba (Africa) is a register tone

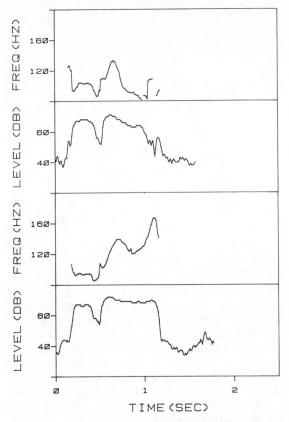

Figure 26.5. Fundamental frequency and sound level contours for "They are going now" (top) and "They are going now?" (bottom).

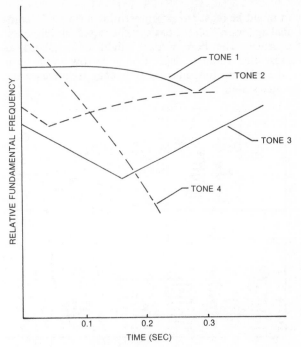

Figure 26.6. Average fundamental frequency contours and durations for the tones of Mandarin. (After Ting, 1971.)

language in which three pitch levels are used.

Cantonese Chinese, Mandarin Chinese, Thai, and Vietnamese are contour-tone languages. Cantonese Chinese uses six contrasting tones, Thai uses five. Mandarin Chinese uses four contrasting tones whose nominal fundamental frequency contours and durations are shown in Figure 26.6. A given phoneme can take on as many as four different meanings depending on which of the four different tones are used. For example, "ma" produced with tone one means "mother"; with tone two it means "hemp"; with tone three it means "horse"; and it means "scold" when tone four is used.

Tonal languages involve considerable processing in the right hemisphere of the brain, whereas atonal languages are processed primarily in the left hemisphere. Tonal features are apparently acquired before segmental features by children learning a tone language. In some languages level tones are acquired before contour tones.

Perceived Personal State

Prosodic features can provide information about the physical, emotional, and "personality" characteristics of a talker. The prosodic features thus help to determine our perception of a talker's "personal state." We are often able to determine the sex of a talker on the basis of fundamental frequency and to a lesser extent on the basis of formant frequencies. The age of a person may be partly apparent on the basis of voice quality. We form our perceptions of an individual's peronality from hearing his speech, often in the absence of visual cues. (Some of the things we will

discuss are not well defined at the present time and should be considered in that light.)

There is evidence that prosodic features are a medium through which the emotional state of a person can be observed. In studies of actual and simulated emotional conditions, it was found that, in general, when a person spoke in anger, durations were shortened and fundamental frequency and sound level were increased relative to normal speech. Conversely, durations increased and fundamental frequency and sound level were decreased for an emotionally depressed talker. Figure 26.7 illustrates average fundamental frequencies for each of three talkers speaking in four emotional situations. Clearly fundamental frequency tends to be lowest for sorrow, increasing in order for neutral, fear, and anger, respectively.

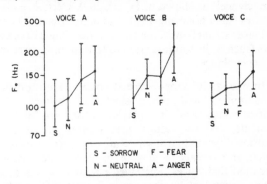

Figure 26.7. Averages and ranges of fundamental frequency for each of three talkers in sorrow, neutral, fear, and anger situations. (From Williams and Stevens, 1972.)

Emotional responses in real life situations is illustrated in the narrow-band speech spectrograms of Figure 26.8 for a radio announcer describing the arrival and crash of the Hindenburg. The top part of the figure might be regarded as emotionally neutral. The middle part might be regarded as representing a state of "emotional agitation" on the basis of the raised fundamental frequency. The bottom part might be regarded as representing a state of depression or sorrow on the basis of the lowered fundamental frequency (relative to the agitated state).

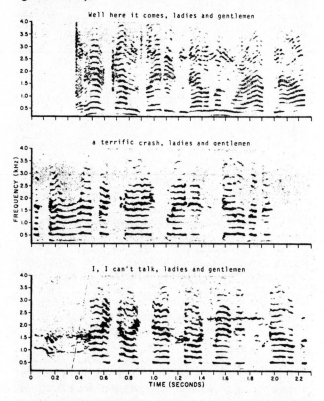

Figure 26.8. Narrow-band spectrograms for a radio announcer talking before (top) and after (middle and bottom) the crash of the Hindenburg. (From Williams and Stevens, 1972.)

Studies have been conducted in which listeners were asked to rate different tape-recorded voices in terms of many pairs of opposites (e.g., kind-unkind). The results of these ratings were then condensed and plotted along two perceptual dimensions labeled "benevolence" and "competence" as shown in Figure 26.9. Certain paired opposites (such as strong-weak and active-inactive) correlate most strongly with the competence dimension and others (such as just-unjust and polite-impolite) correlate with benevolance.

In one study, tape-recorded voices were modified by computers and other electronic devices. The modified voices were then played to listeners for their judgments, and the results were plotted in the two-dimensional perceptual space. Rate of speech, variation of pitch, and average pitch were the three factors varied in the modified voices. Perceived personality was most strongly dependent on changes in speech rate and less dependent on average pitch and variation of pitch. Perceived competence appears to increase and decrease with rate in an almost linear fashion; that is, increased speech rate contributes to judgments of increased competence. Perceived benevolence appears to show an inverted "U" relationship with rate; very high or very low rates produce decreased perceived benevolence, whereas the middle range of speech rate contributes to the

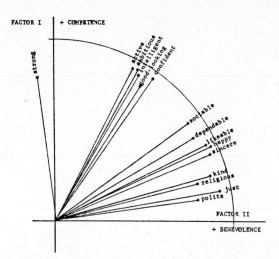

Figure 26.9. "Condensed" patterns for paired-adjective ratings of voices with modified rate of speaking. (From Smith etal, 1975.)

highest perceived benevolence. The effects of speech rate are shown in Figure 26.10 where rate increases around the curve from slowest (4S) to fastest (4F). Increased variation of pitch (i.e., making the speech less monotonous) causes perceived competence and benevolence to increase slightly, whereas decreased variation of pitch (i.e., more monotonous) caused the perception of both to decrease. Changes in average pitch seem to have little effect on perceptions of benevolence and competence.

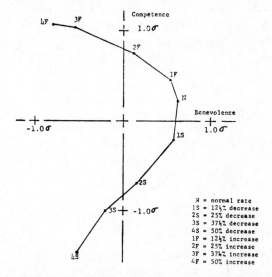

Figure 26.10. Perceived benevolence and competence as functions of speech rate. Nine levels of rate were used varying from 50% lower than normal (4S) to 50% higher than normal (4F). Benenvolence is maximum at normal speech rate. Competence increases with speech rate. (From Smith etal, 1975.)

Exercises

1. Choose a sentence similar to "John drove to the store." Determine different meanings that can be imposed on the sentence by stressing different words. Draw some idealized pitch contours that you might expect to obtain from the different versions.

2. A person who is fluent in a tone language temporarily loses his voicing ability. He resorts to whispered speech. How successful is he likely to be with

the tonal aspects of the language? With the non-tonal aspects?

3. The tone contours shown in Figure 26.11 were produced by a native speaker of Mandarin for a single word. Identify each contour as one of the four Mandarin tones by referring to Figure 26.6. Which contours are obvious? Which are ambiguous? Which mandarin tones are most likely confused?

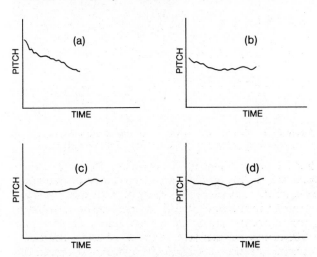

Figure 26.11. Tone contours for each of the four tones of Mandarin.

4. A talker with an "average" speaking rate has a recording of his speech speeded up. How will others perceive his benevolence to change? His competence?

5. Repeat Exercise 4, this time slowing the playback of the recording.

6. A person's speech is made more monotonous. What effect does this have on perceived competence and benevolence?

Demonstrations

1. Stand with your back against a wall and your eyes closed while sounding a steady vowel. Have a friend give your lower chest a push at an unexpected time so you cannot react explicitly. What happens to your fundmental frequency and sound level?

2. Use a sound spectrograph to produce sound spectrograms for selected sentences, as produced normally, in anger, and in depression. What do you observe for fundamental frequency and duration of the angry and depressed versions as compared with the normal?

3. Repeat Demonstration 2, this time measuring fundamental frequency for stressed words in sentences.

4. Repeat Demonstration 2, this time measuring fundamental frequency for various tones of a tone language.

5. Get a friend (perhaps a budding dramatist) to speak the sentence "You always get the same results," while speaking normally and then while simulating anger (or excitement) and depression. For each condition, list the nature of each of the following features: pitch, stress, and rhythm.

6. Get several different people to speak the sentence "You always get the same results," while speaking normally. For each person, list the nature of the following features: pitch, stress, and rhythm.

7. Get a competent speaker of a tone language to

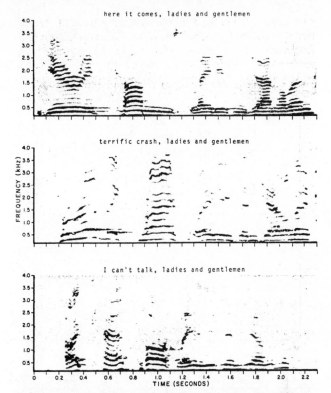

Figure 26.12. Spectrograms for a talker simulating the radio announcer describing the arrival and crash of the Hindenburg. Compare with Figure 26.8. (From Williams and Stevens, 1972.)

illustrate the tones of the language. Note the pertinent features.

Further Reading

Denes and Pinson, Chapter 8

Ladefoged, Chapters 5,10

Pickett, Chapters 5, 11

Singh and Singh, Chapter 8

Fromkin, V.A., ed. 1978. **Tone: A Linguistic Survey** (Academic Press)

Fry, D.B. 1958. "Experiments in the Perception of Stress," Language and Speech 1, 126-152.

Ladefoged, P. 1963. "Some Physiologial Parameters in Speech," Language Speech 6, 109-119.

Lehiste, I. 1979. **Suprasegmentals** (M.I.T. Press).

Melby, A. M., W. J. Strong, E. G. Lytle, and R. Millett. 1977. "Pitch Contour Generation in Speech Synthesis: A Junction Grammar Approach," Am. J. Computational Linguistics, microfiche no. 60.

Smith, B. L., B. L. Brown, W. J. Strong, and A. C. Rencher. 1975. "Effects of Speech Rate on Personality Perception," Lang. Speech 18. 145- 152.

Ting, A. 1971. "Mandarin Tones in Selected Sentence Environments: An Acoustic Study," J. Acoust. Soc. Am. 51, 102.

Williams, C. E., and K. N. Stevens. 1972. "Emotions and Speech: Some Acoustical Correlates," J. Acoust. Soc. Am. 52, 1238-1250.

Audiovisual

1. "Prosodic Features in Speech," (GRP)

2. **The Sounds of Language** (32 min, 1962, IU)

27. Defects in Speech Production and Perception

We will consider only three causes of speech defects in this chapter: cleft palate, impaired vocal folds, and impaired hearing. Each of these defects varies in severity from one case to another, and each has a different effect on the production of spoken language. Impaired hearing, if it is substantial, is probably the most debilitating in terms of its effects on the whole process of spoken-language communication. Thus, we will conclude the chapter with a discussion of sensory aids for the hearing impaired.

Cleft Palate

Speech produced by a person with a cleft palate has what is perceived as a nasal quality. Let us digress a bit at this point and consider what produces a nasal quality for a normal speaker. We have assumed that the vocal tract is a single tube of variable cross-section, as shown in Figure 27.1a. This assumption is valid for most speech sounds, with the exception of the nasal sounds /M/, /N/, and /NG/. The soft palate serves as a valve in the vocal tract to control the sound going into the nasal cavity. For nasal sounds, the soft palate is lowered and the nasal cavity is connected to the rest of the vocal tract. Also, the mouth cavity typically is closed at some point, and we have a tube combination, as shown in Figure 27.1b. The combination of two tubes gives rise to a more complicated set of resonances than those of a single tube. The resulting spectrum for a nasal sound gives the sound its particular quality. (Other speech sounds also have a nasal quality when the soft palate is partially lowered while the vocal tract shape and excitation are otherwise normal for the intended sound. A nasalized sound can be considered to be a mixture of the intended sound and some nasal features.)

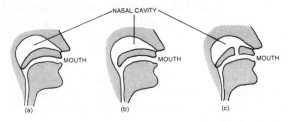

Figure 27.1. Vocal tract configurations showing (a) no nasal tract coupling, (b) nasal tract coupling, and (c) cleft palate giving an ever present coupling to the nasal tract.

A **cleft palate** results when, at birth, the palate fails to fuse and there is an opening from the mouth cavity into the nasal cavity, as shown in Figure 27.1c. Since the nasal cavity is always connected to the vocal tract, the resulting speech sounds have a nasal quality. In addition, a cleft palate talker may not be able to develop sufficient pressure in the oral cavity for the production of stops and fricatives. The usual correction for cleft palate is surgery to close the hole in the palate. This operation is generally successful for the most part.

Impaired Vocal Folds

Hoarseness as a voice quality is symptomatic of vocal fold problems. The vocal folds may become inflamed and swollen which can cause irregularities in their vibration. If the inflammation and swelling are very severe, the ability to produce voiced sounds may be lost entirely for some period of time. Inflammation and swelling of the vocal folds can be caused by infections or abuse such as screaming, and yelling, or prolonged overuse. Usually resting the folds, where necessary, and receiving proper medical attention, are adequate to overcome these problems.

Chronic misuse of the vocal folds—lecturing or singing for extended periods of time, producing higher pitch or loudness levels than appropriate, using the folds when they are in an inflamed condition, or smoking—can give rise to temporary or permanent vocal fold impairment. The impairment may take the form of inflammation and swelling during which time the vocal folds are very susceptible to hemorrhaging. In other cases the impairment may take the form of nodular growths (malignant or benign) on the vocal folds. These usually occur on the edges of the folds that collide and may be due to repeated small scale bleeding. Sometimes these nodules obstruct the folds so they cannot completely close; in other instances they cause the folds to vibrate erratically. Either of these conditions can cause a harsh vocal quality. Sometimes surgery is necessary to remove growths from the vocal folds, and occasionally an entire fold must be removed if the growth is extensive and/or malignant.

Even if a major portion of a vocal fold is removed, it may still function for voice production. If one of the folds is removed entirely, it is often possible to surgically replace it with a stationary structure against which the remaining fold can vibrate. In either of these cases the speech quality will usually be harsh and breathy. When the use of both folds is lost it is often possible to surgically form a constriction with the lower pharyngeal muscles. Air can be trapped in the esophagus, and forced through this constriction, thus causing it to vibrate and interrupt the air flow. This produces so called esophageal speech which is generally weak and not always easy to control, especially when the talker is under emotional stress.

Another method of producing quasi-voiced speech employs an **artificial larynx**, a device for electronically producing a buzzing sound that can be coupled into the vocal tract through openings in the tract or through some of the softer tissues of the tract in the vicinity of the larynx. The vibrating head of the artificial larynx is placed in contact with the external larynx structure and caused to vibrate while the user produces the vocal tract configurations that are appropriate for the intended speech sounds. There is also a mechanical artificial larynx in which air from the lungs passes through an opening in the vicinity of the larynx, through a mechanical vibrator, and is introduced into the corner of the mouth. The user forms the vocal tract shapes that are appropriate for intended speech sounds.

Speech Characteristics of the Hearing Impaired

Scientists at Bell Telephone Laboratories have coined the term "speech chain" to refer to the different levels involved in the speech communication process. They suggest five levels in the speech chain: (1) the linguistic level of the talker (the concept or idea to be communicated), (2) the physiological level of the talker (neural and muscular activity of the speech production mechanism), (3) the acoustical level (speech sounds transmitted from talker to listener), (4) the physiological level of the hearer (mechanical and neural activity of the speech reception mechanism), and (5) the linguistic level of the hearer (the concept or idea received by the listener).

For the normal-hearing child, the development of speech seems to occur so easily and automatically that we are really unaware of the complexity of the process. It is only when we observe the great difficulties in the acquisition of language that are encountered by children with impaired hearing that we can begin to appreciate the intricacies of the language acquisition process. Spoken language is the most common and widely used form of language. Persons without relatively normal speech-communication abilities are placed at a great disadvantage in a society in which spoken language is the basic form of interpersonal communication.

Substantial hearing impairments (particularly when they occur before a child has acquired facility in the use of language) can adversely affect a person in three major areas: reception of spoken language, production of spoken language, and acquisition of language. When the reception of spoken language is impaired, normal language development is hindered. It is possible to create substitutes for spoken language, such as manual ("sign") languages, but these are often not so rich in the features of the language as is the spoken language. Furthermore, in societies in which the spoken language normally serves as the host for language development, substitutes for the spoken language are inconvenient and impractical in many instances even though they may be adequate on other counts.

Some substitute systems for spoken language have proven partially successful for the reception and acquisition of language. However, they have not produced significant successes in speech production for people with impaired hearing. One reason may be that substitute systems do not give the talker any feedback regarding his own speech production. One possible explanation for this is that the substitute systems are not rich enough in spoken language features to provide adequate feedback.

Consideration of feedback systems provides useful insights into the profound effects hearing impairments have on the speech process. A system without feedback implies a collection of several components that function together to produce an output from an input in a predetermined manner. A **feedback system** is a system in which a means has been provided for sensing the output of the system and then modifying the input of the system on the basis of the measured values of the output. A feedback system is illustrated in Figure 27.2. Many systems in our everyday world are feedback systems. When driving an auto on the highway, you are able to keep on the road by means of visual feedback. (Imagine how absurd it would be for a blind person to attempt to drive an automobile.) Even such common tasks as walking and eating require feedback.

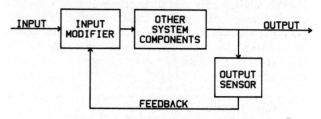

Figure 27.2. Elements of a feedback system.

The output of the speech production system is sensed by the hearing system, and appropriate modifications are made in the control signals being sent to the speech production mechanism. A hearing impairment tends to break the feedback loop of the speech chain and thus to impair speech production because it cannot be monitored. A severe hearing impairment in a very young infant often causes the infant to stop vocalizing altogether—even though the infant may have begun babbling at a normal age. Hearing ability permits a speaker not only to monitor his own speech but also to monitor his speech in relation to the speech of others. A person with normal hearing can continually calibrate and recalibrate his own speech production against that of others, but a person with severely impaired hearing is unable to monitor his own speech—let alone the speech of others.

The effects of a broken feedback loop on speech production can be seen quite clearly by comparing Figure 27.3 with Figure 27.4. Figure 27.3 shows three formant frequencies, fundamental frequency, and sound level graphed as functions of time for the utterance "The leaves will be," spoken by a normal- hearing talker. Figure 27.4 shows similar information for the utterance "The leaves will," spoken by a severely hearing-impaired talker. The time scales are the same in both figures. Note the overall difference in duration in the two cases and the difference in detail of the curves. The differences are presumably due to a broken feedback loop that prevents the hearing impaired talker from monitoring his own speech and from calibrating it against the speech of others.

There is considerable correlation between subjects' audiograms and the confusions they make in speech perception. For example, subjects with high-frequency hearing loss fail to identify noise-like, high-frequency fricative sounds. Speech production errors are highly correlated with speech perception errors; when a subject is unable to perceive a given speech sound reliably he is unable to produce it reliably.

Severely hearing-impaired talkers tend to speak more slowly (as much as one-third, one-half, or even more slowly) than do normal-hearing talkers. They fail to make duration distinctions between stressed and unstressed syllables. Their speaking rhythm is poor and the length and number of pauses is excessive. Ar-

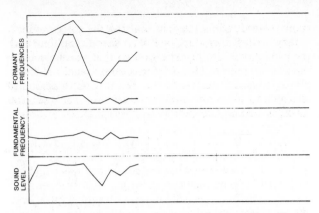

Figure 27.3. Speech parameters for normal-hearing talker. The utterance was "The leaves will be . . . ."

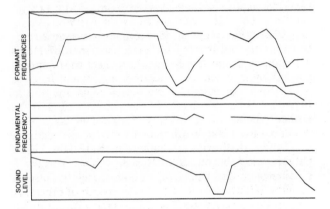

Figure 27.4. Speech parameters for severely hearing-impaired talker. The utterance was "The leaves will . . . ."

ticulatory movements requiring precise timing—such as time before onset of voicing after a plosive burst which differs between voiced and unvoiced plosives—are very troublesome.

The hearing impaired often fail to produce proper formant frequencies for vowel sounds. Distinctions between voiced and unvoiced phonemes are often not made. Consonants are often omitted or substituted. One of the most severe problems occurs for speech sounds in sequence. Nasalization is often inappropriate in amount and time of occurence.

As noted in Chapter 26, intonation may play a very significant role in communicating meaning beyond that of the string of phonemes. Pitch level as used by the hearing impaired is often too high, may have sudden and abrupt breaks, and typically is not varied appropriately. Inappropriate loudness and breathy or harsh voice quality are often typical.

It should again be emphasized that the speech characteristics of the hearing impaired are due to a breakdown of the feedback loop that permits a talker to hear and modify his own speech and to calibrate it against that of others.

Sensory Aids for the Hearing Impaired

The most commonly used device for aiding the hearing impaired is the hearing aid, discussed in Chapter 17. Acoustical amplification is probably the most useful sensory aid for the hearing impaired when there is adequate residual hearing. More sophisticated

hearing aids make it possible to provide frequency selective amplification, frequency compression, and amplitude compression of the speech signal. Some hearing aids have been designed to take part of the high-frequency energy of speech and present it at low frequencies where the hearing-impaired person usually has the most residual hearing. Most of these aids have met with only marginal success, probably due to inadequate design and testing.

The crucial problem is to develop sensory aids that will give hearing impaired persons access to speech. One class of aids may be regarded as mapping devices. Several auditory mapping devices perform spectral analyses, frequency compress the spectrum into a range of from 90% to 25% of the original, and then synthesize the speech for presentation in the limited residual range of the listener. Other devices perform spectral analyses and map the spectrum "tactilely" via vibrators strapped around the waist or attached to the fingertips. Yet other devices map the spectrum for presentation to the visual senses. Some devices convert the spectrum to vocal tract shapes which are presented visually.

An attempt has been made with a number of devices to extract features that complement information available from speechreading. "Cued Speech" (see references) was designed as a manual system in which hand cues would disambiguate the cues seen from the talker. Attempts are being made to automate the system. An eyeglass speechreading aid displays information on voicing, frication, and plosives which are hard to gain through speechreading. One vibrotactile aid presents signals to each of three fingers of a user. The signals relate to nasality, voicing, and frication which are hard to obtain in speechreading.

Other devices have been designed to train hearing-impaired persons on selective speech production deficiencies. These include proper nasalization, adequate production of fricatives, proper pitch level, proper intonation patterns, and so on. A disadvantage of training systems tested to date is that they concentrate on speech features in isolation. Even though the user learns to use the feature in isolation, he is often unable to use it effectively in continuous speech.

With the exception of hearing aids, sensory aids have not yet been tested extensively and so their value is not known. It seems reasonable to conjecture that wearable sensory aids for speech reception will prove more useful than training aids because they become a part of the individual who learns to interpret their outputs in terms of day-to-day encounters with spoken language. The value of these devices should increase with the passing of time because of the subject's greater experience with them. Because the hearing-impaired person's frequency bandwidth for speech reception is substantially reduced as compared to that of a normal-hearing person, it will be necessary to perform partial analyses of the speech signal and present to the user's ear a modified signal whose bandwidth requirements are much less than that of normal speech. Or it may be necessary to develop sophisticated and reliable "speech feature extracting devices" that present signals to the tactual or visual senses if residual hearing is inadequate.

Exercises

1. The speech produced by a person having a cleft palate usually has a nasal quality associated with it. Why?

2. What is the best "cure" for a cleft palate? How is this "cure" accomplished?

3. How does an artificial larynx function to produce quasi-voiced speech? Is the vocal tract used at all when an artifical larynx is used?

4. Describe the components, functions, and inter-relationships at the physiological level in the speech chain for a talker.

5. Describe the components, functions, and inter-relationships at the physiological level in the speech chain for a listener.

6. Describe and discuss some speech features that are available to a listener in the acoustical level of the speech chain that may not be available in some manual means of communication. How important are these features in the perception of speech? In the acquisition of language? In the production of speech?

7. Consider a warm air furnace as a feedback system. Describe the function of each of the following components in the furnace: thermostat, blower, cold air, warm air, gas, firebox. What happens if the thermostat is not working?

8. What are the components of the speech system? What is the feedback in the speech system? What is the input? What is the output? What senses (or measures) the output?

9. What is the role of feedback in singing? In performing on a musical instrument?

10. What musical instruments do you suppose a severely hearing-impaired person can perform on most successfully? Least successfully? Why?

11. There are indications that development of speech in young children proceeds about as rapidly as the development of the speech and speech feedback mechanisms. What happens if the feedback mechanism fails to develop properly?

12. Suppose you have a task of developing a speech reception aid for the severely hearing-impaired. What speech parameters would you attempt to extract from the speech signal? How would you present them to the user? What frequency bandwidth would be required?

13. Compare the speech parameters of normal and hearing-impaired talkers as shown in Figures 27.3 and 27.4. Describe what you observe for relative movements of the formants for the two talkers. What are the differences in the fundamental-frequency contours? Which is more monotonous? What can you say about the relative speech rates for the two talkers? What can you say about the rhythm for the two talkers? Which has the more smoothly-flowing speech?

Demonstrations

1. The role of normal feedback in speech can be demonstrated in the following way. If a talker hears his speech much later than he speaks it (delayed feedback), his speech often becomes somewhat confused and disoriented. This can be demonstrated by having a talker speak into a microphone and then listen to his speech over headphones that seal out the direct sound. A time delay is introduced between the microphone and the headphones. Adjust the gain so that you hear only the delayed speech. Now read several sentences of some printed material and record the result with an auxiliary microphone. Play the auxiliary recording back several times, noting any deviations from normal pitch, stress, and rhythm.

2. Obtain an artificial larynx (usually available through the local telephone office). Place it near your Adam's apple and use it to produce voiced speech. (You may need to experiment a bit to get it positioned optimally.) Perform spectral analyses of your natural and aritficial larynx speech.

3. Plug your ears with your thumbs or a hearing protector to produce "partial hearing loss." Get someone to stand a meter or two in front of and facing you. Have the person speak softly as you carry on a conversation. Then have them face away while conversing. How important is hearing for perceiving speech? How important is sight, especially when there is a partial hearing loss?

Further Reading

Borden and Harris, Chapter 4

Davis and Silverman, Chapters 7, 10, 13-15

Denes and Pinson, Chapters 6-8

Flanagan, Chapter 7

Fletcher, Chapter 19

Levitt, Pickett, and Houde, Parts III-VII, IX

Proctor, Chapters 9, 10

Stevens and Warshofsky, Chapter 7

Calvert, D.R., and S.R. Silverman. 1976. **Speech and Deafness** (Alexander Graham Bell Association).

Ling, D. 1976. **Speech and the Hearing Impaired Child: Theory and Practice** (Alexander Graham Bell Association).

Pickett, J.M., and J. Martony, ed. 1978. **Proceedings of the Research Conference on Speech-Processing Aids for the Deaf** (Gallaudet College).

Stark, R.E. 1977. "Speech Acquisition in Deaf Children," Volta Review **79**, 98-109.

Stark, R.E., ed. 1974. **Sensory Capabilities of Hearing Impaired Children** (University Park Press).

Strong, W.J. 1975. "Speech Aids for the Profoundly/Severely Hearing Impaired: Requirements, Overview, and Projections," Volta Review **77**, 536-556.

Audiovisual

1. Tape recordings of utterances produced by severely hearing-impaired talkers.

2. **The Speech Chain** (19 min, color, 1963, BELL)

3. **Alaryngeal Speech** (21 min, 1966, UKANMC)

4. **Children With Cleft Palates** (29 min, color, 1957, UMICH)

5. **Listen** (30 min, color, 1973, USNAC)

6. **Silent World, Muffled World** (28 min, color, 1966, USNAC)

7. **The Function of the Pathologic Larynx** (24 min, color, ILAVD)

28. Degraded Speech

Almost all speech communication suffers from some kind of degradation. Such degradation is typically due to added noise, limited bandwidth of the transmission system, or peak clipping. Intentionally degraded speech can help us to gain a greater understanding of the speech features that are important in preserving speech intelligibility and talker identity.

Speech Testing

Several kinds of tests are used to assess degraded speech. The most common is an intelligibility test used to measure how much of the original speech information is transmitted under a given set of conditions. Such intelligibility testing will be considered in more detail below. Quality testing is done to determine how natural sounding various speech samples are or whether listeners can identify a talker. When a speech signal has been badly degraded or modified (such as is the case for low frequency speech codes for the hearing impaired), intelligibility and quality tests may not be useful. In such cases discrimination testing can be used to determine whether listeners can detect a difference between two speech-related signals.

In practice, an **intelligibility test** consists of sending a set of speech signals through some kind of speech transmission system, such as a telephone line, a public address system, or even just from one place in a room to another. At the receiving end of the system, listeners are asked to identify the speech stimuli that were sent over the system. An **intelligibility score** is the percentage of the speech signals that were correctly identified. The speech stimuli most often used in intelligibility tests are single words, because single words are easier to deal with than sentences in test situations. The test words typically come from one of the following categories: spondees (two-syllable words), phonetically balanced words (single-syllable words that have occurrences of different phonemes in the same proportions as they occur in the natural language), and rhyming words (words that differ in a single phoneme, such as pat, bat, or mat). Intelligibility tests run with spondees and phonetically balanced lists require the listener to write what he heard and require considerable training of the listeners before the results are reliable. Rhyming lists are typically used with untrained listeners, who are required only to mark the word that was received out of a set of two or more possible words. Rhyme tests can be administered with less effort on the part of the listeners, who do not have to be specially-trained.

The Diagnostic Rhyme Test (DRT) provides an example of an intelligibility test which uses rhyming words. Two rhyming words in which the initial consonants differ by a single "distinctive feature" are used. For example, the words "bat" and "pat" have initial consonants differing only in the voicing feature; /B/ is voiced and /P/ is unvoiced. Listeners hear one of the two words and mark which one they think they heard.

The results must be adjusted for guessing because there is a 50% chance of choosing the word spoken even if the listener could hear nothing. The DRT tests the intelligibility of six speech attributes including voicing, nasality, and sibilation (the presence of noise-like energy). Word pairs illustrating these three attributes are bean-pean, mad-bad, and jot-got, respectively.

Intelligibility tests using only single words eliminate contextual information that is very important in everyday speech communication. However, there is a fairly close relationship between the intelligibility score and the success with which a normal conversation might be expected to be carried on. Intelligibility scores below about 50% indicate that normal conversation would be difficult to carry on. However, scores from 80% to 100% indicate that normal conversation should be very successful; i.e., with the additional information supplied from context virtually all of the conversation should be intelligible to the participants.

Additive Noise

Additive noise is probably the most common degradation of speech in communication systems. For instance, it is not uncommon when using the telephone for speech communication (particularly on long-distance calls) to get a "bad connection" or a "noisy line." Sometimes the talker is in a noisy environment and the rest of the transmission system simply transmits the noise along with the speech. On the other hand, the listener may be in a noisy environment. In any one of these cases, noise (some unwanted signal) is competing with the speech signal. The reduction of intelligibility scores in the presence of noise depends upon the energy in the speech signal relative to the energy in the noise signal. Table 28.1 gives typical DRT values for intelligibility (expressed as a percentage) for different values of signal-to-noise ratio (expressed in dB). A signal-to-noise ratio of 0 dB means that the speech and the noise are equal in sound level. A positive value of say 6 dB tells us that the speech sound level is 6 dB greater than the noise sound level, whereas -6 dB indicates the speech is lower by 6 dB than the noise. Note that when the signal-to-noise ratio is 6 dB or greater we might expect communication to be relatively successful.

Table 28.1. Intelligibility as a function of speech-to-noise ratio.

Speech-to-noise ratio (dB)	Intelligibility (%)
-12	0
-6	15
0	42
6	67
12	84
18	93
24	96
30	98

A segment of waveform for a synthetic diphthong /O/EE/ and its corresponding spectrogram are shown in Figure 28.1. The first and second formant transitions are quite apparent. Figure 28.2 shows a similar waveform and spectrogram for the same /O/EE/ but with an equal amount of noise added. Some feeling for the effect of additive noise can be gained by comparing the two figures.

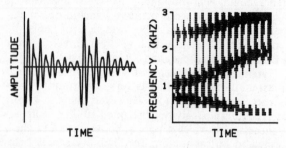

Figure 28.1. Waveform and spectrogram for synthetic diphthong /O/EE/. Two cycles of the waveform for /O/ are shown. The spectrogram is for the complete diphthong.

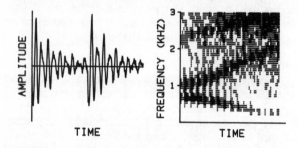

Figure 28.2. Waveform and spectrogram for synthetic diphthong /O/EE/ with added noise.

Bandwidth Limitations

Limited bandwidth is probably the second most common cause of speech degradation. Normal speech may have energy at frequencies ranging from 60 to 10,000 Hz. Most communication systems are not equipped to transmit this wide range of frequencies. (For instance, a telephone system transmits frequencies in a range of about 300-3000 Hz, and typical public address systems are probably poorer.)

It is of interest to know what frequencies must be transmitted in order to achieve good speech intelligibility. This can be determined by taking speech that has an intelligibility score near 100%, passing it through various filters, and determining new intelligibility scores for the speech under the filtered conditions. The effect on the spectrum of passing white noise (a random signal having equal amounts of energy at all frequencies) through three different filter setups is illustrated in Figure 28.3. Note that the low-pass filter leaves the low frequencies in the spectrum but eliminates, or at least reduces, the higher frequencies. The high-pass filter lets high frequencies pass through and eliminates low frequencies. The band-pass filter lets a band of frequencies get through, but it eliminates frequencies that are either lower or higher than the pass band. Table 28.2 illustrates how low-pass and high-pass filtering affect the intelligibility of speech.

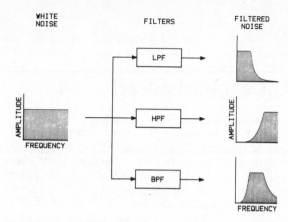

Figure 28.3. Spectra of unfiltered white noise, low-pass filtered noise, high-pass filtered noise, and band-pass filtered noise.

Table 28.2. Intelligibility scores of speech, low-pass filtered or high-pass filtered, at different cutoff frequencies.

Cutoff frequency (Hz)	Intelligibility of low-passed speech (%)	Intelligibility of high-passed speech (%)
100	0	98
200	0	97
500	7	96
800	18	93
1000	26	89
1250	38	85
1500	50	80
1750	64	74
2000	70	65
3000	88	30
4000	93	13
5000	95	5

The waveform for the /O/EE/ of Figure 28.1 is shown in Figure 28.4 after it has been through a band pass filter whose low frequency cutoff is about 500 Hz and whose high frequency cutoff is about 1500 Hz. The corresponding spectrogram also is shown in Figure 28.4. Note that the third formant is missing throughout the spectrogram and that the second formant disappears part way through.

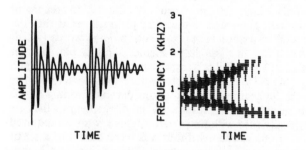

Figure 28.4. Waveform and spectrogram for synthetic diphthong /O/EE/ after band-pass filtering between 500 and 1500 Hz.

Peak Clipping

Many speech transmission systems have limits in terms of the amplitudes of the signals they are able to transmit without producing appreciable distortion. When a signal amplitude exceeds the limits of a system, "peak clipping" may occur. Peaks of the signal that exceed the limits of the system are chopped off. Peaks of the signal that do not exceed the limits of the system pass through the system essentially undistorted. The waveform for the /O/EE/ of Figure 28.1 is shown in Figure 28.5 after it has been peak clipped at a value equal to one-sixth of its peak value. The corresponding spectrogram is also shown in Figure 28.5. Even though distortions are apparent in the spectrogram, the formants are still largely preserved.

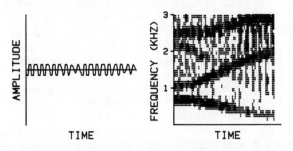

Figure 28.5. Waveform and spectrogram for synthetic diphthong /O/EE/ after peak clipping.

Peak clipping reduces the "naturalness" of speech. This is probably because the clipping process introduces additional amounts of high-frequency energy into the signal. (Any time a waveform is modified so that it changes from one instant to the next in a fashion more abrupt than usual, additional high-frequency energy is introduced.) Although peak clipping reduces the naturalness of the speech signal, it leaves the intelligibility fairly intact.

Center Clipping

Center clipping of a speech signal is of less practical interest than peak clipping because it does not occur in typical speech transmission systems. Portions of the signal that have the smallest amplitudes are clipped out. The waveform for the /O/EE/ of Figure 28.1 is shown in Figure 28.6 after it has been center clipped at a value equal to one-tenth of its peak value. The corresponding spectrogram is also shown in the figure. Distortions of the formants are apparent in the spectrogram.

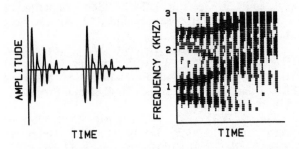

Figure 28.6. Waveform and spectrogram for synthetic diphthong /O/EE/ after center clipping.

Center clipping reduces the naturalness of a speech signal in somewhat the same way as does peak clipping by introducing additional high frequency energy. However, center clipping reduces the intelligibility of a speech signal much more than peak clipping does.

Modified Parameters in Speech Synthesis

Parameter modification is a form of intentional speech degradation. Parameter modification is performed so that the relative importance of different speech features can be assessed with regards to speech intelligibility, speech recognition, speaker identification, and naturalness of speech. There are basically two ways in which parameter modification can proceed: (1) analysis of natural speech and modification of the resulting parameters before synthesis, or (2) synthesis of speech from artificially supplied parameters. Usually the latter procedure is guided, at least in part, by the former procedure; otherwise it is too inefficient.

It is possible with modern speech processing to take naturally-produced speech and analyze it in terms of different sets of time-varying parameters. Some parameters in common use are formant frequencies and amplitudes, energies in different frequency bands, or predictor coefficients. (See Figure 25.6 for an example of time varying formant frequencies and amplitudes obtained from the analysis of natural speech.) The parameter "sets" can be used by speech synthesizers in reconstructing a version of the speech signal. The synthesized speech is in many cases almost as intelligible as the original speech and in some cases almost as natural sounding.

If speech synthesized from a parameter set is virtually indistinguishable from the original speech in terms of intelligibility and naturalness, then it is probably reasonable to assume that the parameter set adequately characterizes the speech. We can then use this parameter set to systematically study various speech features.

Suppose we wish to discover the primary differences between male and female speech. We can start by noting how the typical parameter values differ in the two cases. We would probably find that on the average, fundamental frequencies for female speech are about an octave higher than those for male speech. We would also probably find that formant frequencies of female speech are about 15-20 percent higher on average than for male speech. To test our observations, we might then take male speech and systematically increase the fundamental frequency and/or the formant frequencies and ask listeners to make judgments on whether the speech was produced by a male or a female. Similarly, the fundamental frequency and/or the formant frequencies of different samples of female speech could be adjusted to lower values and a similar listening test could be run. We would probably find that there is considerable overlap in our parameters in terms of speech that is perceived as being spoken by females or males. However, we would probably find that at the extremes of the parameter ranges the judgments are more consistent.

We can test the effect of the fundamental frequency on speech intelligibility by synthesizing speech according to certain speech parameters after modifying the fundamental frequency. We can then run tests to assess what effect, if any, the fundamental frequency has on intelligibility. We might run comparable tests to assess the influence of fundamental frequency on naturalness. The kinds of tests that we might run are almost limitless.

Of practical significance are tests to determine what parts of the speech signal are most important to our perception of plosives, fricatives, vowels, and so on. We might address ourselves to such questions as: What role do formant transitions play in the perception of stop consonants? How necessary is the burst of noise energy to the perception of unvoiced plosives? Do the relative durations of the burst and the interval to onset of voicing have significant influence on distinguishing between voiced and unvoiced plosives? What spectral characteristics are necessary to distinguish between /S/ and /SH/ for example? The answers to these and other questions might be acquired by analyzing natural speech and then systematically modifying various parameters, such as the elimination of formant transitions, elimination of noise bursts, change of the interval from burst to onset of voicing, and so forth.

An alternate procedure to that of analyzing speech, modifying parameters, and synthesizing speech, is to create stylized parameters for controlling a speech synthesizer. For instance, with a three-formant speech synthesizer (similar to the one discussed in Chapter 25) we could investigate the role of formant transitions in the perception of various phonemes. The Pattern Playback machine developed by Haskins Laboratories has been used extensively to investigate some of these questions. Still more modern speech synthesis devices might still be used effectively to investigate these and other questions and to provide refinements to questions already studied.

Exercises

1. What do intelligibility tests measure?

2. What intelligibility score would you expect to obtain for a "good" telephone transmission if its bandwidth is limited to the range 300-3000 Hz?

3. What intelligibility score would you expect to obtain for a high-quality tape recording system if the overall system response is 60-15,000 Hz?

4. What is the most common form of speech degradation in the speech transmission systems in use today? Why?

5. What is probably the second most common form of speech degradation in speech transmission systems? Is this form of degradation a problem in face-to-face transmission?

6. What does a band-pass filter do? If you had an adjustable band-pass filter with a pass band of 1000 Hz, where would you place the pass band to achieve maximum speech intelligibility?

7. Peak-clipped speech is quite intelligible, even though it does not sound natural. What is the difference between intelligibility and naturalness? What does the peak clipping do to formant structure? Refer to Figure 28.5.

8. Center-clipped speech is neither natural sounding nor intelligible. If the claim that intelligibility of speech is largely dependent on an adequate formant structure is true, what does center clipping do to the formant structure of speech? Refer to Figure 28.6.

9. Suppose that with a speech analysis-synthesis system you are able to analyze speech, obtain the formant frequencies and amplitudes, and then modify the control parameters before synthesizing the speech. Suppose you synthesize speech using only one of the formants. You then run intelligibility tests on the synthetic speech. Which formant will you probably find to be most significant in speech intelligibility? Which formant will probably be least significant in speech intelligibility? Support your supposed results. (Hint: Use Tables 24.1 and 28.2.)

10. Suppose for some reason you want to analyze the speech of an adult male and then synthesize it so that it sounds more like that of an adult female. How would you modify the formant and fundamental frequencies?

11. A pulse train (a series of repetitive pulses) having a flat-line spectrum (see Figure 6.10E) is substituted for the white noise in Figure 28.3. It passes through low-pass, high-pass, and band-pass filters. Draw stylized line spectra appropriate for each case.

12. How would the waveform from Figure 28.1 look if it were peak clipped at one-third its maximum value? Sketch the result.

13. How would the waveform from Figure 28.1 look if it were center clipped at one half its maximum value? What would its spectrum be?

Demonstrations

1. Use a variable filter to demonstrate low pass, high pass, and band pass filtered speech

2. Listen to tapes of speech degraded in different ways and write down what you hear. Compare your responses with what was actually said and determine your intelligibility score for each condition. What degradations produce the greatest loss of intelligibility? Why?

3. Have someone speak rhyming words (such as bat, pat, mat, etc.) for you to identify. First do this in quiet surroundings and then with various levels of noise. Do this when you can see the talker's face and then repeat when you cannot see the talker's face.

Further Reading

Denes and Pinson, Chapter 8

Flanagan, Chapter 7

Fletcher, Chapters 6, 15-18

Pickett, Chapters 11, 12

Rossing Chapter 16

Cooper, F.S., P.C. Delattre, A.M. Liberman, J.M. Borst, and L.J. Geerstman. 1952. "Some Experiments on the Perception of Speech Sounds," J. Acoust. Soc. Am. **24**, 597-606.

Keeler, L.O., G.L. Clement, W.J. Strong, and E.P. Palmer. 1976. "Two Preliminary Studies of Predictor-Coefficient and Formant-Coded Speech," IEEE Trans. Acoust., Speech and Signal Processing **ASSP-24**, 429-432.

Voiers, W.D. 1977. "Diagnostic Evaluation of Speech Intelligibility," in Benchmark Papers in Acoustics series (Dowden, Hutchinson, and Ross).

Audiovisual

1. "Degraded Speech: Filtered, Peak Clipped, Center Clipped " (GRP)

2. "Degraded Speech: Elimination of Formants" (GRP)

3. **Science Behind Speech** (8 min, color, 1964, BELL)

29. Machine Processing of Speech

It is of interest to scientists to try to understand natural processes to the extent that they can create substitute means for carrying out the processes. Much of what we will discuss about machine processing of speech falls into this category. Indeed, we often feel that we have a better understanding of natural processes if we are able to create substitutes for them. The four areas of machine speech processing to be discussed in this section are bandwidth compression of speech, machine synthesis of speech, machine recognition of speech, and speaker identification and verification. Bandwidth compression has been of military and commercial interest for several decades. Active interest in the other three areas has grown many fold in the past few years.

Bandwidth Compression

Speech bandwidth compression is of interest any time bandwidth on a transmission channel is limited. This is a primary concern when the transmission is over long distances, such as is the case with transcontinental or transoceanic cables. At first, it seemed that satellite communication would make unlimited bandwidth available at low cost. However, there is growing concern that the total bandwidth available will not be adequate as demands increase. As a result bandwidth compression may be of interest even for satellite communications.

As an example, suppose a particular undersea cable has a channel capacity of 1,000,000 Hz. If an essentially full-bandwidth speech signal (10,000 Hz) is transmitted, only 100 different speech signals can be transmitted concurrently (actually fewer if we allow some "frequency space" as a buffer between contiguous signals). If telephone-quality speech (bandwidth of approximately 3000 Hz) is transmitted, only about 300 different speech signals can be trasmitted concurrently over the same cable. If a means were devised to compress the speech signal without degrading it too much, it would be possible to transmit more speech signals concurrently over the same cable.

Before discussing particular bandwidth compression systems, let's consider digitized speech signals as an alternative to analog speech signals. Digital signals appear to be the way most information will be transmitted in the future. They offer advantages such as error correction and more flexible ways of combining the signals before transmission. With digital signals we have a different, but very convenient, way of talking about bandwidth or data rate.

Suppose we use telephone quality speech as our standard. It has a bandwidth of about 3000 Hz. A roughly equivalent description is to consider telephone-quality speech digitized by means of an analog-to-digital converter. The speech signal is sampled 6000 times per second. (The digitization rate must be equal to or greater than twice the bandwidth.) By allowing 8 bits for each sample we get good quality. Bits are digits in a base 2 number system as opposed to the base 10 system to which we are accustomed. For instance, in base 10 there are ten digits from 0 to 9, but in base 2, there are only two digits, 0 and 1. Eight bits of information means that we can specify any one of 256 (2 to the eighth power) different values for any particular speech sample. Bandwidth or information rate can then be discussed in terms of bits per second. For the example we are considering, we get a rate of 48,000 bits/second (6000 samples/second times 8 bits/sample), which for convenience we round off to 50,000 bits/second as the information rate for telephone quality speech.

Considerable effort has been expended in trying to code speech so that its bandwidth is substantially reduced. Several different voice coder (vocoder) systems have been devised, including channel vocoders, formant vocoders, linear-predictor vocoders, and "phoneme" vocoders. A **channel vocoder** consists of an analysis module (illustrated in Figure 29.1), a transmission channel, and a synthesis part. The analysis module is composed of a set of band-pass filters adjusted so that they cover the speech bandwidth. Each filter lets a signal pass through that consists of energy at the frequencies in its pass band. The signal coming out of the filter is rectified and then low-pass filtered to produce a signal which is a measure of the energy in the channel at a particular instant, varying rather slowly in time. This signal is sampled 40-100 times per second with 3-6 bits per channel. An additional channel is provided for the fundamental voicing frequency. The signals resulting from the analysis are combined and then sent over a transmission line.

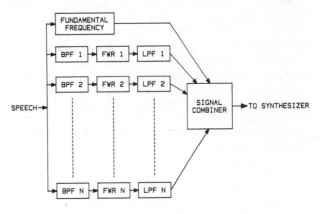

Figure 29.1. Elements of the analysis module of a channel vocoder.

The transmitted signals are received by the synthesis module of the system and used to reconstruct a facsimile of the speech signal. This is accomplished by using the fundamental-frequency signal to control the repetition rate of a pulse generator. The resulting pulse train is input to a set of band-pass filters similar to the ones used in the analysis module of the system. The individual channel signals are used to control the amount of energy coming from each channel in the synthesizer that is to be used to synthesize the speech. The resulting signal is received by the listener.

For the sake of discussion of bandwidth compression, consider a channel vocoder with 20 channels. Assume that each channel is sampled 100 times per second with 6-bit precision, and that the fundamental frequency channel is also sampled 100 times per second with 6-bit precision. The resulting bit-rate is then 21 channels times 6 bits per channel times 100 samples per second or 12,600 bits/second. This represents about one-fourth the bit-rate of digitized telephone-quality speech. This particular vocoder could then compress speech so that four conversations could be transmitted in the same channel capacity as one normal telephone conversation. (Vocoders have actually been built that operate in the range of 1000 to 4000 bits per second and produce reasonable quality speech.)

A **formant vocoder** works on principles similar to those of the channel vocoder. However, the analysis is considerably more sophisticated because formant frequencies and amplitudes must be extracted (see Figure 25.6). At the synthesis end of the vocoder, a pulse train (controlled in frequency by the fundamental frequency signal) is used to excite several formant filters whose frequencies are controlled by the formant amplitude signals. Their outputs are added together to form a facsimile speech signal that is received by the listener. The synthesis module of a formant vocoder is shown in Figure 25.7.

A **linear predictor vocoder** encodes a time-varying spectral envelope, but in a different way from channel and formant vocoders. A channel vocoder as we have seen divides the spectrum into an arbitrary set of fixed-frequency bands and measures the energy in each band. A formant vocoder attempts to individually track the most significant spectral peaks. A linear-predictor vocoder uses a number of parameters (usually about 10) to represent a spectral envelope in terms of a composite of peaks. The peaks are not tracked individually as in a formant vocoder. Each peak depends on all parameters and each parameter affects all peaks. Linear-predictor vocoders are probably the most widely used. They offer a more accurate spectral representation than channel vocoders and they avoid the ambiguities of picking individual spectral peaks which are a problem with formant vocoders. Figure 29.2 illustrates a vowel spectrum obtained by a linear-predictor method and shows how it relates to the actual spectrum.

A **"phoneme" vocoder** represents the ultimate in bandwidth compression. No such vocoder has been successfully implemented to date. In principle it would consist of an analysis module in which the incoming speech would be processed in such a way that the individual phonemes would be extracted. A phoneme code would then be transmitted. At the other end of the system, the message could be typed out to the "listener" or it could be used to control a speech synthesizer to produce a speech signal for the listener. This ultimate compression system could, in principle, reduce the bit-rate required for speech transmission to 20 or 30 bits/second. However, if a typed speech signal is received by the listener, much information that may be of significance

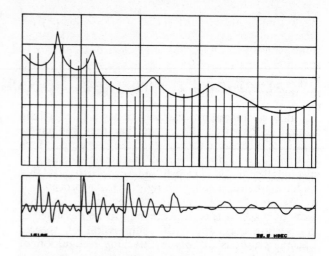

Figure 29.2. Line spectrum (top) obtained from the bracketed single period of a vowel waveform (bottom). The "spectral envelope" (top) is a smoothed spectrum obtained by linear-predictor methods. Frequency is marked off in 1 kHz increments and spectrum level is marked off in 20 dB increments.

is lost. For instance, the listener will receive no acoustical information telling him the talker's identity or emotional state. Clearly, subtleties of the spoken language will be lost. This should remind us that consideration must be given to the desired end results of any speech compression system before any particular system is chosen.

Machine Synthesis of Speech

Talking machines have been of interest for a long time. Sometimes this interest has been motivated by curiosity and sometimes by a real or imagined need for such a device. Early talking machines were mechanical. Later talking machines were built with analog electronic circuitry. Most modern talking machines employ digital electronic circuitry for speech synthesis.

There are many applications of speech synthesis in our modern world. Most applications use machine-produced speech in situations that involve much repetition and that are nonstimulating for a human talker. Examples in this category include time and temperature information, travel schedules and fares, automated inventory, directory assistance, and "talking" consumer goods. In all of these cases information available to a machine is presented as a spoken message.

A second area of application for speech synthesis involves more interaction with the user. Examples in this category include electronic learning aids and talking computers. Speech is the most natural mode of communication for humans. Talking computers capitalize on this by speaking messages to a user. This is especially important for blind users who can key material into a computer, but who have a difficult time receiving computer output. Texas Instruments' "Speak 'n' Spell" was one of the early electronic learning aids which involved user interaction. In one mode of operation a user is asked to spell a word by keying it in. The device then indicates whether the spelling is right or wrong and gives the user a new word to spell or asks the user to try again on words spelled incorrectly.

Two other potentially very useful applications of speech synthesis are in reading machines for the blind and in computer-aided instruction. Both are the objects of continuing research. A reading machine for the blind would consist of a page-scanning device coupled to a speech synthesis system. The page scanner would "read" the various written characters on the page, and the information would then be passed on to a synthesizer. Computer-aided instruction has in the past depended heavily on typed and graphic messages for communication. Since audio communication is so important in the human communication process, it seems reasonable to suppose that a machine capable of "speaking" (and conceivably, of interpreting spoken input) would be better able to "communicate" with a user in a computer-aided instruction environment. Both of these applications imply a fairly extensive dictionary and many sentence structures. Special speech synthesis methods must be employed to meet those requirements.

There are several ways in which information can be made available in spoken form. Pre-recorded messages have been in use for some time. When a request is made, the playback mechanism is actuated and the user receives some kind of recorded message. Prerecorded messages are adequate where a small repertoire of fixed messages is used and when there is limited need for updating. Prerecorded messages are not practical where the message must be extracted from a large volume of data and tailored to a specific user request. Part of the problem is storage limitations. This problem can be resolved in part by parameterizing various parts of the message, thus using bandwidth compression to reduce the amount of storage needed. Furthermore, with parameterized speech it is possible to modify the parameters at word and phrase boundaries to produce more natural sounding speech.

Various methods are used for synthesizing speech. In one method words are spoken in isolation, then digitized and parameterized. For synthesis of phrases and sentences individual words are retrieved from memory and concatenated. Appropriate changes can be made in the prosodic features to produce the desired utterance. In some experiments, concatenated words were found to be as comprehendable to listeners as were naturally spoken words when applied in a system dealing with telephone numbers. Disadvantages of word concatenation are high vocabulary cost and large storage requirements if a large vocabulary is to be used. Furthermore, telephone numbers would seem to be a natural application of word concatenation because they are spoken "quasi-isolated" a digit at a time; more elaborate phrases and sentences would make more severe demands on prosodic features.

Text-to-speech synthesis is a more flexible (and more elaborate) method than word concatenation. In principle it can handle an unlimited vocabulary and unrestrained sentence structure. An utterance (typically a sentence) to be spoken is stored in written form with conventional English spellings. Rules are used to convert a written word into a "phonetic word" of speech sounds. Words that cannot be converted with the rules can be looked up in an exceptions dictionary. Phoneme-related quantities are the usual building blocks in this system. However, when phonemes are concatenated they can take on different values depending on neighboring phonemes. The phoneme /P/, for example, is different in the words "pin" and "spin." Thus, speech sounds are specified by a dictionary of allophones which includes the variations of a phoneme. The phonetic word then exists as a string of allophones. Prosodic information must be available along with the written version of the utterance or it must be supplied by special prosodic rules. The allophone strings and their associated prosodic data are next converted into parameter strings for controlling the speech synthesizer. The synthesizer may employ formant techniques, linear-predictor techniques, or other techniques. When the parameter strings control the synthesizer, it produces an electrical speech signal which can be converted to sound via a loudspeaker or headphones.

In some respects text-to-speech synthesis may be the only true speech synthesis. It gets down to the basics and does not depend on large speech units such as words and phrases spoken by a human. It offers the potential advantage of greater economy for large vocabulary applications and greater flexibility for many applications. But along with all its power and flexibility, text-to-speech synthesis requires detailed specification if quality speech is to result.

Machine Recognition of Speech

Machine recognition of spoken language is a most intriguing challenge. It is probably the most difficult of machine processing tasks. When machine recognition is sufficiently developed, many applications will likely be forthcoming. One such application might be in the area of aids for the hearing impaired, in which a machine recognizer of speech would give the user a written version of the message to work with. Another application might be computer-aided instruction in which a machine recognizer of speech would make it possible for the machine to "listen" and "understand" as well as to speak to the user. Many other actual and contemplated uses of machine recognition of speech include voice actuated typewriters or computer terminals; voice actuated wheelchairs; commercial transactions by voice; voice telephone dialing; voice data entry for materials handling and sorting; conversational and interactive flight information; directory assistance by speaking the spelling of a name. In many of these applications voice entry provides the important bonus of leaving the user's hands free to perform other tasks. Also, speech is the human's most natural and highest capacity modality and no special training is required of the user.

Machine recognition of speech is more difficult than is machine synthesis of speech because the machine must carry a greater burden. In machine synthesis the human listener can compensate for many of the machine's deficiencies and can successfully receive a message with a high degree of intelligibility even when the quality is not as good as that of natural speech. Human speech processors are very flexible in their ability to adapt the processing procedure to accommodate a wide range of speech input; hence the processor's ability to deal with machine-produced speech, along with that of many different talkers, many different dialects, etc. On the other hand, machine recognizers have not yet begun to achieve this human flexibility. A machine geared to accept a rather limited vocabulary for one person may have a difficult time if it tries to accept the same vocabulary from a different talker. Many of the ambiguities and inconsistencies that we tend to ignore in everyday speech communication must be dealt with in one way or another in the machine recognizer.

One finds that it is not possible to perform an acoustical analysis of running spontaneous speech and extract a phoneme string that corresponds clearly to the "intended" phoneme string. When we speak spontaneously we tend to produce a speech signal that carries no more information than is necessry for the task at hand. When we are conversing with a friend on a well-known subject, we tend to be more careless in our phoneme production than if we are conversing with a stranger on a less well-known subject. We are able to make the necessary adjustments in our reception and perception of speech almost automatically. We capitalize on our knowledge of the subject and the vocabulary to provide contextual clues and supply missing or distorted phonemes. Indeed, in conversations of this nature we can often anticipate what friends are going to say before they have completed saying it. The acoustical signal then serves to confirm what we had anticipated, and a few missing or distorted phonemes or even a few distorted words do not impair the effectiveness of our conversation. However, if the subject of conversation is suddenly changed, even with a friend, the success of the communication is very often reduced until each one has been reoriented to the new conversation.

Let us assume that machine recognition of speech is to be attempted in a way that parallels human recognition of speech. Humans receive speech sounds through their ears. Some kind of time-dependent spectrum analysis is carried out in the cochlea. These time dependent spectral parameters pass via the auditory nerve to the brain where speech features are extracted. Strings of speech features are classified as utterances by comparison to speech information previously stored in the brain.

A machine speech recognizer receives speech signals via a microphone which, first of all, may limit the bandwidth. Background noise and inadverdant speaker-related noise (such as heavy breathing and "ums" and "ahs") become part of the signal and create problems later. The speech signal is usually spectrum analyzed in a time dependent way. This may be in the form of band pass filtering (with about 20 channels), narrow band filtering (with about 400 channels), or linear predictor analysis (with about 10 coefficients). Usually, measurements are also made of fundamental frequency and sound level as they vary in time. These spectral, fundamental frequency, and sound level parameters can be used in a number of ways.

There are two basic recognition strategies, but each has many variations in terms of details and constraints. One recognition strategy deals with isolated words; the other deals with connected words. One constraint that can be imposed with either strategy is to do "talker-dependent" recognition in which the recognizer is trained to recognize the speech of "allowed" talkers only. For talker independent recognition, a broad, representative range of talkers may be used to train the recognizer. Other constraints that can be imposed are vocabulary sizes and difficulty; number and complexity of sentence structures; and number and complexity of meanings as restricted by the tasks the system will handle.

Probably the most common strategy used for isolated word recognition is template matching. Templates of appropriate time-varying parameters are created for the words in the vocabulary in a "training" process. When a spoken word is to be recognized a template of its time varying parameters is created. The unknown template is matched against templates stored in the system and the closest match is chosen as the word spoken. It is comparatively easy to define the beginning and ending of a word when spoken in isolation. However, the template of an unknown word must be "time warped" with stored templates to provide the best possible matches. Simple stretching or compressing of a template can account to some extent for different rates of speaking. However, a word spoken by different talkers or even repeated by the same talker will have different relative durations of its syllables. Dynamic time warping in which different segments of word templates are adjusted to each other must be used to provide the best possible match between two templates.

Isolated word recognition by template matching is usually confined to situations involving vocabularies of a few tens to a few hundreds of words; otherwise the number of templates becomes too large. It is practical for isolated digit recognition and other isolated word recognition tasks such as computer commands or machine control.

Word spotting is a technique of searching for occurrences of a given word in continuous speech. It is related to isolated word recognition in that a template is created for the desired word. The template is then compared against all portions of a continuous speech signal to find where it matches.

Continuous speech recognition is more difficult than isolated word recognition because it is difficult to segment the speech parameters into meaningful units. Word boundaries, syllable boundaries, and phoneme boundaries are all ill-defined. Questions arise as to what phonological labels to use—allophone, phoneme, diphone, syllable, or word. Imagine that speech features are extracted from the speech parameters, that segmenting is done, and that the segment is given a phoneme label or labels. A string of phonemes can be used to hypothesize words which can be checked against a stored dictionary. Alternative word strings can be checked against allowed sentence structures and the task environment. The "best" word string presumably represents the intended sentence. When inconsistencies are encountered at one level (parameter, phoneme, word, grammar, or task) interaction with a lower or higher level to select other alternatives may resolve the inconsistencies. Sometimes prosodic information such as fundamental frequency maxima can be used to provide a framework with which to carry out the recognition process.

One approach to machine recognition of continuous speech assumes the form of a task-oriented problem. The task is defined in terms of the vocabulary, the grammar, and the meaningful sentence types. Systems being developed using this approach typically have vocabularies of around 1000 words, use some simplified grammar, and have tasks that involve limited numbers of meaningful sentences. Computer Chess is an example of a task-oriented problem.

As an ultimate application, we might think of a machine interpreter which would serve to accept one spoken language and produce another spoken language. This machine interpreter might consist of a machine recognizer (to convert spoken language into written language), a machine translator (to convert one written language into a different written language), and a machine synthesizer (to convert written language into spoken language). A sufficiently advanced version of a machine interpreter would then permit two talkers, whose native languages are different, to carry on a conversation with each other. Such a system is very speculative at this point, but it is exciting to contemplate and work toward such a goal.

Talker Indentification and Verification

We know as a practical matter that we can identify several different people quite reliably from their voices. Sometimes we accomplish this identification by noting the particular words and sentences a person uses. However, even when two different persons say the same thing, we are still able to differentiate between them. Presumably we use a person's prosodic features for identification. In other words, it is not so much a matter of what people say as how they say it that is important in identification.

There are two categories of the general talker recognition problem. One is the identification of a talker out of a population of many talkers. The other is the verification of a talker as the one claimed. One difficulty encountered by any talker identification system is that as the number of possible talkers increases, there is increased likelihood of incorrect decisions.

The talker identification problem has been of considerable interest in forensic situations where establishing guilt is at issue. There has been much discussion of using speech spectrograms for talker identification. In analogy to fingerprints, the term "voiceprint" has been coined to refer to speech spectrograms. Fingerprints are well defined and stable for an individual and two sets of prints can be identified as belonging to the same person with confidence. Speech spectrograms, on the other hand, are susceptible to many degradations. The bandwidths of a speech signal may be different for two spectrograms from the same talker. Voice production is susceptible to health conditions and emotion which affect the spectrograms. In forensic situations the person to be identified is typically uncooperative which makes consistency of speech samples difficult to obtain. In some instances spectrograms of other talkers show more similarity to the spectrogram of a talker than do spectrograms of repetitions by the same talker. Figure 29.3 illustrates a situation where two spectrograms from the same talker (upper) may be viewed as less similar than two spectrograms from two other talkers (bottom). At this point in time it appears that spectrographic evidence is not sufficiently reliable for most legal purposes because of the great variation that can occur in the acoustical signal.

Talker verification imposes less stringent demands

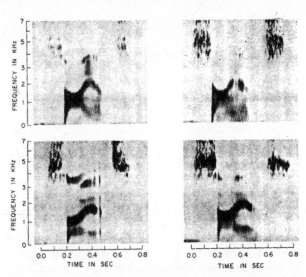

Figure 29.3. Four spectrograms of the word "science." Three of the spectrograms are from three different talkers. One is from a repetition by one of the talkers. (From Bolt etal, 1970.)

than talker identification. Talker verification has been proven successful enough to be used in several practical applications, including authorizations in banking and credit, voice transactions from remote locations, in limiting access to confidential information, and in security applications. Voice verification is based on previous speech data obtained from a cooperative talker under controlled conditions. The talker typically claims a certain identification by keying in a code or by some similar means. Voice verification is used as an additional check on identity. As an example, imagine the following situation: A selected group of 30 persons has access to a company's confidential files. When any authorized person wishes to gain access to the files he must properly identify himself. One way of doing this is by typing in a code. However, to increase the security of the files an additional verification of the person's identity is required. The person is requested to repeat some standard sentence; the sentence is then analyzed and compared to previously accumulated data. If the analysis results compare within a sufficiently low tolerance, the person is accepted as the same person whose code was typed.

Several different parameters extracted from the speech signal are used, usually in some combination, for talker verification. Parameters include long time average speech spectra, fundamental frequency contours, sound level contours, formant frequency contours, and linear-predictor coefficient contours. Contours of some parameter versus time are often dynamically time warped against stored template patterns to produce best possible fits. When the fit is sufficiently close, as measured by some previously specified criterion, the talker is verified. Otherwise the talker may be rejected or asked to repeat the verification procedure. As an idealized example, consider the three stylized fundamental-frequency contours shown in Figure 29.4. The first and second contours are standards for two different talkers. The final contour is from an unknown talker, but is presumed to be the contour for one of the the two talkers. The dashed lines represent the superposition of the third contour onto the first and second contours. You can see that the unknown contour fits the second contour better than it does the first contour. If the fit is sufficiently close, a judgment might be made that the unknown is in fact the second talker.

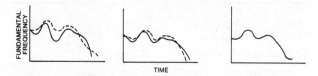

Figure 29.4. Stylized fundamental frequency contours for three talkers.

Practical talker verification systems are in use. Error rates for rejection of legitimate talkers or acceptance of impostors are typically only a few percent. Impressionists seem to pose slightly more difficult problems than other impostors. Error rates are typically reduced as more parameters are included in the verification procedure.

Exercises

1. Assume that you have a channel vocoding system similar to the one described in the text. Suppose your system has 16 channels, each of which is sampled with 4-bit accuracy 100 times each second. Suppose the fundamental frequency is sampled 100 times each second with 6-bit accuracy. What is the total bit-rate of your vocoder in bits/second? What percentage of the bit-rate of natural speech is this? (Use the data rate of 50,000 bits/seconds suggested in the text.) Suppose that you sample each of the signals in your vocoder only 40 times each second instead of 100 times. What would be the bit-rate? What is this as a percentage of the bit-rate of natural speech?

2. Assume that you have a formant vocoding system similar to the one described in the text. Suppose your system transmits information on the frequencies and amplitudes of three formants and on the fundamental frequency. Suppose that each of the 7 signals is sampled 100 times each second, with 6 bit accuracy. What is the bit-rate of this formant vocoder? What is this as a percentage of the bit-rate of normal speech?

3. Assume that you have a phoneme vocoder similar to the one described in the text. The system involves the use of 64 phonemes. How many bits are required to represent this set of 64 phonemes? Suppose the phonemes are transmitted at an average rate of 10 phonemes per second. What is the bit-rate of the system? What is this as a percentage of the bit-rate of normal speech? What has been lost in achieving this bandwidth reduction?

4. Consider the phonemes in Table 23.1. Which ones can be represented by a single allophone for speech synthesis? Which ones require two? Which ones require three or more?

5. You are given the task of creating a telephone number information device. The digit strings are already available. You are to develop a speech synthesizer that will accept codes for written digits at its input and produce synthetic spoken digits at its output. You decide that the best kind of speech synthesizer for the project is an isolated word, three-formant speech synthesizer. Draw stylized control signals (three formant frequencies) for the synthesizer in Figure 29.5 for

the digit string "9165". (Refer to Figure 29.6 to get formant frequency data.) How natural would you expect the resulting speech to sound? How intelligible?

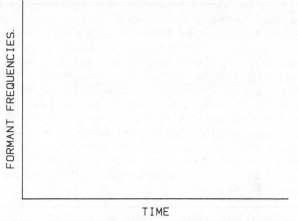

Figure 29.5. Stylized formant frequency contours are to be drawn on this figure.

6. The claim is made that machine synthesizers of speech are easier to build than are machine recognizers of speech. Give reasons as to why this may or may not be true. Consider the flexibility of the human vs. the machine in synthesis and recognition.

7. A digit recognizer consists of an acoustical analyzer that extracts the three lowest formant frequencies from the speech signal. When the signal is voiced, the formant is represented with a solid line; when unvoiced, with a dashed line. The 20 sketches in Figure 29.7 represent stylized outputs of an acoustical analyzer for each of the spoken digits (zero, one, two, three, four, five, six, seven, eight, nine) as spoken by each of two talkers. You are to be the rest of the "recognition machine." Based on your knowledge of formant patterns for various phonemes and voiced-unvoiced information of various phonemes, label each sketch with an appropriate digit label.

8. The top four tracings in Figure 29.8 show pitch period as a function of time for four talkers. The bottom two tracings in the figure show pitch period for two talkers who may or may not be one of the original four talkers. In your judgment, is either of these talkers one of the first four? Give reasons for your judgment.

Demonstrations

1. Listen to speech before and after it has been through a vocoder.

2. Synthesize speech using a synthesizer available on a small computer. If possible, experiment with the effects of controlling different parameters.

3. Prepare spectrograms of someone saying unknown digits in isolation. "Recognize" the spoken digits on the basis of your knowledge about phonemes. Copy, randomize and then use the spectrograms in Figure 29.6 if you cannot prepare your own.

4. Get several people to speak the same short sentence two or more times. Obtain spectrograms, fundamental frequency contours, or other representations of the sentences and "recognize" which were produced by the same person.

5. Observe how difficult it is to "understand"

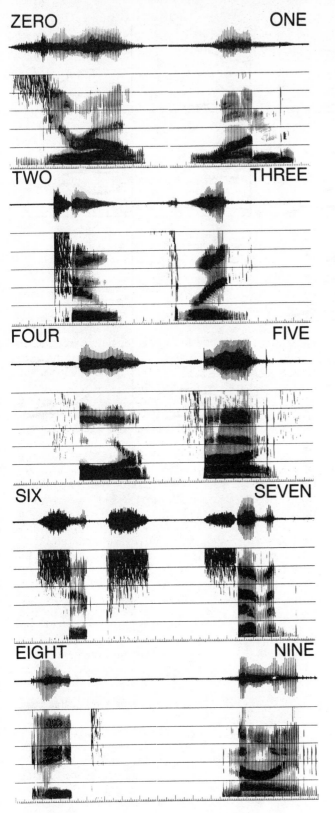

ZERO ONE

TWO THREE

FOUR FIVE

SIX SEVEN

EIGHT NINE

Figure 29.6. Spectrograms and waveforms for the spoken digits "zero" through "nine." (Courtesy of Kay Elemetrics Corp.)

Further Reading

Denes and Pinson, Chapter 8

Flanagan, Chapters 5, 6, 8

Flanagan and Rabiner

Ladefoged, Chapter 8

Rossing, Chapter 16

Pickett, Chapters 11,12

Schafer and Rabiner

Atal, B.S. 1976. "Automatic Recognition of Speakers from Their Voices," Proc. IEEE **64**, 460-475.

Bolt, R. H., F. S. Cooper, E. E. David, P. B. Denes, J. M. Pickett, and K. N. Stevens. 1970. "Speaker Identification by Speech Spectrograms: A Scientist's View of Its Reliability for Legal Purposes," J. Acoust. Soc. Am. **47**, 597-612. See also J. Acoust. Soc. Am. **54**, 531-537.

Dixon, N. R. and T. B. Martin, ed. 1979. **Automatic Speech and Speaker Recognition** (IEEE Press).

Doddington, G. E., and T. B. Schalk. 1981. "Speech Recognition: Turning Theory to Practice," IEEE Spectrum **18** (Sep), 26-32.

Elphick. M. 1981. "Talking Machines Aim at Versatility," High Technology **1** (Sep/Oct), 41-48.

Elphick, M. 1982. "Unravelling the Mysteries of Speech Recognition," High Technology (March/April), 71-78.

Flanagan, J. L. 1972. "The Synthesis of Speech," Sc. Am. **226** (Feb), 48-58.

Frantz, G. A., and R. H. Wiggins. 1982. "Design Case History: Speak and Spell Learns to Talk," IEEE Spectrum **11** (Feb), 45-49.

Lea, W. A., ed. 1980. **Trends in Speech Recognition** (Prentice-Hall).

Levinson, S. E., and M. Y. Liberman. 1981. "Speech Recongition by Computer," Scientific Am. **244** (Apr), 64-76.

Melby, A. M., W. J. Strong, E. G. Lytle, and R. Millet. 1977. "Pitch Contour Generation in Speech Synthesis: A Junction Grammar Approach," Am. J. Computational Linguistics, microfiche no. 60.

Songco, D. C., S. I. Allen, P. S. Plexico, and R. A. Morford. 1980. "How Computers Talk to the Blind," IEEE Spectrum **17** (May), 34-38.

Strong, W. J. 1967. "Machine-Aided Formant Determination for Speech Synthesis, "J. Acoust. Soc. Am. **41**, 1434-1442.

Audiovisual

1. **The Human Voice . . . and the Computer** (IEEE Soundings Tape 70-S- 04)

2. **The Speech Chain** (19 min, color, 1963, BELL)

3. "Synthetic Vowels from Vocal Model" (GRP)

when someone abruptly changes the topic of conversation. This is especially apparent when someone unknown to you calls on the telephone and begins talking about an unfamiliar topic.

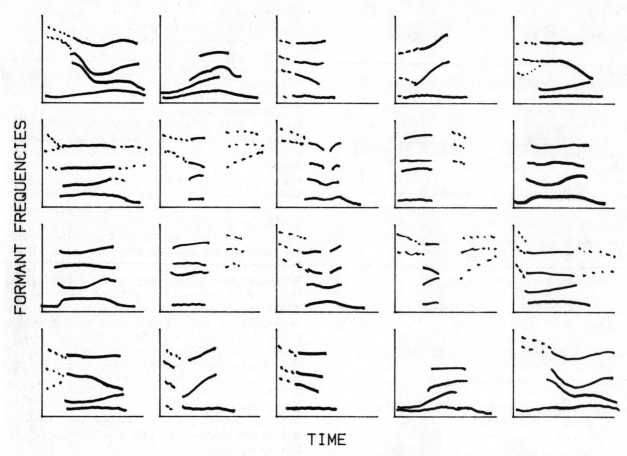

FORMANT FREQUENCIES

TIME

Figure 29.7. Stylized spectrograms for the ten digits "zero" through "nine" as spoken by two talkers.

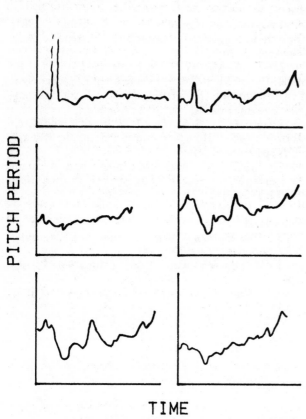

PITCH PERIOD

TIME

Figure 29.8. Pitch period contours for four known (upper and middle) talkers and two unknown (bottom) talkers. (From Atal, 1976.)

198

30. The Singing Voice

As noted earlier in Chapter 23, the vocal folds produce periodic puffs of air which are filtered and modified as they pass through the vocal tract. The air flow through the glottis is periodic (but not sinusoidal) and consists of many harmonic components. The frequency of oscillation of the vocal folds is determined primarily by their effective mass and tension and is for the most part independent of the vocal tract shape. The formant frequencies of the vocal tract in relation to the fundamental voicing frequency determine which voicing harmonics are emphasized.

Several phenomena occur in singing that are not otherwise present in speech production. Pitch vibrato (modulation of vocal fold fundamental frequency) is one of these and is discussed in Chapter 31. Tone color and loudness modulations accompany pitch modulation and all affect the quality of the resulting tone. There are some shifts in vowel quality in singing that result from vocal tract configurations differing from those used in normal speech production. A lowering of the larynx results in a lengthening of the vocal tract and a lowering of the formant frequencies, which in turn cause the "darker" vowel quality of "covered singing." We will now consider other significant features of the singing voice, including vocal registers, singing formant, formant tuning, and vocal power and efficiency.

Vocal Registers

In our discussion of speech production, no specific mention was made of vocal register because speech is normally produced within a single register. When singing, however, some two or more registers are commonly used. (The reader is referred to the references at the end of this chapter for discussions on the multiplicity of registers.)

A **vocal register** is characterized by one of the modes or combinations of modes in which the vocal folds vibrate. The mass and spring vocal fold model discussed in Chapter 23 is too simplified to accurately model the singing voice. It works reasonably well for speech modeling although even there it fails to capture significant features of the vocal fold motion because it does not treat wave phenomena in vocal fold motion. The mass and spring vocal fold model also fails to account for motion of the vocal folds in the direction of air flow through the larynx, and is inherently incapable of representing some of the more string-like behavior of the vocal folds observed in high-pitched singing. A many-mass model of the vocal folds more accurately represents some of these features because it can show wavelike, single-mass, or double-mass behavior. We will discuss briefly two registers that lie toward the extremes of the male voice. There are roughly comparable registers for the female voice.

We will consider only the major muscular adjustments made to configure the vocal folds for voice production. The interarytenoid and cricarytenoid muscles serve to bring the vocal folds together. These muscles control the initial glottal width which affects fundamental frequency and glottal air flow. Two opposing sets of muscles—vocalis and cricothyroid—are the primary determiners of fundamental frequency and vocal register. The vocalis muscle is the muscular part of each vocal fold. Tensing the vocalis muscles tends to shorten the vocal folds, raising the fundamental frequency. Tensing the cricothyroid causes the thyroid cartilage to rotate relative to the cricoid cartilage, lengthening the vocal folds. (To explore this, place a fingertip firmly in the small opening between the thyroid and cricoid cartilages at the bottom of the Adam's apple. Now vocalize at a low pitch and then glide to a high pitch. Note that your fingertip tends to get squeezed out of the opening as it is made smaller.) **Modal register** ("chest" or "heavy") is characterized by an active vocalis muscle which maintains the length of the folds short and increasing tension of the cricothyroid muscle from low to high frequencies. **Falsetto register** ("head" or "light") is characterized by a lax vocalis muscle and by increasing cricothyroid tension that increases vocal fold length from low to high frequencies. The ligament is the primary vibrating portion of the vocal folds in the falsetto register.

The modal register is used for low fundamental frequency production and is the register commonly used in speech production. A combination of vocal fold modes typical of this register is shown in the upper row of Figure 30.1. The top view (a) shows the characteristic oval glottal opening produced when the two ends of the fold are fixed and the middle portion moves. This is essentially a half-wavelength mode. The end view (b) shows that the upper and lower portions of the fold tend to vibrate out of phase with each other. This is essentially a half-wavelength "thickness mode" with upper and lower portions of the fold free to move relative to each other. The edge view (c) shows the fold moved upward in the direction of airflow. Again, this is a half-wavelength mode in the vertical direction, with the ends fixed. Note that all of these modes are vibrating together so that the motion of the vocal folds is rather complex.

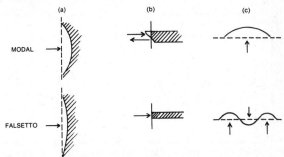

Figure 30.1. Vocal fold modes typical of modal and falsetto registers: (a) superior view looking from above; (b) coronal view looking along the vocal fold; (c) sagittal view looking edge-on. (After Titze and Strong, 1975.)

The modal register is characterized by the following features: (1) The vocal folds are comparatively short and thick, with relatively large mass per unit length. (2) Vocal fold tension is comparatively low. (3) Vibration amplitude of the folds is large. (4) The vocal folds tend to close completely over part of the vibratory cycle. (5) The signal produced by airflow through the glottal opening tends to be rich in harmonics. (6) The energy supplied by the lungs is converted rather efficiently into sound energy.

The falsetto register is used for high fundamental frequency production. A combination of modes typical of this register is shown in the lower row of Figure 30.1. The top view (a) shows a characteristic oval opening. The vibration amplitude is smaller than in the modal register. The mode shown is half-wavelength, although sometimes higher modes involving several half-wavelengths may be involved. The end view (b) shows upper and lower portions of the cord moving together as a unit, as opposed to the modal register where they vibrate out of phase. Note that the fold is much thinner. The half-wave vertical mode shown for modal register is also the primary vertical mode for falsetto register. There may also be some of the three half-wave vertical mode shown in the lower row of Figure 30.1c involved in the falsetto register. Keep in mind that these modes are vibrating together.

The following features are characteristic of the falsetto register: (1) The vocal folds are longer (up to 30%) and thinner, with a smaller mass per unit length. (2) Longitudinal vocal fold tension is comparatively high. (3) Vibration amplitude of the folds is small. (4) The vocal folds tend to lack complete closure during any part of the vibratory cycle. This results in increased airflow and a more breathy quality. (5) The signal produced by glottal air flow tends to be poorer in higher harmonics. (6) The conversion of "lung energy" into sound energy is less efficient than in the modal register.

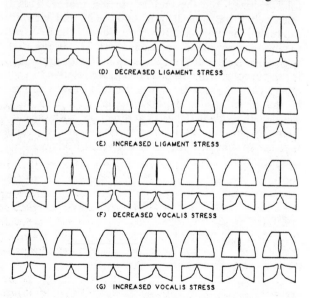

Figure 30.2. Computer generated vocal fold movements in horizontal and vertical cross sections for four adjustments of vocal fold parameters. (From Titze and Talkin, 1980.)

Some of the features described above have been studied in computer simulation with a many-mass vocal

fold model. The simulation verifies some of these features and demonstrates that the modal register exhibits natural frequencies that are lower than those of the falsetto register. Figure 30.2D illustrates vocal fold motion (top view and end view) for decreased ligament tension and nominal vocalis tension—which corresponds to low chest register. Note the large motions of the folds and their two-part motion in the vertical direction. Increased ligament tension (Figure 30.2E) corresponds roughly to a falsetto adjustment. Note the much smaller motions, the thinner folds, and the single part motion in the vertical. Figure 30.2F and Figure 30.2G illustrate simulations for decreased and increased vocalis tension with nominal ligament tension.

Fundamental frequency varies directly with vocalis tension and ligament tension and inversely with vocal fold length. Glottal air flow varies directly with glottal length and subglottal pressure and inversely with ligament tension. Some of these effects can be seen in Figure 30.3 which shows glottal air flow for the conditions of Figure 30.2.

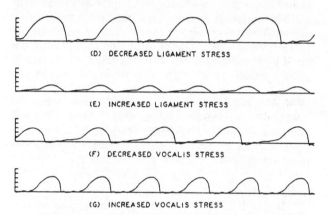

Figure 30.3. Glottal air flow waveforms for the four adjustments of the vocal fold model corresponding to Figure 30.2. (From Titze and Talkin, 1980.)

Singing Formant

A first-class opera singer can be heard above relatively high sound levels from an orchestra and in relatively large halls. This is apparently due to the **singing formant**, a strong peak in the spectrum at about 3.0 kHz for sung vowels. The spectra of a spoken and sung /OO/ (as in cool) are shown in Figure 30.4. The dashed lines in the figure represent the spectrum of a typical orchestra averaged over long periods of time. If a tenor produced a spectrum comparable to that of ordinary speech (Figure 30.4a), his song would be masked by the orchestra. However, the spectrum produced by a trained singer (Figure 30.4b) differs from that of normal speech primarily because of the singing formant. The singing formant is in a frequency region where the sound level of the orchestra is reduced, which enables the singer to be heard above the orchestra.

The origin of the singing formant appears to be an extra formant in the singer's vowel spectrum. We noted in Chapter 24 that frequencies in the vicinity of formants are emphasized. Furthermore, when two formants lie close together in frequency the enhancement is even more marked for frequencies in the vicinity of either of the two formants. This can be seen in the vowel

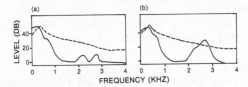

Figure 30.4. Solid line shows spectra for (a) spoken /OO/ and (b) sung /OO/. Dashed line shows time-averaged spectrum of orchestra. (After Sundberg, 1974.)

spectra in Figure 24.10, where frequencies around 2.5 kHz are emphasized because of the proximity of the second and third formants in the vowel /EE/ (as in "heat") and where frequencies around 1.0 kHz are emphasized because of the proximity of the first and second formants in the vowel /O/ (as in "hot").

The introduction of an additional formant in the vicinity of the third and fourth formants would result in significant enhancement of the energy at about 3.0 kHz. This phenomenon has been demonstrated in models of the larynx-pharynx system more detailed than we have considered. We have considered the larynx-pharynx system to consist of a single tube of smoothly-varying cross-section terminated at the vocal folds (see Figure 24.2). Anatomical evidence indicates that this is an oversimplified model, especially when the larynx is lowered for singing. A more realistic model is one in which the vocal folds connect to the pharynx via a small cavity (the laryngeal ventricle, or sinus Morgagni) and a narrow tube, as seen in Figure 30.5. The width of the pharynx is much larger than the narrow tube, which results in an acoustical "mismatch" where they join. Because of this acoustical mismatch the small cavity and narrow tube can be considered a resonator in its own right. Its resonance frequency furnishes the extra formant at about 3.0 kHz when the larynx has been lowered and the pharynx widened as noted.

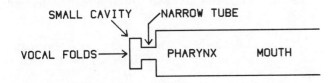

Figure 30.5. Model of "singer's" vocal tract. (After Sundberg, 1974.)

Formant Tuning

The first formant ranges from about 300 to 900 Hz for female vocal production, depending on the vowel being produced. When the singing voice is producing very high fundamental frequencies (as would be the case with sopranos), it is possible that the fundamental frequency will not lie close to a formant frequency. This can result in substantial loss of amplitude (and hence loudness), since the further a harmonic is from a formant the less its enhancement. Furthermore, the loudness of the tones produced would fluctuate depending on the relation of the fundamental and first formant frequencies.

The singer's solution is to move the first formant frequency to match more closely to the fundamental frequency being produced, especially when the fundamental lies above the first formant. This is accomplished by

varying the vocal tract shape via increased jaw and lip opening and by shortening the vocal tract via lip retraction. Figure 30.6 illustrates the combined effects of these two adjustments for the vowel /OO/. The vocal tract shape in the upper part is for a soprano's spoken /OO/. The spectral pattern shows how a fundamental frequency of 500 Hz lies above the resulting first formant—resulting in low amplitude. The tract shape in the lower part is for a sung /OO/. The increased jaw opening and lip retraction have raised the first formant so that it coincides with and reinforces the fundamental.

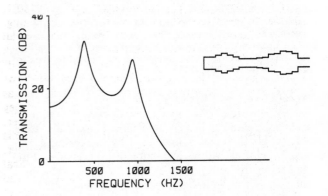

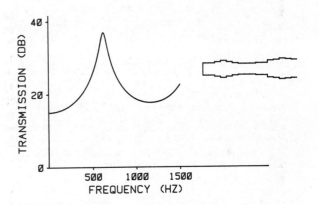

Figure 30.6. Multi-tube vocal tract models and corresponding spectra for spoken /OO/ (upper) and sung /OO/ (lower).

Formant tuning is used by singers when they produce a fundamental frequency that would lie above that of the first formant. When fundamental frequencies lying below that of the first formant are used, singers use a more speech-like vocal tract shape. Figure 30.7 illustrates first and second formant frequencies and fundamental and second harmonic frequencies for the sung vowel /OO/. Formant tuning clearly comes into play as the frequency of the fundamental or second harmonic would exceed that of the first or second formant, respectively. This minimizes the loudness variation for different vowels sung at different pitches.

Formant tuning to the fundamental might be expected to result in the loss of vowel intelligibility. It probably does, but in singing, tonal power is usually of more importance, with intelligibility of lesser concern. The intelligibility of high pitched vowels is decreased in any event because there are too few harmonics to "define" the formants.

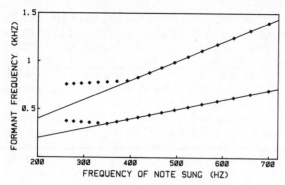

Figure 30.7. Formant tuning to match fundamental and second harmonic of note sung. The lines represent the frequencies of the fundamental and second harmonic. The symbols represent the frequency of the first and second formants. Formant tuning is used when the fundamental frequency would normally exceed the first formant frequency. (After Sundberg, 1975.)

Vocal Power and Efficiency

As previously discussed, fundamental frequency can be increased by increasing ligament or vocalis stress. Glottal air flow can be increased by increasing glottal length, glottal width, and subglottal pressure. Fundamental frequency and glottal air flow vary in a monotonic way with their controlling parameters.

Radiated vocal power depends on alternating glottal air flow and subglottal pressure. Vocal efficiency is the ratio of radiated vocal power to the vocal source power. Vocal power and efficiency do not vary in a monotonic way with vocal fold configurations. There are certain optimum vocal fold configurations that maximize vocal power and efficiency. Vocal power can be varied from a few tens of microwatts to a few tens of milliwatts—a dynamic range of about 30 dB. Vocal efficiency can be varied from about 0.1% to about 10%. The trained singer is able to achieve the optimum configurations so as to produce more vocal power more efficiently than is possible when speaking.

Exercises

1. Have a trained tenor illustrate the difference between his spoken and sung vowels.

2. Have a trained soprano illustrate the effects of formant tuning for high pitches. Note any differences in loudness and vowel quality.

3. Have a trained singer produce a tone at a given pitch in different registers.

4. Produce something akin to the "darker" vowels of "covered" singing by singing a vowel at a given pitch under normal circumstances and then with a short length of tubing surrounding the lips. This lengthens the vocal "tract" and lowers the formant frequencies. (Of course in actual covered singing the length of increase of the tract is at the other end of the tract, in the vicinity of the vocal folds.)

Demonstrations

1. Do Exercise 1 in conjunction with a spectrum analyzer. Compare your results to Figure 30.4.

2. Do Exercise 2 in conjunction with a spectrum analyzer. Compare your results to Figure 30.6.

3. Do Exercise 3 in conjunction with a spectrum analyzer.

4. Do Exercise 4 in conjunction with a spectrum analyzer.

Further Reading

Backus, Chapter 11

Benade, Chapter 19

Flanagan, Chapters 2, 3

Hall, Chapter 14

Proctor, Chapters 1, 5, 7, 8

Rossing, Chapter 17

Seashore, Chapters 4, 20

Large, J., ed. 1973. **Vocal Registers in Singing** (Mouton and Co.).

Smith, L.A., and B.L. Scott. 1980. "Increasing the Intelligibility of Sung Vowels," J. Acoust. Soc. Am. **67**, 1795.

Sundberg, J. 1974. "Articulatory Interpretation of the 'Singing Formant'," J. Acoust. Soc. Am. **55**, 838-844.

Sundberg, J. 1975. "Formant Technique in a Professional Female Singer," Acustica **32**, 89-96.

Sundberg, J. 1977. "The Acoustics of the Singing Voice," Scientific American **236** (Mar), 82-91.

Sundberg, J. 1979. "Perception of Singing," STL-QPSR **1**, 1-48.

Titze, I.R. 1973, 1974. "The Human Vocal Cords: A Mathematical Model, Parts I and II," Phonetica **28**, 129-170; **29**, 1-21.

Titze, I.R. 1979. "Variations of Pitch and Intensity with Pre-phonatory Laryngeal Adjustments," Current Issues in Linguistic Theory **9**, 209-215, H. Hollien and P. Hollien, ed.

Titze, I.R. 1979. "The Concept of Muscular Isometrics for Optimizing Vocal Intensity and Efficiency," J. Res. in Singing **2**.

Titze, I.R. and D.T. Talkin. 1979. "A Theoretical Study of the Effects of Various Laryngeal Configurations on the Acoustics of Phonation, J. Acoust. Soc. Am. **66**, 60-74.

Vennard, W. 1967. **Singing: The Mechanism and Technic**. (Carl Fischer, Inc.)

Audiovisual

1. **Function of the Normal Larynx** (21 min, color, 1956, ILV)

VI. Musical Acoustics

Courtesy of College of Fine Arts and Communications, Brigham Young University.

31. Production and Perception of Musical Tones

Many of the physical properties that affect tone production are peculiar to a particular class of instruments (e.g., strings or winds) and even to particular instruments within a class (e.g., violin, viola, cello, double bass). However, some general features of tone production and perception apply to a broad range of instruments. We will discuss some of these features under the headings of families of musical instruments, spectral characteristics, transients in musical tones, dynamics of performance, uncertainty in musical tones, and perception of isolated tones.

Families of Musical Instruments

Three broad classes of musical instruments may be considered: mechanical, electronic, and electromechanical. Mechanical instruments employ a mechanical source of excitation, such as blowing, bowing, plucking, or striking, to supply energy to a mechanical resonator. The resonator, such as an air column or a string, largely determines the frequencies produced by the instrument. In the case of non-percussive instruments, the resonator typically interacts with the source of excitation to control the rate at which the source supplies energy. Part of the energy in the resonator is converted into sound by radiation from tone holes or an instrument body. There is no use whatsoever of electronics in sound production by mechanical instruments.

Electronic instruments employ electronic oscillators as sources of excitation. In lieu of a resonator interacting with the source, they employ electronic filters to modify the spectra produced by the oscillators. All electronic instruments employ a transducer, such as

a loudspeaker, to convert electrical energy into acoustical energy and to radiate it as sound.

Electromechanical instruments employ both mechanical and electronic elements. In some cases they employ a transducer to convert a pressure signal in a wind instrument into an electrical signal, which is modified and amplified before being radiated by a loudspeaker. In other cases the motion of a mechanical vibrator is transduced and the resulting signal passed through electronic filters to simulate an instrument's body response. Examples in the latter category are electric guitar, electronic piano, electronic carillon, and electronic violin.

There are several ways in which mechanical instruments can be categorized into families. The term "family" suggests similar features among the members of an instrumental family, with size being a major difference among instruments within that family. Physical similarities that might be considered include the type of source (e.g., blowing action or bowing action), type of resonator (e.g., string, cylindrical tube, or conical tube), and radiator type (e.g., violin body, trumpet bell, clarinet tone-holes, or drum head).

Table 31.1 shows one way in which instruments might be assigned to families on a physical basis. The first seven families shown in the table are all mechanical instrument families. The first four are non-percussion families in which energy is continually replenished. Three of the mechanical instrument families are percussion families in which a tone is produced with a single input of energy: a striking mallet, a striking hammer, or the plucking of a string. Once excited, the vibrations are allowed to die away in a manner determined by the

Table 31.1. Families of musical instruments

Family	Excitation	Resonator	Radiator	Examples
Lip reed	Blown lips Blown vocal folds	Conical/cylindrical tube Variable shaped tube	Bell Mouth	Trumpet Voice
Mechanical reed	Blown single reed Blown double reed	Cylindrical tube Conical tube	Tone holes Tone holes	Clarinet Oboe
Air reed	Blown air jet	Conical/cylindrical tube	Tone holes End of tube	Flute Organ
Bowed string	Bowed string	String and body	Body	Violin
Plucked string	Plucked string	String and body	Body	Guitar
Struck string	Struck string	String	Soundboard	Piano
Percussion	Struck membrane Struck bar Struck plate	Membrane Bar Plate	Membrane Bar Plate	Tympani Xylophone Cymbals
Electronic	Electronic oscillators	Electronic filters	Loudspeaker	Synthesizer
Electromechanical	Plucked string	String	Loudspeaker	Electric guitar

resonator. The last two families in the table employ electronics in some way.

Spectral Characteristics

If, in fact, members of an instrumental family are large- and small-scaled versions of one another, it is reasonable to suppose that the tones they produce will be frequency-scaled versions of each other; low-frequency characteristics exhibited by a large instrument in a family will be observed at higher frequencies for the smaller instruments of a family. In other words, the spectra measured for large and small instruments in a family should be approximately frequency-scaled versions of each other. There is evidence to support this idea, particularly for some instrumental families. The example for average spectra of trumpet, trombone, and tuba shown in Figure 31.1 demonstrates this quite convincingly.

by vocal tract resonances (see Chapters 24 and 25). More recent evidence from analyses of musical instrument tones indicates that the spectral peak idea is applicable to many nonpercussive musical instruments. Spectral peaks can be determined for musical instruments by plotting the spectra for each of many different tones in a chromatic sequence. The spectra are superimposed on each other, as shown in Figure 31.2 for violin tones, and a curved line is drawn through them to "represent" the average spectrum over all tones. Examples of spectral envelopes obtained in this way for bassoon and flute are shown in Figures 31.3 and 31.4. The spectral peaks of a spectral envelope, then, represent the most intense frequencies for the instrument, although the partials of individual tones may deviate substantially from the average. The instruments of a particular family might be expected to exhibit spectral peaks with similar shapes, although having different

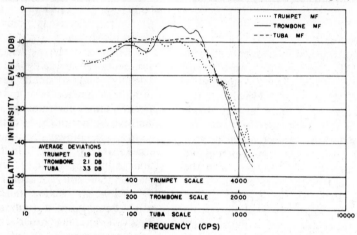

Figure 31.1. Trumpet, trombone, and tuba spectral envelopes scaled in frequency. (From Luce and Clark, 1967.)

One interesting way of describing an instrument's spectral characteristics is in terms of peaks in its spectral envelope. The idea of spectral peaks, called formants, has been used extensively to describe the vowel spectra of speech. The formants of vowel sounds are produced

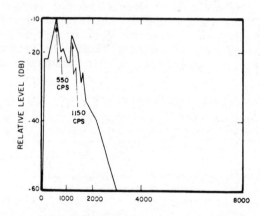

Figure 31.3. Spectral envelope for bassoon. (From Strong and Clark, 1967.)

relative frequencies, as seen for the brasses in Figure 31.1. The spectral peaks may also shift in frequency as a result of loud versus soft playing. This has been noted for flute tones, in which the spectral peaks shift to higher frequencies for the more loudly blown tones.

A few comments need to be made about the spectral peak representation of musical instruments. Spectral peaks for musical instruments are due to such things as cutoff frequencies, body resonances, radiation properties, and excitation mechanisms. They tend to be few

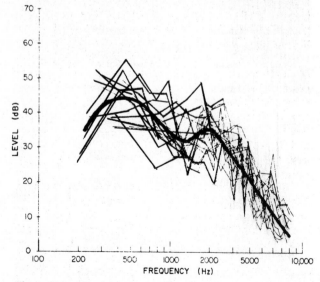

Figure 31.2. Superimposed mezzo-forte violin tone spectra with "average spectrum" line drawn in. (After Beauchamp, 1974.)

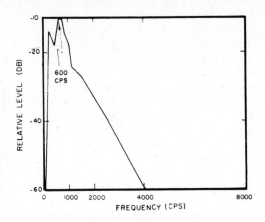

Figure 31.4. Spectral envelope for flute. (From Strong and Clark, 1967.)

in number (typically one or two) and rather broad (see Figures 31.1-31.4). They represent gross spectral properties of the instrument and cannot be tied explicitly to tube resonances as can the formants in speech production. As noted previously, spectra obtained for an instrument are generally sensitive to microphone placement, and a given spectral envelope will not always be representative of the given instrument's average envelope, as seen in Figure 31.5. These spectra were obtained at five different microphone positions for a violin in an anechoic chamber.

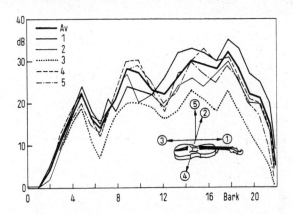

Figure 31.5. Long-time average spectra (LTAS) from recordings of a violin made in an anechoic chamber with the microphone placed in each of five different directions as indicated. An average for the five directions is also shown. (From Jansson, 1976.)

Insofar as the notion of spectral peaks is applicable, the average characteristic frequencies of an instrument can be specified in terms of fixed-frequency bands of emphasis, which means that partials lying within peak regions will be the strongest. Thus, as the fundamental frequency is varied, different partials will be emphasized as they lie within a peak region. The spectral peak description appears to be more representative of what actually happens than does a description of fixed relative harmonic amplitudes (see Chapter 14).

Transients in Musical Tones

A certain amount of time is required to "start" musical tones and to "stop" them once they are started. The idealized waveform shown in Figure 31.6 illustrates these features. The part of the waveform labeled "at-tack" is that part associated with the starting up, or onset, of the tone. The part labeled "decay" is that part of the waveform associated with the shutting down, or ending, of the tone. The "steady state" is the part of the waveform between the attack and decay; it may be fairly steady or it may be quite unsteady.

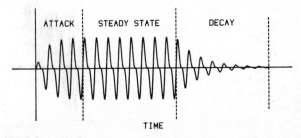

Figure 31.6. Idealized waveform illustrating attack transient, steady state, and decay transient.

Percussive tones would not be expected to exhibit a steady state since they are excited with an impulse, and once the excitation has ceased the tone begins to decay. Nonpercussive tones, on the other hand, would be expected to exhibit all three features—attack, steady state, and decay—with a steady state that is determined by how long the excitation (blowing or bowing) is applied.

From the physical point of view it might be expected that the large instruments of a particular family would have longer attack times than the smaller instruments of the same family (e.g., the double bass longer than the violin, the bassoon longer than the oboe). This should be so because the disturbances produced in a larger instrument must travel longer distances before standing waves can be set up (e.g., the longer strings of a double bass vs. the shorter strings of a violin, the longer tube of a trombone vs. the shorter tube of a trumpet). By the same argument, it might be expected that the high-frequency tones on some instruments would have shorter attack times than the lower-frequency tones because the instrument is made shorter by fingering.

In addition to the size of an instrument, its relative energy losses should influence the attack and decay times. This should be so because musical instruments store energy, and it takes time for the energy to be stored when the instrument is being excited and time for the store of energy to be exhausted when the excitation is shut off. A low-loss system takes a long time to store or exhaust its energy supply, whereas a high-loss system stores or exhausts its energy supply more quickly (see Chapters 4, 5, and 20). Thus, higher-loss instruments, such as the double reeds, would be expected to exhibit shorter attack times than lower-loss ones, such as the clarinet.

Attack and decay times can also depend on the style of playing and the particular performer. However, we cannot give a very firm description of this dependence. There is some problem in defining the duration of an attack. It might be defined as the time from the beginning of the signal until some percentage of the final amplitude has been reached. It could also be defined in terms of the slope of an amplitude envelope over some portion of the beginning of the signal. Measured attack times based on 50 percent of the final amplitude are reproduced in Figures 31.7 through 31.9. The figures basically exhibit the features we have talked about: decreasing attack times for higher-frequency tones, and longer attack times for large instruments in a family. (The strings seem to violate this convention, perhaps because playing style influences attack time more in the string family.)

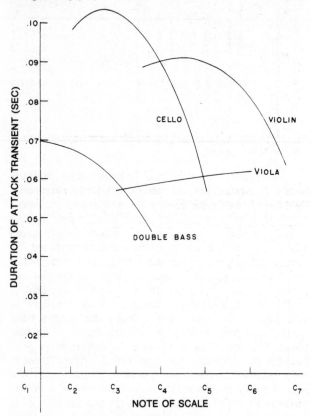

Figure 31.7. Duration of the attack transients of bowed string instruments. The smooth curves shown were drawn to provide a best fit to original discrete points which are not shown. (From Luce and Clark, 1965.)

Dynamics and Performance

We now consider some acoustical features that result when a performer plays at different dynamic markings (i.e., pp, mf, ff). In addition to the obvious increase in loudness produced when an instrument is excited more strongly, there is reason to expect that the tone color may change. It is a fairly general rule that any time a vibrating system is driven more vigorously, the high-frequency modes gain more energy than do the low-frequency modes. This should be true in part because when vibrators are weakly excited their motion tends to be smoother and more sinusoidal, which should produce strong low-frequency partials and rather weak high-frequency partials. On the other hand, a strongly driven vibrator may exhibit abrupt changes in its motion (e.g., as when a clarinet reed beats against the mouthpiece), which will result in a spectrum much richer in high-frequency energy. Spectra observed for loud tones are typically much richer in high-frequency energy than those observed for soft tones.

The dynamic range of instruments as a function of

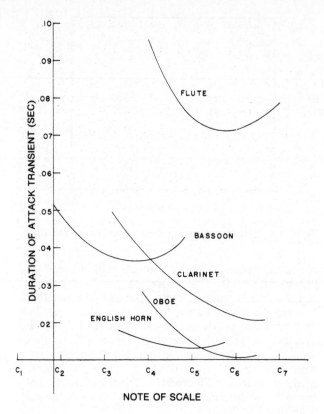

Figure 31.8. Duration of the attack transients of woodwinds. (From Luce and Clark, 1965.)

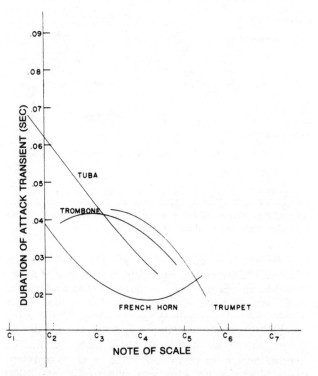

Figure 31.9. Duration of the attack transients of brass instruments. (From Luce and Clark, 1965.)

different dynamic markings is also interesting. For nonpercussive musical instruments the average dynamic range is about 11 dB between the dynamic markings of pp and ff. The woodwinds exhibit dynamic ranges smaller than the average, while the strings and brasses

208

exhibit a greater range, as shown in Figure 31.10. Dynamic range and dynamic level are also dependent to some extent on whether the instrument is performing in its low range or high range, with the differences being larger for double bass, flute, and French horn than for the other instruments. Details of this can be seen in Figures 31.11 to 31.14.

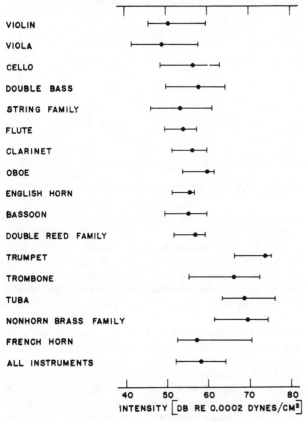

Figure 31.10. Average levels of scales played pianissimo (left-most, short, vertical bar), mezzoforte (dot), and fortissimo (right bar) for various instruments. (From Clark and Luce, 1965.)

Uncertainty in Musical Tones

Uncertainty in musical tones is often desirable and makes them more interesting and exciting. For instance, a typical pipe organ, with its pipes not perfectly in tune (in an exact mathematical sense), sounds more interesting than an electronic organ in which the oscillators are mathematically in tune with each other. Uncertainty may be inherent in the tone production process or it may be introduced by the player as a desirable ornament. We will consider the following tonal uncertainties: vibrato, tremolo, and choral effects.

Vibrato is an ornamental frequency modulation usually produced intentionally by the performer. Vibrato is used most extensively in singing and in performing on the bowed string instruments. Vibrato is accomplished on the bowed strings by alternately shortening and lengthening the string with a low-frequency rolling motion of the finger stopping the string. The partials produced are higher in frequency when the string is short and lower when the string is long. Associated with the frequency modulation is a spectral modulation that can be quite pronounced. Figure 31.15 illustrates large

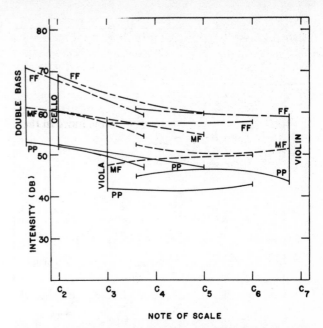

Figure 31.11. Intensity levels of scales of bowed string instruments played pianissimo (pp), mezzoforte (mf), and fortissimo (ff) as functions of the note played. The smooth curves shown represent original discrete points not shown. (From Clark and Luce, 1965.)

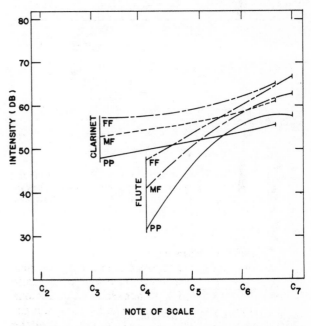

Figure 31.12. Intensity levels of scales of the clarinet and flute played pianissimo, mezzoforte, and fortissimo. (From Clark and Luce, 1965.)

spectral differences between the "low" and the "high" parts of a vibrato tone. Other vibrato tones might be expected to have similar behavior, while differing in extent. An amplitude modulation usually accompanies vibrato, but it is not significantly perceptible. The amount that the frequency changes from the average frequency (0.6% to 9%) and the rate of the frequency changes (about 5-8 complete changes per second) are typically dependent on the particular performer and on the musical situation.

Tremolo is an ornamental amplitude modulation

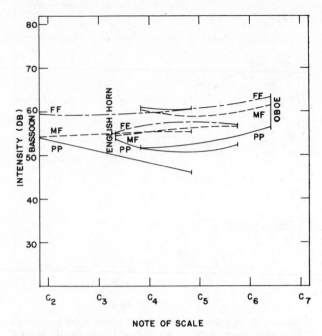

Figure 31.13. Intensity levels of scales of the double reed instruments played pianissimo, mezzoforte, and fortissimo. (From Clark and Luce, 1965.)

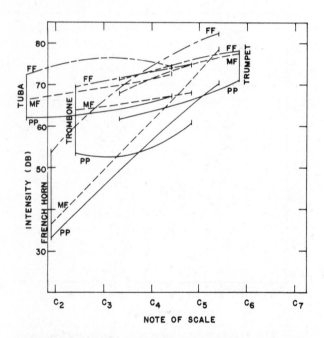

Figure 31.14. Intensity levels of scales of the brass instruments played pianissimo, mezzoforte, and fortissimo. (From Clark and Luce, 1965.)

usually produced intentionally by a performer. It is the ornament used most often with the flute (which is an air reed instrument), where it can be produced by modulating the blowing pressure. Tremolo rate (5 to 8 changes per second) is similar to that for vibrato, but the amount of amplitude modulation can be as much as 80% of the amplitude. Again, as with vibrato, tremolo is accompanied by a spectral modulation and, to a lesser extent, by a frequency modulation.

When two or more instruments produce musical tones together, **choral effects** appear which are not pre-

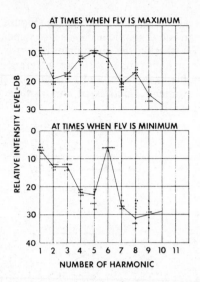

Figure 31.15. Harmonic structure for violin tone E played on the A-string at times when the frequency level is maximum and minimum during a vibrato cycle. (From Fletcher and Sanders, 1967.)

sent when a single instrument is performing. Choral effects probably arise unintentionally for the most part, although performers may invoke them or exaggerate them for particular musical reasons. Suppose, for example, that two performers are asked to produce a note in perfect time synchronism and with perfect intonation. Even if they can accomplish this task, the result will be different from that of increasing the sound level of a single instrument; there will be two spatially separate sound sources which give rise to a different perception. The performers will be unable to play in perfect time synchrony, resulting in a slight staggering of the attacks of the two instruments. This will further modify the percept of the instruments playing together as opposed to the sound of a single instrument amplified. Staggered attacks can also occur on an organ when the performer depresses the keys nonsynchronously, or when different ranks of pipes have different attack times. Two performers will be unable to perform perfectly in tune (in a mathematical sense) and this will result in beating between neighboring partials in the combined tones, further coloring the percept. Organ pipes that are not perfectly in tune (mathematically) exhibit these same features. It has been found that electronically synthesized organ tones can be made to closely resemble the tones of a pipe organ by detuning them slightly (about 0.3-1.2%) and adding the detuned signal to the original.

Perception of Isolated Tones

We have noted previously that the perceptual attributes of tone color and tone quality are multidimensional in that they depend on more than one physical dimension (see Chapter 14). In one study of perceptual scaling of musical timbre (including both tone color and tone quality), 16 isolated instrumental tones having the same pitch, loudness, and duration were used. Listeners were asked to rate the similarity of all possible two-tone pairs. The similarities were then plotted in a three dimensional physical space with the first dimension related to spectral energy distribution. The second dimension related to synchronicity in the attack times of the higher partials and spectral fluctuation within the tone. The third dimension related to low-amplitude, high-frequency energy in the attack.

Along with the physical bases discussed for families of instruments, there is experimental evidence which demonstrates the existence of perceptual families of instruments. Tones which had been speeded up relative to the original recording speed were presented to listeners, who were then asked to identify the instruments producing the tones. In this way the sound of the viola was changed to the sound of the violin, and so on. The results are in agreement with the classifications in Table 31.1, as the following perceptual families were found: string (violin, viola, cello, double bass), brass (trumpet, trombone, French horn, tuba), and double reed (oboe, English horn, bassoon).

Perceptually, the tone color of nonpercussive musical instruments is a weak function of the intensity with which they are played, even though the relative amounts of energy at low and high frequencies may be quite different. This has been demonstrated in experiments in which the loudness of tones produced at different dynamic markings was equalized before presenting the tones to musically competent listeners. The listeners were able to differentiate the soft tones from the loud tones on the basis of tone color rather than volume.

Attack transients of non-percussive tones have been shown to be important in the perception of musical tones. Even short segments of a tone that include the attack transient enable a listener to identify a musical instrument much more accurately than a substantially longer segment of a tone from which the attack has been deleted. One reason for the aural significance of the attack is the great spectral evolution that takes place during the attack. The spectrum at the beginning of the attack is usually characterized mostly by low-frequency partials, but by the end of the attack, high frequencies have been added (see Figure 31.16). The decay of nonpercussive tones is probably of subordinate importance, whereas the decay of percussive tones is likely to be almost all-important.

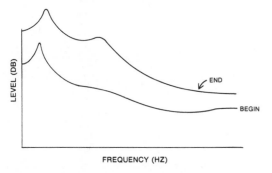

Figure 31.16. Spectra measured at the beginning and ending of an oboe attack transient. (Courtesy of J. D. Dudley.)

Exercises

1. In what ways can the musical instruments be organized into families other than the way shown in Table 31.1?

2. The curve in Figure 31.4 is the spectrum envelope for a flute. What might be a reasonable estimate for the spectrum envelope of a "bass flute" if it performs an octave lower?

3. What might be a reasonable spectrum envelope for an oboe if it performs an octave higher than a bassoon? (See Figure 31.3.)

4. A trumpet has a spectral peak centered at 1000 Hz. Which partial (by number) will be the strongest when the trumpet is sounded with a fundamental frequency of 250 Hz? 300 Hz? 500 Hz? 1000 Hz?

5. Which instruments (percussive or nonpercussive) are likely to exhibit the longer attack times? Why?

6. What tones (percussive or nonpercussive) have little or no steady state?

7. What determines the amount of time required for the attack? For the decay?

8. What determines the length of the steady state?

9. Put the following instruments in order of increasing attack time and give typical attack times (when possible): violin, guitar, piano, clarinet, oboe, saxophone.

10. Determine the attack time, the steady state time, and the decay time for the idealized waveform envelope shown in Figure 31.6.

11. Which of the instruments shown in Figure 31.10 has the greatest dynamic range? The smallest?

12. Which instruments have dynamic ranges that increase with increasing frequency? That decrease?

13. Which instruments exhibit the greatest dynamic level (i.e., are the most powerful)? Which the least? Do these results agree with your experience?

14. The spectra in Figure 31.16 can also be viewed as showing the average spectra of an instrument blown softly and loudly. Which curve corresponds to each condition?

15. A vibrato produces frequency variations of 1% above and 1% below the average frequency. What are the highest and lowest frequency values for each of the first five partials if the fundamental frequency is 200 Hz?

16. Similar organ pipes each have spectra exhibiting 15 partials. If the nominal fundamental frequency of each pipe is 200 Hz, what beat frequencies are produced by adjacent partials when the two pipes are sounded together if they are 0.5% out of tune?

Demonstrations

1. Display attack and decay transients on an oscilloscope.

2. Perform spectral analyses for tones produced at different dynamic markings.

3. Have players demonstrate vibrato, tremolo, and choral effects with their instruments.

Further Reading

Roederer, Chapters 1-5

Seashore, Chapters 4, 8, 9, 17-20

Winckel, Chapters 3, 6, 7, 10

Backus, J. 1973. "Acoustical Investigation of the Clarinet," Sound **2**, 22-25.

Beauchamp, J. W. 1974. "Time-Variant Spectra of Violin Tones," J. Acoust. Soc. Am. **56**, 995-1104.

Beauchamp, J. W. 1975. "Analysis and Synthesis

of Cornet Tones Using Nonlinear Interharmonic Relationships,'' J. Audio Eng. Soc. **23**, 778-795.

Clark, M. 1959. ''Several Problems in Musical Acoustics,'' J. Audio Eng. Soc. **7**, 2-4.

Clark, M., and D. Luce. 1965. ''Intensities of Orchestral Instrument Scales Played at Prescribed Dynamic Markings,'' J. Audio Eng. Soc. **13**, 151-157.

Clark, M., D. Luce, R. Abram, H. Schlossberg, and J. Rome. 1963. ''Preliminary Experiments on the Aural Significance of Parts of Tones of Orchestral Instruments and on Choral Tones,'' J. Audio Eng. Soc. **11**, 45-54.

Clark, M., and P. Milner. 1964. ''Dependence of Timbre on the Tonal Loudness Produced by Musical Instruments,'' J. Audio Eng. Soc. **12**, 28-31.

Clark, M., P. Robertson, and D. Luce. 1964. ''A Preliminary Experiment on the Perceptual Basis for Musical Instrument Families,'' J. Audio Eng. Soc. **12**, 199-203.

Fletcher, H., E.D. Blackham, and D.A. Christensen. 1963. ''Quality of Organ Tones,'' J. Acoust Soc. Am. **35**, 314-325.

Fletcher, H., and L.D. Sanders. 1967. ''Quality of Violin Vibrato Tones,'' J. Acoust. Soc. Am. **41**, 1534-1544.

Fletcher, N.H. 1975. ''Acoustical Correlates of Flute Performance Technique,'' J. Acoust. Soc. Am. **57**, 233-237.

Grey, J.M. 1977. ''Multidimensional Perceptual Scaling of Musical Timbres,'' J. Acoust. Soc. Am. **61**, 1270-1277. (See also J. Acoust. Soc. Am. **62**, 454-462 and **64**, 467-472.)

Jansson, E.V. 1976. ''Long-Time-Average Spectra Applied to Analysis of Music. Part III: A Simple Method for Surveyable Analysis of Complex Sound Sources by Means of a Reverberation Chamber,'' Acustica **34**, 275-280. (See also Acustica **34**, 15-19, 269-274 and Acustica **42**, 47-55.)

Luce, D., and M. Clark. 1965. ''Durations of Attack Transients of Nonpercussive Orchestral Instruments,'' J. Audio Eng. Soc. **13**, 194-199.

Luce, D., and M. Clark. 1967. ''Physical Correlates of Brass-Instrument Tones,'' J. Acoust. Soc. Am. **42**, 1232-1243.

Patterson, B. 1974. ''Musical Dynamics,'' Scientific Am. **231** (Nov), 78-95.

Strong, W., and M. Clark. 1967. ''Synthesis of Wind-Instrument Tones,'' J. Acoust. Soc. Am. **41**, 39-52.

Strong, W., and M. Clark. 1967. ''Perturbations of Synthetic Orchestral Wind Instrument Tones,'' J. Acoust. Soc. Am. **41**, 277-285.

Audiovisual
1. ''Real and Synthetic Organ Tones'' (GRP)
2. ''Speeded and Slowed Brass Tones'' (GRP)

32. Lip Reed Instruments

The lip reed, or brass, instruments include the trumpet, trombone, tuba, and French horn. The common physical features of brass instruments are a cup-shaped mouthpiece, sections of cylindrical and conical tubing, valves or slides to change the length of the tubing, and a flared output end called a bell. These features can be observed in Figure 32.1, where it can also be seen that the tubing is coiled to make the instruments more compact. The air column contained within the tubing is excited by the vibrating lips of the player. The tone quality of a brass instrument depends on the shape of the mouthpiece, the shape and size of the tube, and the flare of the bell. We will establish some general requirements for a brass instrument and then discuss a trombone prototype to illustrate how the bell and the mouthpiece are used to meet these requirements. The role of slides and valves will then be considered. Radiation properties and sound spectra, including mute ef-

fects, of the brasses will be discussed. Finally, brief historical notes on the brass instruments will be given.

General Requirements

Production of sound in a brass instrument is somewhat analogous to sound production by the voice. Vibrating lips excite the brass bore; vibrating vocal folds excite the vocal tract. Both the lips and vocal folds produce pressure antinodes at the nearly-closed ends of the tubes which they excite. The lips and the vocal folds are both comparatively massive and subject to conscious muscular control. Most of the sound is radiated from a single opening—the bell for the brass and the mouth for the voice—where a pressure node exists.

In the vocal mechanism, the vocal tract has little influence on the vibration frequency of the vocal folds. The vocal tract makes only weak "suggestions" to the vocal folds in the form of standing waves, as its formant

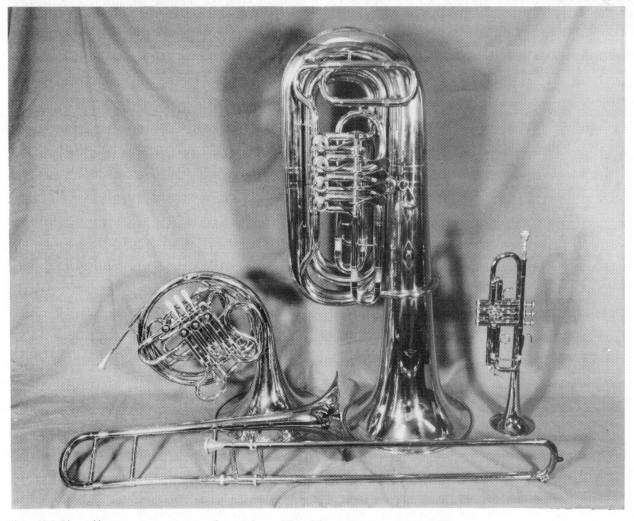

Figure 32.1. Lip reed instruments: trumpet, trombone, tuba, and French horn. (Courtesy of G. R. Williams.)

frequencies are not harmonic and its formant amplitudes are comparatively low. The vocal folds do not respond to the tract "suggestions" because of their mass and conscious muscular control. Hence, the vocal folds vibrate at a frequency determined primarily by their mass and tension, and not influenced much by the vocal tract.

In the brass instruments, the tube and flaring bell have considerable influence on the vibration frequency of the lips. The tube makes strong "suggestions" to the lips in the form of standing waves because of its harmonically related natural mode frequencies and large amplitude natural modes. Muscular control of the lips permits their response to and cooperation with the suggestions from the tube. Hence, the lip reed can, in effect, choose one of several available tube resonance frequency combinations to excite. There is usually a considerable frequency variation possible near a resonance, as a player can "lip" the note up or down a semitone or so by tensing or relaxing the muscles controlling the lips.

Let us look briefly at the lip valving function to gain some insight into its behavior and how it might interact with the wave in the tube. Assume that we start with the lips stretched taut but closed as shown in Figure 32.2a. The pressure of the air in the mouth exerts a force against the lips, which causes them to blow open, as shown in Figure 32.2b. When the lips reach their maximum opening (Figure 32.2c), the air flow through the opening is quite large. Because the lips still form a constriction, the speed of the air is increased, thus causing a pressure reduction by Bernoulli's Law (see Chapter 3). Because of the reduced pressure, and because of the tension on the lips, the lips begin to close, as shown in Figure 32.2d. The lips continue to close until they resume their original position, where the entire process begins again.

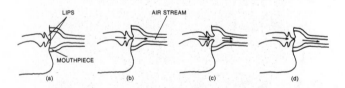

Figure 32.2. Lip position and airflow through lip opening at successive times in a lip vibration cycle. (After Olson, 1967.)

Some interesting points regarding the production of a pulsating air stream can be made. The area of the opening between the lips often varies in a nearly sinusoidal way. The air pulses would be almost sinusoidal and would have little harmonic content if they were to vary as the lip opening does. However, there are other means (discussed in Chapter 23) by which the air pulses are enriched with a good deal of harmonic content. For one thing, the air flow through the lips does not vary directly as the lip opening because the resistance of the opening to air flow does not vary directly with opening area. In addition, the standing waves in the tube give rise to a fluctuating pressure in the mouthpiece which affects the air flow. These two features are responsible for producing a series of air pulses having a rich harmonic content, even when the lip opening area varies almost sinusoidally.

As just noted, the amplitude of the standing pressure wave in the tube of a brass instrument helps determine the air flow through the lip opening, and thus influences the harmonic content of a tone. The amplitudes of the natural modes of the tube help to determine the amplitude of the standing waves and thus the strength of the tube suggestions. Large amplitude natural modes must exist in order to give strong suggestions to the lips.

The air flow through the lip opening will tend to be periodic and thus composed of harmonically related partials. Strong standing waves will be produced in the tube, when its natural mode frequencies are harmonic and coincide with those of the air flow. Furthermore, when the tube resonance frequencies are harmonic, the standing waves of all modes can act together to produce the largest possible composite standing wave. Thus, **harmonically-related tube modes enable the tube to exert substantial control over the lips.** As we follow the evolution of the trombone in the next sections we will look for response curves that have large amplitude peaks with frequencies in the ratios 1, 2, 3, 4, and so on.

Bell Function

We have chosen the trombone as our prototype brass in order to illustrate significant functions of various parts of the instrument. The trombone consists of a mouthpiece attached to a cylindrical U-shaped tube which is doubled back on itself and connected to a tapered section of tube. This tapered section flares into a bell. The total length of this tube is about 270 cm.

For instructional purposes, let us consider the evolution of the trombone from a closed-open cylindrical piece of pipe. The response of such a pipe is shown in Figure 32.3. The resonance frequencies are

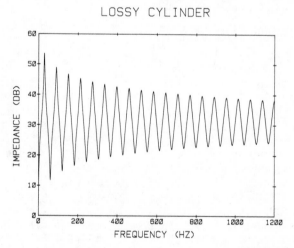

Figure 32.3. Input impedance response curve for cylindrical tube. (Courtesy of J. D. Dudley.)

shown in the second column of Table 32.1. Note that these frequencies are as expected for a cylindrical closed-open tube. The third column gives frequency ratios of each mode, which are normalized with the third mode set equal to three. In this way we can readily see the relative harmonicity of the modes. We note that

Table 32.1. Resonance frequencies (Hz) and their ratios for various idealized approximations to a trombone.

Mode	Pipe		Pipe plus bell		Trombone	
	Frequency	Ratio	Frequency	Ratio	Frequency	Ratio
1	29.5	0.58	35.7	0.62	36.5	0.62
2	90.4	1.79	107.4	1.86	109.3	1.87
3	151.6	3.00	173.3	3.00	175.3	3.00
4	212.9	4.21	231.9	4.01	233.3	3.99
5	274.4	5.43	292.4	5.06	292.3	5.00
6	335.9	6.65	355.9	6.16	352.3	6.03

the relative frequencies are related approximately as odd harmonics. However, we want our instrument to have all harmonics and so the higher modes must be lowered relative to the lower modes. This can be accomplished in part through the addition of a bell, which appears longer at high frequencies than at low ones. Figure 32.4 is a response curve for the pipe and bell combination. Columns 4 and 5 in Table 32.1 give mode frequencies and ratios. The ratios are now much closer to an all-harmonic relationship than those for the pipe alone.

POWER REFLECTION AT BELL INLET

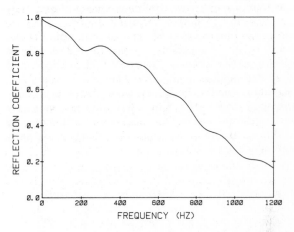

Figure 32.5. Power reflection coefficient at the point where the trombone bell is attached to a cylindrical tube. (Courtesy of J. D. Dudley.)

CYLINDER + BELL

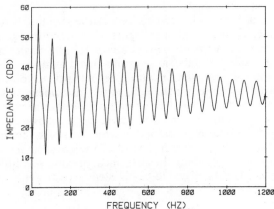

Figure 32.4. Input impedance response curve for cylindrical tube with trombone bell attached. (Courtesy of J. D. Dudley.)

The addition of the bell has several other effects. First, it helps the trombone produce a louder and clearer tone by increasing the sound radiation. Second, it not only changes the resonance frequencies of the tube, but it reduces the number of resonances present. Beyond a certain cut-off frequency, the pressure waves in the instrument are not reflected by the bell to produce standing waves, but leak out through the bell, thus reducing the number of resonance peaks and their overall amplitude. This can be seen by comparing Figures 32.3 and 32.4 for a cylindrical tube with and without a bell. This can be understood in terms of the power reflection coefficient shown in Figure 32.5. The **power reflection coefficient** is a measure of what fraction (as a percent) of the power coming from the lips is reflected back toward the lips at the point where the bell is attached to the cylindrical tubing. In one sense it can be regarded as a measure of how much of a "barrier" the bell is at different frequencies. At low frequencies the bell does not let much energy escape, while at high frequencies most of it can escape. The larger the bell of an instrument, the lower its cutoff frequency and the fewer the higher harmonics. This is illustrated in Figure 32.6 with trumpet

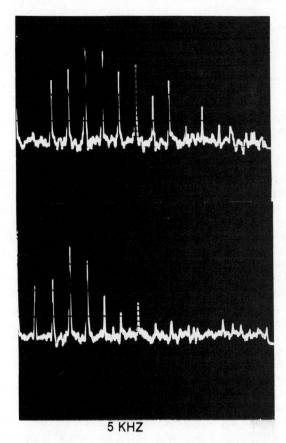

5 KHZ

Figure 32.6. Radiated spectra of trumpet (upper) and cornet (lower).

215

and cornet spectra. Although the cornet spectrum is very similar to the trumpet spectrum, harmonics above the seventh are usually not observed in cornet spectra, while they are present in trumpet spectra. This is because the larger bell and conical shape of the cornet reduce the higher partials.

Mouthpiece Function

Two further adjustments are necessary to make the pipe and bell combination useful as a musical instrument. The natural mode frequencies need to be made more harmonic and their amplitudes need to be increased. Both of these requirements can be met by adding an appropriate mouthpiece and tapered mouthpipe at the lip-end of the instrument. When this is done the response curve shown in Figure 32.7 is obtained and related data appear in columns 6 and 7 of Table 32.1. The modes from the third to the eleventh are very nearly harmonic. The second is a little flat and the first is very flat in relation to higher modes. In addition, the mouthpiece and mouthpipe increase the response peak amplitudes considerably.

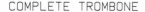

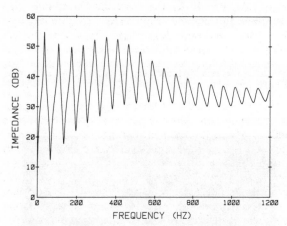

Figure 32.7. Input impedance response curve for trombone with slide in. (Courtesy of J. D. Dudley.)

If a musician uses this instrument to play a note having a frequency of 114 Hz (nominally B2ᵇ), the fundamental of the note will coincide with the second response peak of Figure 32.7. The player's vibrating lips produce a set of harmonics; these harmonics will be reinforced by the response peaks which are multiples of 114 Hz, namely peaks 4, 6, 8, 10, and so on. If the player vibrates his lips at the frequency of peak 3, response peaks 3, 6, 9, and 12 will function in a similar manner. As a trombonist blows harder, the harmonic content of the tone increases and more response peaks are utilized, resulting in a brighter and louder tone.

Another feature of interest with our trombone is the "privileged notes" made possible by the cooperation of several modes of the tube. We consider for illustration the response curve shown in Figure 32.7. Suppose we wish to produce a tone having a fundamental frequency lower than the second mode. We could try to produce a note using the first mode as the fundamental, but we would find it to be weak and unstable because the higher modes are not harmonically related to it.

Another possibility would be to produce a note at a frequency half that of the second mode, where there is no resonance peak. However, there are resonance peaks at 109 Hz, 175 Hz, etc. which are nearly integer multiples of 58 Hz. If the second harmonic produced by the vibrating lips is made to coincide with the 109 Hz mode, the third harmonic with the 175 Hz mode, and so on, a fundamental frequency of 58 Hz will be possible even though there is no mode present to encourage it. A "**privileged note**" in this sense is a note produced with a fundamental frequency with which no natural mode coincides. It is made possible because higher air-flow harmonics coincide with other tube modes; these other tube modes need to have a nearly integer relationship with the desired fundamental frequency if they are to cooperate effectively in the production of a privileged note. The "privileged frequencies" associated with a collection of modes can be determined by seeing what frequency differences occur between pairs of modes. In the case of the trombone if we consider the modes having frequencies of 109, 175, and 233 Hz, the frequency difference of 58 Hz between nearest neighbor modes is a privileged frequency.

It is of some interest to look at the standing wave patterns in our trombone. Figure 32.8 shows the standing pressure wave of the first mode. It has a single antinode and exhibits a pressure maximum at the mouthpiece end. The pressure does not penetrate as a sinusoid to the end of the bell, so the instrument looks shorter for this mode than its measured length as noted in the previous section. The power curve (shown as a dashed line) decreases smoothly from the lip end to the bell end of the instrument. Most of the power input at the lip end is lost along the instrument because very little is present at the bell end to be radiated.

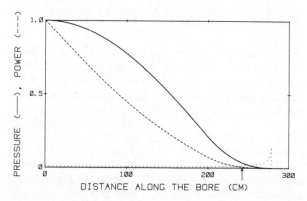

Figure 32.8. Standing pressure wave (solid line) and power (dashed line) for the first mode of a cylindrical tube and trombone bell. Profile of tube and bell appears at the bottom. (Courtesy of J. D. Dudley.)

Figures 32.9 and 32.10 illustrate standing waves for a high amplitude mode (mode five in Figure 32.7) and a low amplitude mode (mode eleven in Figure 32.7). Mode five exhibits five antinodes. (The antinodes have all been plotted as pressure magnitudes. In an actual standing wave pattern alternate antinodes would be alternately positive and negative.) Modes five and eleven penetrate farther into the bell than mode one.

The instrument appears longer to modes five and eleven than to mode one, so they are flattened more as noted in the previous section. More power arrives at the bell to be radiated for these modes than for mode one. The pressure amplitude decreases as the bell flares and increases as the tube is constricted in the mouthpiece region as can be seen in Figure 32.10.

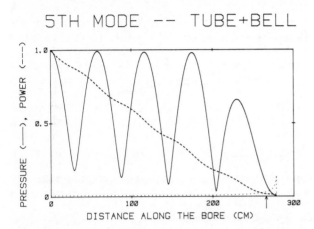

5TH MODE -- TUBE+BELL

Figure 32.9. Standing pressure wave and power for the fifth mode of a cylindrical tube and trombone bell. (Courtesy of J. D. Dudley.)

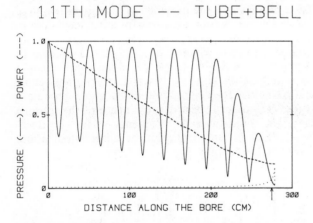

11TH MODE -- TUBE+BELL

Figure 32.10. Standing pressure wave and power for the eleventh mode of a cylindrical tube and trombone bell. (Courtesy of J. D. Dudley.)

Slides and Valves

The trombone we have designed so far is actually a type of bugle. It can play notes corresponding to its modal frequencies, but that is all. From Table 32.1 we see that these notes would have fundamental frequencies of 109, 175, 233 Hz, and so on. In order to make the instrument more useful musically, we must be able to play many additional notes within this range. This is accomplished for the trombone by adding a slide which can be used to increase the length of the tube. Every time the length of the instrument is changed, a new set of resonances, based on a different fundamental frequency, is produced. Since the slide can be extended arbitrarily, an infinite number of variations is possible. Seven standard slide positions have been found to be adequate, as not all possible variations can be

distinguished. Each new position gives an increase in length of approximately 6% over the preceding position as can be seen in Figure 32.11. When the trombone slide is not extended (first position), a B-flat or one of its harmonics can be played. Since each consecutive position lowers the pitch by one semitone, the seventh position corresponds to a fundamental frequency of E.

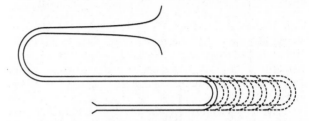

Figure 32.11. Slide positions of the trombone.

For instruments other than the trombone, piston or rotary valves are used to couple in predetermined lengths of tubing. One valve controls a length of tubing equal to about 6% of the instrument length and thus lowers the frequency a semitone. The two other valves control tubing to lower the frequency nominally two and three semitones. By using various combinations of valves, six frequency lowerings (from one to six semitones) are possible. Problems arise with valved instruments which are less critical when they occur in slide instruments. Position one of the trombone can add an effective length of 6% to the trombone with the slide in. Position two can add an effective length of 6% to position one, and so on. Suppose one valve adds an effective length of 6% to the trumpet with no valves depressed. Suppose a second valve adds an effective length of 12% to the trumpet with no valves depressed. When the second valve is used in conjunction with the first valve, the fixed amount of tubing it controls will add an effective length of less that 12% to the trumpet with one valve depressed. Various compromises are made in the tube lengths controlled by the valves and additional valves are sometimes used to alleviate the problem.

Whenever abrupt changes in tubing diameter occur, turbulence is produced and acoustical energy is lost. The discontinuities at valves in trumpets and tubas or where the different-sized pieces of tubing meet in a trombone will give rise to such turbulence losses, especially at high dynamic levels. Carelessness in the construction of an instrument will intensify this problem.

Brass Spectra

The spectrum of a brass instrument is dependent on the dynamic level at which the instrument is blown. Internal spectra measured at the mouthpiece will exhibit little harmonic development when produced at low playing levels, and increasing harmonic development as the playing level is increased. Stylized internal spectra are shown in Figure 32.12 for three dynamic levels.

An external brass spectrum measured with a microphone at some point outside the bell depends on the internal spectrum and how efficiently different frequencies are radiated. We can imagine, for example, that some internal spectrum has been produced and that

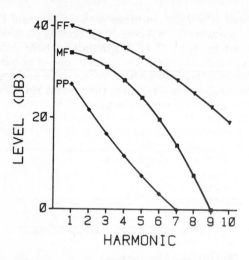

Figure 32.12. Stylized internal brass spectra for pianissimo (pp), mezzoforte (mf), and fortissimo (ff) tones. (After Beauchamp, 1980 and Benade, 1976.)

as the associated waves travel along the tube they encounter the bell barrier shown in Figure 32.5. Low frequencies are mostly reflected back into the tube and are not radiated efficiently. Radiation efficiency might be obtained by taking 100% minus the power reflection curve. It would provide a high frequency boost up to some upper limit. The external spectrum (see Figure 32.13) thus depends on the internal spectrum and radiation efficiency. It will exhibit relatively more high frequency energy than the internal spectrum, as seen by comparing Figures 32.12 and 32.13.

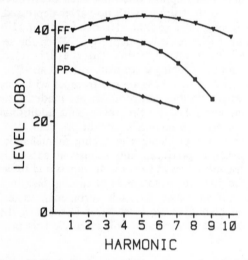

Figure 32.13. Stylized external brass spectra for pianissimo, mezzoforte, and fortissimo tones.

Mute Function

We have seen that the bell of a brass instrument modifies the natural mode frequencies and forms a selective radiation barrier on a tube to which it is attached. A mute inserted in the large end of a bell modifies the bell properties, which in turn modify the radiation barrier produced by the bell. Figure 32.14 illustrates a straight mute inserted into a bell. All mutes, including cylindrical, cup, and Harman, produce their own specific modifications in a brass response curve. A

modified response curve changes the playing properties of an instrument and results in a different radiated spectrum.

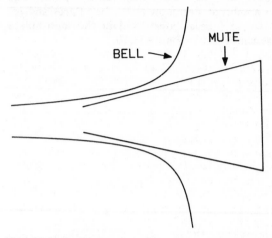

Figure 32.14. Straight mute inserted into a brass instrument bell.

The Trumpet and Tuba

The coiled tube of a trumpet is cylindrical for about the first two- thirds of its 140 cm (uncoiled) length. The rest is conical, with a very slow rate of flare (except for the last 25 cm, which flares rapidly to form the bell). Three piston valves are provided to change the length of the air column. A response curve for a trumpet is shown in Figure 32.15. It bears a striking resemblance to our trombone curve of Figure 32.7. The effects of the bell, mouthpiece, and mouthpipe are apparent. Modes 5 through 7 are the largest in the trumpet response curve and all modal frequencies are scaled about an octave higher. The overall spectral prominence of the trumpet occurs at about twice the frequency as for the trombone, which is consistent with the observations about Figure 31.1 in Chapter 31.

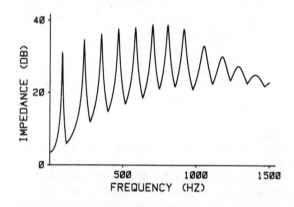

Figure 32.15. Input impedance response curve for a trumpet with open valves. (After Backus, 1976.)

We see from the response curve that we have a musically useful series of resonance frequencies, with the exception of the lowest mode which is badly out of tune with the others. This is of limited consequence, however, because the actual lowest resonance (E2) is not used in practice. The remainder of the resonances form an approximately harmonic series based on a fun-

218

damental (B2-flat) which does not exist as a mode. When we compare the harmonic series based on B2-flat with the resonances of the trumpet we find that, with the exception of the first mode, the trumpet resonances are very close to the harmonic series. The playable frequencies are all approximately integer multiples of a fundamental (B2-flat) which is not an actual trumpet resonance, and which might be termed a nonexistent fundamental. The so-called "pedal tone" in the trumpet is the octave below the first useful mode (B3-flat) and hence is identical to the nonexistent fundamental. The pedal tone is produced as a privileged note, even though there is no actual resonance present, by vibrating the lips at a frequency of 116 Hz. The upper harmonics of the lips are integer multiples of 116 Hz (232, 348, etc.), which very nearly coincide with the resonances of the trumpet, thus reinforcing the vibration. In other words, the presence of harmonics 2, 3, 4, etc. in the tone will cause the lip and air column vibrations to repeat at the fundamental frequency of 116 Hz.

The coiled tube of the tuba is a slowly flaring cone which terminates in a large bell, the entire (uncoiled) length being about 720 cm. The tuba usually has four valves, with their associated plumbing, in order to give the player greater flexibility and ease of performance.

The French Horn

The French horn consists of a coiled tube about 360 cm in length, with a slow conical flare terminated in a very large bell. The instrument is equipped with three valves. The French horn differs from other brass instruments in one important aspect. The trumpet and trombone use the first eight or nine resonance modes for playing, whereas the horn uses the series up to an octave higher, as far as the sixteenth resonance. This requires that the higher resonances be pronounced and distinct, which is accomplished by having more of the French horn tube conical and less of it cylindrical. A response curve for a French horn is shown in Figure 32.16. Note that, since a more conical tube was used, the fundamental frequency coincides more closely to the nonexistent fundamental.

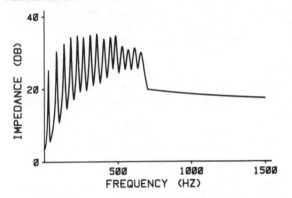

Figure 32.16. Input impedance response curve for a French horn. (After Backus, 1976.)

Since the horn typically plays in a frequency region where the modes are very close together, the player has considerably more flexibility than do players of other brass instruments. It was discovered empirically, that by placing one's hand in the bell of the instrument, one

could lower the pitch by an amount that depends on how far the hand is inserted. This technique was developed to a point where a complete chromatic scale was possible over the upper part of the instrument's range. When a hand is inserted in the bell, the area is constricted, which increases the acoustic mass at the end of the instrument, thus lowering the resonance frequencies. In addition to lowering the frequencies, the number of harmonics is usually increased as the hand is inserted, thus making it possible to play in resonance modes which are not sufficiently well-developed without the hand insertion. A comparison of Figures 32.17 with Figure 32.16 shows the effect of inserting the hand into the bell of a French horn.

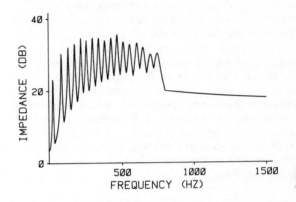

Figure 31.17. Input impedance response curve for a French horn with player's hand inserted. (After Backus, 1976.)

When the hand is thrust as far as possible into the bell, a tone of quite different quality, called the "stopped tone," is produced. The stopped tone has a pitch which is a semitone higher than the normal horn tone. The apparent rise in pitch is caused by the large increase in acoustic mass as the hand is inserted. The mass actually increases so much that each mode is lowered in frequency until it is within a semitone of the original frequency of the mode below it. A given note is then played in the next higher mode with the horn stopped, the original mode now having gone much too flat. This can be demonstrated by following a note down in pitch as the hand is inserted. The note can be followed down to its new low pitch, but if the player tries to hold his pitch, the note jumps to the next higher mode, which is a semitone sharper than the unstopped mode.

Historical Notes

In addition to the principal orchestral brass instruments which we have discussed, the cornet, flugelhorn, bugle, baritone, euphonium, and sousaphone— other related brass instruments—are also used in marching bands and symphony orchestras. All of these instruments share similar physical features and acoustical properties, and have varying amounts of cylindrical and conical tubing to provide harmonic modes for interaction with the lips.

The ancestry of our modern brass instruments dates back to two prototypes: basically conical animal horns and basically cylindrical cane trumpets with a funnel-like end. An example of the first prototype is the

Jewish shofar made of ram's horn (see Figure 32.18A) which has been in use for over three thousand years. It is played only on the second and third harmonic and thus produces two crude but awe-inspiring sounds about a fifth apart. The Roman lituus (Figure 32.18B) is an example of an early trumpet. These instruments were employed for festive and ceremonial occasions; no "music" was played with them, as only one or two notes could be produced.

By the 15th century a type of trumpet arranged into a long coil was in common use as a hunting horn. The coiled shape (see Figure 32.18C) reduced the size of the instrument and made it possible for a mounted horseman to carry the horn on his shoulder during a hunt. Although hunting horns and trumpets were utilized more often than were their predecessors for musical purposes by playing the notes of the harmonic series, they were severely handicapped by the limited number of tones available. Three different developments ensued, all intended to obtain a complete melodic scale from one instrument. The earliest development was the invention of the telescopic slide, which only worked on the cylindrical instruments. The sackbut, a type of medieval trombone with a smaller bell and narrower tube, existed in all sizes from alto to contrabass. Only the descendents of the tenor instrument survive in our modern trombone.

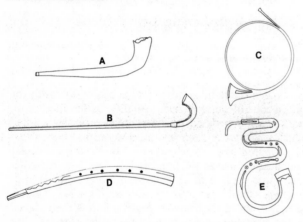

Figure 32.18. Some ancestors of modern brass instruments: (A) Jewish shofar, (B) Roman lituus, (C) hunting horn, (D) treble cornett, and (E) serpent. (After Baines, 1969.)

The second method of obtaining a melodic scale was the use of fingerholes, such as those employed in woodwind isntruments. The cornetto family members (no relation to the modern cornet) were hollow conical wooden tubes, each with six fingerholes and a thumbhole (Figure 32.18D). The bass version, the "serpent" which originated in the late 16th century, was so named because its 8 foot wooden tube was folded into a serpentine form, as shown in Figure 32.18E. These interesting instruments, a hybrid of woodwind and brass technologies, have no surviving descendents among our modern instruments.

The third method of obtaining melodic scales evolved considerably later. By 1650 the helical hunting horn, which orginated in France, was being used occasionally for musical purposes. Its musical development took place in Germany, however, where it evolved into a

slender conical tube coiled in one or more circles and terminating in a rapidly flared bell. This "French horn," the true parent of the modern instrument, required a separate horn for each different key. By the 18th century, as orchestras grew in size, a set of tubes of differing lengths (called crooks) were inserted in the horns to change the length of the instrument. By adding or removing pieces of tubing one instrument could play in many different keys. Around the middle of the same century, an innovative horn player experimenting with mutes discovered that a hand inserted into the bell lowered the pitch. Although different sets of harmonics could be generated merely by inserting the hand to various depths, a considerable change of tone color and loudness accompanied the stopped tones. This problem as well as the deficiencies of the crook tube system were corrected by the invention of the valve action in 1815. By the end of the nineteenth century it was commonplace for musicians to use the combination of valve and hand technique used today. The final development in the long history of the French horn occurred during the early years of this century. French horn players were required to play in a high register, where the response peaks are closely spaced. By using a higher-pitched horn, the same tones could be produced on peaks of lower modes. Once again the musician needed two separate horns: the older horn in F, and a horn in B-flat for more modern music. The problem was solved by the invention of the "double horn," a combination of F and B-flat instruments with a thumb valve which allows the player to switch instantly from one horn to the other.

The modern bugle and the 19th century ophicleide are both descendents of the "bugle horn," literally "horns of a young bull" ("bugle" in old French). The ophicleide was invented in the early years of the 19th century as a bass bugle. Since it was considerably louder, the ophicleide soon replaced the serpent as the bass of the brass section in large orchestras. By 1900 the ophicleide had likewise been replaced by the more sonorous sounding tuba. The 19th century also saw the bugle evolve into a valved version called the flugel horn, while the unvalved post horn (used by 18th century postmen to announce their arrival at country manors) became the conical, but valved, cornet.

Exercises

1. How can a bugle, with no valves (and hence a single length), play tones at more than one fundamental frequency?

2. Determine several "privileged note" frequencies from the trumpet response curve of Figure 32.15.

3. Refer to the trumpet response curve in Figure 32.15. Draw lines to indicate the partials in the spectrum for a tone of 393 Hz (G4). Draw similar lines indicating the partials for a tone having a frequency of 262 Hz. For 524 Hz. How do the spectra of these tones compare?

4. The trumpet is similar to the trombone except that it is much shorter. What effect does this have on natural mode frequencies?

5. What effect do different shaped mouthpieces have on brass tones?

6. If the trumpet has only three valves (giving eight different length combinations) how can it play

tones at 30 different frequencies?

7. How is it possible to produce the pedal note of a brass instrument in which the lowest natural mode frequency lies several semitones below the pedal note frequency?

8. What is the effect of a leaky valve in a brass instrument?

9. Why is it possible to sing all notes (within a certain range) without changing the length of the resonator (your vocal tract), while musical instruments only produce frequencies close to a natural mode frequency?

10. Explain why vocal fundamental frequency is determined by the vocal fold properties, whereas brass frequencies depend on lip and tube properties.

11. Answer the following questions by referring to both spectra of Figure 32.6. How many harmonics are present in each spectrum? What are the approximate frequencies of the harmonics? What is the fundamental frequency and to what musical note does it correspond?

12. A trumpet with no valves depressed is used as a bugle. What fundamental frequencies can be produced? To what notes do these correspond? Comment on the stability of the various notes played pianissimo and fortissimo.

13. Suppose that the valve controlled tubing on a trumpet is adjusted so the first valve produces an exact two semitone lowering, the second one semitone, and the third three semitones. How much will the frequencies differ between a tone using valves one and two and a tone using valve three? If all three valves are depressed, how sharp will the resulting tone be relative to an intended six semitone lowering?

Demonstrations

1. Play on a cylindrical tube directly and then with a brass mouthpiece.

2. Get a good brass player to demonstrate features described in the text.

Further Reading

Backus, Chapter 12
Baines, Chapters 11, 12
Benade, Chapter 20
Hall, Chapter 13
Kent, Part II
Rossing, Chapter 11
Backus, J. 1976. "Input Impedance Curves for the Brass Instruments," J. Acoust. Soc. Am. **60**, 470-480.

Backus, J., and T. C. Hundley. 1971. "Harmonic Generation in the Trumpet," J. Acoust. Soc. Am. **49**, 509-519.

Baines, A. 1976. **Brass Instruments: Their History and Development** (Faber and Faber, Ltd.).

Bate, P. 1966. **The Trumpet and Trombone** (W. W. Norton).

Beauchamp, J. W. 1975. "Analysis and Synthesis of Cornet Tones Using Nonlinear Interharmonic Relationships," J. Audio Eng. Soc. **23**, 778-795.

Beauchamp, J.W. 1980. "Analysis of Simultaneous Mouthpiece and Output Waveforms of Wind Instruments," Audio Eng. Soc. 66th Convention, paper 1626(C-3).

Benade, A. H. 1973. "The Physics of Brasses," Scientific Am. **228** (Jan), 24-35.

Benade, A. H., and D. J. Gans. 1968. "Sound Production in Wind Instruments," Ann. N.Y. Acad. Sci. **155**, 247-263.

Fletcher, N. H. 1979. "Excitation Mechanisms in Woodwind and Brass Instruments," Acustica **43**, 63-72.

Martin, D. W. 1942. "Lip Vibrations in a Cornet Mouthpiece," J. Acoust. Soc. Am. **13**, 305-308.

Martin, D. W. 1942. "Directivity and Acoustic Spectra of Brass Wind Instruments," J. Acoust. Soc. Am. **13**, 309-313.

Morley-Pegge, R. 1960. **The French Horn** (Ernest Benn)

Pratt, R. L., S. J. Elliott, and J. M. Bowsher. 1977. "The Measurement of the Acoustic Impedance of Brass Instruments," Acustica **38**, 236-246.

Pyle, R. W. 1975. "Effective Length of Horns," J. Acoust. Soc. Am. **57**, 1309-1317.

Stauffer, D. 1954. **Intonation Deficiencies of Wind Instruments** (Catholic Univ. of Am. Press).

Webster, J. C. 1949. "Internal Tuning Differences due to Players and the Taper of Trumpet Bells," J. Acoust. Soc. Am. **21**, 208-214.

Young, R. W. 1967. "Optimum Lengths of Valve Tubes for Brass Instruments," J. Acoust. Soc. Am. **42**, 224-235.

Audiovisual

1. **Sounds in Your Ears—Brass Instruments** (3.75 ips, 2 track, 30 min, UMAVEC).

2. **Horn, Part I** (21 min, color, 1968, OPRINT)

3. "Speeded and Slowed Brass Tones" (GRP)

4. "Basic Trumpet Acoustics" (REB)

33. Mechanical Reed Instruments

The mechanical reed instruments are one subclass of the woodwind group. (The other subclass will be considered in Chapter 34.) Mechanical reed instruments employ a cylindrical tube or a conical tube. The air-valving is accomplished by a mechanical reed, which alters the steady air stream from the player's lungs and causes the air column in the tube to begin oscillating. Unlike the brass instruments, however, the vibration frequency of the mechanical reeds is controlled more by the tube. The mechanical reed instruments can be classified by the shape of the tube and by the type of reed used. The two types of tube are cylindrical and conical; the two types of reed are single and double. We will consider three of the instruments shown in Figure 33.1 as representative of their respective families: clarinet (single reed and cylindrical tube), oboe (double reed and conical tube), and saxophone (single reed and conical tube). We will establish some general requirements for a mechanical reed instrument and then explore the roles of tube shapes, tone holes, and register holes in a prototypical clarinet. Radiation properties and woodwind spectra will then be considered. Some discussion will be given on specific instrumental families. Finally, brief historical notes on the mechanical reed woodwinds will be given.

General Requirements

Some insight into sound production in mechanical reed instruments can be gained by contrasting them with the brasses. A vibrating mechanical reed modulates the air flow that excites the woodwind air column; vibrating lips modulate the air flow that excites the brass air column. The mechanical reeds are blown inward by the mouth pressure and tend to shut off air flow. The lips are blown outward by the mouth pressure. Both the

Figure 33.1 Mechanical reed instruments: clarinet, oboe, saxophone, and bassoon. (Courtesy of G. R. Williams.)

mechanical reed and the lips nearly close the ends of the tubes which they excite; pressure antinodes exist at these ends. The mechanical reed is light and subject to limited player control whereas the lips are massive and subject to conscious muscular control. Most of the sound is radiated from several open tone holes in a woodwind where pressure nodes exist. The radiation is entirely from the bell in the brasses.

In the brasses, the air column in the tube sends strong suggestions to the lips because of its large amplitude and harmonically related response peaks. The massive lip reed, under conscious muscular control, can select which of several tube resonance combinations it will excite.

In the mechanical reeds, the air column in the tube also sends strong suggestions to the reed because of its large amplitude and harmonically-related response peaks. However, in this case there are typically fewer response peaks acting in cooperation. The largest peak typically "captures" the reed and largely determines its vibration frequency, as the reed is light and subject to limited player control. This makes it very susceptible to tube suggestions.

Our mechanical reed system may be seen as an air supply under pressure, a reed mechanism for converting an otherwise steady air flow into a pulsating flow, and a resonant air column. If the reed were driven by some external device, it could be made to vibrate at all frequencies, at all blowing pressures, and into any tube. The vibratory motion of the reed will die out unless sufficient power is supplied to offset the reed losses. Power can be supplied to or extracted from the reed by any forces acting upon it. The blowing pressure in the player's mouth will supply power to the reed as the reed closes, but will extract about an equal amount of power as the reed opens. (Mouth pressure always pushes in on the reed. When the reed closes, it moves in the direction of the applied force and so gains power. When the reed opens it moves opposite the direction of the applied force and so loses power. Refer back to Chapter 4 to explore these ideas for a simple vibrator.) Hence, the mouth pressure acting on the reed supplies no net power. The source of power to keep the reed in motion comes from the oscillating pressure inside the mouthpiece, which is caused by the resonating air column. When the mouthpiece pressure is positive it will push the reed open; when it is negative it will pull the reed closed. If phase relations are right, both actions supply power to the reed because the force and the displacement are in the same direction. A Bernoulli force can act between the tip of the reed and the mouthpiece. It is not very important for clarinet-type instruments when the reed motion is small, but may be important for large single-reed motions and for double reeds.

The reed and air column must extract sufficient power from the constant blowing pressure of the player if oscillations are to be maintained. If sufficient power is supplied to the reed to maintain its vibration, oscillatory flow into the air column will occur. The amount of power supplied to the reed by a given air column oscillation depends on the amplitude of the corresponding response peak. It also depends on how much of that frequency has been introduced into the tube by the air flow through the reed opening. As with the brasses, when the tube resonance frequencies are harmonic the standing waves of all modes can act in phase to supply the largest amount of power to the reed.

Arguments similar to those for brasses can be made to describe the origin of the harmonic content of the mechanical reeds. Figure 33.2 illustrates single and double reeds used on the mechanical reed woodwinds. The area of the opening between the reed and the mouthpiece (or between double reeds) usually varies in a nonsinusoidal manner. The resistance to air flow does not vary directly with reed opening, so the flow does not vary directly with the opening. These features are responsible for producing a series of air pulses rich in harmonics.

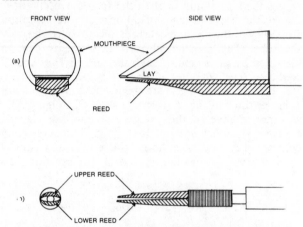

Figure 33.2. Reed and mouthpiece for (a) clarinet and (b) oboe.

As with the brasses, we look for tubes whose response curves have harmonically related peaks. Peak height is not quite so critical as with the brasses because of the lighter mechanical reeds. Since fewer peaks are involved in cooperative action, we look for tubes for which the response curves have all harmonics (1, 2, 3, 4) or odd harmonics (1, 3, 5, 7).

Bore Shapes

Since the mechanical reed instruments are closed at the reed-end, we explore various possible tube shapes which are closed at one end to see which ones satisfy our requirements. Four possible tube shapes and their response curves are shown in Figure 33.3. Their natural mode frequencies and frequency ratios to the first mode are shown in Table 33.1. The cylindrical tube (a) gives response peaks at odd integral multiples of the first peak. The conical tube (c) gives response peaks at all integral multiples of the first. The negative flare tube (b) and the positive flare tube (d) give response peaks at nonintegral multiples and so are not useful for musical purposes. Furthermore, cylindrical and conical tubes maintain their odd harmonic and all harmonic modes, respectively, even when shortened by open tone holes and modified by open register holes.

Tone Hole Function

The simple tubes just considered would be of limited musical value because they could only produce a single tone (as in the case of an organ pipe) or two or

Table 33.1. Natural mode frequencies and frequency ratios of lowest four modes for cylindrical (a), decreasing flare (b), conical (c), and increasing flare (d) tubes.

Mode	Mode Frequency (Hz)				Ratio to first mode			
	(a)	(b)	(c)	(d)	(a)	(b)	(c)	(d)
1	136.4	208	255.6	281	1.00	1.00	1.00	1.00
2	409.2	477	512.0	520	3.00	2.29	2.00	1.85
3	682.0	748	771.0	767	5.00	3.60	3.02	2.73
4	954.9	1019	1032.0	1021	7.00	4.90	4.04	3.63

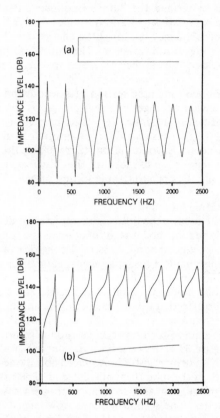

Figure 33.3. Prototype mechanical reed wind instrument tube shapes and their response curves: (a) cylindrical, (b) negative flare, (c) conical, and (d) positive flare.

three tones through reed adjustments. Some means must be provided to change the length of the tube; this is accomplished by the use of tone holes.

The presence of open tone holes serves to shorten the tube and thus to raise the natural mode frequencies. The presence of closed tone holes makes the bore rougher. Both give rise to changes in natural mode frequencies and to increased losses caused by air flow past discontinuities. The tube must be adjusted so that combinations of open and closed tone holes for each note produce nearly harmonic modal frequencies.

At the open end of a tube, the power reflection coefficient is large (Figure 33.4) so that most of the wave is returned to the reed-end of the instrument. This results in large and regularly spaced peaks in the response curve, as seen in Figure 33.5. If a tube is terminated with a series of open tone holes, its response curve is modified relative to that of a simple tube. Figure 33.6 shows the response curve for a tube like that of Figure 33.5 but with the addition of a section of tubing with eight open tone holes. The response curve looks about the same to the third peak but is changed

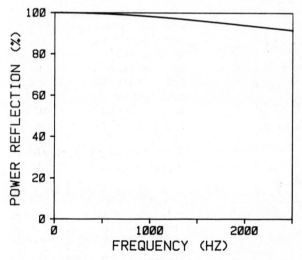

Figure 33.4. Power reflection coefficient at open end of cylindrical tube.

drastically beyond that. This can be understood by looking at the power reflection coefficient (Figure 33.7)

225

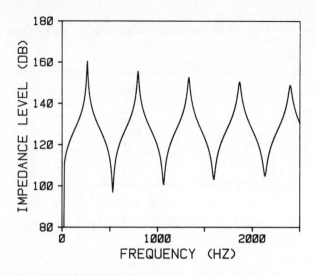

Figure 33.5. Input impedance response curve for cylindrical tube.

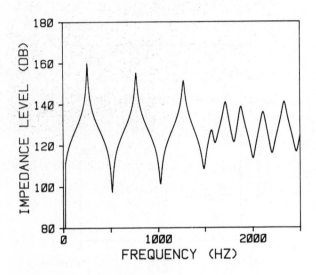

Figure 33.6. Input impedance response curve for cylindrical tube terminated with a series of open tone holes.

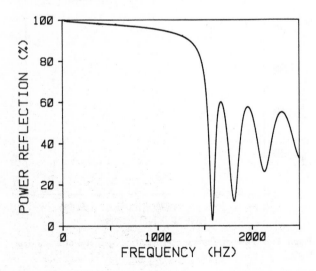

Figure 33.7. Power reflection coefficient at point where series of open tone holes is attached to cylindrical tube.

at the point where the open hole section of tube is added to the original piece of tube. At low frequencies the curve looks much like the comparable curve (Figure 33.4) for the open end of a tube. However, at a frequency near 1500 Hz the open-hole portion of tubing ceases to be like a barrier and higher frequencies are not reflected, but travel right into the open-hole section of tube.

The frequency at which waves begin to travel into an open hole section of tubing, rather than being reflected, is termed the **cutoff frequency**. Peaks in a response curve are small and irregularly spaced beyond the cutoff frequency because little energy is reflected to set up strong standing waves. The cutoff frequency is determined by the number, size, and spacing of the tone holes.

The open tone hole section of a woodwind plays a role somewhat analogous to that played by the bell for brasses. Each represents a reflecting barrier at low frequencies which gives rise to large amplitude peaks in the response curve. Beyond the cutoff frequency the barrier becomes smaller and most of the wave energy propagates into the open tone-hole section or bell.

Register Hole Function

In the brasses the lips can be adjusted to determine which of the tube modes is to function as the fundamental for a particular tone. Less reed control is available with the mechanical reeds; a register-hole mechanism is used to modify an instrument's response curve so that a higher tube mode will control the reed. For a register hole to be effective it must decrease the amplitude and shift the frequency of the first mode. Figure 33.8 shows a response curve for a cylindrical tube with no register hole. Figure 33.9 shows a response curve for a similar tube with an open register hole placed about one quarter the length of the tube from the reed end. The first mode is now misaligned and so cannot cooperate with the other modes. The second mode is now larger than the first and is better aligned in frequency with higher modes. The second mode now serves to define the fundamental frequency of vibration and receives cooperation from modes five and eight. In the case of this cylin-

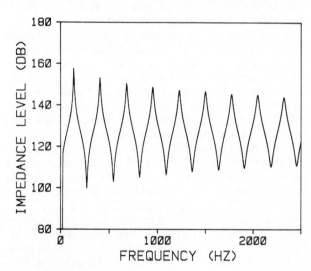

Figure 33.8. Input impedance response curve for a cylindrical tube with no register hole.

226

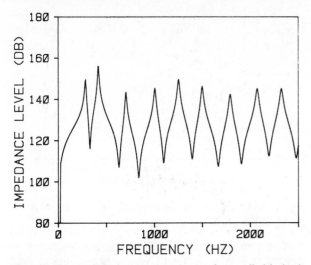

Figure 33.9. Input impedance response curve for a cylindrical tube with an open register hole.

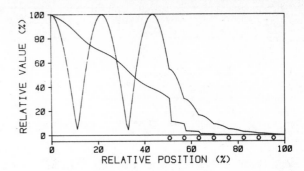

Figure 33.11. Standing pressure wave and power for the third mode of a cylindrical tube with open tone holes.

drical tube, use of the register hole gives rise to a tone which is a twelfth above the original because the second mode has a frequency three times that of the first. In conical tubes the register hole gives rise to tones an octave higher because the second mode frequency is twice that of the first.

Radiation Properties

Radiation from brasses is through the bell. Radiation efficiency and the directionality of the radiation both increase with frequency. Radiation from woodwinds can be from any of several open tone holes. Radiation from an open hole is dependent on the properties of the tone hole and the amount of the acoustic wave propagated to the position of the hole. The cutoff frequency largely determines the radiation pattern of an instrument. Below cutoff radiation occurs primarily from the first few open tone holes, as can be seen by looking at the power distribution for modes one and three in Figures 33.10 and 33.11. (The discontinuity in the power distribution at the position of the first open hole for mode one provides an estimate of radiation loss from the hole relative to other losses in the tube. Adding the discontinuities at holes one and two gives an estimate of power lost to radiation of about 9% relative to a power loss of 91% in the bore. Relative radiation losses for the third mode are about 27%. The estimates

for radiation loss are all too large because losses in the tube are underestimated.) Above cutoff radiation occurs from all open tone holes in the series, as can be seen in the power distribution for mode five in Figure 33.12. (Relative radiation losses for the fifth mode are 55%.) Much of a woodwind's characteristic sound is determined by its cutoff frequency; the lower the cutoff frequency, the darker the tone color for a given mouthpiece and reed configuration.

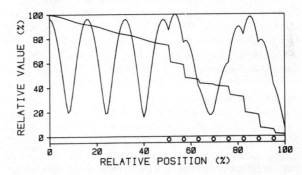

Figure 33.12. Standing pressure wave and power for the fifth mode of a cylindrical tube with open tone holes.

The sound one hears from a woodwind, of course, is not the internal standing wave in the tube, but the radiated sound. The bell of the instrument is not the radiating source of the sound except when all the tone holes are closed; the bell can be removed with little consequence to the instrument, except for the lowest few notes.

Clarinet Family

The clarinet provides a relatively simple example of a vibrating reed producing oscillations in an air column. Since the reed opening is quite small, the mouthpiece end of the instrument is at a pressure antinode. Let us follow a pressure pulse down a hypothetical clarinet tube which we assume to be completely cylindrical, but with a clarinet mouthpiece attached. When the reed opens, a positive pressure pulse is admitted into this "tubinet," as shown in Figure 33.13A. The positive pressure pulse travels down the tube until it encounters the open end (bell end on an actual clarinet). Since the open end must be a pressure node, the reflected wave is a negative pulse which then travels back along the tube toward the reed (Figure 33.13B). When this negative pulse encounters the reed, it tends to pull the reed closed and is reflected at the reed end as a negative pulse which

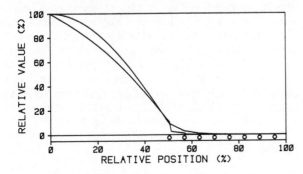

Figure 33.10. Standing pressure wave (smooth curve) and power (curve with sharp discontinuities) for the first mode of a cylindrical tube with open tone holes. Positions of open tone holes are shown at the bottom of the figure.

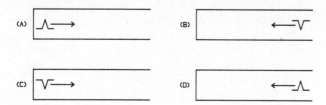

Figure 33.13. Pressure pulse propagation and reflection in a "tubinet."

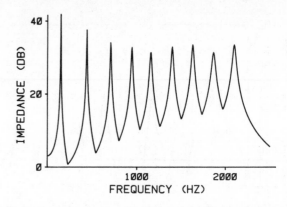

Figure 33.14. Input impedance response curve for E3 of clarinet. (After Backus, 1974.)

again travels toward the open end (Fig. 33.13C). Upon encountering the open end, the pulse is reflected as a positive pulse which travels back toward the reed. When this pulse encounters the reed, it now opens the reed, thus admitting another positive pressure pulse, and the motion repeats.

Since it takes the pulse four trips along the tube before one cycle is completed, the wavelength of this vibration must be four times the length of the tube, a situation we have come to expect for closed-open tubes. Furthermore, since only odd harmonics are allowed in closed-open tubes, the frequencies present will be odd multiples of the fundamental. If we measure the input impedance of our tubinet using the method described in Chapter 10, we obtain a result similar to the curve shown in Figure 33.8. The resonance frequencies (peaks) of the curve are those frequencies where oscillations are favored. (Note the odd multiples of the fundamental frequency, f_1.) Likewise, the places where weak oscillations of the air column take place are the valleys, which occur at even multiples of f_1. When the shape of our perfect cylinder is altered in any way, such as by adding the bell, the resonance peaks are shifted in frequency. A slight enlargement at the open end will move the peaks upward in frequency, but the lowest peak will be moved the most and each consecutive peak will be shifted less. Conversely, a flaring at the mouthpiece end will displace the peaks downward, but the higher-frequency peaks will be affected more than the lower ones. In general terms, any alteration at the reed end has the opposite effect from the same alteration at the open end.

Now consider the resonance peaks of an actual clarinet for the lowest note (E3). Naturally the effect of the bell will be appreciable here, as can be seen in Figure 33.14. The effect of the bell is to raise the fundamental frequency, thus making the other modes less than harmonic multiples of the fundamental. Also, note that the resonance frequencies depart more and more from the odd harmonics as we go to higher frequencies. The eighth harmonic, for instance, coincides more nearly with a resonance than either the seventh or ninth. Consequently, the eighth harmonic would be quite predominant in a tone with this fingering. In general, actual clarinet tones show the odd-harmonic structure up to about the cutoff frequency; beyond this frequency even and odd harmonics are present to about the same extent.

Now, consider the clarinet reed itself. When the reed is played without the dominating influence of the instrument tube, a high-pitched squeal of 1500-3000 Hz results. The parameters which control the frequency at which this reed vibrates are the stiffness of the reed and the mass of the reed. When the tube is coupled to the

reed, the pressure inside the mouthpiece varies (as we noticed in Figure 33.13) in a complicated manner and controls the reed vibration frequency. When the clarinet reed is not vibrating, the tip of the reed is located about 1 mm from the mouthpiece. When the instrument is played, the lower lip and the air pressure in the mouth reduce this distance to about 0.5 mm. When the blowing pressure is low, so as to produce a soft tone, the reed vibrates about its equilibrium position without ever touching the mouthpiece. When the blowing pressure is increased, the amplitude of the reed vibration increases until it begins to beat against the mouthpiece. For fortissimo tones, the reed is in contact with the mouthpiece for approximately one-half of each cycle. Figure 33.15 shows the motions of a clarinet reed for soft, medium, and loud tones.

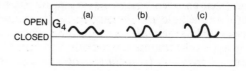

Figure 33.15. Clarinet reed openings as a function of time for (a) soft, (b) medium, and (c) loud tones. (After Backus, 1963.)

Note that the vibration for soft tones is essentially sinusoidal. As the blowing pressure increases, more and more of the top of the sinusoid is clipped off (by the reed banging into the mouthpiece). Clipping a sinusoid in this manner adds more and more high-frequency energy; in the extreme case, the sinusoid is clipped to such a great extent that it almost becomes a square wave. (You may recall from Chapter 11 that square waves have a large number of odd harmonic frequencies present.) Figure 33.16 shows how the higher frequency harmonics increase relative to the low frequencies as the clarinet is played at higher dynamic levels. Note that, in going from a dynamic level of pp all the way to ff, the change in amplitude of the low-frequency components is small. Rather, one increases loudness by increasing the number and amplitudes of the high-frequency components.

Double Reed Family

The oboe, English horn, and bassoon have bores which, to a first degree of approximation, may be represented as cones. The cones, of course, must be

228

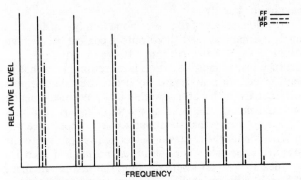

Figure 33.16. Clarinet spectra for pp, mf, and ff dynamic markings. (Data from Backus, 1963.)

truncated slightly so that the reed may be put into place, as shown in Figure 33.17. A cylinder closed at one end has resonances which are odd-integer multiples of the fundamental. Enlarging the open end or contracting the closed end of a cylinder moves the resonances upward in frequency. If a cylinder is converted into a cone by expanding the open end and contracting the closed end, all of the resonances move up in frequency by an equal amount. The fundamental resonance f_1 moves up to the position of the former valley, $2f_1$. Likewise $3f_1$ moves upward to $4f_1$, etc., as can be seen by comparing Figure 33.3a with Figure 33.3c.

We note that two things have happened. A cone having a length equal to the cylinder has its fundamental resonance an octave higher at $2f_1$ rather than at f_1. Secondly, the resonances of the cone are seen to be multiples of the new fundamental frequency ($2f_1$). If we let $2f_1 = f'$; then the resonances of the cone are given by f', $2f'$, $3f'$, etc. Since these are the same resonance frequencies as for a cylinder open at both ends, we can say that in some respects a cone acts like an open-open cylindrical tube. Since the oboe and bassoon are basically conical instruments, they will have resonances which are multiples of the fundamental.

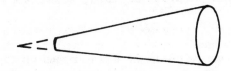

Figure 33.17. Truncated cone as an approximation to double reed instrument tube.

Real instruments can only be approximately represented as a cone. In actuality the simple cone is modified in the following ways: (1) the cone is truncated so that the reed can be attached; (2) the tube consists of several truncated cones joined end to end, but each conical to a different degree; (3) there is a noticeable discontinuity between the end of the tube and the reed staple; (4) there are closed tone holes; and (5) the reed cavity has a larger diameter than the bore. The results of these sequential changes are summarized in Table 33.2. They consist of the following: (1) the resonances are slightly stretched; (2) the fundamental resonance is lowered, but the upper resonances are raised slightly; (3) all the resonance frequencies are raised slightly (because the area at the top of the cone is decreased slightly); (4) all the resonance frequencies are lowered slightly (by increasing the effective cross-sectional area); and (5) the high-frequency resonances are lowered much more than the lower ones, which somewhat compensates for the stretching of the higher modes due to the truncation.

Measured and calculated oboe impedance curves are shown in Figure 33.18. The agreement between the two curves is quite good up to the cutoff frequency of about 1300 Hz. For a good, clear tone, several well-aligned peaks are required. Note that, B4, whose response curve is shown in the figure, will be somewhat weak because only two large peaks are present.

By being able to compute the input impedance for each note, one could presumably improve a mediocre oboe by predicting what changes should be made to the tube and to the tone holes, thus saving experimentation with an actual instrument. Since the input impedance can now be predicted with reasonably good accuracy, it may be possible to optimize the placement and size of each tone hole, as well as the size and shape of the tube, so as to produce an instrument having excellent tone quality throughout its entire range. The answer to this question must await the result of future research.

The spectrum of the oboe is rather interesting. For many notes which are played, harmonics with frequencies near 1000 Hz and 3000 Hz are emphasized, while those with frequencies near 2000 Hz are attenuated. The spectrum shown in Figure 33.19 is for an oboe playing a note with a fundamental frequency of 250 Hz at a dynamic marking of mezzo forte. When a note of higher frequency is played, the entire spectrum changes so that frequencies near 1000 and 3000 Hz are emphasized, regardless of whether they are 1st, 2nd, 3rd, 4th, etc., partials of the spectrum. Likewise, the bassoon spectrum emphasizes those partials near 500 and 1150 Hz

Table 33.2. Natural mode frequencies (Hz) of different approximations to an oboe. (From Plitnik, 1972)

One joint	Five joints (continuous)	Discontinuity at staple	Closed holes	Reed cavity	Losses
256.0	241.0	246.5	234.0	230.0	230.0
513.0	517.0	523.0	504.0	481.0	480.5
772.0	793.0	788.0	762.0	708.0	707.5
1033.0	1058.5	1049.0	1003.0	938.0	938.0
1296.0	1302.0	1289.0	1244.0	1174.5	1175.5
1561.0	1557.0	1543.0	1496.0	1388.5	1391.5
1827.5	1827.5	1814.5	1750.5	1620.5	1626.5
2095.0	2102.0	2093.0	2018.0	1860.0	1868.0
2364.0	2372.5	2369.5	2281.0	2123.5	2134.5
2633.0	2638.0	2640.5	2535.0	2384.0	2397.0
2903.0	2906.0	2915.0	2800.0	2631.0	2645.0

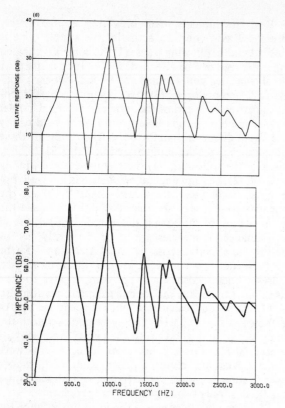

Figure 33.18. Input impedance respone curves for note B4 on an oboe. The upper curve is an experimentally measured curve; the lower curve is calculated numerically. (From Plitnik and Strong, 1979.)

and suppresses those near 1000 Hz. The dashed curve drawn over the spectral lines in Figure 33.19 shows a spectral envelope which represents the average manner in which the harmonics vary in amplitude over many notes. Two possible explanations for the emphasis of certain frequency regions in both oboe and bassoon spectra are the cutoff frequency and the manner in which the reed vibrates.

Now consider the effect of playing an oboe at dynamic markings of pianissimo, mezzo forte, and fortissimo. As with the clarinet, the oboe reed action is more abrupt for higher levels, thus adding more high-frequency energy to the spectrum. The dashed curves in Figure 33.19 show how the spectral envelopes of an oboe vary with dynamic level. Note that the high-frequency components increase substantially as the dynamic playing level increases, while the lower frequencies change much less.

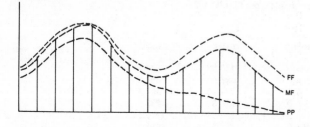

Figure 33.19. Oboe line spectrum for a 250 Hz tone played at a dynamic marking of mf. The dashed curves show approximate spectral envelopes for tones played at dynamic markings of pianissimo, mezzo forte, and fortissimo. (After Plitnik, 1972.)

Saxophone Family

Each member of the saxophone family consists of a single mechanical reed fitted onto a conical tube which ends in a slight flare. Although the saxophone is a conical tube instrument, its action is more similar to that of the clarinet. In one detail, however, the saxophone is different from the other mechanical reed instruments: the diameter of the tube at the reed end is relatively large. This means that the input acoustic impedance will be fairly small; the reed will be less tightly coupled to the tube than the reeds for the other mechanical reed instruments. With a looser coupling, the saxophone reed, although still controlled by the tube, will have a somewhat faster attack than its cousins. (It does not take as long for the vibrating reed to reach its steady state.)

The response curve for the note C6 of an alto saxophone appears in Figure 33.20. The cutoff frequency is quite apparent. We can see that there are fewer resonances than for the other conical instruments. Also, the two lowest resonances are, over most of the instrument's range, exactly an octave apart. This instrument will thus be observed to have a fairly uniform tone color throughout its range.

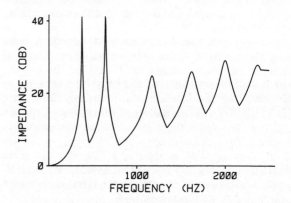

Figure 33.20. Input impedance response curve for note C4 on a saxophone. (After Backus, 1974.)

Historical Notes

Although the history of reed instruments is a long one, the common woodwind instruments of the mechanical reed variety are of more recent origin than is often realized. The oboe was invented during the latter part of the seventeenth century, and although it possessed only three simple keys, it was much closer to the present instrument than to earlier instruments of the same name. The bassoon, which developed at about the same time as the oboe, was conceived as a bass to the oboe (sounding a twelfth lower in pitch), but with a tonal character all its own. Because of its large size, the bassoon is bent back on itself and the fingerholes are drilled at angles to avoid impossible wide finger spacings. The forerunner of the English horn was the tenor oboe, but the English horn soon established its own unique tone color and can no longer be considered a tenor oboe. The globular bell on the English horn has a profound visual effect but influences tone color only in the lowest notes.

The crumhorn and shawm families (Figure 33.21) may be regarded as ancestors of our modern double reeds. The crumhorn is an enclosed reed instrument in which the reed is placed in a chamber and not held directly in the player's mouth. The player blows air into the chamber which drives the reed. Reeds of this type are used in pipe organs and bagpipes. The shawm uses a large double reed held between the player's lips, which gives the player considerably more control over the sounding properties of the instrument. It may be regarded as a recent ancestor of the oboe.

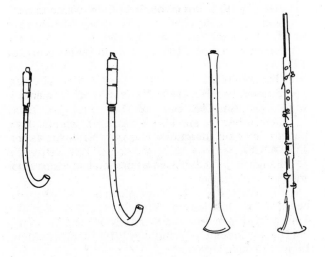

Figure 33.21. Some ancestors of modern double reed instruments. Treble and tenor crumhorns appear on the left and treble and tenor shawms on the right. (After Baines, 1969.)

The clarinet was invented at the beginning of the eighteenth century, some forty or fifty years after the invention of the oboe. The precursors to the clarinet, although using a cylindrical tube, had a very limited range—the clarinet was invented when it was realized that a register key could be added to extend the scale into the second register (the twelfth). The clarinet descended from the chalumeau which was a cylindrical tube, single reed instrument, which lacked the frequency and dynamic range characteristic of the modern clarinet.

The saxophone is most often used in bands and jazz groups; it does not appear often in a symphony. It was invented by Adolphe Sax during the mid nineteenth century as a reed instrument loud enough for a military band. It has no obvious ancient ancestry.

An important phase in the development of the mechanical reed instruments occurred during the nineteenth century, when more keys were added and the key mechanism itself underwent considerable change and improvement. The oboe tube also underwent some minor changes; the bassoon was completely reworked because of the unevenness and instability of some of the notes in the scale. By the end of the nineteenth century the mechanical reed instruments had become more-or-less standardized into the familiar forms of today.

Exercises

1. A cylindrical tube 34 cm in length is sounded with a single reed. What are the approximate frequencies of the first three modes?

2. Repeat Exercise 1 for a conical tube 34 cm in length.

3. Compare and contrast the bell of a brass, the open tone holes of a woodwind and the open end of a tube as barriers to low and high frequency waves. (Compare Figures 32.5, 33.7, and 33.4.)

4. What will the fundamental frequency be for a tone produced with the response curve shown in Figure 33.8?

5. Repeat Exercise 4 with Figure 33.9.

6. The same open tone hole tube was used for Figures 33.6, 33.7, 33.10, 33.11, and 33.12. How do the peak heights in Figure 33.6 depend on the power reflection in Fig. 33.7?

7. Refer to Exercise 6. How do the peak heights in Figure 33.6 depend on wave penetration into the open tone hole section in Figures 33.10, 33.11, and 33.12?

8. Refer to Exercise 6. The same open hole tube appears longer to mode 3 shown in Figure 33.11 than to mode 1 shown in Figure 33.10. Justify this statement by comparing wave penetration into the open tone hole section of tube.

9. Is the spectrum of a reed instrument primarily due to reed type (single or double) or tube type (cylindrical or conical)?

10. Refer to Figure 33.13 to answer the following questions. Will the reed tend to open or close at (C)? At (A)?

11. Suppose, that the tube in Exercise 1 is lengthened. How will this affect the number of reed vibrations in each second? If the tube is shortened?

12. What effect do the open tone holes have on the vibration frequency of the reed? Does the size of these holes make any difference?

13. What effect do the closed holes have on the overall spectrum of the instrument? How does this affect the tone quality?

14. What is the function of the register hole in a clarinet?

15. What is the function of the register hole in an oboe?

16. Draw the spectrum of an oboe playing a note of frequency 500 Hz and compare it to Figure 33.19. Draw the spectrum for a note of 1000 Hz.

17. Draw the spectrum of a clarinet (see Figure 33.16) playing a note of frequency 250 Hz and compare to Figure 33.19.

18. From the information given in the text, construct (approximately) spectral envelopes for a bassoon playing at dynamic markings of pianissimo, mezzo forte, and fortissimo.

19. Sketch the approximate spectral envelope for a bassoon playing mezzo forte tones. Then draw, on the same graph, the spectrum for a bassoon playing notes having frequencies of 100 Hz, 250 Hz, and 500 Hz.

20. The spectra shown in Figure 33.22 resulted from clarinet and saxophone tones. What is the fundamental frequency of each if the range shown is from 0 to 5000 Hz? Which spectrum exhibits primarily odd harmonics, particularly at lower frequencies? Does the odd-harmonic spectrum occur because of reed type or

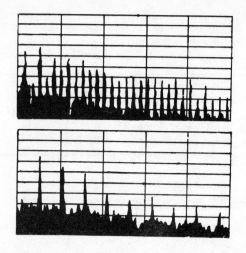

Figure 33.22. Spectra for a clarinet tone and a saxophone tone.

tube type? Why are both odd and even harmonics present with about equal amplitudes at higher frequencies?

Demonstrations

1. Build a slide clarinet consisting of one cylindrical tube which slides over a smaller cylindrical tube equipped with a clarinet mouthpiece.

2. Use a clarinet mouthpiece on cylindrical, conical, negative-flare and positive-flare tubes.

3. Display spectra of various mechanical reed tones.

4. Decide on the gross features of a simple wind instrument that you wish to construct. By consulting appropriate sources determine the physical properties of your instrument (e.g., length, diameter, tone hole positions, tube shape) so that it will produce the desired frequencies. Construct the instrument. Take the instrument into the laboratory and make measurements to determine how well you have achieved your design goals.

Further Reading

Backus, Chapter 11
Baines, Chapter 10
Benade, Chapters 21, 22
Culver, Chapters 12, 13
Hall, Chapter 13
Kent, Part III

Rossing, Chapter 12

Backus, J. 1961. "Vibrations of the Reed and the Air Column in the Clarinet," J. Acoust. Soc. Am. **33**, 806-809.

Backus, J. 1963. "Acoustical Investigation of the Clarinet," Sound **2**, 22-25.

Backus, J. 1974. "Input Impedance Curves for the Reed Woodwind Instruments," J. Acoust. Soc. Am. **56**, 1266-1279.

Baines, A. 1963. **Woodwind Instruments and Their History** (W. W. Norton).

Bate, P. 1956. **The Oboe**, 3rd ed. (W. W. Norton).

Benade, A. H. 1959. "On Woodwind Instrument Bores," J. Acoust. Soc. Am. **31**, 137-146.

Benade, A. H. 1960. "On the Mathematical Theory of Woodwind Finger Holes," J. Acoust. Soc. Am. **32**, 1591-1608.

Benade, A. H. 1960. "The Physics of Woodwinds," Sci. Am. **203** (Oct), 145-154.

Benade, A. H., and D. J. Gans. 1968. "Sound Production in Wind Instruments," Ann. N.Y. Acad. Sci. **155**, 247-263.

Langwill, L. G. 1965. **The Bassoon and Contrabassoon** (W. W. Norton).

Plitnik, G. R. 1972. "The Calculation of Input Impedance for Double-reed Wind Instruments, and the Time Variant Analysis and Synthesis of their Tones using Digital Computer Techniques" (Ph.D. dissertation, Brigham Young University).

Plitnik, G. R. and W. J. Strong, 1979. "Numerical Method for Calculating Input Impedances of the Oboe," J. Acoust. Soc. Am. **65**, 816-825.

Rendall, F. G. 1971. **The Clarinet**, 3rd ed. (W. W. Norton).

Strong, W., and M. Clark. 1967. "Synthesis of Wind-Instrument Tones," J. Acoust. Soc. Am. **41**, 39-52.

Strong, W., and M. Clark. 1967. "Perturbations of Synthetic Orchestral Wind Instrument Tones," J. Acoust. Soc. Am. **41**, 277-285.

Audiovisual

1. **Woodwind Instruments** (3.75 ips, 2 track, 30 min, UMAVEC)

2. **Clarinet** (10 min, color, 1968, SF)

3. **The Clarinet, Part 1** (20 min, color, 1965, OPRINT)

4. **The Clarinet, Part 2** (17 min, color, 1965, OPRINT)

34. Air Reed Instruments

The lip reed and mechanical reed instruments discussed in the preceding chapters depend on the alternate opening and closing of a physical reed to interrupt air flow. Air-reed instruments depend on the action of an air stream as it moves toward a sharp edge attached to a resonator to provide an oscillating air flow. Representative air-reed instruments shown in Figure 34.1 are the organ flue pipe, the recorder, and the flute. We will establish some general requirements for an air-reed instrument. The air stream and resonator interaction will be considered as it relates to a blown bottle and to an ocarina. The organ flue pipe will be used as an example to introduce the roles of the resonator response and air stream parameters. The role of tone holes will be considered in a discussion of the recorder. The flute family, the most versatile of the air reeds, will then be discussed. Effect of wall materials will be considered briefly. Finally, brief historical notes will be given.

General Requirements

The air flow into brass and mechanical reed woodwinds is controlled by the opening and closing of the lips or by reeds acting as a valve. The valve is pressure-controlled in the sense that oscillatory pressures from the tube drive it open and closed. Sound is radiated primarily from the bell or tone holes and not from the reed, where excitation occurs. In air-reed instruments we encounter a flow-control mechanism for controlling air flow. The end of an air-reed instrument which has been excited by blowing is basically open. Oscillating air flow (rather than pressure) from the air column controls an air stream and governs its flow into and out of the instrument. Sound is radiated from the hole at which excitation is applied, as well as from other openings such as tone holes.

The response curve peaks for brasses and reed woodwinds are essentially pressure peaks, and so are

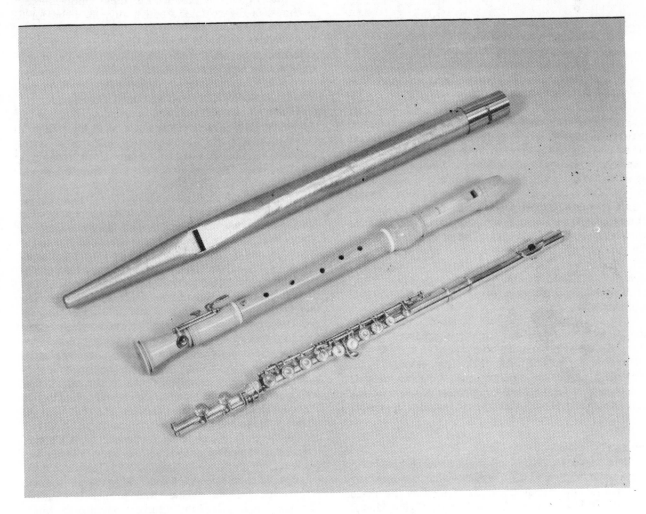

Figure 34.1. Air reed instruments: organ flue pipe, recorder, and flute. (Courtesy of G. R. Williams.)

233

important for pressure controlled reeds. Response curve valleys are essentially pressure minima, but flow maxima. For flow controlled excitation we look for response curves for which the valleys show all harmonics or odd harmonics. The greater the depths of the valleys, the stronger the control they exert on the air stream.

Air Stream and Resonator Interaction

When a stream of air is directed through a narrow slit toward a sharp edge attached to a resonator, the air stream does not simply divide into two parts, but alternately flows into and out of the resonator, as shown in Figure 34.2. The air stream causes the air in the resonator to vibrate, and the resonator in turn controls the air stream. This results in the sequence shown in Figure 34.2 for a pipe resonator. Possible pipe modes are determined by whether the pipe is open or closed. Pipes associated with air-reed instruments may be regarded as open at the blowing end where pressure nodes and air flow antinodes exist. Pipes open at the other end are termed "open pipes" while pipes closed at the other end are called "closed pipes."

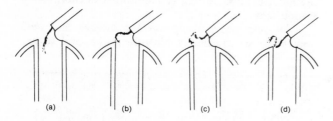

Figure 34.2. Interaction of an air stream with a pipe open at both ends. (After Coltman, 1968.)

Blowing on a bottle provides a simple example of air-reed excitation. The bottle is a Helmholtz resonator; its natural frequency depends on air volume and size of opening. If we position and aim our lips properly, and blow with a correct pressure, the bottle will sound at its natural frequency. When this happens, the air in the bottle is alternately compressed and expanded; flow in the opening is alternately into and out of the bottle at the bottle's natural frequency. The air stream from our lips is controlled by the alternating air flow at the opening. In this way the vibrating air in the bottle extracts energy from the air stream to offset losses, which maintains its motion. If the air stream is removed, the oscillations die out.

An ocarina is a bottle with tone holes and a fipple mouthpiece. A fipple mouthpiece consists of a flue (basically a flat pipe) and a sharp edge in a fixed configuration. Air travels through the flue, emerges as an air stream, and strikes the sharp edge. Opening the tone holes makes several natural frequencies possible for the ocarina and makes it useful for producing simple melodies.

Organ Flue Pipes

An organ flue pipe functions in a manner similar to that shown in Figure 34.3. The flue, which produces the required narrow stream of air, is formed by the space between the lower lip and the languid (see Figure 35.4a). A sound is produced when this slit of air impinges on

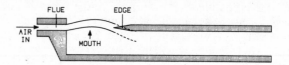

Figure 34.3. Air stream and pipe combination showing the growth of the air stream oscillation from the flue to the edge.

the upper lip, as described previously. The final tone of each pipe is greatly affected by the process of voicing, whereby the height of the mouth and the width of the slit are adjusted. The pipes are tuned by adjusting the length of the pipe at the upper end.

As mentioned earlier, we look for pipes for which the response curves have harmonically related valleys. Looking back to Figure 33.3 we see that both cylindrical pipes and conical pipes satisfy this requirement. Even the other two pipe-shapes in Figure 33.3 seem to satisfy it. The valleys are not shifted drastically, as are the peaks. Another pipe not shown here is a cylindrical pipe closed at its far end, which would exhibit valleys where the peaks appear in Figure 33.3a.

We can go through an argument similar to that associated with Figure 33.13 for the clarinet to gain some insight into the behavior of open-open and open-closed flue pipes. For our discussion we will use a simplified organ flue pipe with a fipple "mouthpiece" as shown in Figure 34.3. Acoustically, the behavior is the same as for an actual organ flue pipe (see Figure 35.4a). Suppose that a positive flow (into the pipe) is produced at the edge at the left end of the pipe. It will travel to the right end where it will reflect as a positive flow. It will return to the left end as a positive flow, and then be amplified if the air stream is flowing into the pipe. The vibration period will be the time required for the wave to travel twice along the tube. If the right end of the tube were closed, the positive flow would reflect as a negative flow and two round trips (four trips along the tube) would be required before the motion would repeat.

The air stream must be adjusted to interact properly with the pipe; it is initially disturbed by pipe oscillations as it emerges from the flue. This disturbance grows much larger as the air stream moves from the flue to the edge. The initial disturbance is in the form of a small "push" or "pull" to cause the air stream to flow out of or into the pipe, respectively. The disturbance which arrives at the edge is in the form of a large flow out of or into the pipe (see Figures 34.2 and 34.3). The time required for the air stream disturbance to travel from the flue to the edge is determined by the distance between the flue and the edge (called the cutup), and the speed with which it travels. The speed of travel of the disturbance is about 0.4 times the speed of the air stream, which in turn depends on blowing pressure. The disturbance on the air stream must arrive at the edge with the proper phase to enhance the air motions in the tube. If it arrives too early it will cause the tone to have a higher frequency, while late arrival will result in a lower frequency.

Several parameters can be adjusted to achieve desired tonal characteristics and proper intonation. Changing the length of the cutup or changing the blowing pressure will change the disturbance travel time and thus the intonation. If the travel time is decreased too much, the air stream disturbance may not be able to interact with the first pipe mode. It may interact with the second or higher pipe modes in a process known as **overblowing**.

Increasing the size of the flue will change the amount of air flowing and thus the loudness. The air stream can be aimed above or below, or centered on the edge. Centering will give rise to symmetrical motion, which will favor the odd harmonics. Offsetting the air stream will give rise to nonsymmetrical motion, which will favor all harmonics. **Nicking** is the process of cutting narrow, vertical grooves along the front edge of the languid and the inner side of the lower lip, at right angles to the flue. Nicking in an actual organ pipe tends to produce less turbulence and less of the resulting chiff during attack. Some other features will be considered in Chapter 35.

Recorder Family

An organ pipe is used to sound a single tone. A small organ pipe could be made more versatile by blowing into it with the mouth. The blowing pressure could be controlled in this way to permit overblowing to the second, third, and higher modes. However, this instrument would have very limited usefulness for melodic purposes. The addition of tone holes as done with the mechanical reeds makes the instrument useful for melodic purposes.

The recorder family, which includes five members of different sizes, is a family of mouth blown instruments with tone holes. The recorder uses a fipple mouthpiece; it suffers from inflexibilities due to a fixed flue size, a fixed flue-to-lip distance, and a fixed air stream direction. It has a limited dynamic range because loudness is controlled by blowing pressure, and as we have seen, an appreciable increase in blowing pressure gives rise to overblowing. However, the recorder has seven finger holes and a thumb hole, which permit a chromatic scale over one octave when used in the proper combinations. Additional octaves can be obtained by overblowing with a repeat of the lowest octave fingerings.

Normally an instrument with tone holes is fingered so that all holes from the mouthpiece end to some point along the tube are closed, and all holes beyond that point are open. However, for instruments such as the recorder with too few tone holes to produce a chromatic octave, "forked" or "cross" fingerings are used to provide the additional notes. The recorder, with its nominal seven tone holes, needs to employ several cross fingerings within the first octave and even more in the higher octaves. Cross fingering consists of closing one or more normally open tone holes beyond some open hole. Certain modal frequencies can be shifted substantially by having a closed tone hole beyond one or more open holes, as the wave is not terminated completely at the open hole and so "sees" and is sensitive to conditions beyond the open tone hole. (See, for example, the wave penetration of mode three in Figure 33.11.)

Flute Family

Members of the flute family used in the modern symphony orchestra are the flute, the piccolo, and the alto (or bass) flute. The flute is typical of the family and consists of a metal tube about 66 cm in length, with an inside diameter of about 1.9 cm. An oval embouchure (mouth hole) is placed near one end (see Figure 34.1). The flute has about 16 tone holes, which can be opened and closed by a series of pad-covered keys; these serve to change the effective length of the instrument. The instrument behaves very much like an organ flue pipe or a recorder for the production of individual notes.

The natural frequencies of a cylindrical flute are approximately determined by the distance between the embouchure and the first open tone hole, and the speed of sound in air. The sound is affected by the air temperature and the relative proportions of carbon dioxide present, as in other wind instruments. Although we consider the flute to be a simple tube open at both ends, it is considerably more complex than this. At the mouth end the hole is partially covered by the player's lip, and at the fingered end the opening (for all but the lowest note) is an open tone hole with a key pad directly above it. Furthermore, the small volumes added by the closed tone holes and the portion of the tube beyond the open tone holes all have an effect on the flute resonances. The flute tube does not "end" at the position of the first open tone hole. The tone hole and the opening between the hole and the key covering it have much smaller areas than the area of the main tube. A short length of a small tube behaves like a longer length of a large tube. The **end correction** of a tone hole is its equivalent length of main tubing. The effective length of the main tube is its actual length plus any end corrections.

End corrections at the embouchure hole are more complicated because the open area of the embouchure hole is varied from note to note. For producing higher notes, the airstream needs to travel from the lips to the embouchure edge in less time. One way of accomplishing this is to shorten the lip-to-edge distance. Figure 34.4 illustrates typical positions to which the lips

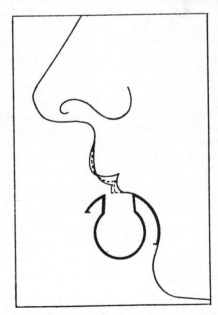

Figure 34.4. Lip positions for three notes on the flute: withdrawn for C4, intermediate for E5, and advanced for C6. (From Coltman, 1972.)

are moved for different notes. When the lips are moved forward the embouchure opening is made smaller and the end correction is made larger. This would make a higher note sound flat in relation to a lower note if no compensation were made. However, the flute is designed so that the higher modes are made sharp in relation to the lower modes for a given embouchure opening (see Figure 34.5A). The higher modes are flattened when played with a smaller embouchure opening which results in the notes being relatively in tune (see Figure 34.5B).

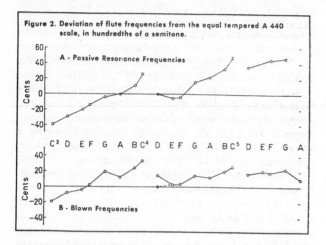

Figure 34.5. Deviation of flute passive resonance frequencies and blown frequencies from equal tempered A-440 scale. (From Coltman, 1972.)

The sharpening of higher modes in relation to lower ones is accomplished by tapering the head joint so that it becomes smaller toward the cork. The cork position in relation to the embouchure hole also can be adjusted to help give proper relationships among the various note frequencies, and thus help to provide proper intonation.

The great versatility of the flute as compared to a fixed-mouthpiece instrument like the recorder is due largely to the flexibility which the player has in exciting the instrument at the embouchure hole. The player can vary blowing pressure, lip-to-edge distance, lip opening, and air stream-to-edge angle. The air stream disturbance travel time from lips to embouchure edge should be about one-half period of the tube fundamental. The travel time is governed by blowing pressure and lip-to-edge distance. If pressure were used exclusively, the higher notes would be much too loud in relation to the lower notes because of the greater total air flow. If lip-to-edge distance were used exclusively, problems of intonation and loudness would arise because of the great variation in embouchure opening. Accomplished players typically use both pressure and lip-to-edge distance in combination. Figure 34.6 shows how blowing pressure is increased with frequency. Figure 34.7 shows how lip-to-edge distance is decreased with frequency.

The dynamics of flute playing depend largely on the amount of air flow. This can be adjusted by varying the lip opening, by varying the blowing pressure, or by a combination of the two. There are limits to the adjustments that can be made for dynamic purposes while

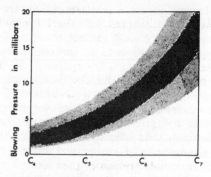

Figure 34.6. Range of blowing pressures used for different notes and dynamic levels by a selection of flute players. (From Fletcher, 1974.)

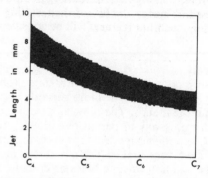

Figure 34.7. Lip-to-edge jet lengths used by a selection of flute players. (From Fletcher, 1974.)

sounding a particular note. Apparently flute players consistently increase the lip opening to produce greater air flow for louder outputs. Lip opening is typically decreased at higher frequencies (Figure 34.8) to partially offset the increase of air flow due to higher pressures, and thus maintain similar dynamic levels at all frequencies. This is only partly successful, as seen in the increased dynamic levels for high notes of the flute in Figure 31.12.

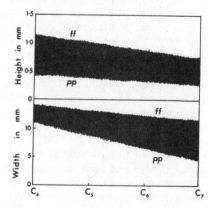

Figure 34.8. Dimensions of lip opening as a function of note played and dynamic level. The opening is elliptical in shape. (From Fletcher, 1974.)

It is important to have several open tone holes beyond the one which determines the fundamental frequency, so that the tube is effectively terminated; otherwise an appreciable standing wave can exist beyond the first open tone hole. This is dramatically demonstrated for two alternate fingerings of A6, which are known to produce very different playing characteristics. Figure

236

34.9 shows the standing wave pattern used for the fundamental when A6 is fingered with the D# key open. There is some penetration of the wave into the open tone-hole region, but the tube is approximately terminated at the beginning of the five open holes. This A6 is quite easy to sound. Figure 34.10 shows the standing wave pattern when A6 is fingered with the D# key closed. There is extreme penetration of the wave into the open tone-hole region. This A6 is very difficult to sound. These standing wave patterns can be understood in terms of the power reflection beyond the cutoff frequency at the open-hole portion of the tube. When the D# key is closed, the power reflection is almost zero for the mode shown in Figure 34.10. Too little power is available to produce a large standing wave in the closed portion of the tube, and so the tube does not make adequate suggestions to the air stream.

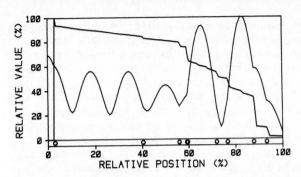

Figure 34.10. Standing pressure wave and power for a flute with alternate A6 fingering. The D# tone hole (near a relative position of 80%) is closed as compared with Figure 34.9.

The tone of the flute is less rich in upper harmonic development than are oboe and trumpet tones. However, flute tones are far from being nearly pure sinusoids, as is often suggested. Figure 34.11 shows spectra for three different flute tones, produced by four players. Note that the spectra are characterized by a spectral peak around 550 Hz for forte tones, and at a lower value for piano tones. Note also that the forte tones show a significantly greater upper partial development than the piano tones.

Effect of Wall Material

The material from which pipes are constructed has long been considered to exert considerable influence on the tone produced. For hundreds of years organ builders have constructed bright-toned pipes from tin, while lead was used to produce a duller sound. Most of the pipes were constructed from a mixture of lead and

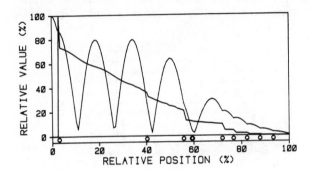

Figure 34.9. Calculated standing pressure wave and power for a flute fingered for A6. Open tone holes are shown at the bottom.

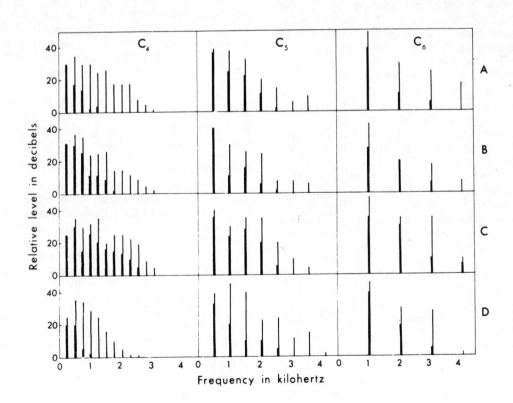

Figure 34.11. Spectra of flute tones produced by four players (A, B, C, and D) at dynamic levels of forte and piano. (From Fletcher, 1975.)

tin, but it was believed that the higher the percentage of tin, the brighter the tones. Recent scientific experiments indicate that, not only does the wall material not affect the tone quality of thick-walled pipes, but the vibrations of the material are too small even to be measured. (However, some measurements indicate that wall material may play a role for thin walled pipes.) The flue pipes of an organ may be either open or closed; both wood and metal are used in constructing these pipes. As you recall, an open flute will have all harmonics present, while a stopped flute will contain primarily odd harmonics. In the next chapter we will present considerably more detail on the pipe organ as an instrument and the tonal qualities of various organ pipes.

Tones of flutes made of very thin material, or having very thin walls at positions of antinodes, may depend to some extent on wall material and wall thickness. Tonal properties of a silver, a copper, and a wood flute were tested in a series of experiments. Musically trained listeners were unable to significantly distinguish their tones. Flutists were unable to distinguish among the playing properties of the three flutes.

Historical Notes

Instruments of the air-reed family date back to antiquity. The earliest of these family members were probably small pipe or bamboo instruments. Sometime in the early history of these instruments it was noticed that small finger holes along the bore of the instrument allowed one to change the pitch by uncovering the holes, thus effectively shortening the length of the tube. The primitive instruments utilized six holes; by successively uncovering these holes one could play a diatonic scale. Instruments with six finger holes and one key were the standards of the orchestra for about 100 years from the mid seventeenth century. Additional tone holes and more elaborate keying mechanisms were added to give the flute greater versatility.

Theobald Boehm, the man who would bring the flute to its modern form, was born at the end of the eighteenth century. He began his amateur playing career on a six-hole, one-key, conical tube flute. He made two landmark innovations in flute design. The first, announced in 1832, was a greatly elaborated and improved key mechanism for the conical tube flute. The second, announced in 1847, was a redesign of the flute. This design used a cylindrical metal body with large, key-covered tone holes. The modern flute is the Boehm 1847 model with some refinements. The history of Boehm's innovations on the flute includes some interesting stories about competition among contemporaries in the flute world.

The recorder was developed and refined nearly in parallel with the flute. However, it was supplanted in the orchestra by the flute, probably because of the flute's greater power and versatility. Recorders have recently come back into popular use.

Exercises

1. What is the fundamental frequency of an open-open tube if its length is 200 cm? What are its modal frequencies? Do these correspond to peaks or valleys in the response curve? Why?

2. Repeat Exercise 1 for an open-closed air-reed tube 200 cm long.

3. Are flue pipes in the organ open-open or open-closed? What harmonics do they produce? Why?

4. When an open-open flue pipe sounding a frequency of 220 Hz is overblown, what frequency do you hear? When an open-closed flue pipe having a frequency of 220 Hz is overblown, what frequency do you hear? Explain why the result is different in these two cases.

5. For organ flue pipes, the slit-to-edge distance is kept as short as possible (for a given blowing pressure) without overblowing, in order to produce a brighter sound. Why does a shorter slit-to-edge distance produce a brighter sound?

6. If an organ flue pipe is voiced on a particular wind pressure, will the pipe speak properly on half that pressure? Why or why not?

7. A flutephone spectrum for a tone of about 800 Hz is shown in Figure 34.12. Does it exhibit characteristics of an open pipe or a closed pipe? How can you tell?

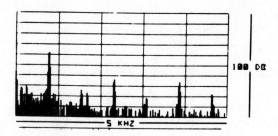

Figure 34.12. Flutephone spectrum.

8. A flute can be considered an open-open cylindrical pipe. What harmonics are present? Explain.

9. Does the playing frequency of a flute increase or decrease with an increase of temperature? Explain.

10. If a flutist inhaled helium gas and then played a note on the flute, would it sound at its normal pitch? If not, how would it differ? Why?

11. If the material from which organ flue pipes are made does not affect their sound, why do wooden and metal pipes sound quite different? Could they be constructed to sound identical? Explain.

Demonstrations

1. Make a "flute" from a piece of 1.5 cm diameter plastic pipe about 60 cm long. Cut an embouchure of about 1.2 cm diameter near one end. Insert a cork or plug about 2 cm from embouchure hole center. Play it.

2. Attach a clarinet mouthpiece to a piece of tubing similar to that of Demonstration 1. Compare the pitches and tonal qualities of the "flute" and "clarinet."

3. Compare tonal qualities by blowing small open and stopped organ pipes.

4. Experiment with a set of adjustable metal organ pipes. (Sargent Welch #3280, for example.)

Further Reading

Backus, Chapter 11
Baines, Chapter 10
Benade, Chapter 22
Hall, Chapter 12
Kent, Part III

Olson, Chapters 5, 6

Rossing, Chapter 12

Wood, Chapter 8

Backus, J., and T. C. Hundley. 1966. "Wall Vibrations in Flue Organ Pipes and Their Effect on Tone," J. Acoust. Soc. Am. **39**, 936-945.

Boehm, T. 1964. **The Flute and Flute Playing** (Dover).

Coltman, J.W. 1968. "Acoustics of the Flute," Physics Today **21** (Nov), 25-32.

Coltman, J. W. 1971. "Effect of Material on Flute Tone Quality," J. Acoust. Soc. Am. **49**, 520-523.

Coltman, J. W. 1972. "Acoustics of the Flute," The Instrumentalist **26** (Jan) 36-40 and (Feb), 37-43.

Fletcher, N. H. 1974. "Some Acoustical Principles of Flute Technique," The Instrumentalist **28** (Feb) 57-61.

Fletcher, N. H. 1975. "Acoustical Correlates of Flute Performance Technique," J. Acoust. Soc. Am. **57**, 233-237.

Fletcher, N. H., and L. M. Douglas. 1980. "Harmonic Generation in Organ Pipes, Recorders, and Flutes," J. Acoust. Soc. Am. **68**, 767-771.

Lyons, D. H. 1981. "Resonance Frequencies of the Recorder (English Flute)," J. Acoust. Soc. Am. **70**, 1239-1247.

Audiovisual

1. **The Flute** (22 min, color, OPRINT)

35. The Pipe Organ

If not the king of instruments, the pipe organ may at least be considered the dinosaur of the musical instrument world; it has become (after the baroque era) an un- wieldy giant which may have outlived the climatic conditions in which it prospered. The pipe organ is an assemblage of hundreds of resonant tubes of various

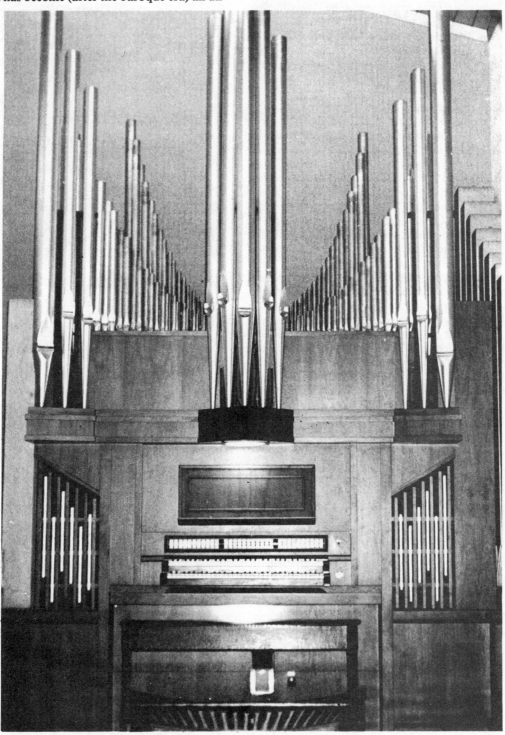

Figure 35.1. The organ at Christ Lutheran Church, LaVale, Maryland. (Designed by G. R. Plitnik.)

shapes and sizes (see Figure 35.1), each pipe producing only one pitch and one tone quality. The majority of the pipes found in organs are of the flue pipe variety, although about 20 percent are usually of the mechanical reed variety. The pipes sit on a wind chest containing a set of valves, each of which admits air into the appropriate pipe when the valve is energized from a large modern console. Each desired tone quality consists of a set of pipes, graduated in size (and hence pitch) through the entire manual or pedal division. In this chapter we will consider several types of organ action; then we will detail the workings of an organ reed pipe. We will next classify organ stops into families and relate the families to pipe scales. We will then consider the speech of organ pipes, and an analysis of the spectral qualities of pipes representative of each tonal family. We will conclude with a brief historical section.

Construction and Action

The typical modern pipe organ consists of a large variety of flue and mechanical reed pipes controlled by two or more manuals and a pedalboard. Each manual and the pedalboard usually control separate groups of pipes; the organ is in reality a composite of several organs. The basic components of a pipe organ, diagrammed in Figure 35.2, are a blower to provide wind, a power supply (for electric actions), regulators to maintain constant pressure, wind chests (which hold the pipes), the pipes themselves, shutters to control the volume (swell division), and a console to control it all. The upper manual, called the **Swell**, controls the pipes which are enclosed behind a set of movable shutters (see Figure 35.3). When the shutters are open the sound can escape; by closing the shutters (by means of the swell pedal), the organist can reduce the volume considerably. The lower manual controls the pipes of the **Great** organ; these pipes are not usually enclosed and so are heard at the same loudness.

The wind chest contains a set of valves which, when opened, will admit air into the appropriate pipe. There

Figure 35.3. Interior of the swell chamber of a pipe organ. (Frostburg State College organ built by G. R. Plitnik. Photo courtesy of G. Orvis.)

are three types of windchest actions in use today. The oldest type, used (in some form) for nearly a thousand years, is the tracker, or mechanical, action. No power supply is needed with this action because the keyboards are connected directly to the chests by a complicated set of mechanical linkages called "trackers." The second type of action, developed in the early decades of this century, is called electropneumatic because small eletromagnets are used to exhaust pneumatic bellows. The final type of action is the direct electric action; in this system an electromagnet opens a pipe valve directly. Figure 35.4 is a representation of each type of action. Inspection of the figure shows that, for the tracker action, air is admitted into a pipe only when two conditions are met. First, a long slider bar under each set of pipes must be pulled to the "on" position. The sliders run perpendicular to the page in the figure; there is a separate slider for each pipe, representing different ranks. Second, a valve (called the pallet) must be pulled down by pressing the key. For the type of electropneumatic action shown, air is admitted to the pipe when the air in a thin leather pouch valve is exhausted. This happens via the lifting of the armature disc by the energized electromagnet. The direct electric valve, the simplest mechanism, opens whenever the armature is attracted by the energized magnet.

When considering the different types of action, we have not considered the effect of the stop keys. The **stops** are small tablets, usually located above the top

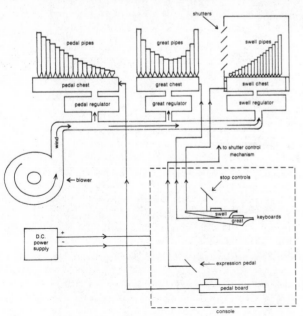

Figure 35.2. Schematic diagram of a typical pipe organ.

manual, or drawknobs alongside the manuals. Each stop controls a set of pipes called a **rank**. The pipes are graduated in pitch so that there is one pipe for each manual (or pedal) key, but all the pipes of one rank possess the same tone quality. Different ranks of pipes possess different tone qualities and may occur at different pitches. When one key is depressed, one pipe will sound for every stop which is on. Thus, the keyboard controls various sets of pipes which can be selected according to the tone quality, volume, or pitch desired.

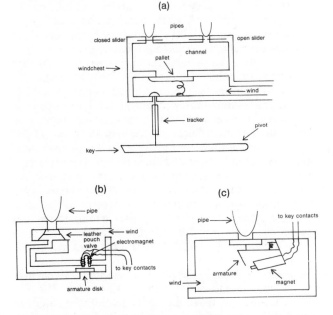

Figure 35.4. Windchest actions: (a) simplified tracker, (b) electropneumatic, and (c) direct electric.

Organ Reed Pipes

Having examined the basic construction of the organ, let us now direct our attention to the pipes. Organ pipes consist of two basic types: flue pipes (air reed) and reed pipes (mechanical reed). In both types the tone is the result of a vibrating air column; the difference is in the method used to excite the air column. We have already considered (in Chapter 34) the operation of an organ flue pipe. The reed pipe differs from the flue pipe in several important ways: (1) a vibrating brass reed modulates the air stream, (2) the energy is applied at the closed end of the pipe, and (3) all of the air passes through the pipe. Although the organ reed is a mechanical reed instrument, the air column of the resonator does not control the reed as it does for mechanical reed woodwinds. Rather, the vibrating organ reed is influenced by the resonator in much the same way as in the lip-reed instruments. Figure 35.5 shows the essential features of organ flue and reed pipes. The sound-producing section of the reed pipe is the curved brass tongue (reed), which beats against the shallot at a certain frequency. The shallot and reed are held in place in the lead block by a small wooden wedge, the entire arrangement being enclosed by the boot. The shallot is a brass tube, closed at the lower end and flattened on the reed side. A long, narrow hole in the flat side leads through the boot to the resonator, which is

the only route by which air can escape from the boot. A sliding wire spring presses the tongue firmly against the shallot, which changes the vibrating length of the tongue and can thus be used for tuning. When air enters the bottom of the boot it begins to escape through the opening under the curved reed tongue. But, by Bernoulli's Law, the reduced pressure of the rushing air pulls the tongue against the shallot, thus closing the opening and interrupting the escape of air. Since the tongue is elastic, however, and a positive pressure pulse soon returns from the resonator, the reed again pulls away from the shallot, allowing more air to escape. The entire process is then repeated. The puffs of air are admitted to the resonator, thus setting up a standing wave. The lower end of the resonator will be a pressure antinode. If the resonator is cylindrical, the odd harmonics will be favored, as with the orchestral clarinet. A conical resonator will, as discussed in Chapter 33, reinforce all harmonics; for this reason most reed pipe resonators are conical.

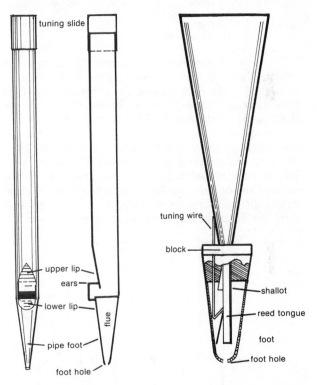

Figure 35.5. Parts of organ pipes: two views of flue pipe at left and reed pipe at right.

Although the natural frequencies of the brass tongue and the resonator are generally about the same, there is a certain degree of latitude. When the tongue is lengthened it not only vibrates at a lower frequency but, being more flexible, also vibrates with greater amplitude. The sound of the reed becomes louder and richer through production of more upper harmonics. As the pitch can be brought back to its orginal value by changing the length of the resonator, we have a method with which to regulate the tone of a reed pipe. If we want to change the loudness without any alteration of tone quality, we have to modify the curve of the tongue (a job which requires a considerable amount of skill).

Reed pipes are extremely sensitive. The curve of the

tongue must be so perfectly graduated that it will roll smoothly down the shallot. Nonuniformities of the tongue, or dust on the shallot, will cause an unpleasant sound when the reed is played.

Classification of Organ Stops

The separation of organ stops into flue pipes and reed pipes forms the basic classification of the instrument's tonal resources. Within each of these main divisions there are several important tonal families, as summarized in Table 35.1, which we will consider in more detail.

Table 35.1. Classification of Organ Stops.

Flue Tone		Reed Tone	
Diapason	Open Diapason	Classic	Chorus
	Principal		Regal
Flute	Open	Romantic	Solo
	Stopped		Effect
String	Imitative		
	Non-Imitative		

The word **diapason** is a Greek word meaning "through all." Many centuries ago not all ranks went from the bottom to the top of the keyboard, but were incomplete sets. The diapason, however, was a rank that went "through all" the notes of the keyboard—there was a pipe for each key. The diapasons are cylindrical open-open pipes having a tone which is the most characteristic of the organ. This tone, quite unlike that produced by any other instrument, forms the basis of the organ ensemble (rank upon rank of pipes having different tone colors and pitches). We can consider the diapason family to be composed of two dif-

ferent groups: the open diapason and the principal. Although both types are cylindrical open-open pipes, they are quite different in function and purpose. The tone quality of an open diapason is "round" and "smooth," with relatively few harmonics present. The tone quality of a principal is brighter than an open diapason because there are more partials present in the tone. The open diapason sound is typical of organs built around 1900, while the newer (and also older) instruments used a principal chorus. We could summarize the difference by saying that the open diapason chorus has typically been used to add "gravity" and volume to an organ ensemble, while the principals have been used to add clarity and brightness. Figures 35.6a and 35.6b show an open diapason and a principal pipe.

The **flute** stops are those whose tone quality somewhat resembles the orchestral flute, although this is not their purpose. Organ flutes come in three different shapes (cylindrical, conical, and rectangular), which occur in each of the two tonal groups (open and stopped). The open flutes (see Figures 35.6c and 35.6d) have a quality which is most similar to an orchestral flute, as would be expected since both instruments produce a tone by the same method and both have an open-open tube resonator. Since the open flute is an open-open resonator, all harmonics are present in the spectrum; typically, however, only the first three or four harmonics are prevalent. Sometimes the open flute pipes are made "harmonic." In this case, a small hole is drilled in the center of the pipe (approximately mid-way between the mouth and the tuning slide), which produces a pressure node at the center of the pipe. The pipe is thus caused to speak an octave higher; the length of the pipe is the same as the wavelength of the fundamental.

The stopped flutes, with their emphasis on odd par-

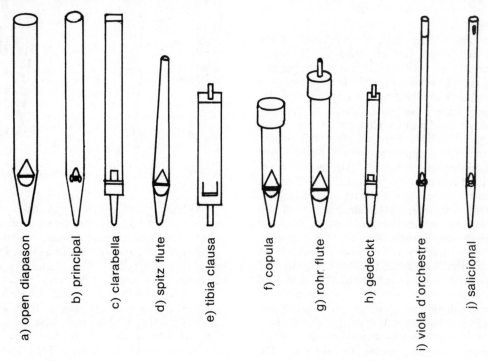

Figure 35.6. Flue pipes: (a-b) diapason; (c-d) open flute; (e-h) stopped flute; (i) imitative string; (j) non-imitative string.

tials, have relatively few harmonics in the spectrum. Typically, the stopped flutes are of rectangular cross-section, fitted at the end with an adjustable stopper (for tuning purposes). An illustration of wooden stopped flutes, the tibia clausa and the gedeckt, are shown in Figures 35.6e and 35.6h. A capped metal flute, the copula, is pictured in Figure 35.6f. The metal cap is adjustable so that the pipe can be tuned. One type of stopped flute, the rohr flute, consists of a cylindrical pipe fitted with an adjustable cap, but with an open-open tube of small diameter placed in the center of the cap. Although the large stopped pipe emphasizes the odd partials, the addition of the small "chimney" increases the amplitude of some of the upper partials. The length of the chimney is usally chosen so that the fifth or sixth partial is emphasized. A rohr flute is pictured in Figure 35.6g.

The **string** family of organ stops slightly resembles the tone produced by the string section of the orchestra. Their tone quality can best be described as soft, but rich in harmonic development. In many examples, the upper harmonics have a greater amplitude than the fundamental. The imitative strings actually attempted to imitate their orchestral counterparts, as much as this was possible with flue pipes, by producing a strident, "buzz-saw-like" tone. Such names as cello and viole d'orchestre are common for the imitative string stops; the viole d'orchestre is pictured in Figure 35.6i. On the other hand, the nonimitative strings make no attempt to imitate orchestral instruments. Their tone tends to be placid and bland compared to the imitative strings, due to less harmonic development. The salicional, pictured in Figure 35.6j, is the best know example of this type of string pipe.

The reed pipes can be divided into two tonal families: the classic and the romantic. Classic reeds are the typical reeds found on most pipe organs today. The **chorus reeds** may be used for playing chords (ensemble), or for a bright solo voice (individual notes). When several ranks of chorus reeds (usually of different pitches) are present on the same manual, a reed chorus of great brilliance and beauty is created. In general, chorus reeds on the organ serve the same purpose as the brass instruments of an orchestra. Several typical chorus reeds are shown in Figures 35.7a-d. Notice that each of these chorus reeds utilizes either a single or a double cone as a resonator.

The regal reeds, typically found only in larger pipe organs as a supplement to the chorus reeds, are characterized by being softer than the chorus reeds and

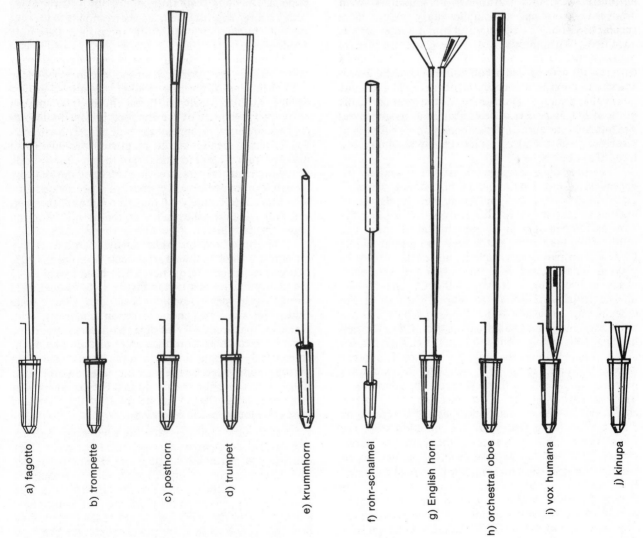

a) fagotto b) trompette c) posthorn d) trumpet e) krummhorn f) rohr-schalmei g) English horn h) orchestral oboe i) vox humana j) kinupa

Figure 35.7. Reed pipes: (a-d) chorus reeds; (e-f) regals; (g-h) solo reeds; (i-j) special effect.

by often having very little fundamental energy but many harmonics. The attenuated fundamental occurs because fractional length resonators are often used. (In some cases the resonator may be only one-eighth or one-sixteenth of the fundamental wavelength.) Illustrations of regals are given in Figures 35.7e-f.

The romantic reeds are of two varieties: the solo reeds and the special effect reeds. The **solo reeds** are analogous to the orchestral woodwinds, as can be seen from their names: clarinet, orchestral oboe, English horn, etc. Two examples of solo reeds are pictured in Figures 35.7g-h. The **special effects reeds** are to the romantic reeds group as the regal reeds are to the classic reed group. Just as the regals can be used to color a baroque flue chorus, so can the special effect reeds be used to color romantically voiced flue-stops. The special effect reeds do not imitate any known sound; they produce an entirely unique, though not very useful, tone color. Like their distant cousins, the regals, the special effect reeds use fractional length resonators to produce a wealth of harmonics with an attenuated-fundamental. The two best known special effect reeds, pictured in Figures 35.7i and 35.7j, are the vox humana and the kinura. The vox humana (human voice) has been a popular organ sound for over 300 years, although it is not currently in vogue. Although its sound has been described in various uncomplimentary ways, it is capable of yielding a very beautiful sound, under certain conditions. When massive chords are played on the vox humana on a large cathedral organ with the tremolo engaged, the effect is remarkably similar to the ethereal sound of a choir humming in the distance. The kinura, invented as a special effect sound for the cinema organs of the 1920's, is neither as beautiful nor as useful as the vox humana. Because of the extremely short cylindrical resonators, the sound is reminiscent of an angry bee trapped in a bottle.

The preceding classification scheme is entirely independent of pitch; on an actual pipe organ, ranks of pipes are present at many different pitch levels. An 8' manual stop is one whose lowest pipe has a pitch the same as an open-open pipe approximately 8' long. The third C from the bottom of the keyboard is then middle C. A 4' stop plays one octave higher while a 16' stop plays an octave lower. A 2' manual stop plays a pitch which is two octaves above middle C. The higher-pitched stops were found to add brightness to the ensemble by artificially adding upper partials. All the pitch levels mentioned so far can be grouped under one heading: foundation. **Foundation stops** change the octave of a note without changing the note itself; in other words, "C" pipes, although at different pitches, always play from middle C on the keyboard. **Mutation stops**, on the other hand, do change the pitch characteristic—the pitch sounded has a different name from the key being depressed. The mutation stops are necessary to fill in the gaps in the harmonic series created by foundation stops of different pitches. As an example, suppose we depress middle C on the keyboard. When an 8' stop is turned on, a pitch of middle C (C4) sounds. By adding a 4' stop, a pitch one octave higher (C5) is added; this corresponds to the pitch of the second harmonic of middle C. By adding a mutation stop (2 2/3'), the pitch G5 is added to our ensemble; this is

also the pitch of the third harmonic of C4. Next we add a 2' stop, which adds C6, the pitch of the fourth harmonic of C4. For the higher partials of the harmonic series, a mixture stop (also a mutation) is usually provided. The mixture stop immediately adds from 2 to 6 ranks of pipes, each rank corresponding to some higher partial of C4. Mixture pipes are generally of diapason quality as they, along with their lower- pitched relatives, help form part of the diapason chorus.

Each group of pipes shown in Table 35.1 can be found at 16', 8', and 4' pitch levels. Flutes and diapasons are also typically present at higher pitch levels, while flutes, diapasons, and chorus reeds are sometimes also found at the 32' pitch level.

Pipe Scales and End Correction

The **scale** of a rank of pipes refers to the ratio of diameter to length for the lowest-pitched pipe of the rank. Large-scale pipes (with a large diameter) have a dominating fundamental frequency and few upper harmonics. Small-scale pipes (small diameter in relation to length) have considerably more harmonic development. If we arrange the families of open-open flue organ pipes according to scale (from large to small), we get the following arrangement: wide scale flute, open diapason, principal, medium scale flute, non-imitative string, imitative string. This list, then, also corresponds to arranging the families in order of increasing harmonic development.

When the air within an open-open flue pipe is set into vibration, the maximum air motion does not occur right at the end of the pipe. Rather, because of an impedance mismatch between the air in the tube and the outside air, the node of the enclosed pressure wave occurs a short distance beyond the pipe end. The same effect is much greater at the mouth of the pipe. The distance from the end (or mouth) of the pipe to the node of the fundamental pressure wave is termed the **end correction**. For an open-open cylindrical pipe, the end correction is approximately 0.6 times the radius at the open end and 2.7 times the radius at the mouth. For the higher partials the end correction is somewhat less.

The end correction explains why a large scale flue pipe must be shorter than a small scale flue pipe in order to produce the same frequency. Three open flue pipes of the same pitch are pictured in Figure 35.8: a small scale string, a principal of medium scale, and a large scale flute. Because the end correction increases with diameter, the three different pipe resonators support the same wavelength and produce notes of the same pitch. Interestingly enough, the resonators of some reed pipes must be made longer as the scale increases, in order that notes of the same pitch be produced. The explanation of this phenomenon lies in the behavior of the truncated cone commonly used in reed pipes.

Let us now consider how the pipe scale explains why small-scale pipes encourage, and large-scale pipes inhibit, upper harmonic development. The amplitude of a harmonic present in a pipe depends on how strongly it is encouraged by the pipe. When the diameter of a pipe is small in comparison to the wavelength of a given harmonic, most of the wave is reflected at the open end of the pipe, resulting in a large amplitude for the harmonic. However, when the wavelength of a given har-

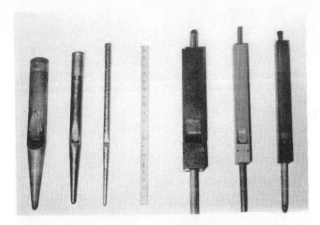

Figure 35.8. Large, medium, and small scale flue pipes: open metal pipes on the left and closed wood pipes on the right. (Courtesy of G. Orvis.)

monic is smaller than the pipe diameter, most of the wave radiates out the end of the pipe (see Chapter 12) and not much of the wave is reflected to produce a large amplitude for the harmonic. For a large-scale pipe, only the first few harmonics are reinforced by the pipe, as the higher frequency modes are radiated efficiently. For a small- scale pipe, the higher frequencies are radiated less efficiently than in a large-scale pipe, so that higher frequency modes can be set into large amplitude vibration. Therefore, the upper partials are more strongly developed in small-scale pipes because of reinforcement by the pipe.

Speech Characteristics

In the previous section, we classified organ stops in several different tonal families based primarily on the tone color of the pipes. There are other important characteristics, however, which determine the overall effect of any given rank of pipes. In addition to the loudness and efficiency of a rank, the nature of the attack is very important. We will consider the attack time (which determines how promptly a pipe speaks) and the number of attack transients (which determines the smoothness of the attack). Since a vibrating reed emits air in a series of puffs, there is little wasted energy. Flue pipes, on the other hand, are much less efficient because there is a continuous stream of air through the pipe. This hissing "wind noise" is often audible in the tone of the pipe.

The attack time of a reed pipe is usually one-half to one-fifth the attack time of a flue pipe of the same pitch. In a flue pipe the tone "builds up" as the air within the pipe begins to vibrate. A reed pipe attains its full vibration almost immediately, as it is dominated by the vibrating brass tongue. In almost all nonpercussive instruments, the low tones have longer attack times than higher tones, as we discovered in Chapter 31. Organ pipes are no exception to this rule. As you descend in pitch through any rank, the attack time increases. Experiments indicate that for any rank of flue pipes the attack time is approximately proportional to the wavelength. Also, the greater the mass of air contained within a pipe, the larger the attack time will be; particularly long attack times can be observed for 16' or 32' open- open pipes of large scale. The brass tongues of 16' or 32' chorus reeds sometimes have a small weight at-

tached (see Figure 35.9) in order to increase the mass, and thus lower the frequency, without changing the stiffness or length of the tongue. The extra mass lengthens the attack time substantially, sometimes to such an extent that the usefulness of the stop is greatly reduced. One of the authors of this text owns a 32' chorus reed with weighted tongues in which the attack time of the lowest notes exceeds two seconds. Obviously such a stop

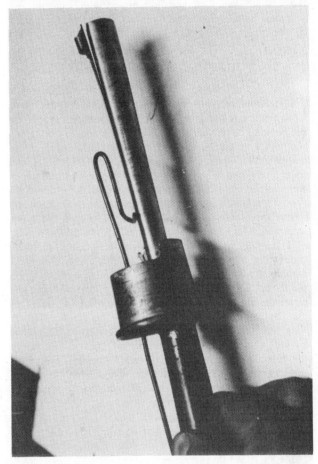

Figure 35.9. Weighted brass tongue of a reed pipe. (Courtesy of G. Orvis.)

can be used effectively only for slowly moving pedal passages.

The nature of the attack of a flue pipe is dependent upon various details of the design and the voicing of the pipe. A small-scale imitative string pipe, for example, is voiced to yield an abnormally full complement of upper partials. The techniques used tend to encourage the pipe to speak its second harmonic rather than to speak the fundamental. A cylindrical "bridge" placed in front of the mouth, parallel to the flue opening, has been found to eliminate this problem without adversely affecting the desired spectrum. The voicing device which has the greatest influence on the nature of the attack is termed "nicking." Nicking is the process of cutting narrow, vertical grooves along the front edge of the languid and the inner side of the lower lip, at right angles to the flue. Nicking increases the width of the air stream, securing a softer, smoother attack with fewer transients. When flue pipes have been voiced with heavy nicks, we say that the voicing is "romantic." When only very light nicking (called "feathering") is used, the voicing is baroque, because nicking was an unknown technique during the baroque era. Since pipe organs are generally located in buildings with fairly long reverberation times, and since the pipe organ is played primarily as a legato instrument, baroque voicing of flutes and principals is considered to be highly desirable. The "percussive consonant" preceding the "vowel" of steady-state sound helps to separate succeeding tones and adds a clarity to the voice lines of polyphonic music which cannot be achieved with heavily-nicked pipes. In the performance of romantic and popular music it is not essential to separate voices; rather, a mass of tone color is desired. The smooth attacks of heavily-nicked pipes are conducive to producing such an effect.

In recent decades there has been considerable debate about whether the chest mechanism which causes a pipe to speak has any effect on the sound produced by the pipe. In particular, builders of mechanical action instruments have long maintained that the electropneumatic action has a deleterious effect on the speech of pipes. In one study, two identical sets of pipes were voiced to sound the same on two different chest actions: a tracker and an electropneumatic. The chests were constructed by the M.P. Moller Co. and the pipes were voiced by their head voicers. Although the results of this experiment were somewhat inconclusive, some interesting observations can be made. First, the chest mechanism definitely influenced the attack portions of the tone. This was tested, quite simply, by reversing the pipes on the tracker and the electropneumatic sides of the chest. Perhaps the most interesting observation is that the voicers, by deft manipulation of the pipe mouth, were able to eliminate the differences caused by the chest. In summary, it appears as though the chest mechanism does influence the sound, but a skilled voicer can compensate for this difference.

Analysis of Tonal Qualities

We will now consider the steady-state spectra of typical representatives of the pipe families given in Table 35.1. While such spectra are useful in understanding the physical basis of the tonal families and in comparing members of the same family, we must remember not to attach undue importance to these spectra. As we have seen, the steady-state spectrum is only one factor in the overall tonal effect of any pipe. Furthermore, the spectrum measured for an open-open pipe will vary considerably, depending on whether the recording microphone is placed near the mouth, near the open end, or some distance from the pipe. Since organ pipes are almost never heard in close proximity, we have recorded the sound of various pipes at a distance of 2 to 4 meters (in their natural environment) so that an overall spectrum, characteristic of the pipe, would be received. Figure 35.10 illustrates spectra of the typical flue pipes shown in Figure 35.6. Figure 35.11 illustrates spectra of the typical reed pipes pictured in Figure 35.7. These spectra were recorded in many different environments but, unless otherwise stated, the note being analyzed is middle C. A brief discussion of these spectra will follow. Although each spectrum shows the relative proportions of each of its ingredients, direct comparisons cannot be made between spectra since they were not recorded under uniform conditions or at the

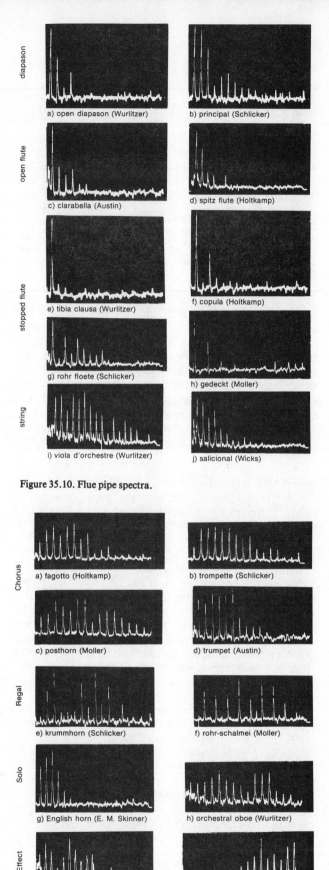

diapason

a) open diapason (Wurlitzer)

b) principal (Schlicker)

open flute

c) clarabella (Austin)

d) spitz flute (Holtkamp)

stopped flute

e) tibia clausa (Wurlitzer)

f) copula (Holtkamp)

g) rohr floete (Schlicker)

h) gedeckt (Moller)

string

i) viola d'orchestre (Wurlitzer)

j) salicional (Wicks)

Figure 35.10. Flue pipe spectra.

Chorus

a) fagotto (Holtkamp)

b) trompette (Schlicker)

c) posthorn (Moller)

d) trumpet (Austin)

Regal

e) krummhorn (Schlicker)

f) rohr-schalmei (Moller)

Solo

g) English horn (E. M. Skinner)

h) orchestral oboe (Wurlitzer)

Effect

i) vox humana (Austin)

j) kinura (Wurlitzer)

Figure 35.11. Reed pipe spectra.

same sound level.

Comparing the open diapason spectrum (Figure 35.10a) with the spectrum of the principal (b), we see that the difference in scale and voicing is obvious. The open diapason has only four harmonics, while the principal shows nine or ten. The spectra of (c) and (d) are for two open flutes, the clarabella and the spitz flute. The clarabella (c) is a fairly large-scaled rectangular flute. You may notice that the spectrum is not too different from that of the open diapason, except that the clarabella gives more emphasis to the fundamental. The spitz flute is tapered (see Figure 35.6d), which helps increase the strength of the second partial without decreasing the first partial, and produces a brighter sound. The next four spectra are various types of stopped flutes which, of course, emphasize the odd harmonics. The tibia clausa (e) is a very large-scale flute; we should not be surprised to discover that the spectrum shows only the fundamental (the third harmonic is of smaller amplitude than even the second harmonic). The copula (f) is a capped metal flute of somewhat smaller scale; hence we note the prominent third harmonic. The gedeckt (h) is of even smaller scale; we see that the fifth harmonic is now apparent, although small. The rohr flute (g) is basically a stopped flute, which is apparent in the emphasis of odd partials. The small chimney is usually of such a length as to emphasize the fifth or sixth harmonic. In this case, the fifth harmonic has been made somewhat more prominent. Finally, the spectra of two string pipes are presented. The viole d'orchestre (i), which is of a very small scale (the diameter of the 8' pipe is less than 4 cm), shows many harmonics, the first six being especially prominent. It is also interesting to note that the second and third partials are both somewhat stronger than the fundamental. This is because the pipe is first voiced to overblow and to sound the second partial; a small wooden dowel (the beard) is then fitted near the mouth to cause the pipe to again sound the fundamental. This process gives emphasis to the second and third partials, which is deemed to be desirable for a "realistic" orchestral string effect. The salicional spectrum (j), on the other hand, is somewhat bland in comparison to the orchestral string. Although there are more harmonics present than for a small-scale flute, the lessened harmonic development results in a more refined tone.

The spectra of reed pipes given in Figure 35.11 are for chorus reeds. The first three, all built within the past 10 years, show a high development of partials with the fundamental being somewhat attenuated. The trompette (b) is voiced to be somewhat brighter than the fagotto (a), although they have very similar tonal qualities. The posthorn (c) is voiced to be extremely bright and loud; the spectrum shows sixteen prominent partials and there are undoubtedly quite a few more beyond the range of the spectrum analyzer. The trumpet (d) was made about fifty years ago when chorus reeds were larger in scale and were not voiced to be as bright as is the general practice today. This is obvious from the fact that the harmonics higher than the sixth are not very prominent. The krummhorn (e) not only shows a reduced fundamental, but the fundamental is missing entirely. Since the krummhorn consists of a cylindrical resonator (albeit of very small scale), we would expect

to find mostly odd harmonics. Except for the very prominent second harmonic and the missing fifth harmonic, this is essentially true. The rohrschalmei (f) (which is C5 instead of C4) consists of two cylindrical tubes of approximately the same length. We note that, although all partials are present in the spectrum, the fourth harmonic is the most prominent. The English horn (g) consists of a small-scale conical resonator with a bulge on the end, which resembles the bulge of its orchestral prototype. Since the end is essentially closed, with a relatively small opening for the sound to escape, the upper harmonics are greatly attenuated. One will notice, however, that the second and third partials are very prominent. The orchestral oboe (h) has a very small-scale capped resonator. This stop, although fairly soft, is voiced to have many harmonics of nearly equal intensity present. The vox humana (i) utilizes a capped cylindrical resonator having a length of only one-eighth that of a full-length conical resonator. The resonator, being very short, does not control the reed, but rather serves to emphasize certain frequency regions and attenuate other regions in a manner similar to the way the vocal tract modifies the vocal fold energy. The kinura (j) consists of an extremely short conical resonator (8 cm long at 8' C), which exerts no influence on the reed but serves somewhat as an impedance-matching device from the reed to the outside air. That it fails miserably in this regard for the low partials is evidenced by the fact that the first eight partials are absent. The higher partials are present, including many which are beyond the range of the spectrum analyzer. You may have noticed that for reed pipes there is almost no limit to the type of spectrum (and its associated tonal quality) which can be produced by combining unusually shaped resonators with certain types of shallot.

The characteristic ensemble effect of pipe organs, which has not yet been satisfactorily reproduced by their electronic imitators, occurs when more than one set of pipes play at the same time. Even though the pipes may have been tuned to the same frequency, they will be very slightly out of tune. This gives rise to beating among the various partials and adds warmth to the tone. The ensemble or choral effect is produced by beating when more than one voice is sounded. The ensemble effect is missing on most electronic organs because the voices are exactly in tune; there is no beating, and the tone lacks the warmth of a pipe organ tone (see Chapter 31).

It is not uncommon, on many organs, to have a rank of pipes which is purposely tuned slightly sharp. This stop, usually called the voix celeste, is a member of the string family. When it is played with a similar string (such as the salicional), very noticeable beats are produced, giving a warm sound reminiscent of the orchestral strings. Although the string family, because of the many partials, is best adapted to this purpose, very beautiful flute celestes have also been produced.

A vibrato is produced in organ stops by a mechanical device called the tremolo. The tremolo (for pipe organs) creates a periodic fluctuation of the air pressure in the windchest. Since the pressure is alternately increasing and decreasing around its average value, a fluctuating pitch change is introduced into the tone.

Historical Notes

Although the origins of the pipe organ are obscure, a prototype organ was probably fashioned from a set of pan pipes blown by a bellows. The first mention of the instrument in recorded history appears about 250 B.C. in Egypt, with the hydraulis, invented by Ctesibius. The hydraulis used water pumped into a large covered clay jar, which caused the entrapped air to sound a set of pipes. The Greeks made refinements on the hydraulis, and by the time of the Roman Empire, hydraulises were being used in large numbers for temples and theaters.

The first pneumatic organ, employing a bellows to provide the wind pressure, appeared in the 3rd century A.D. In 400 A.D. St. Jerome described a pneumatic organ in Jerusalem with 12 bronze pipes, 2 elephant skins, and 15 blacksmith's bellows. The organ apparently could be heard a mile away.

Until the 11th century flue pipes were either open diapasons made of copper or brass, or stopped flutes constructed of wood. All pipes in one rank, however, were of the same diameter. Accordingly, the bass end of each rank was clear and bright because of the small scale, while the large scaled trebles produced dull tones. The compass of these organs was usually sixteen to twenty notes played by pulling out sliding rods. Since each slide (note) controlled some dozen pipes, the entire instrument could contain several hundred pipes. It has been known for some time that ranks sounding octaves and fifths added to the brightness of the instrument, so these were present in abundance. During the 13th and 14th centuries the slider mechanism was replaced by a set of primitive keys (actually levers), each "key" measuring 3 to 5 inches wide, 2 inches thick and at least 1 foot long. To activate the pipes, the lever had to be depressed about 1 foot, using a clenched fist to press or strike them. The organist who performed on these unwieldly instruments was known as the "pulsator organorum."

Reed pipes were in use during the 14th century. By the end of this century, second and occasionally third sets of keys were being provided. During the 15th century the organ grew to a considerable size and power, and "stops" were invented for shutting off parts of the instrument. By the 16th century the pipe organ had, more or less, achieved its present form. The keys had been reduced to "fingerable" form, the large pedal keys were present, and slider chests were being used so that sets of pipes could be added or silenced merely by moving the stop knob. During the Baroque Era, the Germans became the preeminent organ builders, constructing beautifully crafted instruments with a variety of tone colors and choruses for synthetic tone building. During the 18th century the swell shutters were added to help give more flexibility to the instrument.

The next great technical advance in the art of organ building came at the end of the 19th century when an eccentric English engineer, Robert Hope-Jones, first used electricity to control a pipe organ. He apparently demonstrated his new invention in a very dramatic way. Since an electrically controlled organ does not require that the console be in close proximity to the pipes, Hope-Jones rebuilt the old organ in the rear balcony of St. John's Church in Birkenhead, England so that he could play it from a remote console placed in the chur-chyard among the tombstones. While the invention of the electric action changed the future of the pipe organ, mechanical action instruments continued to be built and are still popular. Although the electric action organs have considerably greater flexibility, advocates of the tracker action claim that the mechanical linkage puts the musician in more direct and intimate contact with his instrument. Whether or not this "advantage" compensates for the disadvantages of mechanical action instruments is as yet an unresolved question.

Exercises

1. Compare the advantages and disadvantages of a tracker action and an electric action pipe organ. (Hint: Consider such items as cost, flexibility, and keyboard touch.)

2. What would be the effect on the tone of a reed pipe if a thin sheet of soft leather were put on the shallot face where the tongue rests? Explain.

3. Several different types of shallot are used in reed pipes. What difference in volume and tone would you expect for a shallot face with a large opening compared to one with a smaller opening? Explain.

4. Chorus reeds are sometimes found with half-length conical resonators (i.e., an 8' pitched pipe will have a resonator approximately 4' in length). What effect would using a half-length resonator have on the tone quality? On the amplitude of the fundamental? On the perceived pitch?

5. Would you expect a typical open diapason pipe to be nicked or unnicked? Why? What about a typical principal pipe?

6. Compare an open diapason and a non-imitative string pipe. Consider scale, harmonic development, strength of the fundamental, and loudness.

7. How would you expect the tone color of a very large-scale open flute to compare to the tone color of a very large-scale stopped flute? Explain.

8. Pipe organs almost always have a diapason chorus and/or a flute chorus, but seldom does one find a complete string chorus. Suggest some reasons for this.

9. The purpose of a diapason chorus is to reinforce the harmonics of the 8' diapason stop. Would large-scale open diapasons or the smaller-scale principals fulfill this purpose better? Explain.

10. In small pipe organs the 16' pedal pipes are almost always stopped wooden pipes. Why?

11. Which families of organ stops have the greatest upper harmonic development? The least? Why?

12. An organ stop called the French horn is a fairly loud reed with a very strong fundamental and little harmonic development which does not blend well in ensemble. With which family would you classify this stop? Why?

13. The translation of the Latin words vox humana is "human voice." The typical vox humana utilizes a one-eighth length cylindrical resonator which is partially closed on top. How is this similar to the human vocal apparatus? How is it different?

14. An organ stop labeled "32' Resultant" is often found in the pedal division of small or medium-sized organs. The bottom octave of this stop is a 16' tone and a 10 2/3' tone (the fifth above) played simultaneously. By referring to the discussion of dif-

ference tones in Chapter 15, explain how a 32' pitch is produced by this effect.

15. For an imitative orchestral string pipe would it be more effective to use nicking or not? Explain.

16. Will a diapason and a string pipe of the same pitch have exactly the same length? Explain.

17. Why is a voix celeste an effective stop to help produce an orchestra-like quality of string tone?

18. In order to increase the loudness of an organ ensemble it is more effective to add stops of higher pitch than to add more 8' stops. Explain why this is true.

Demonstrations

1. Blow various single pipes with a laboratory air supply. Adjust the pressure to an optimum value and to values above and below. Make any adjustments possible on the pipe.

2. Visit a pipe organ installation and have someone explain and demonstrate the workings of the organ and the tonal families.

Further Reading

Baines, Chapter 2

Hall, Chapter 12

Rossing, Chapter 14

Anderson, P.G. 1969. **Organ Building and Design** (Oxford University Press).

Barnes, W.H. 1964. **The Contemporary American Organ**, 8th ed. (J. Fisher).

Bonavia-Hunt, N., and H.W. Homer. 1950. **The Organ Reed** (J. Fisher).

Fletcher, H., E.D. Blackham, and D.A. Christensen. 1963. "Quality of Organ Tones," J.

Acoust. Soc. Am. **35**, 314-325.

Fletcher, N. 1976. "Jet-drive Mechanism in Organ Pipes," J. Acoust. Soc. Am. **60**, 481-483.

Ingersler, F., and W. Frobenius. 1947. "Some Measurements of the End Corrections and Acoustic Spectra of Cylindrical Open Flue Organ Pipes," Trans. Danish Acad. Tech. Sciences **1**.

Jones, A.T. 1941. "End Corrections of Organ Pipes," J. Acoust. Soc. Am. **12**, 387-394.

Kinnaugh, N.T. 1980. "Initial Transients in Organ Flue Pipes: Computer Analysis of the Effects of Two Different Wind Chest Actions" (Unpublished M.S. Thesis, Brigham Young University).

Klotz, H. 1961. **The Organ Handbook** (Concordia Publishing House).

Nolle, A.W., and C.P. Boner. 1941. "Harmonic Relations in the Partials of Organ Pipes and of Vibrating Strings," J. Acoust. Soc. Am. **13**, 145-148.

Nolle, A.W., and C.P. Boner. 1941. "The Initial Transients of Organ Pipes," J. Acoust. Soc. Am. **13**, 149-154.

Audiovisual

1. **Instruments Using a Bellows** (3.75 ips, 2 track, 30 min, UMAVEC)

2. **Organ, Part 1** (19 min, color, OPRINT)

3. **Organ, Part 2** (22 min, color, OPRINT)

4. **Vibrating Strings and Air Columns** (30 min, 1957, EBEC)

5. **Wind and Percussion Instruments** (30 min, 1957, EBEC)

6. "Pipe Organ Tones of Figures 35.10 and 35.11" (GRP)

36. Bowed String Instruments

The vibrating string, which was discussed in Chapter 10, is the basis of this chapter. Strings may be made to vibrate by bowing, plucking, or striking. The acoustical effects of excitation by bowing form the content of this chapter. We will state some general requirements for bowed string instruments, and examine the role of the strings and the bowing action. In addition, we will consider the influence of the body and other factors on the spectra produced by vibrating strings. Finally, we will discuss recent developments of the violin family and give some brief historical notes.

General Requirements

The members of the bowed string family are the violin, viola, cello, and double bass shown in Figure 36.1. They all have strings approximately fixed at both ends as their primary vibrators. The string plays a role similar to that of the air column in a wind instrument. It determines the frequencies that will be present, which are approximately all harmonics. This would be expected for a fixed-fixed string. The bowing action supplies energy to the string to replenish the energy that is lost and to keep the string vibrating. The waves on the string interact with the bow to determine when it will catch onto and break loose from the string. In this way the string determines when the bow will add energy, similar to the way a tube controls a reed.

A string is an inefficient radiator of sound because of its small area. It must be coupled to a body with a large area to provide efficient radiation. The body will impose its own resonance features as it radiates the string energy. The body of a bowed string instrument must have resonance properties which will enhance the desired frequencies for that particular instrument.

Construction of the Violin

The violin is representative of bowed string instruments. A sketch of the main parts of the violin body is given in Figure 36.2. The strings which provide the original vibration are attached to adjustable pegs at one end and to the tail piece at the other end. The vibrations of the strings are transmitted to the body of the instrument through the bridge. Directly under the feet of the bridge, on the underside of the top plate, are the bass bar and the sound post. The bass bar stiffens the top plate, and the sound post transfers some of the vibration energy to the back plate so that it also vibrates. The fingerboard provides a support against which the strings can be stopped, to shorten them. The holes in the top plate, called the f-holes, enable the air inside the body to

Figure 36.1 Bowed string instruments: violin, viola, cello and double bass. (Courtesy of G. R. Williams.)

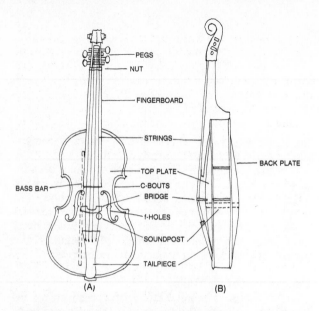

PEGS
NUT
FINGERBOARD
STRINGS
BACK PLATE
TOP PLATE
C-BOUTS
BASS BAR
BRIDGE
f-HOLES
SOUNDPOST
TAILPIECE

(A)   (B)

Figure 36.2. Construction aspects of a violin: (A) top view and (B) side view.

participate in the vibration (as a Helmholtz resonator). The graceful curves in the sides of the violin called the C-bouts, enable the violinist to bow the outside strings without rubbing the side of the instrument.

The four open strings of the violin are G3 (G below middle C), D4, A4, and E5. The unstopped strings have their effective length for vibration defined between the bridge of the instrument at one end and the nut at the other. By stopping the strings with the fingers of the left hand, one can obtain a range of over three octaves.

Role of the Strings

The laws governing the frequencies of the various modes of a vibrating string have been discussed in Chapter 10. In review, the frequencies of the modes of a vibrating string are (1) inversely proportional to the wavelengths of the modes, which in turn are proportional to the length of the string, and (2) directly proportional to the wave speed on the string, which in turn is proportional to the square root of the tension on the string and inversely proportional to the square root of the mass per unit length of the string. Thus, for any given material the mass per unit length depends on the square of the diameter of the string, and so thicker strings give lower frequencies. Also, doubling the length of the string lowers its modal frequencies by an octave, while halving the length of the string raises them by an octave. To raise the frequencies of a stringed instrument, one need only increase the tension, or decrease the tension to lower the frequencies. This is typically done in tuning all stringed instruments. The bass strings on violins, harps, and pianos have a larger diameter than the treble strings, which produces lower pitches without requiring strings of inordinate length.

Assuming a uniform string of constant cross-sectional area (and thus constant mass per unit length) and constant tension, the natural modes of a vibrating string all have integer multiples of one-half wavelength along the string's length, as shown in Figure 10.1. Note that the odd-numbered modes have antinodes (displacement maxima) at the center of the string, while the even

modes have nodes (zero displacement) at the center of the string. Now suppose that we bow the string at its center. Any natural modes with nodes at the center of the string will not be excited because the point where we bow the string is a point of maximum vibration, or an antinode. Thus, odd modes are excited and even modes are not excited on a string bowed at its center. Suppose a string is excited (by bowing, plucking, or striking) at a point one third of its length from one end. It can be seen that any partial having a node one third of the distance from the end of the string is not possible; hence the third, sixth, ninth, etc. modes are not excited. In general, the partials that are missing are those having a node at the point of excitation.

In practice, the bow has a finite width and is not maintained at a fixed position relative to the end of the string, which means that particular partials will not be completely missing from the spectrum. However, the position of the bow in relation to the end of the string will have a strong influence on the spectrum, and the tonal quality can be made quite different by changing the bow position on the string. Any given partial will be made to sound stronger as the bow is placed more closely to an antinode of the mode associated with the partial; it will sound weaker as the bow is placed closer to a node of the associated mode.

Action of the Bow

When a horsehair bow is drawn across a violin string, the frictional force exerted on the string by the bow displaces the string until the restoring force in the string is great enough to overcome the frictional force of the bow, and the string snaps back toward its rest position. Frictional forces of the bow then displace the string again, and the process is repeated periodically, resulting in vibration of the string, as shown in Figure 36.3.

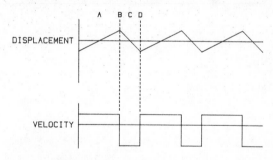

DISPLACEMENT

VELOCITY

Figure 36.3. Idealized displacement and velocity waves in a string. The string is moving with the bow in region A; the string begins slipping at point B; the string is slipping past the bow in region C; the bow "captures" the string at point D. (After Schelleng, 1974.)

The period of time the string is free from the bow is substantially shorter than the period of time the bow is displacing the string. (These time periods could be equalized by bowing the string at its center.) It is also apparent from Figure 36.3 that the string does not merely return to its equilibrium position, but overshoots this position—a characteristic of oscillatory motion. The time required for the string to slip back is governed by the sliding friction between the string and the bow; since sliding friction is always less than static friction, the return time is shorter than the time required for displacement. The motion of the string is then a com-

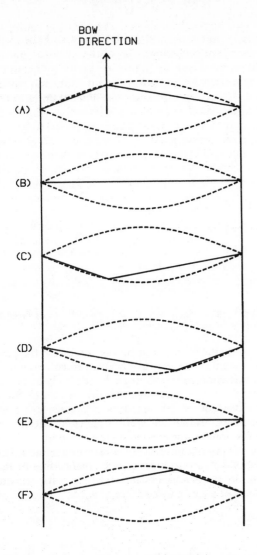

BOW
DIRECTION

(A)

(B)

(C)

(D)

(E)

(F)

Figure 36.4. Shape of violin string at different times during one cycle of vibration. String slips from bow at (A) and catches bow at (C). Compare Figure 36.3. (After Schelleng, 1974.)

displacement of the string occurs at the bow position. The negative displacement results in a positive restoring force on the string, in the same direction as the bowing force. These combined forces cause the string to stick to the bow again. The downward bend travels (Figure 36.4D) to the far end of the string where it is reflected (Figure 36.4E) as an upward bend, which travels (Figure 36.4F) toward the bow. When this upward bend arrives at the bow (Figure 36.4A), it causes a negative restoring force on the string. The negative restoring force is opposite to and greater than the bow's frictional force. This causes the string to break away from the bow again and the motion repeats. In this way the standing waves on the string cause the string to stick to the bow and to break away from the bow at periodic intervals.

Although the frequency of the tone emitted by bowing does not depend strongly upon the bowing force, the quality of the emitted sound is dependent upon the force. It is also dependent on the position of the bow on the string and the speed of the bow. The bow force, however, is more effective in modifying tone than is the bow speed; increasing the bow force tends to increase the intensity of the higher harmonics. The point at which the bow is applied to the string also greatly modifies the tone quality. Bowing close to the bridge increases the number of higher harmonics, while bowing farther away from the bridge reduces the higher harmonics, giving the sound a softer quality. The amplitude of the string vibration at a particular bowing position depends on how rapidly the bow is drawn over the string—the greater the bow speed, the louder the resulting sound. The amplitude of vibration, however, does not depend on bowing force.

For a constant bowing speed, there is a maximum bow force and a minimum bow force which will give the required frequency. The range of acceptable forces depends on the distance of the bow from the bridge, as shown in Figure 36.5. Note that the distance between maximum and minimum bow force increases as the distance from the bridge increases, and that there is a point (very close to the bridge) where the maximum and minimum forces are equal. When the maximum force is exceeded, the discontinuity in the string—due to the traveling waves—can no longer provide the necessary force to trigger the slipping. A raucous, erratic vibration results. If less than the minimum force is provided, the static friction is insufficient to hold the string for its full displacement and the string slips sooner, thus replacing the fundamental tone by the octave.

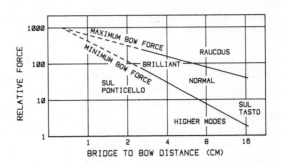

Figure 36.5. Range of acceptable bow forces as related to bow position for a cello bowed at a speed of 20 cm/s. (After Schelling, 1973.)

plex type of vibration which will contain many harmonics of the fundamental frequency. We might expect the period of the waves on the string to depend on the bowing force exerted by the hand. In actuality the period of the string is remarkably constant over a wide range of bowing forces. The explanation is found in the time it takes the sharp bend on the string to travel to the near end of the string and then to the far end and back to the point of excitation. The appearance of the string at different instants in the cycle is shown by the straight line segments in Figure 36.4. If the A-string is bowed for example, the bend makes 440 complete trips per second to the near end, then to the far end, and finally back to the excitation point (since the frequency is 440 Hz). The bend moves as if it were enclosed by the dashed curves in the figure. For "up bowing," shown by the arrow, the direction of circulation of the bend is toward the left end. When the bow changes from "up bow" to "down bow," the direction of circulation of the bend also changes. The upward bend of Figure 36.4A travels to the near end where it is reflected (Figure 36.4B) as a downward bend. When this downward bend arrives back at the bow (Figure 36.4C), maximum negative

As mentioned previously, as the bow moves away from the bridge, the proportion of higher harmonics decreases so that the tone has the gentle characteristic called **sul tasto** ("bow over the fingerboard"), and a wide degree of latitude exists between maximum and minimum bow forces. As one bows closer to the bridge, larger bow forces are required and there is a less generous latitude between maximum and minimum force, but the tone is greatly enriched by the increase of the higher partials. Experience is necessary in order to bow successfully in this region; amateurs find it prudent to confine their bowing to the region closer to the fingerboard. As one attempts to bow even closer to the bridge, the bow force mounts even more and the fundamental tone disappears, leaving only a swarm of high harmonics; composers call this the **sul ponticello** ("bow over the little bridge").

Influence of the Body

A vibrating violin string transfers most of its energy to the body of the instrument through the bridge; a small amount of energy is radiated directly by the string, but without the violin body the sound would be extremely weak. The body of the violin, however, modifies the sound of the strings because it has a complex set of resonances. Thus, radiated sound is the sound produced by the strings as "filtered" by the body. The violin body behaves in basically three different ways at different frequencies. At low frequencies, as the string vibrates left to right, the body is set into motion so that when the top plate is depressed, the back rises and vice versa. This is known as the breathing mode because the air is driven into and out of the body through the f-holes; the resulting broad-band resonance is called the air resonance. Higher frequencies are amplified by a mode in which the top plate executes a rocking motion about the centerline. The resonance associated with this mode is known as the primary "wood resonance." Finally, several modes at high frequencies control the instrument's brightness. The small patch of wood between the f-holes is lightweight enough to respond to very rapid vibrations; this region of the violin radiates the high frequencies. The bridge also has a resonance at about 3000 Hz.

There are several methods for determining the resonance frequencies of the violin body. In one method the violin is bowed in semitone intervals over its entire range to produce the loudest tone possible. When the intensity level (in dB) is plotted as a function of the note, prominent peaks, corresponding to the body resonances, are observed. One such loudness curve for the violin is displayed in Figure 36.6. The letters at the bottom indicate the open strings. The air resonance is labeled A, the main wood resonance is labeled W, and the lowest prominent peak is labeled W' (for wood prime). The W' peak is not an actual body resonance, but is a "subharmonic" of the main wood resonance; it arises from the reinforcement of the higher partials of the low tones by the main wood resonance. Even though the fundamental of the tone is not reinforced, the ear hears an overall increase in loudness because of the increased levels of the higher partials. Note that for a good instrument (in this case a Stradivarius), the peaks are uniformly spaced, with the air resonance falling ap-

proximately on the second (or D) string and the main wood resonance occurring close to the third (or A) string. The wood prime resonance is, by its very nature, always an octave lower than the main wood resonance, so that it reinforces those tones near the lowest open string (the G-string).

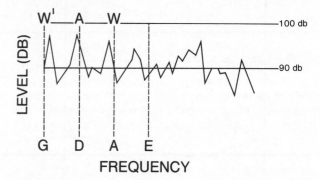

Figure 36.6. Violin loudness curve showing sound level as a function of the note played. (After Hutchins, 1973.)

Some methods for determining resonance frequencies use machine bowing of the instrument to obtain more consistent results; others excite the body by means of external sources and measure the response. Some methods use optical sensing devices to measure body displacements. Hologram interferometry has been used to provide detailed pictures of various body modes. Chladni patterns are useful for illustrating some of the most important lower modes of body plates. Figure 36.7 illustrates nodal patterns of modes 2 and 5 for top and back plates of a violin before assembly.

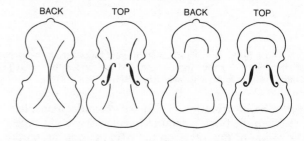

Figure 36.7. Chladni patterns showing nodal lines for second mode (left) and fifth mode (right) of violin back and top plates. (After Hutchins, 1977.)

In one of the most widely used methods for determining violin body resonance frequencies a small coil is attached to the bridge and excited by a loudspeaker magnet, or a small magnet is attached to the bridge and excited by a coil. In this way the bridge and the body are driven sinusoidally over a range from low to high frequencies. The resulting response curve, shown in Figure 36.8, is useful for defining the acoustical properties of the violin. The response peak labelled A0 (just below 300 Hz) and the peak labelled T1 (just below 400 Hz) correspond to the air and main wood resonances of the loudness curve for a different instrument than the one shown in Figure 36.6. The main body resonances appear too low to produce good quality violin tones.

Research done on many violins shows that high-quality violins have the resonances placed so that the air and main wood resonances reinforce partials in the

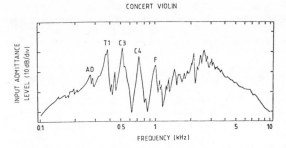

Figure 36.8 Input admittance response curve of a violin. (From Moral and Jansson, 1982.)

vicinity of the D and A strings, respectively. The air resonance (which results from the air contained in the body) can be adjusted by changing the air volume of the body and the size of the openings (f-holes) in the body. The main wood resonance (which is the lowest fundamental mode of the coupled top and back plates) can be adjusted by proper construction of the plates. One successful construction procedure involves tuning the second modes (see Figure 36.7) in the unassembled top and back plates to within about 1.4% of each other. The fifth modes (see Figure 36.7) in the two plates are tuned to within less than two semitones of each other. Modes 2 and 5 should be about an octave apart and of large amplitude. The average frequency of the top and back plate fifth modes should be about seven semitones below the desired main wood resonance of the assembled instrument, as coupling the plates via the ribs and soundpost stiffens the system and raises the frequencies.

Recent research of two kinds has further clarified the desirable resonance properties of violin bodies. Good correlation has been found between violins having good tonal quality and the following physical attributes: (1) high amplitude peaks for body resonances, labelled T1, C3, and C4 in Figure 36.8, (2) uniformity of peak heights for these resonances, and (3) a steep upward slope of the response curve from 1500 to 3000 Hz (see Figure 36.8). The response curve shown in Figure 36.8 was measured on the bridge at the G-string position. A response curve can also be measured on the bridge at the E-string. The similarity of these two response curves between 1500 and 3000 Hz is another indicator of good tonal quality. An electronic violin with a single broad resonance peak at about 3000 Hz can produce realistic violin tones on the three upper strings. The G-string tones lack realism because there are no electronic air and main wood resonances.

Additional Factors in Violin Tone

Recently other violin characteristics have been investigated to determine their influence on tone quality. The type of wood and how the wood is cut in relation to the grain both play important roles in the final results. The effect that aging has on violins is being studied; the preliminary results indicate that aging may affect the tone. It appears that playing a violin can eventually improve the tone quality to some extent. One reason is that continued playing loosens the glue joint between the top plate and the sides, thus making the top plate more flexible (and hence more responsive). It has been discovered that the purfling—a narrow groove cut around the outside edge of the top and back plates, with thin strips of wood glued inside—increases the flexibility of the plates as the glue ages and begins to crack slightly.

The excellent tonal quality of the Stradivarius violins has often been attributed to the varnish, the formula of which was a secret shared only by the master craftsmen. Recent acoustical investigations have shown, however, that while the response peaks of the violin are lowered slightly after the varnish has been applied, the change is so small that it cannot ordinarily be detected by ear. Furthermore, analyses of varnish scrapings from old violins show that the composition is very much like that of good modern varnishes. The main function of the varnish seems to be to seal the pores and thus to protect the wood.

If the most prominent body resonance of a stringed instrument is placed too close to a string resonance, a "wolf" tone can result from an interaction between the two resonances. The wolftone is a harsh beating between the two vibrations.

Some problems may arise if stiff strings are used on bowed string instruments. Stiff strings give rise to inharmonic partials which do not arrive back at the bow "in phase." Inharmonic partials do not effectively trigger the slipping and sticking of the string on the bow. Ranges of useable bow pressure, intonation, and strength of high frequency partials can all be restricted because of inharmonic partials.

The basic string motion has already been described in the section on action of the bow. There are additional aspects of string motion which may be important in violin performance. The direction in which a string is bowed—either horizontal or with a vertical component—can affect how it couples energy to the bridge and in turn how the bridge couples energy to the body. The string can roll as it is displaced by the bow. While the string is sticking to the bow, the bow divides the string into two segments. Each of these segments can exhibit its own set of characteristic frequencies; ripple or sinusoidal motions sometimes appear superimposed on the basic triangular wave. A source of aperiodicity, which appears to be musically important, arises from unequal slipping of the bow hairs during the sticking phase. It gives rise to a series of "spikes" superimposed on the basic wave, which produce audible noise under playing conditions.

A final feature of interest is the radiation patterns of string instruments. Figure 36.9 illustrates violin radiation characteristics in the horizontal direction at different frequencies. One can observe that the radiation is not very directional at low frequencies (200-500 Hz) and becomes quite directional at high frequencies (2000-5000 Hz). This is consistent with the discussion in Chapter 12, although violin radiation patterns are more complex than loudspeaker patterns.

Recent Developments

In order to determine the placement of the main wood resonance in a finished instrument, violin makers traditionally used "tap tones;" i.e., they tapped the unassembled front and back plates and listened to the pitch produced. Correctly judging the resonances of the top and back plates was part of the art of violin making. Recent research has made the determination of the free plate resonances accessible to scientific analysis. By sup-

257

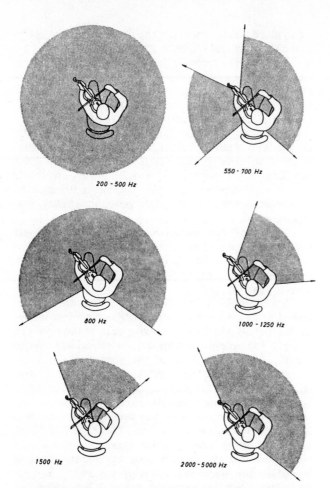

200 - 500 Hz
550 - 700 Hz
800 Hz
1000 - 1250 Hz
1500 Hz
2000 - 5000 Hz

Figure 36.9. Primary radiation patterns of violin in different frequency ranges. (From Meyer, 1972.)

Table 36.1. Data on new bowed string family of instruments. Length scaling is relative to conventional violin (mezzo). Multiply length ratios by 36 to get lengths in cm. (Data from Hutchins, 1967.)

Instrument	Strings	Body length ratios	
		Theoretical	Actual
Contrabass	EADG	6.0	3.60
Small bass	ADGC	4.0	2.92
Baritone	CGDA	3.0	2.42
Tenor	GDAE	2.0	1.82
Alto	CGDA	1.5	1.44
Mezzo	GDAE	1.0	1.07
Soprano	CGDA	0.75	0.87
Treble	GDAE	0.50	0.75

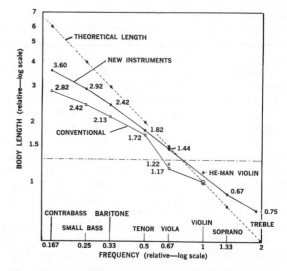

Figure 36.10. Scaling factors for old and new bowed string families. (From Hutchins, 1967.)

porting the plate at its nodes and driving it with a variable oscillator, the response of the plate as a function of frequency can be plotted. As noted earlier, good violins result when adjacent peaks of a plate are properly tuned, and when the front and back plates have peaks which alternate in frequency. Then, when the instrument is assembled, the main wood resonance is about seven semitones above the average frequencies of the tap tones from the front and back plates. Thus, by accurately predicting the tap tones and properly placing them in frequency, it is possible to place the main wood resonance where it will do the most good for the tone of the instrument.

The violin design became more or less optimized under the skilled hands of the Cremona craftsmen, but the same development was not achieved for the other members of the string family. The air and wood resonances in violas and cellos are generally too high for excellent tone quality to be achieved. With this in mind, Carleen Hutchins applied the principles of good violin design to designing and constructing a family of eight stringed musical isntruments covering a frequency range from E1 to E6. If the instruments were linerarly scaled to the violin, the large bass would have a body length of 2.1 m (six times that of the violin) and the small treble violin would be only about 16 cm long. Table 36.1 and Figure 36.10 give the relative body lengths chosen for the instruments, as well as their theoretical lengths. The actual scaling makes the body of the smallest instrument

27 cm long, and the large bass body 130 cm long. The philosophy of design was to then determine the other dimensions so that the air resonance and main wood resonance would occur near the two central strings, respectively. In order to get the air resonance of the treble violin high enough without using very small ribs, a set of six holes was made in the ribs. The viola was redesigned to have a longer body length (52 cm as opposed to the standard 41 cm), which meant that small players must play it vertically on the floor like a cello. The new baritone instrument is similar to the cello in length but has slightly longer strings. The tone quality of the new baritone instrument, however, has been judged superior to that of the conventional cello. The small bass is a newly constructed instrument; it is slightly smaller than the conventional bass but has a desirable tone and is easy to handle. The new contrabass is 1.3 m high and, although somewhat hard to handle, has an exceptional depth of tone. The overall evaluation of these instruments is that they are musically successful. There is a lack of musical literature for some of them, however, which limits their popularity.

Another area of research is the application of electronics to string instruments in order to simulate the acoustical properties of the body with electronic filters. The sound is produced by bowing the strings in the usual fashion, but string vibrations are picked up by electromechanical transducers, fed to the body-simulating filters, and finally radiated by a loudspeaker mounted either externally or internally in the instrument body. Electronic string instruments will be discussed further in Chapter 40.

Historical Notes

The precise origin of bowed string instruments is one of music's most intriguing enigmas. Although stringed instruments are ancient, no records before the ninth century mention bowing. Yet, by the tenth century bowed "fiddles" were in common use all across Europe, Asia, and North Africa. Around the year 1550 the violin, in the form we know it today, emerged from several immediate ancestors which flourished during the renaissance. One of these ancestors, the lyra da braccio (see Figure 36.11), was remarkably similar to the violin in body outline. It was slightly larger, however, and utilized seven strings. Five stopped strings ran along the fingerboard, while two unstopped strings were placed along the side of the neck piece. It is often assumed that the viol family members, because they preceded the violin by about 100 years and have now become obsolete, were the ancestors of the violin. Actually, the viols (see Figure 36.12) represent a separate class of instruments with different construction and stringing.

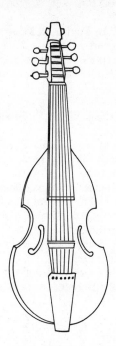

Figure 36.12. Sketch of a viol. (After Baines, 1969.)

Figure 36.11. Sketch of the lira da braccio. (After Baines, 1969.)

The violin developed rapidly in Italy, aided by Andrea Amati who, in the mid-sixteenth century, founded the great Cremona school of violin makers. Within 150 years of his death, his descendents and their pupils (notably Antonio Stradivari and Guiseppe Guarneri) had brought the art of violin making to such a state of perfection that their instruments are only now being equalled. By the end of the seventeenth century, Antonio Stradivari had developed a classical model of violin which became the standard for succeeding generations. Around the year 1800, the violin underwent its last important transformation to provide the greater power and brilliance needed to fill the larger music halls. The changes included greater string tension, a higher bridge, a sloped fingerboard, and greater playing length of the strings, all of which resulted in further internal changes in the instrument.

Exercises

1. Some partials are (approximately) not present in a string tone. As an example, consider wave 6 from Exercise 8 of Chapter 11 as a string bowed at its center. What partials are present? What partials are absent?

2. What partials may be missing for a string bowed 1/4 the distance from one end? For 1/5? For 1/6? For 1/7?

3. Why is the spectrum of radiated sound for violins different from the spectrum of the string waveshape?

4. Where is the violin bowed? What partials (approximately) are missing?

5. Name other instruments with characteristics similar to those of the violin.

6. What are the approximate frequencies of the air resonance and the main wood resonance of the violin?

7. What would the air and main wood resonance frequencies be for the new treble violin, which is scaled an octave higher than the violin? For the new tenor violin, which is scaled an octave lower?

8. How do the standing waves on a violin string affect the interaction between the string and bow? How do inharmonic partials affect the interaction?

9. Why does shortening a string by fingering increase its frequency?

10. If you owned a violin of poor quality and research showed that the wood resonance was located too high, what would you do to lower the frequency of the resonance? How would you modify the instrument in order to change the frequency of the air resonance? (Better get expert help if you actually do it.)

11. In the viola and cello, the air and wood resonances are generally three or four semitones higher with respect to the frequencies of the open middle strings than for the violin. Can you think of any reasons why this might be expected? What can you predict about the tone quality of these instruments?

12. Point B of Figure 36.3 corresponds to Figure 36.4A. To what part of Figure 36.4 does point D of Figure 36.3 correspond? To what parts of Figure 36.4 does region A of Figure 36.3 correspond? Region C?

Demonstrations

1. "One String Violin," Freier and Anderson, page S-23.

2. Perform spectral analyses of violin tones produced with different bowing speeds and different bow positions.

3. Have a competent player demonstrate vibrato.

4. Bow the D-string of an inexpensive violin with masking tape covering the f-holes. Then repeat the bowing as the tape is removed.

5. Produce violin tones with and without a small mass attached to the bridge.

6. Damp different parts of the violin body with your fingers while different tones are being produced.

Further Reading

Backus, Chapter 10

Benade, Chapters 23,24

Hall, Chapter 11

Rossing, Chapter 10

Arnold, E.B. and G. Weinreich. 1982. "Acoustical Spectroscopy of Violins," J. Acoust. Soc. Am. **72**, 1739-1746.

Farga, F. 1961. **Violins and Violinists** (Frederick A. Praeger).

Gorrill, W. D. 1973. "Viola Tone Quality Study Using an Instrument with Synthesized Normal Modes," J. Acoust. Soc. Am. **54** 311(A).

Hacklinger, M. 1978. "Violin Timbre and Bridge Frequency Response," Acustica **39**, 323.

Hutchins, C. M. 1962. "The Physics of Violins," Scientific Am. **207** (Nov), 78-93.

Hutchins, C. M. 1967. "Founding a Family of Fiddles," Physics Today **20** (Nov), 23-27.

Hutchins, C. M., 1973. "Instrumentation and Methods for Violin Testing," J. Audio Eng. Soc. **21**, 563-570.

Hutchins, C. M., ed. 1975. **Musical Acoustics, Part I: Violin Family Components** (Dowden, Hutchinson, and Ross).

Hutchins, C. M., ed. 1975. **Musical Acoustics, Part II: Violin Family Functions** (Dowden, Hutchinson, and Ross).

Hutchins, C.M. 1977. "Another Piece of the Free Plate Tap Tone Puzzle," Catgut Acoust. Soc. Newsletter **28**, 22.

Hutchins, C. M. 1981. "The Acoustics of Violin Plates," Sc. Am. **245** (Oct), 172-186.

Hutchins, C. M. 1983. "A History of Violin Research," J. Acoust. Soc. Am. **73**, 1421-1440.

John, R. 1977. "Musical String Vibrations," The Physics Teacher **15**, 145-156.

Lawergren, B. 1980. "On the Motion of Bowed Violin Strings," Acustica **44**, 194-206.

Mathews, M. V. 1982. "An Electronic Violin with a Singing Formant," J. Acoust. Soc. Am. **71**, S43.

Mathews, M. V., and J. Kohut. 1973. "Electronic Simulation of Violin Resonances," J. Acoust. Soc. **53**, 1620-1626.

McIntyre, M. E., and J. Woodhouse. 1978. "The Acoustics of Stringed Musical Instruments," Interdisciplinary Sc. Rev. **3** (2), 157-173.

McIntyre, M. E., R. T. Schumacher, and J. Woodhouse. 1981. "Aperiodicity in Bowed-String Motion," Acustica **49**, 13-32.

Meyer, J. 1972. "Directivity of the Bowed String Instruments and Its Effect on Orchestral Sound in Concert Halls," J. Acoust Soc. Am. **51**, 1994-2009.

Moral, J. A. and E. V. Jansson. 1982. "Eigenmodes and Quality of Violins," J. Acoust. Soc. Am. **71**, S42.

Schelleng, J. C. 1968. "Acoustical Effects of Violin Varnish," J. Acoust. Soc. Am. **44**, 1175-1183.

Schelleng, J. C. 1973. "The Bowed String and the Player," J. Acoust. Soc. Am. **53**, 26-41.

Schelleng, J. C. 1974. "The Physics of the Bowed String," Scientific Am. **230** (Jan), 87-95.

Audiovisual

1. **Bowed String Instruments and Zithers** (3.75 ips, 2 track, 30 min, UMAVEC)

2. "Real and Synthetic Bowed String Tones" (GRP)

3. **The Cello, Part 1** (24 min, color, OPRINT)

4. **The Cello, Part 2** (24 min, color, OPRINT)

5. **The Violin, Part 1** (22 min, color, OPRINT)

6. **The Violin, Part 2** (26 min, color, OPRINT)

7. **The Great Violin Mystery** (30 min, NOVA 813, Station WGBH)

37. Plucked String Instruments

We have already considered two classes of vibrators called "forced" and "free." The bowed strings discussed in the last chapter are forced vibrators in which energy is continually being added to the string by the bow. The instruments to be described in this chapter are free vibrators in which energy is supplied to the string by plucking and the vibration is then free to die out. We will consider general requirements for plucked string instruments and some properties of a plucked string. Guitar, banjo, harp, and harpsichord will be considered as examples of different types of plucked string instruments. The chapter will conclude with brief historical notes.

The plucked string instruments may be divided into four classes: harps, lyres, lutes, and zithers. Harps are constructed in triangular form (Figure 37.1) from three main components: (1) the sound board, the upper edge of which leans on the player's shoulder, (2) the neck which contains the tuning pins, and (3) the fore-pillar, which supports the neck against the string tension. Strings of varying length are strung between the sound board and the neck. Lyres have strings running from a yoke supported by two arms to a tailpiece attached to a resonator. Although the strings are of equal length, they are tuned to different pitches. Examples are the box lyre constructed with a box-like resonator such as the ancient Greek cithara (Figure 37.2), and the bowl lyre using a skin covered tortise shell as the resonator. Lutes consist of a neck or fingerboard, and a body which acts as a resonator. The strings are attached at the lower end of the body, cross a bridge, and travel up the neck to a tuning peg. Although the word "lute" is a generic term describing a class of widely varying instruments such as the guitar, the Indian sitar, and the banjo, a specific instrument called the lute (Figure 37.3) was quite popular during the 16th and 17 centuries. The strings of the zither are stretched across and parallel to the body, which serves as a broad-band resonator. Although there are many different types of zither, one familiar type is the board zither, consisting of a flat wooden resonator with some three dozen strings attached. The four or five strings nearest the performer can be stopped on a fretted finger board to play melodies. The right hand plucks these strings with a plectrum while the left hand plucks the other strings as accompaniment. Other examples of zither instruments are the harpsichord (Figure 37.4), the autoharp, and the Japanese koto.

Figure 37.1. A harp. (Courtesy of G. R. Williams.)

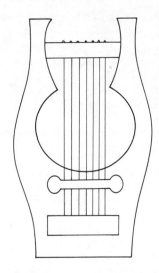

Figure 37.2. Sketch of a box lyre. (After Baines, 1969.)

General Requirements

The plucked strings, as one class of percussive strings, share many common features with bowed strings. A string fixed at both ends is the primary vibrator, all partials are present (with some exceptions), most of the

Figure 37.3. Guitar, banjo, and lute. (Courtesy of G. R. Williams and D. B. Farnsworth.)

Figure 37.4. Harpsichord. (Courtesy of G. R. Williams.)

energy is radiated by the body, and the body imposes its own resonance features on the radiated sound. The common feature of the plucked strings is excitation by plucking of strings.

As discussed previously, the string plays a very important role in determining when the bow adds energy to the string of the bowed string instruments. Having a string with nearly harmonic partials is necessary, so that the slipping and sticking action of the bow is triggered adequately. Inharmonic partials are of less consequence for a plucked string as the string motion is not used to control the plucking action. The plucking action excites the various modes of the string, which are then allowed to decay. The resulting tone quality depends on the plucking position, the nature of the plucking device, and the relative inharmonicity of the natural string fre-

quencies.

As noted in the preceding chapter, a string is an inefficient radiator and must be coupled to an effective radiating body to form a useful instrument. The bodies of guitars and banjos, and the soundboards (or soundboxes) of harps and harpsichords impose their own resonance properties as they radiate the string energy to the air. Guitar and banjo bodies are designed to enhance certain frequencies; harp and harpsichord soundboards are designed for a more uniform response.

The Plucked String

When a string is plucked there is no periodic driving force to produce harmonic partials. Consequently, the partials may be slightly inharmonic because the string is free to vibrate. Plucked string tones have short attack times, no steady state, and variable length decays; they have this in common with the struck string tones. Since we are dealing with a string fixed at both ends, it will have natural modes as shown in Figure 10.1. Which modes are excited depends on the point of plucking and the nature of the plucking device. When a hard, narrow plectrum is used for plucking, a sharp bend in the string results. A sharp bend is an abrupt change in string shape and many higher partials are required to produce it. Hence, "hard and narrow plucking" gives rise to many high frequency partials. On the other hand, when a soft, broad finger is used for plucking it produces a more gradual bend in the string. "Soft and broad plucking" results in a reduced number of higher partials.

The spectrum produced on a plucked string depends on where the string is plucked. An example of the spectrum for a guitar string plucked one-quarter the length of the string is shown in Figure 37.5. Note that the fourth partial is absent. The spectrum can be understood in part by referring to the top of Figure 37.6 where we see a string plucked at one-fourth its length. Below this we see modes one through four for a string fixed at both ends. The amount of each mode required to build up the actual string shape depends on how close

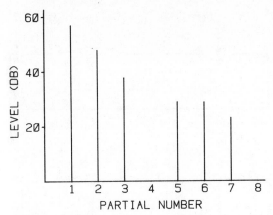

Figure 37.5. Spectrum of a guitar string plucked at one-quarter its length.

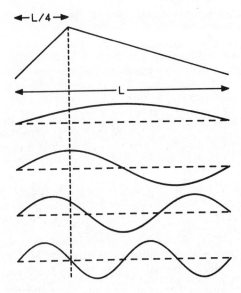

Figure 37.6. Excitation of first three modes of a string plucked at one-quarter its length. The fourth mode is not excited because it has a node at the plucking position.

the plucking point is to an antinode of that mode. When the plucking point lies at a node, as it does with the fourth mode, that mode is not excited.

The motion of a plucked string after its release is illustrated as the sum of two traveling waves moving in opposite directions, shown by the dashed lines in Figure 37.7. The dashed lines show the traveling waves and the solid line represents the sum of the two at any instant. The sequence of waves in Figure 37.7 shows the traveling waves and their sum (the observed wave) at successive eighth cycle intervals for half a cycle.

The Guitar and Banjo

There are several different families of guitars. The classical (Figure 37.3) and flamenco guitars are strung with nylon strings; the flattop and archtop guitars are strung with steel strings. The steel-stringed guitars typically produce louder sound than the others. Various bracing schemes (Figure 37.8) are used for the top and back plates of guitars to make them more rigid and to couple them to bridge vibrations more effectively.

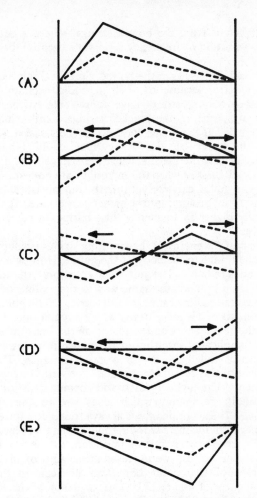

Figure 37.7. Motion of a plucked string after its release. The string shape (solid line) is the sum of the traveling waves (dashed lines) at any instant. Successive eighth cycle intervals are shown in A-E.

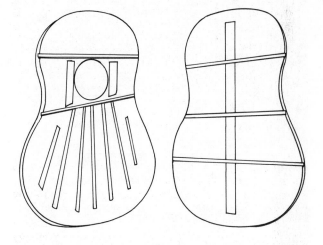

Figure 37.8. Sketch of bracing scheme for top and back of a guitar. (After Baines, 1969.)

The modern classical guitar has six strings tuned to E2, A2, D3, G3, B3, and E4. The strings are terminated at one end by the bridge and at the other end by the nut. The fingerboard is fitted with frets, which facilitate the production of definite frequencies. A player can produce one of the definite frequencies by stopping the string against the corresponding fret. This arrangement is quite different from the bowed strings, where a continuous variation of frequency is possible because there are no frets.

Forced strings, as in the bowed violin, exhibit harmonic spectra because of their periodic excitation, whether or not the strings have appreciable stiffness. Free strings, such as the plucked strings, exhibit harmonic partials if the strings have no stiffness and if the end points are exactly fixed. They may exhibit inharmonicity among the partials when the strings have appreciable stiffness or when the end points are not exactly fixed. The guitar exhibits very small, but measurable, amounts of inharmonicity; the higher partials are usually slightly higher in frequency than harmonic partials would be.

Most of the energy of the vibrating strings is coupled via the bridge to the body of the guitar, from which the sound is radiated. The guitar body "colors" the string spectrum in much the same way as the violin body influences the radiated spectra of its strings. The guitar body consists of a wooden cavity with a circular opening called the rose. The wooden plates of the instrument have a complicated spectrum of resonances; the lowest resonance is called the W (or wood) resonance. The placement of this resonance is crucial for well-designed instruments. The enclosed cavity and opening constitute a Helmholtz resonator, which gives rise to the air resonance. The air resonance is always found at a lower frequency than the main wood resonance, and its placement (in frequency) is just as important as the placement of the lowest wood resonance. Approximate resonances of the guitar body can be seen in the response curve of Figure 37.9. The frequency of the air resonance (about 90 Hz) is determined by the volume of air enclosed and the size of the rose.

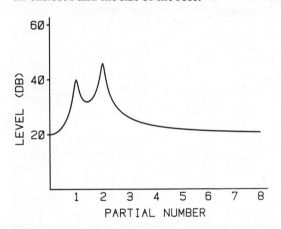

Figure 37.9. Approximate first two resonances of a guitar body. (After Caldersmith, 1978 and Firth, 1977.)

For the air mode, the air alternately moves in and out of the guitar body via the rose, and the air inside the body is alternately compressed and expanded. As air flows into the guitar body to excite the air resonance, the resulting compressed air in the body pushes outward on the top plate as shown in Figure 37.10A. In this case the air flow into the rose and the air flow resulting from the bulging of the top plate are out of phase and interfere destructively with each other, reducing the external sound pressure level.

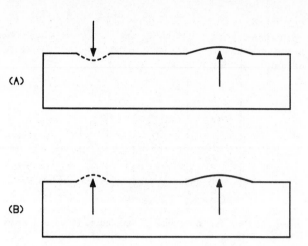

Figure 37.10. (A) Air motion at air resonance of guitar. (B) Air motion at first wood resonance of guitar. (After Rossing, 1982.)

The first wood resonance is characterized by an in-phase motion of the whole lower part of the top plate, to which the bridge is attached. This resonance occurs at a frequency near 180 Hz. The phase relationships of the air inside the body are, at this frequency, such that air flows out of the rose at the same time the top plate bulges outward as shown in Figure 37.10B. In this case the two flows are in phase and interfere constructively, increasing the external sound pressure level.

The expected radiated spectrum for a guitar tone can be obtained by adding the string spectrum of Figure 37.5 and the response curve of Figure 37.9. (Addition of the two is possible because they are expressed as signal levels.) Figure 37.11 shows the resulting radiation spectrum for a tone with a fundamental of about 84 Hz. Note that the second partial is increased in amplitude relative to the first because the wood resonance is larger than the air resonance.

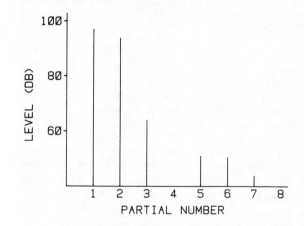

Figure 37.11. Idealized radiated guitar spectrum.

Plucked string sounds are characterized by an abrupt attack transient followed by an extended decay transient. The information in the decay transient is important in our perception of the tone. The time-evolving spectrum shown in Figure 37.12 illustrates some of these features of the decay transient for the guitar. Note that time increases from back to front in the figure, frequency increases from left to right, and sound level appears in the vertical. The guitar tone spectrum exhibits almost harmonic partials of rather long duration.

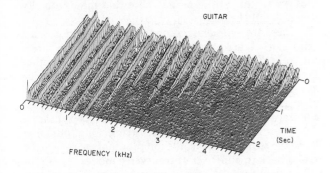

Figure 37.12. Time evolving guitar spectrum. (Courtesy of E. P. Palmer and I. G. Bassett.)

The banjo is another plucked string instrument. It has some features in common with the guitar, but its bridge is attached to a membrane rather than a top plate. A time-evolving spectrum of the banjo is shown in Figure 37.13. Its partials die out more rapidly than those of the guitar, which may account for its twangy tone character.

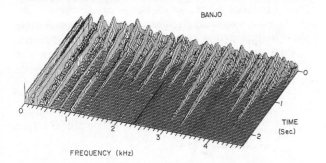

Figure 37.13. Time evolving banjo spectrum. (Courtesy of E. P. Palmer and I. G. Bassett.)

The Harp

The harp (Figure 37.1) is a plucked string instrument quite different in many respects from the guitar and banjo. It has no fingerboard and no frets. Its strings lie nearly in the vertical, with their upper ends supported by the neck. The lower ends of the strings are terminated in a soundboard, which forms the top of a soundbox enclosure.

The 47 harp strings are tuned in a diatonic scale (corresponding to the white notes on a piano). The strings are not fingered to produce higher notes, as with the violin and guitar. There are seven pedals on the concert harp, which activate string shortening mechanisms. Each of the seven pedals controls one set of key strings (C, D, E, etc.) and has three positions to produce a natural or a sharp. The mechanism which does this is located in the neck of the instrument, where the strings are supported by two sets of rotatable pins as shown in Figure 37.14. When the levers are in one position, the pins do not engage the strings and their full length is free

to vibrate. One lever combination rotates the upper set of pins so they engage, which shortens the strings enough to raise the pitch a semitone. Another lever combination rotates both sets of pins and raises the pitch two semitones. Thus two consecutive strings may be tuned to the same pitch, making a very rapid repetition of identical tones possible by using alternate strings. It is also possible to tune the entire instrument to a diminished seventh chord, so that when the fingers run over all the strings, only the tones of this chord will be present. The harpist can pluck the strings at different positions in order to produce different tone qualities. When a string is plucked at the center, odd harmonics are produced. As the plucking point is moved closer to the soundboard, a brighter tone is apparent.

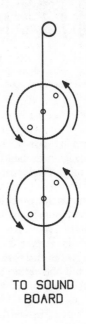

Figure 37.14. String shortening mechanism for harp. Activation of upper part of mechanism raises tone by a semitone. Activation of lower part of mechanism raises tone by two semitones.

Harp soundboards have been studied with many of the methods used for studying the violin and guitar. Response curves and Chladni patterns are two of the methods used. Figure 37.15 shows Chladni patterns for a soundboard to which straining and cover bars have been added. These bars play a role somewhat similar to the bass bar in the violin, and make the soundboard more rigid. We see that the lowest modes are due to standing waves along the length of the soundboard, as the accumulation of powder at nodal positions is across the soundboard. The 583 Hz mode shown in the figure has some standing waves across the soundboard. Similar patterns exist for a completed soundboard attached to the soundbox. In addition, a new mode appears, due to combined vibrations of the whole instrument. The action of various strings on the harp depends on the response of the soundboard to that string. Strings attached to a responsive soundboard position produce loud tones. Those attached at positions of small response produce weaker sounds.

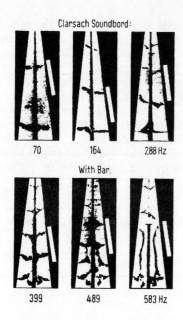

Clarsach Soundbord:

70   164   288 Hz

With Bar.

399   489   583 Hz

Figure 37.15. Powder patterns of six modes of barred harp soundboard. (From Firth, 1977.)

The Harpsichord

The harpsichord (Figure 37.4) employs freely vibrating plucked strings attached to a radiating body, as do the guitar and harp. Each string is used to produce a single note. The instrument is played from a keyboard which is used to actuate the plucking mechanism. The plucking action is more nearly invariant than that used with the guitar and harp. A simplified plucking mechanism is shown in Figure 37.16. When a key is depressed, a jack on its far end is raised. A plectrum attached to the jack pulls the string upward and releases it. The string is left free to vibrate while the key remains depressed. When the key is released, the plectrum (which is hinged to "open" upward) falls past the string and a damper contacts the string.

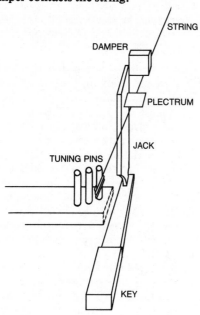

Figure 37.16. Simplified harpsichord action.

The string spectra for the harpsichord are governed by the same rules as those for the guitar and harp. The position of plucking and the nature of the plectrum determine missing partials and relative strengths of high frequency partials. There is relatively little inharmonicity of the various string partials, but that inharmonicity gives a slightly bell-like quality to the sound.

There are two polarizations or directions in which a string can vibrate transversely. If we choose one direction to be that in which the string is plucked, the other direction can be chosen to be perpendicular. Each of the two polarizations has its own set of modes which will have similar, but not identical, modal frequencies. When a string moves transversely, some of its transverse energy is used to lengthen the string longitudinally. When the string shortens, the longitudinal energy can be passed back to either set of transverse modes. This provides a way for both sets of transverse modes to be set into vibration, even though only one is initially excited. Beating of the decaying tone results when the corresponding modes of the two polarizations are slightly out of tune with each other.

The soundboard of the harpsichord is roughly triangular in shape, as shown in Figure 37.17. Ribs, shown by dotted lines in the figure, are glued to the underside. The figure illustrates some of the strings and their termination by the nut at one end and the bridge at the other. The position of the jacks (and thus of the plucking) also are shown. The soundboard usually forms the top of a box, which has a rose opening in its bottom. The resonance frequencies of the soundboard should be distributed evenly, so as to serve all strings. An even distribution is achieved by partitioning the soundboard into various-sized regions with the rib placements.

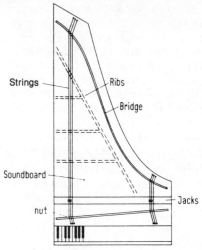

Figure 37.17. General arrangement of mechanical features in a harpsichord. (From Fletcher, 1977.)

Tonal coherence among the strings may be achieved by controlling the spectrum, decay time, and sound level. Because the ear is most sensitive to frequencies of up to 3000 Hz, it is important to have energy present up to this frequency for all strings. In referring to Figure 37.17, we see that the lower note strings are excited closer to one end, which will result in more harmonic development. But with lower fundamentals, more harmonics must be developed to get partials lying in the 3000 Hz range. Sound level and decay time are controlled by string thickness. Decay time is generally longer in the bass of the harpsichord, and is about four times as long for the lowest string (about 50 Hz) as for the highest string (about 1500 Hz). The decay time is controlled by scaling the string diameters from about 0.6 mm at the low end to about 0.2 mm at the high end. Brass is used as string material in the low end, and steel is used in the high end.

Historical Notes

It seems likely that the musical bow, one of the earliest musical instruments, evolved because of some musical quality of the hunting bow. Musical bows are of great antiquity (although still in common use by certain aboriginal tribes), as can be ascertained from the famous cave drawings of Les Trois Freres, France. There among the paintings of bison, elk, and other animals is a painting of a man playing a musical bow, with his mouth used as the resonator. As the popularity of the musical bow spread over the ancient world, accounts of it appeared in the legends of ancient Japan, Greece, and Africa. Gourds, boxes, and even shallow pits in the ground were used as resonators.

At some time in antiquity the flexibility of the musical bow was extended by adding a second string. Compound musical bows were constructed in a variety of ways; it was but a small step from the rigid compound bow to the bow harp, shown in Figure 37.18. The bow harp, which appeared in Egypt and Sumer about 5000 years ago, has its strings inserted into a box-like resonator, often covered with a layer of dried skin. About 4000 years ago the angle harp (Figure 37.18) appeared in ancient Egypt and Assyria. Angle harps differ from bow harps in that they have a sharp elbow-like angle, more strings, and sometimes no resonator. The more familiar triangular frame harp (Figure 37.18) originated in Europe during the Middle Ages. The pillar added between the neck and resonator strengthened the frame, allowing more strings and at greater tension. The early frame harps were small (1 m high) and portable, but they gradually increased in size and complexity. By the end of the 17th century, a set of mechanical hooks was being used to shorten the strings by a semi-tone. This device was the direct precursor of the modern harp mechanism described earlier in this chapter.

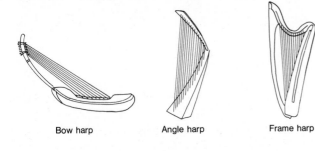

Figure 37.18. Three different kinds of harp. (After Baines, 1969.)

Bow harp   Angle harp   Frame harp

The box lyres, like the early harps, date back at least 5000 years. Several box lyres have been found in the archaeological diggings at the royal cemetary of Ur. They range throughout the Middle East and are still found, virtually unchanged in form, in Ethiopia. The "harp" used by the biblical King David was undoubtedly a box lyre similar to the Greek cithara.

Although no one knows when the first lute was constructed, there is ample evidence that long-necked lutes existed in Mesopotamia and Egypt during the third century B.C. The easiest way to build a long-necked lute is to drive a long stick through a leather-topped gourd and then attach strings. The short-necked lute, the ancestor of the modern lute family, seems to have orginated near Samarkand in Central Asia around the 8th century B.C. "El'ud" (the Arabic name for the lute) became, in the Middle East, the 'ud, a type of lute still popular throughout the Mediterranean region. The anglicized version became the lute of the 16th and 17th centuries. The Spanish version, called the qitara, became the guitar.

According to legend, the ch'in, a seven-stringed Chinese zither, was invented by Fu-hsi in 2900 B.C. From that time until the present, it has been very important to Chinese scholars. In western culture the zither instruments are descendants of the psaltery, which evolved from the harp during the 10th century A.D. Psalteries might be described as harps with a flat sounding board and two long bridges to couple the strings to the sounding board. From the 12th century on, the instrument was plucked with a pair of quill plectra. Sometime during the Renaissance a keyboard version of the psaltery, called the harpsichord, was constructed. By the 16th century, the harpsichord had evolved to a form which changed little during the next two centuries. At this time it was the principal keyboard stringed instrument. During the first half of the 18th century the harpsichord attained its fullest development — musically and technologically. During the latter part of the 18th century the harpsichord was replaced by the pianoforte, but harpsichords have recently regained popularity and many new instruments are being built.

Exercises

1. Why are violin partials harmonic when bowed but guitar partials are slightly inharmonic?

2. What would happen to the violin partials if the instrument were plucked instead of bowed? Why?

3. Fill in Table 37.1 with the frequencies of harmonic guitar partials and the "stretching" of the real partials from their harmonic values.

Table 37.1. To be completed as Exercise 3.

Partial	Real guitar	Harmonic guitar	Stretching of guitar partials
1	145	145	
2	292		
3	436		
4	—		
5	728		
6	871		
7	1018		
8	—		
9	—		
10	—		

4. The spectra shown in Figure 37.19 resulted from a violin string plucked at its midpoint and again at

a distance one-eighth the length from the bridge. Which spectrum corresponds to which plucking? Can you approximate the air and first body resonances from the spectra? Refer to Chapter 36.

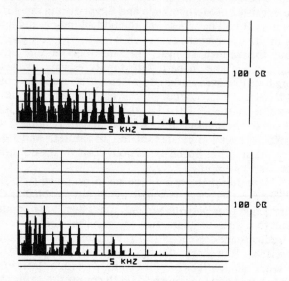

Figure 37.19. Spectra resulting from a violin string plucked at distances one-half and one-eighth its length from the bridge.

5. What is the approximate frequency of the air resonance for the guitar?

6. Superimpose the guitar body resonance curve given in Figure 37.9 on the spectrum of guitar strings given in Figure 37.20 to determine the spectrum of the radiated sound.

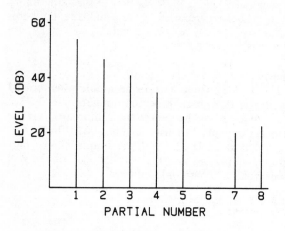

Figure 37.20. Spectrum of a guitar string plucked at a distance one-sixth its length from the bridge.

7. The modern harp has 47 strings tuned diatonically (white notes on a piano). How many octaves does it cover? If its lowest note is C1, what is its highest note?

8. If a harp string is 600 mm long, by how much must the rotating pin mechanism shorten the string to produce a note a semitone higher?

9. Explain why the tonal quality of a plucked harp string changes for different plucking positions.

10. Refer to Figure 37.17 and estimate the relative position along the string where the plectrum acts for low-, mid-, and high-pitched strings. What partials will be weak in each case?

11. Name other plucked string instruments.

Demonstrations

1. Perform spectral analyses of guitar tones when a string is plucked at different points.

2. Perform spectral analyces of guitar tones with and without the rose hole covered.

Further Reading

Backus, Chapters 10, 13

Benade, Chapters 7, 9, 18

Hall, Chapter 9

Olson, Chapters 5, 6

Rossing, Chapters 10, 14

Caldersmith, G. 1978. "Guitar as a Reflex Enclosure," J. Acoust. Soc. Am. **63**, 1566-1575.

DiLavore, P., ed. 1975. **The Guitar** (McGraw-Hill).

Firth, I. M. 1977. "On the Acoustics of the Harp," Acustica **37**, 148-154.

Firth, I. M. 1977. "Physics of the Guitar at the Helmholtz and First Top Plate Resonances," J. Acoust. Soc. Am. **61**, 588-593.

Fletcher, N. H. 1977. "Analysis of the Design and Performance of Harpsichords," Acustica **37**, 139-147.

Audiovisual

1. **Plucked String Instruments** (3.75 ips, 2 track, 30 min, UMAVEC)

2. **The Guitar** (23 min, color, OPRINT)

38. Struck String Instruments

We have previously discussed forced strings and one class of free strings. The instruments to be discussed in this chapter represent a second class of free strings in which energy is supplied to the string by striking, and the vibration is then free to die out. We will consider general requirements for struck string instruments. The hammered dulcimer and clavichord will be considered as examples of small struck string instruments. The piano will be considered in some detail because of its great importance and wide use. Consideration of the piano will include the piano action, inharmonicity in strings, string motion polarization, coupled string motion, and piano tone quality. Brief historical notes will conclude the chapter.

General Requirements

The struck strings, as one class of percussive strings, share many features with bowed and plucked strings. A string fixed at both ends is the primary vibrator, all partials may be present, and most of the energy is radiated by the soundboard. Three examples of struck string instruments are the hammered dulcimer, the clavichord, and the piano. The common feature of these instruments is that of string excitation by striking.

As with plucked strings, struck strings are able to exhibit any inharmonicity that may exist among their modes. Plucking gives a string an initial displacement but zero velocity. The subsequent motion of the string results from the initial displacement. A very simplified view suggests that striking gives a string an initial velocity but zero displacement. The subsequent motion of the string results from the initial velocity. Struck strings typically have more high frequency energy than plucked strings. The tonal quality depends on striking position, the nature of the striking device, and the relative inharmonicity of the string modes.

Strings of the struck string instruments require a soundboard for efficient radiation of sound. Ideally, the soundboard would provide a uniform response for each string so that tonal quality would be uniform across the range of the instrument. The fairly uniform sound pressure levels produced by different strings indicate that a uniform response has been achieved.

The Hammered Dulcimer

The hammer dulcimer (Figure 38.1) is a multiply stringed instrument having a flat soundboard. The strings are struck with small hammers. The hammered dulcimer is not to be confused with the plucked "dulcimore," nor with the "dulcimer" mentioned in the Old Testament, which actually refers to a type of bagpipe. In the construction of a contemporary hammered dulcimer, the treble strings are arranged in a course with three strings per note. A treble bridge divides each note in the treble course into two tuned

Figure 38.1. Hammered dulcimer. (From G. R. Plitnik collection. Photo courtesy of G. Orvis.)

lengths a fifth apart, with a 3:2 ratio of string lengths. The strings slope away from the central bridge and a bass course of strings, with two strings per note, travels below the treble bridge to a raised bass bridge on the right side. The diagram in Figure 38.2 summarizes the main parts of the hammered dulcimer. The strings are tuned by turning a pin with a tuning hammer. Since the soundboard provides a broad-band wood resonance the decorative soundholes serve no acoustical function.

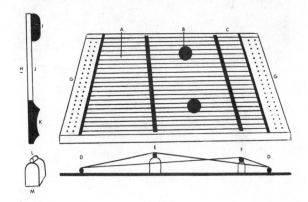

Figure 38.2. Hammered dulcimer components. A. Soundboard. B. "Sound holes." C. Frame. D. Side bridge. E. Treble bridge. F. Bass bridge. G. Pin block. H. Hammer. I. Head of hammer. J. Shaft of hammer. K. Handle of hammer. L. Bridge cap. M. Individual type bridge (detail).

As the left and right part of each string is tuned a fifth apart, some notes may be played from two different positions, giving the player additional flexibility. There are no dampers on the dulcimer, so each string continues to sound after being struck. The resulting cacaphony could obliterate the melody; different methods have evolved for damping the strings. The dulcimer players of the French Court were known to have let their long, laced sleeves drag across the strings as they played, thus providing a simple damping mechanism. The concert cymbalom is equipped with felt damping bars hooked to a pedal. Damping is usually not used with dulcimers played in the United States, as the sustained "ringing" is considered part of the charm of the instrument. When the player wishes to dampen an individual string, however, he presses it with the side of his hand.

One characteristic effect of the hammered dulcimer is the roll, or tremolo, accomplished by holding the hammer loosely and letting it bounce quickly and repeatedly on the string. This technique, which gives a distinctive sparkle to hammered dulcimer music, may be used to emphasize a note, or to prolong a sustained note.

The Clavichord

The clavichord is pictured in Figure 38.3; a top view is shown in Figure 38.4. The clavichord has a string and key arrangement in which the keys lie almost at right angles to the strings, unlike those of the piano and harpsichord. The strings are terminated by hitch pins at their left ends and by the bridge pins at their right ends when no keys are depressed. The bridge is attached to a small soundboard.

Figure 38.3. Clavichord (Courtesy of G. R. Williams.)

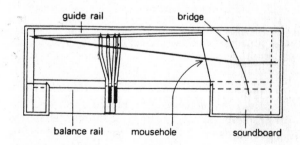

Figure 38.4. Main structural features of a clavichord as seen from above. The strings shown are the tenth pair, there being twenty pairs in all. (From Thwaites and Fletcher, 1981.)

The keys pivot on the balance rail so that, when a key is depressed, a brass tangent attached to the other end of the key mechanism strikes a string. The clavichord action is shown in Figure 38.5. While the key remains depressed, the length of string between the tanget and the bridge pin vibrates. The damper keeps the length of string between the tangent and the hitch pin from vibrating. When the key is released, the tangent breaks contact with the string and the damper causes all string motion to decay rapidly. The striking action in the clavichord is rather different from that of the piano. In the piano the hammer strikes the string and then falls away from the string, letting it vibrate. The sounding length of the string is between two pins. In the clavichord, the tangent strikes the string and remains in contact with the string during the vibration. The sounding length of the string is between the tangent and bridge pin.

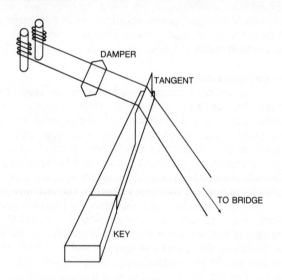

Figure 38.5. Simplified clavichord action.

The clavichord is very sensitive to player touch because the key-controlled tangent remains in contact with the string. The tangent rises several millimeters to contact the string, but the string can be deflected several millimeters after contact. The player can stop the string deflection by the feel of the key and thus control tone quality.

Several of the clavichord strings can be struck by more than one tangent and so can produce more than one note. This is possible because the tangent defines one end of the sounding length of the string; a playing range of about four octaves is thus obtainable with 20 strings. The brass strings are scaled partly by length and partly by diameter to cover the required frequency range.

For most notes, a pair of strings can produce louder tones. These strings are coupled to each other at the bridge, which results in the interesting coupled-string phenomena discussed later in this chapter. We simply remark here that two strings tuned to almost the same frequency can interact with each other via the bridge; this interaction produces beating and affects the rate of decay.

There is an enclosed air cavity between the sound-board and the body, which has an opening to the outside. As might be expected, there are air resonances and soundboard resonances in the clavichord as in other stringed instruments. We have treated air and wood resonances for other instruments as being isolated from each other. However, this treatment is not valid for the low air and wood resonances of the bodies of many stringed instruments. When an air and wood resonance are coupled (and interact strongly with each other), two combined resonances result, neither of which can be identified as "all air" or "all wood." These two resonances are farther apart in frequency than the air and wood resonances from which they originate.

The response curve of the clavichord soundboard is quite non-uniform at low frequencies. However, the sound level produced by all notes in the instrument's range is fairly uniform, except for some dropping at low frequencies. Apparently, note fundamentals that occur at soundboard response minima, and are thus not enhanced, have their higher partials enhanced by response maxima at higher frequencies.

The Piano Action

The primary vibrators of the piano are strings struck by hammers. The length, tension, density, and stiffness of the strings determine the frequencies of the various modes of vibration. The sound of the piano strings is mostly radiated by the soundboard, with the strings transmitting their vibrations to the soundboard through the bridge. The soundboard frequency response is uniform over the frequency range of the instrument.

The strings of the modern piano are made of steel wire; there are three strings per note for most of the piano's range. Since the strings of the top octaves would be too short if a constant tension were maintained for all strings, the strings of the upper octaves shorten by a ratio of about 1.9 to 1 for each octave, rather than 2 to 1. Also, the smaller strings have reduced diameters and slightly increased tensions. At the bass end of the piano the opposite problem exists: the strings would become inordinately long. Decreasing the tension or increasing the diameter of the strings would solve the problem. Decreasing the tension, however, would result in poor tone quality. Using a wire of larger diameter would also affect the tone quality, due to the increased stiffness of the wire. The solution is to increase the bass strings in mass, without an objectionable increase in stiffness, by wrapping the steel wire with fine copper or iron wire. The string layout in the modern baby grand piano can be seen in Figure 38.6. Notice how the bass strings are overstrung in order to save space and to bring them closer to the center of the soundboard. In order to maintain string tension in the modern grand piano, the tuning pin is driven into a pinblock consisting of about a dozen cross-grained layers of hardwood.

The moving parts of the piano involved in striking the string are called the action. Nearly all modern grand piano actions are descendants of the original action invented in 1709 by Cristofori; modern actions are more complex than the original, however, consisting of about 7000 separate parts. A comparison of Cristofori's action and that of the modern grand piano is presented in Figures 38.7 and 38.8. In the Cristofori action, (Figure 38.7) the key (1) is seen to pivot at point (2). The escapement or jack (3) then actuates the intermediate level (4), which delivers an impulse to the hammer (5). The back check (6) prevents the hammer from bouncing off the mechanism and hitting the string a second time. In the modern grand action (Figure 38.8), the key (1) pivots about point (2), raising the capstan (3), which causes the wippen (4) to rotate about point (5). The rotating wippen raises the jack (6), which is pivoted at the other end of the wippen. The upper end of the jack pushes on the roller (7), which is attached to the hammer shank (8), thus causing the hammer (9) to rise toward the string. Just before the hammer encounters the string, the lower end of the jack strikes the jack regulator (10), which causes the jack to rotate so that its upper end moves out from the under the roller, no longer pushing on it. Since inertia carries the hammer toward the string, the hammer rebounds and falls back until the roller rests on the repetition lever (11). The back-check (12) prevents the hammer from striking the string a second time. When the key is lifted slightly, a spring pulls the jack back under the roller, so that pressing the key again repeats the note. Thus, the grand piano action allows for rapid repetition of notes without having the key return to its starting position. This is why the grand action is superior to the upright action. A careful consideration of the action of an upright piano will show that a note cannot be repeated unless the key returns to its starting position.

The sustaining pedal allows the player to raise all of the dampers off all of the keys, so that all tones are sustained. The soft pedal, operated by the left foot, shifts the entire action (in a grand piano) to the right or to the left so that the hammer strikes only two of the three strings. In the earlier pianos, when there were only two strings per note, the soft pedal caused the action to strike only one of the strings; hence the term una corda, which still indicates use of the soft pedal in musical scores. In upright pianos, the soft pedal moves the hammers closer to the strings, thus reducing the volume of sound produced by decreasing the distance the hammer travels before encountering the strings.

Inharmonicity in Percussive Strings

In this section and the two following sections we consider three topics mentioned briefly in earlier chapters on strings. The three topics— inharmonicity of string modes, string motion polarization, and coupled string motion—are conveniently considered in relation to piano strings, where they have been studied more extensively. The first two topics are important to some extent for all string instruments. Coupled string motion is important in instruments using more than one string per note, the piano being a prime example.

We have assumed that strings are nearly perfectly flexible and that the only restoring force is the tension applied to them. Actual strings have some stiffness which provides restoring force (in addition to the tension), slightly raising the frequency of the string partials. This additional restoring force is greater for the higher partials of the string, which means that the higher partials are increased in frequency by a larger fraction than are the lower partials. Calculations indicate that the frequency increase is related to the square

Figure 38.6. Cutaway view of studio grand piano. Note "overstringing" of the bass strings. (Courtesy of Baldwin Piano and Organ Co.)

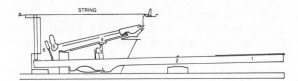

Figure 38.7. Cristofori piano action.

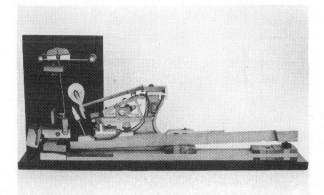

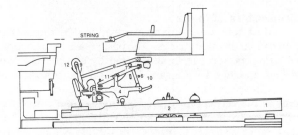

Figure 38.8. Modern grand piano action.

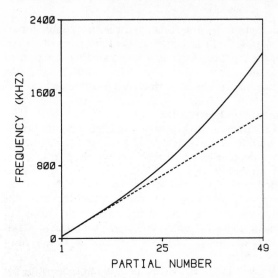

Figure 38.9. Inharmonicity of real piano tone partials (solid line) compared to harmonic partials (dashed line). (Data from Fletcher etal, 1962.)

of the partial frequency; consequently, the second harmonic will be shifted four times as much as the fundamental. Figure 38.9 shows the stretching of the partials for the piano tone. The fact that the partials of the piano tone become successively sharper has an important bearing on the normal piano tone. Piano tones were synthesized and played to a jury consisting of both musicians and nonmusicians, to study effects of inharmonicity. When the synthetic tones consisted of all harmonic partials, they were unconvincing and described as "lacking warmth," even though the attack and decay of the waveforms duplicated those of real piano tones. When tones were synthesized with inharmonic partials, however, the jury could usually not distinguish the real tones from the synthetic tones.

While a certain amount of inharmonicity is desirable in piano tone, too much inharmonicity ruins the tone. It has been found, for example, that the bass tones on the piano are better if the strings are made longer. When short, thick strings are used for the bass notes of a piano (as in the typical spinet piano), the inharmonicity is much greater because of the greater stiffness of the strings. Since the fundamental of low-frequency piano tones has very little energy, our ears rely on the difference tones produced by the upper partials to determine the pitch. The increased inharmonicity of short bass strings, however, causes the partials to be

stretched to such an extent that a single clear difference tone is no longer produced. The tone is then rough-sounding with poorly-defined pitch. This is why spinet pianos sound poor in the bass range, while the notes of a grand piano in the same range are more clear and sonorous.

The stretching of piano tone partials plays an important role in piano tuning. Since pianos are tuned by eliminating beats between octaves and fifths, the octaves will all be stretched slightly. The piano tuner, when tuning an octave, will eliminate beats between the second partial of the lower tone and the first partial of the upper tone. Since the second partial of the lower tone is slightly stretched, however, the octave will be stretched slightly from twice the fundamental frequency of the lower note when the beats are eliminated. The piano is not tuned to equal-temperament, but is stretched somewhat from this tuning. Figure 38.10 illustrates the tuning deviations from equal-temperament for a small piano. A piano tuned exactly to equal temperament sounds flat in the upper register.

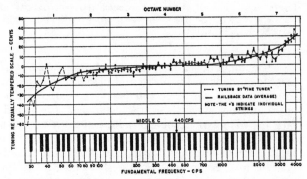

Figure 38.10. Tuning deviations from equal temperament for a small piano. (From Martin and Ward, 1961.)

The actual amounts of inharmonicity in a piano depend on how big the "stiffness force" is, relative to the "tension force" for restoring the string to its rest position. For piano strings, the stiffness force is a significant fraction of the tension force, and very noticeable inharmonicity results. For guitar strings (and most other strings), the stiffness force is a much smaller fraction of the tension force than for piano strings, and the resulting inharmonicity is much less pronounced.

String Motion Polarization

Consider a single piano string lying horizontally. Its sounding length is defined at one end by a tuning pin and at the other end by the bridge, which is attached to the soundboard. Most of the string energy that is radiated goes from the string via the bridge to the soundboard, and then into the surrounding air. The hammer excites the string in the vertical direction, but will excite some horizontal motion as well, due to hammer irregularities or hammer motion that is not completely vertical.

We will now discuss two possible polarizations (or directions) of transverse motion on a piano string. One of these is vertical in the direction of an "ideal" hammer strike, and the other is horizontal at right angles to the vertical, as shown in Figure 38.11. The string in the figure is seen from an end view as it passes over the bridge. The vertical and horizontal string motions will experience different responses from the bridge and soundboard.

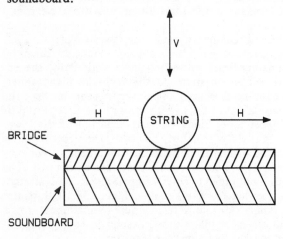

Figure 38.11. Vertical (V) and horizontal (H) polarizations of transverse string motion. The string is seen resting on the bridge which lies on the soundboard in this end view.

The soundboard can be viewed as a large, stiff diaphragm lying in the horizontal plane. It is designed to radiate sound via its up and down (vertical) motions. A piano string moving vertically will "see" the soundboard as willing to accept and radiate energy via its own vertical motion. This efficient radiation will result in the rapid decay of vertical string motion. The piano soundboard is not very responsive to pushes and pulls in the horizontal direction. Hence, a string moving horizontally will "see" the soundboard as unwilling to accept and radiate energy. This inefficient radiation will result in long lasting horizontal string motion.

Piano tones are characterized by rapidly decaying initial sound and more slowly decaying later sound as seen in Figure 38.12. This is consistent with the vertical and horizontal string motions just discussed. Suppose that the vertical motion is excited by the hammer with

10 times more amplitude than the horizontal motion. Sound due to the vertical motion decays rapidly, which accounts for the initial rapid decay shown in the figure. After the initial rapid decay, sound due to the horizontal motion dominates; it decays more slowly, which accounts for the final slower decay shown in the figure.

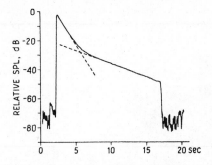

Figure 38.12. Time variation of relative sound pressure level from a single string. (From Weinreich, 1977.)

Coupled String Motion

Most notes on the piano are produced by the excitation of two or three strings. The basic features of coupled string motion can be illustrated by considering two strings attached to the same point on the bridge. Imagine that the strings are perfectly in tune and moving only in the vertical direction. Normally, one would expect them to be excited in phase by the hammer so that they move together. Another possibility would be to have them move out of phase with each other so that, as one moves up, the other moves down. The rate at which they give energy to the bridge is drastically different in the two cases, which affects tonal quality.

We first consider the in-phase motion of the two strings, in which they move up and down together. The two strings acting on the bridge will produce twice the force that one alone would produce, resulting in the bridge moving with twice the velocity. Recall from Chapter 3 that power loss (the rate at which energy is lost) is the product of force times velocity. This means that two strings will give four times the power to the bridge that one string would give. After being struck, two strings would start with twice the energy of one string, but they would give it up twice as fast. Hence, the decay time of two strings moving in phase will be much shorter than that of one string moving alone.

Now consider the out-of-phase motion of two strings. When one moves up, the other moves down so that, as one applies a positive force to the bridge, the other applies a negative force. The sum of the two forces is zero, the bridge velocity is zero, and the strings supply no power to the bridge. If there were no other losses, the strings could vibrate indefinitely. In an actual piano the two strings would not be perfectly in tune. If they were started in phase, their initial decay would be rapid, but as they got out of phase (due to being out of tune) the decay would become more gradual. If the strings were not excited with identical amplitudes and motions, the situation would be even more complicated.

A piano tuner can adjust the tuning of coupled strings, and may use this as a method to control the decay rates of initial and later sound, maintaining some uniformity across the range of the piano.

Piano Tone Quality

We have mentioned that a partial having a node at the point of excitation is expected to be missing or very weak. The hammers of the piano are arranged so that they strike their respective strings at a distance of about one-seventh the length of the string. It has frequently been asserted that this is done in order to eliminate the seventh harmonic from the tone. However, recent research has shown that the seventh partial is present with an intensity comparable to the other partials (see Figure 38.15). Since the hammers do not strike the string at a point, but rather make contact with a certain length of string, the above conditions for missing partials do not apply exactly. Also, the hammer remains in contact with the string for a fraction of the whole vibration period. During this time, vibrations occur in the short length of string between hammer and tuning pin; the frequencies of these vibrations are equal to those of the missing partials. One-seventh the length of the string has been designated as the point where the hammer strikes, as the resulting tone quality is good.

To achieve a clear sound on the piano, it is important that the hammer heads be properly voiced; that is, they must be smooth and uniformly shaped. The degree of hardness or softness of the hammers is usually dictated by personal tastes. The harder the felt, the more brilliant the sound while hammers which become too soft result in dull sound. The hardness of the felt heads is generally increased by applying a hot iron to the surface. Pricking the surface with needles will make the head softer, but piano tuners caution amateurs not to attempt these changes without expert guidance.

It has often been asserted that, apart from the force of the hammer impact, the way that a pianist strikes the key is extremely important to the resulting tone. Experiments have shown, however, that when the force of the stroke is not changed, the tonal quality does not change. This was dramatically demonstrated in a set of experiments where every tone played by a skilled pianist was recreated by a cushioned weight falling on the piano key. A comparison of the waveforms for the tones produced in each case showed them to be identical. Although no amount of pressing or wiggling of the piano key can affect the tone quality (since the performer is out of contact with the hammer at the moment of impact), the pianist can control the speed of his finger in striking the key. The greater this speed, the greater the force of hammer impact on the string. This increases the development of upper partials, resulting in a richer and louder tone. Thus, a pianist can control the quality of sound only by controlling the speed with which the keys are struck.

A second factor is that the different notes of the sounded chord may be struck with different speeds, giving a different tonal quality to the chord as a whole. Also, the notes of the chord may be initiated at different instants of time, thus affecting the resulting sound.

A third factor in piano tone may be the noise of the hammer striking the string and the noise of the finger striking the key. The so-called "hard touch" may be due to a great deal of surface noise resulting from the finger striking the key from a height. Finally, the piano tone may be influenced at cessation by the manner in which the pianist releases the note. A sudden dropping of the felts may dampen the vibrations differently than releasing the key gradually, although more research is needed to ascertain if this effect is really perceptible.

A discussion of piano tone quality would be incomplete without mentioning the possible differences between pianos produced by different manufacturers. One investigation has found that comparable pianos by different manufacturers do indeed exhibit different spectra, as shown in Figure 38.13. Even more surprising is the fact that a grand piano and a spinet piano by the same manufacturer have remarkably similar spectra in the middle and upper range, although the dynamic range of the spinet is much more limited.

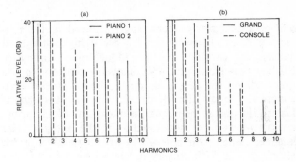

Figure 38.13. Comparison of piano tone spectra for: (a) two different makes; (b) grand and console of same make. (After Culver, 1956.)

The tone of the piano is without a steady state, but consists of a brief attack and a long decay, as illustrated in Figure 38.14. Furthermore, the spectrum of piano tones is very dependent upon pitch, the lower tones having many more partials than the higher notes. This is clearly shown in Figure 38.15, where the spectra of seven C's on a studio piano are displayed. Note the decrease in the number of harmonics for the higher notes, and also that the upper harmonics decrease more rapidly in sound level for the higher notes than for the lower. One study gave an acceptable range of about a 2-dB decrease per partial for low tones to as much as a 40-dB decrease per partial for high tones. It is also apparent that the seventh partial is present in most of these tones.

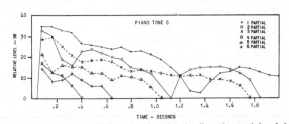

Figure 38.14. Relative level versus time for the first six partials of the piano tone C. (From Fletcher et al, 1962.)

Historical Notes

The direct ancestor of the hammered dulcimer is the psaltery, discussed in the previous chapter. One can

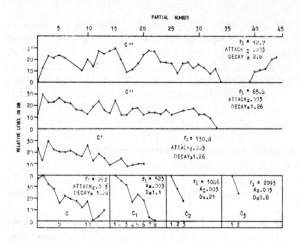

Figure 38.15. Average partial structure of a studio piano. (From Fletcher et al, 1962.)

imagine a medieval psaltery player discovering the unique sounds produced by striking the strings with small mallets. A carving in the Manchester Cathedral (England), built about 1450, shows a psaltery in the trapezoidal form of the dulcimer, being played with hammers. Today the dulcimer, known throughout Eastern Europe as the cimbal, is one of the most important non-keyboard instruments in the region. The dulcimer came to the new world with the early English settlers; by the 19th century dulcimers were so popular that factories were producing them in large quantities. During the early years of the 20th century, their use declined steadily, although they remained popular in Appalachian America. Today, with the renewed interest in American folk music, the hammered dulcimer has been "rediscovered."

The development of the piano began in 1709 when Bartolomeo Cristofori, an Italian harpsichord maker, invented a key-actuated hammer with an escapement mechanism, which allowed the hammer to bounce away from the string after striking. Loud or soft tones could be created by striking the key lightly or heavily—as opposed to ordinary harpsichords, which produce tones at constant loudness, regardless of the force exerted by the finger. Cristorfori called his invention "a harpsichord with loud and soft." The invention came to be known as the piano forte, from piano ("soft") and forte ("loud").

By 1726 Cristofori had invented all the essentials of the modern pianoforte mechanism, but his instrument had one severe disability: the frame was not strong enough to withstand the string tension required to give musical effect. Consequently, the instrument was not popular for almost 70 years; composers and musicians showed little interest. About 1770, however, Johann Stein developed what came to be known as the German action. Also, about this time some composers began playing, and thus helping to popularize, the pianoforte. Johann Christian Bach was one of the foremost. It was left to Mozart, however, to start the pianoforte on its next stage of evolution. During the latter part of the eighteenth century he evolved a style of keyboard composition for the piano which could not be adequately performed on a harpsichord. Meanwhile, Cristofori's

double action was being produced in England. It is the standard piano action used today, having replaced the German action.

The year 1800 was a major turning point in the development of the piano. About this time Beethoven began composing his piano pieces, which made increasing dramatic demands upon the instrument. In the early years of the nineteenth century, the development of the pianoforte centered around increasing the string tension and designing frames to withstand the greater forces thus required. In 1788 an English builder named Stodart experimented with metal reinforcing arches to help counteract the string tension, and in 1808 Broadwood began using metal bracing to strengthen the treble region, where the tension is greatest. The greatest development along these lines occurred in the United States when Alpheus Babcock of Boston patented a complete cast-iron frame, able to withstand greatly increased tension. About the same time, the demand for a more powerful response from the piano led to the use of steel strings, which replaced the earlier iron and brass strings. In 1826 Henri Pope introduced the use of felt, rather than leather, for hammer coverings; the felt retained its resiliency better than leather. By 1851 Broadwood had produced the first completely iron-framed grand piano, and in 1855 the Steinway metal frame (which has been the model for all successful frames ever since) was introduced. By 1859 Steinway had also perfected the system of overstringing, in which the bass strings run at an angle to the treble strings, thus shortening the piano case and also distributing the tension more evenly over the frame.

Meanwhile, the keyboard had been extended to over six octaves (C1 to E7) by 1840 and had become standardized at seven octaves (A0 to A7) by 1855. By this time the piano had essentially achieved its modern form. The double action had evolved into the double escapement, or repetition action, which is the direct ancestor of all modern grand piano actions. The middle pedal, which allows a given chord to be sustained, was invented in 1862 by C. Montal; the soft and sustaining pedals had come into almost universal use by 1790. Soon after 1880 the three top notes, which extended the piano compass to C8, were adopted; by 1885 the grand piano had achieved its modern form in all particulars.

The upright and spinet pianos came into vogue after 1890, partly to conserve space and partly to reduce production cost and increase sales. The piano continues to evolve, primarily in terms of materials for construction, with plastic, teflon, nylon, and aluminum being used in place of earlier materials.

Exercises

1. Where is the piano string struck? What partials should be missing? Are they? Why?

2. What are the resonance characteristics of a piano soundboard? How does this compare to the violin and guitar?

3. Suppose a piano were built with all strings made of the same material, having the same diameter and subjected to the same tension. If the top note (C8) had a string length of 8 cm, compute the length of the lowest C string (C1). Is this practical? Why?

4. How long would the string in Exercise 3 have

to be if mass per unit length were increased by a factor of four? If the tension were reduced by a factor of four? If both of the above changes were made?

5. When a chord is played on the piano while the sustaining pedal is depressed, the tone is richer than when the same chord is played without using the sustaining pedal. Explain.

6. Fill in Table 38.1 with the "harmonic" piano frequencies and the stretching of the real partials from the harmonic values. How does the stretching of piano partials compare with that of guitar partials in Exercise 37.3?

Table 38.1. To be completed as Exercise 6.

Partial	Real piano	Harmonic piano	Stretching of piano partials
1	393	393	
2	785		
3	1180		
4	1577		
5	1976		
6	2380		
7	2796		
8	3174		
9	—		
10	—		

7. If a piano string is struck by a narrow hammer at a point one-seventh its length, the seventh partial should be missing. Why? If the hammer remains in contact with the string for awhile, vibrations will exist in the short segment of string whose length is one seventh the length of the whole string. What will its fundamental frequency be, relative to that of the whole string? Does this help to explain the presence of the missing partial?

8. Name other struck string instruments.

Demonstrations

1. Place your finger on a piano key and, keeping contact between finger and key, depress the key with maximum speed. Now, continually reduce your speed until depressing the key no longer produces a tone. Describe what is taking place in the action when this occurs.

2. Try placing weights on a piano key to determine the force necessary to depress it. Do different keys on the piano require different weights? If possible, repeat the experiment on different pianos of different makes and draw some conclusions.

3. Try striking a piano string with a small mallet at different points and in different directions. How does the tone change?

Further Reading

Backus, Chapter 13

Benade, Chapters 8, 9, 17, 18

Culver, Chapter 11

Hall, Chapter 9

Kent, Part I

Olson, Chapters 5, 6

Rossing, Chapter 14

Blackham, E. D. 1965. "The Physics of the Piano," Scientific Am. **99** (Dec), 88-99.

Fletcher, H. 1964. "Normal Vibration Frequencies of Stiff Piano Strings," J. Acoust. Soc. Am. **36**, 203-209.

Fletcher, H., E. D. Blackham, and R. Stratton. 1962. "Quality of Piano Tones," J. Acoust. Soc. Am. **34**, 749-761.

Hart, H. C. 1934. "A Precision Study of Piano Touch and Tone," J. Acoust. Soc. Am. **6**, 80-94.

Kuerti, A. 1973. "What Pianists Should Know about Pianos," Clavier (May-June), 12-22.

Loesser, A. 1954. **Men, Women, and Pianos** (Simon and Schuster).

Martin, D. W. 1970. "A Concert Grand Electro-piano," J. Acoust. Soc. Am. **47** (Part 1), 131(A).

Martin, D. W., and W. D. Ward. 1961. "Subjective Evaluation of Musical Scale Temperament in Pianos," J. Acoust. Soc. Am. **33**, 582.

McFerrin, W. V. 1971. **The Piano—Its Acoustics** (Tuners Supply Co.)

Thwaites, S., and N. H. Fletcher. 1981. Some Notes on the Clavichord," J. Acoust. Soc. Am. **69**, 1476-1483.

Weinreich, G. 1977. "Coupled Piano Strings," J. Acoust. Soc. Am. **62**, 1474-1484.

Weinreich, G. 1979. "The Coupled Motions of Piano Strings," Sc. Am. **240** (Jan), 118-127.

Audiovisual

1. "Tonal Comparison of Pianos" (GRP)

2. **The Piano, Part 2** (21 min, color, OPRINT)

3. **Cybalom in Hi-Fi** (Janos Hosszu, Period Records RL 1912).

4. **The Hammered Dulcimer** (Front Hall Records FHR-01)

5. **Mountain Melodies: Tunes of the Appalachians** (Whorley Gardner Stereo OL-3-7-2)

39. Percussion Instruments

The plucked and struck string instruments discussed in the last two chapters are percussion instruments. Each has a string as its primary vibrator. The percussion instruments to be considered in this chapter may be divided into four classes according to their vibrator type: membranes, bars, plates, and bells. They can be further subdivided into definite pitch percussives (those producing tones with a definite pitch) and indefinite pitch percussives. We will consider general features of percussion instruments and then the membrane, bar, plate, and bell families will be discussed.

General Features

Percussion instruments are generally excited by striking. The mallets and sticks used for striking can be soft or hard and narrow or broad (in a two-dimensional sense). Hard, narrow mallets and sticks will produce high harmonics. Soft, broad ones will produce fewer high harmonics. The place at which striking occurs will partly determine which modes are excited most strongly. Many of the percussion instruments have two-dimensional vibrators, and so will have nodal lines rather than nodal points. When striking occurs at some point on a nodal line, the associated mode will only be slightly excited. A given mode will receive little excitation when a mallet or stick contacts the vibrator for an amount of time greater than the period of the mode.

Although most struck strings have natural frequencies that are nearly harmonic, piano strings do exhibit a significant amount of inharmonicity because of their stiffness. It might be expected that the partials of bars, plates, and bells will exhibit even greater inharmonicity because their restoring forces are entirely due to stiffness. Although an ideal two-dimensional membrane is quite flexible and has no inharmonicity due to stiffness, the partials are still inharmonic. This is due to the complicated standing wave patterns on a circular membrane. Vibrating plates and bells have inharmonic modes caused both by stiffness (such as exhibited by vibrating bars) and their two-dimensional structure (such as exhibited by circular membranes).

The percussive strings discussed in the last two chapters required a body or soundboard to provide efficient radiation. The vibrating surfaces of the percussion instruments discussed in this chapter generally have large vibrating areas, and so radiate the sound directly. Resonators are used with some vibrating bars to enhance the radiation.

Membrane Percussives

Membrane percussives employ as their vibrator a membrane stretched in two dimensions. Two-dimensional waves are set up in the membrane, which can result in complex standing wave patterns. The membrane percussives may be classified into three basic types, depending on their body type. The first type has some kind of tube for its shell (or body), and one or two drum heads. These shell drums produce a sound of indefinite pitch. Common examples are the bass drum and the snare drum (Figure 39.1). The second type uses a kettle or hemispherical bowl with a single membrane covering the top. The most common example of this type is the kettledrum (Figure 39.2), an instrument of definite pitch. The final group consists of an animal skin stretched on a circular hoop frame with no body. The tambourine is the best known example of these frame drums.

Figure 39.1. Bass drum and snare drum. (Courtesy of G. R. Williams.)

The kettledrum (timpanum) consists of a circular membrane stretched tightly over the open sides of a large, hemispherical bowl. The player sets the membrane into vibration by striking it with a felt-covered stick. The drum also has a mechanism for changing the tension on the membrane, which changes the frequency of vibration. The kettledrum is the only stretched membrane percussion instrument which has a tone of definite pitch. This pitch is determined by four factors: (1) the diameter of the membrane, (2) the volume of the bowl, (3) the tension on the membrane, and (4) the density (mass/area) of the membrane. The kettledrum is usually struck at a distance several centimeters from one edge, so that many natural modes of vibration are excited.

279

Figure 39.2. Kettledrums. (Courtesy of G. R. Williams.)

Table 39.1. Frequency ratios of the first six modes to that of the second mode for circular membrane "instruments." (Data from Rossing, 1977.)

Mode	Membrane alone	Membrane and air	Actual kettledrum
1	0.63	0.80	0.85
2	1.00	1.00	1.00
3	1.35	1.50	1.51
4	1.45	1.83	1.68
5	1.67	1.92	1.99
6	1.84	2.33	2.09

because it radiates sound efficiently. Thus, any excitation of the lowest mode will die out quickly and will contribute only an initial "thump" to the sound, as shown in the time-evolving spectrum in Figure 39.4.

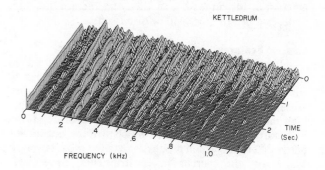

Figure 39.4. Time evolving kettledrum spectrum. (Courtesy of I. G. Bassett.)

Many different modes of vibration are possible on a two-dimensional membrane. We considered some of these for a square membrane in Chapter 10; only circular membranes are used for musical percussion instruments, however. We will now consider their natural modes. The eight lowest-frequency natural modes for a circular membrane with a fixed circumference are shown in Figure 39.3. The circular and diametrical lines represent nodal lines of zero displacement for the membrane. We note that the entire membrane moves as a unit in the first mode. It moves as two opposing parts in the second mode, and as four parts in the third mode. The fourth mode involves the motion of two opposing parts, concentric with each other. The fifth mode involves six parts separated by diametrical nodal lines and so on. The natural frequency ratios for the lowest six modes of a circular membrane are listed in the second column of Table 39.1. The frequency ratios show how the frequency of each mode is related to that of the second mode. We note that, in general, the modes bear a very inharmonic relationship to one another.

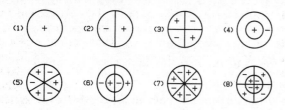

Figure 39.3. Eight lowest frequency modes of a circular membrane. The (+) and (-) signs show relative directions of motion.

The kettledrum has a definite pitch even though the membrane modes are inharmonically related. We find part of the reason for this by considering the effect that the air in the "kettle" has on the different modes. Imagine the excitation of the lowest mode in which the membrane moves as a unit. The motion of the membrane will tend to compress and expand the air in the "kettle," which will result in more stiffness for the mode and thus a higher frequency. However, the lowest mode is of limited importance for the kettledrum

The second mode is the first musically useful mode. Because the membrane moves as two opposing parts, this mode does not tend to compress and expand the air inside, but rather to push it back and forth from one side of the "kettle" to the other. The air in this case adds mass to the vibrator, thus decreasing the frequency. This mass loading affects most of the higher modes, but to a much smaller extent than it affects the second mode. So, the second mode is lowered in frequency by a greater factor than are the higher modes, which results in the frequency ratios shown in column 3 of Table 39.1 for an air-loaded membrane. (We assumed for the calculations, that the air loading increased the effective membrane surface density by 60%, 29%, and 20% for modes 2, 3, and 5, respectively. Modes 4 and 6 were assumed to be unaffected.) Frequency ratios actually measured for a kettledrum are shown in the last column. A comparison of the two ratios shows that the frequency-ratio differences between a plain membrane and an air-loaded membrane are due to increased mass resulting from the air loading.

The frequency ratios for the kettledrum are seen to be more "musically interesting" than the corresponding frequency ratios for a plain membrane. Note in the last column of Table 39.1 that, if we ignore the first mode, the frequency ratios of the second and fifth modes are roughly 1.0 and 2.0, which correspond to the fundamental and second harmonic partial. Furthermore, the frequency ratios of modes 2, 3, and 5 are about 1.0, 1.5, and 2.0. This is just an appropriate collection of frequencies to be the harmonics of a missing fundamental of 0.5. Either of these combinations gives rise to the perception of a definite pitch which we expect from a kettledrum. Apparently the fundamental, having a relative value of 1.0, is the one commonly associated with the pitch of the instrument.

The bass drum does not produce sounds of definite pitch, and its vibrations are more complex than those produced by the kettledrum. Not only does the instrument have the complex circular modes, but the two heads are tuned to slightly different frequencies, which results in some beating in the tone. When a drummer strikes one head, the vibrating air inside the drum causes the other head to vibrate; the vibrations of the two heads alternately add constructively and destructively. The time-evolving spectrum shown in Figure 39.5 illustrates the inharmonic modes of the bass drum. The effects of beating can be seen with some of the modes. (Note that time increases from front to back in the figure.)

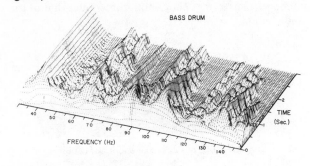

Figure 39.5. Time evolving bass drum spectrum. (From Fletcher and Bassett, 1978.)

An additional feature of interest with the bass drum is that, when it is struck hard, all of its natural frequencies are a few percent higher than normal at the beginning of the tone. This is because the large displacement caused by the hard blow produces an increase in the effective tension on the membrane, which results in an increase of all the natural frequencies.

Drum heads are usually made of calfskin, but in recent years manufacturers have been making plastic heads which are not as affected by temperature and humidity changes. The tone of these heads, however, is different, and some players do not care for them.

Percussive Bars

Instruments belonging to the percussive bar family include the glockenspiel, marimba, xylophone, vibraphone, chimes, and triangle. "Bars" are their vibrating element, in which the restoring force is due to internal stiffness of the bar and not to any externally-applied tension. The vibrating bars are one-dimensional vibrators, somewhat analogous to strings. (There are also torsional, longitudinal, and edge-wise modes not considered here.) When we discussed piano strings in Chapter 38, we observed that the partials were stretched and inharmonic because of the stiffness of the string. We might expect a much more exaggerated inharmonicity for vibrating bars.

The shapes of the first four modes of a uniform bar, free at both ends, are shown in Figure 39.6. Cor-

responding natural mode frequency ratios are listed in column two of Table 39.2. We note that the frequency ratios are very inharmonic, which results in a sound of indefinite pitch. However, several things can be done to adjust the frequency ratios and to produce a sound of definite pitch.

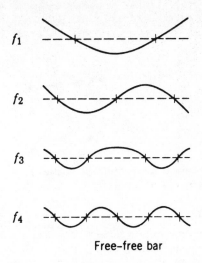

Free–free bar

Figure 39.6. Shapes for the first four modes of a uniform bar free at both ends.

Table 39.2. Frequency ratios for first six modes of bar vibrators. (Data from Rossing, 1976.)

Mode	Uniform Bar	Glockenspiel	Xylophone	Chime	
1	1.00	1.00	1.00	1.00	
2	2.76	2.71	2.97	2.79	
3	5.40	5.15	5.90	5.45	
4	8.93	8.43	—	8.95	(2.0)
5	13.34	12.21	—	13.20	(2.95)
6	18.64	—	—	18.16	(4.06)

The glockenspiel, or orchestra bells (Figure 39.7), employs nearly uniform rectangular metal bars, as shown in Figure 39.8a. The bars are supported by a felt rail at the nodal points for the lowest mode of vibration. Since the nodes of the lowest mode do not correspond to the nodes of the higher partials, the higher modes die out rather quickly, leaving the fundamental mode to linger on. Consequently, no adjustment of the higher mode frequencies is necessary. The pitch is associated with the frequency of the lowest mode. The higher modes contribute primarily to the initial decay transient of the instrument. The decaying partials can be seen in Figure 39.9, which shows a time-evolving spectrum for a glockenspiel note.

The marimba and xylophone (Figure 39.10) are close relatives. Both employ hard wooden bars (such as rosewood) as their vibrating elements. The xylophone, however, is always higher pitched and of shorter compass than the marimba. The use of wood as opposed to metal results in the characteristic "mellow" sound of these instruments. Since there is more internal friction in wood than in metal, the vibrations of a wooden bar die out much more rapidly. The wooden bars are made thinner in the middle than at the ends by undercutting, as shown in Figure 39.8b. Thinning a bar at a nodal

281

Figure 39.7. Glockenspiel. (Courtesy of G. R. Williams.)

Figure 39.8. Bar types: (a) uniform; (b) undercut.

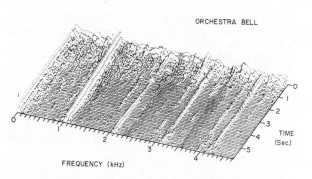

Figure 39.9. Time evolving glockenspiel spectrum. (Courtesy of I. G. Bassett.)

Figure 39.10. Xylophone. (Courtesy of G. R. Williams.)

point produces little change in frequency. Thinning a bar at antinodal points can decrease the frequency of the bar by decreasing the stiffness if the thinning is done in a region of maximum bending (e.g., mid-point of bar for mode 1). If the thinning is not at a bending point, the thinning reduces the mass and raises the frequency. Thinning results in a modification of the nodal points of the various modes, and in a significant modification of the frequency ratios of the modes. Typical frequency ratios for a xylophone appear in column 4 of Table 39.2. Note that the second mode frequency is about three times that of the first mode, and that the second mode, when sounding with the first, will produce a pitch associated with the first mode. The xylophone and marimba also have stopped cylindrical resonators (located beneath the bar) which are tuned to the fundamental frequency, enhancing the sound. The time-evolving spectrum of a typical xylophone tone appears in Figure 39.11. Note that the fundamental is the strongest component and that the next most prominent component has a frequency about three times that of the fundamental.

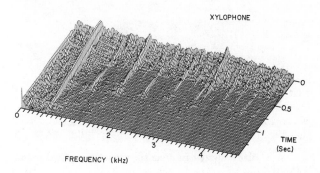

Figure 39.11. Time evolving xylophone spectrum. (Courtesy of I. G. Bassett.)

Two factors contribute to the fact that the prominent second partial has a frequency of almost exactly three times the fundamental frequency. First, the frequency of the second partial has been raised in relation to the first partial frequency by undercutting the bar. Second, the capped resonator, being an open-closed tube, supports resonances only at odd multiples of the fundamental frequency. Similar spectra for marimba tones show that the second and third prominent partials are tuned to frequencies nominally four and nine times that of the fundamental.

The vibraphone uses undercut metal bars in which the second mode is tuned approximately to the octave of the first mode. A unique feature of the vibraphone is a set of motor-driven discs which rotate over the resonator. By alternately opening and closing the top of the resonator, a tremolo effect is produced.

Chimes (or tubular bells) use cylindrical metal tubes as their vibrating elements. A plug is placed in one end of the tube, which modifies some of the natural frequencies. Frequency ratios for a chime are listed in column 5 of Table 39.2. If we give the fourth mode a relative value of 2 (in parentheses), we see that the fifth and sixth modes have relative values nearly equal to 3 and 4, respectively. These three modes produce frequencies with relative values of 2, 3, and 4, resulting in a perceived strike tone pitch with a relative value of 1 (i.e., an octave lower than the fourth mode). The first three modes contribute primarily to tonal quality, but not to the perceived pitch. We can observe these features in the time-evolving A4 chime spectrum shown in Figure 39.12. Note that the first partial dies out rapidly, the second partial lasts longer, and the third component is very long-lasting. Note that the fourth, fifth, and sixth partials have frequencies of about 880 Hz, 1300 Hz, and 1750 Hz, which result in a "pitch frequency" of 440 Hz.

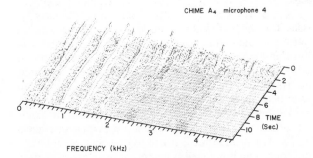

CHIME A₄ microphone 4

Figure 39.12. Time evolving chime spectrum. (Courtesy of H. Fletcher and I. G. Bassett.)

Of the smaller percussion instruments, the triangle is one of the most important. It consists of a steel rod bent into a triangular shape, suspended by a piece of leather to allow maximum vibration of all the modes. Since many of the inharmonic modes are excited at significant levels, triangles are indefinite pitch instruments.

Percussive Plates

Vibrating plate instruments, such as the cymbals (Figure 39.13) and tam-tam (Figure 39.14) are the two-dimensional extension of a vibrating bar and are also indefinite in pitch. In many respects, vibrating plates combine the complexities of membranes (due to two-dimensional wave motion) and bars (due to stiffness). If a solid circular plate is clamped at its center, thus making the center a nodal point, and then bowed on the edge, it emits a tone with many inharmonic partials. The plate vibrates in a series of segments, the simplest being when there are four segments and four nodal lines (producing the fundamental). The series of sketches in Figure 39.15 shows this type of vibration and several higher modes of vibration for such a plate. These are known as Chladni figures, named after the man who first obtained such figures by placing sand on the plate; the vibration makes the sand accumulate on the nodal lines, thus making them visible.

Chladni also formulated several laws of vibrating plates, which state that (1) the frequencies of two like-shaped plates which have the same nodal patterns vary according to the thickness of the plates, and (2) the frequencies of two plates of the same thickness and same nodal pattern vary as the inverse square of the diameters. Thus, for any particular mode of vibration, doubling the plate thickness doubles the frequency and doubling the diameter lowers the frequency by a factor of 0.71.

Figure 39.13. Cymbals. (Courtesy of G. R. Williams.)

Figure 39.14. Tam-tam. (Courtesy of G. R. Williams.)

Gongs are large circular sheets of metal having a deep bent rim and a central "knob" or dome. Gongs are tuned to a definite pitch and are often used in tuned sets, such as the bonang of Java and the cheng lo of China. When struck near the center, a gong produces an initial tonal burst which decreases in intensity, then slowly builds to a peak after which it decays for a considerable time. Some oriental gongs produce upward or downward glides in pitch after being struck.

The tam-tam, often confused with a gong, is made of thinner metal, has a shallow rim and no dome (Figure

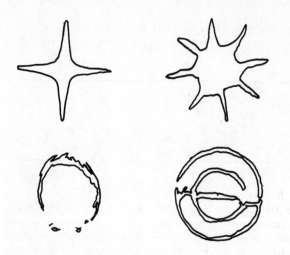

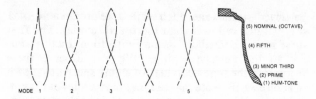

Figure 39.16. The first five modes of a carillon bell. (After Grutz-macher etal, 1966.)

Figure 39.15. Chladni patterns for a circular metal plate. (After Rossing, 1977.)

39.14). Tam-tams are most often of indefinite pitch and large size. The best of the large Chinese tam-tams have been made by a secret but very expensive process for centuries. The instruments have traditionally been engraved with a dragon; the number of claws proclaim the rank of its owner. The ingenious steel drums recently introduced in Trinidad and other Caribbean Islands are a set of tuned tam-tams. Sections of the bottom of a large oil drum are hammered into a set of concave bulges tuned to the notes of a scale. Instruments of treble and bass pitch are then played together as a "steel band."

Bells

Bells are acoustically related to vibrating plates, but they are curved and mass-loaded in the center. Like vibrating membranes and vibrating plates, bells have modes of vibration with two principal kinds of nodal lines: circular (which circle around the body of the bell) and meridian (which start at a lower edge, pass vertically up the bell to the top and down to the lower edge opposite their starting point). The number of vibrating segments of the bell is always even, with alternate segments moving in opposite directions. When the bell vibrates in four segments, the fundamental tone, called the **hum tone**, is produced. The fundamental frequency of the bell (as for a plate) varies inversely with the diameter and the mass—that is, the larger and heavier the bell, the lower the fundamental frequency.

After the bell has been cast, the various partials are tuned by machining metal from the inside surface with a lathe. By knowing the vibration pattern of the bell and carefully thinning the bell at the appropriate place, one can bring the bell into tune with itself. Figure 39.16 shows vibration patterns of the first five modes, as well as the areas on the interior surface which are important for tuning each mode. (The nodes shown in these patterns are circular nodes that pass completely around the bell.) When the bell is thinned along a nodal line, the frequency of the associated mode is affected very little. When the bell is thinned at antinodal points, the frequency of the associated mode may be increased or decreased (in a manner similar to bars), depending on

whether the mass or stiffness for a particular mode is affected.

There are four meridian nodes and no circular nodes for the bell hum tone, meaning that alternate quarters of the bell are moving outward and inward. Its vibrational pattern is shown in Figure 39.17A, in which nodal lines are represented by dashed lines for the side and overhead views. The next mode is called the prime mode because it has a frequency close to the so-called "strike tone." The prime mode, which has four meridian nodes and one circular node, is shown in Figure 39.17B. The prime tone is tuned to one octave above the hum tone. The hum tone is perceived as a lower-pitched hum following the initial strike tone. The next mode above the prime is called the third because it is tuned to sound a minor third above the prime tone. The next mode is termed the fifth because it is tuned a perfect fifth above the prime. Both the third and fifth have six meridian nodes. The next highest mode is the octave (or nominal), which has twice the frequency of the prime tone. The octave has eight meridian nodes. The strike tone, mentioned earlier, has traditionally been associated with the perceived pitch of the bell. The strike tone is thought to be determined by the octave and two additional upper modes: the twelfth and the superoctave. These three modes are tuned to a 2:3:4 frequency ratio, and the resulting strike tone is adjusted to a frequency close to that of the prime tone.

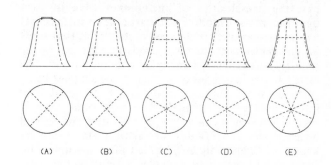

Figure 39.17. Sideviews and overhead views showing nodal lines (dashed) for first five modes of a bell.

It has long been considered that the pitch associated with the bell is that of the strike tone. The initial subjective perception of pitch is then reinforced by the well-tuned prime tone. Table 39.3 lists the measured frequencies (and frequency ratios) for the first five modes of three carillon bells. The carillon (Figure 39.18) located on the Brigham Young University campus, was built by the I. T. Verdin Company in 1975. A time-evolving spectrum of the A4 bell is shown in Figure 39.19 for comparison. From the spectrum and Table

Table 39.3. Frequencies and frequency ratios for first five modes of each of three carillon bells. (Courtesy of I. G. Bassett.)

Mode name	Frequency			Frequency Ratio		
	A3	A4	A5	A3	A4	A5
1 hum	220.0	439.2	881.0	1.00	1.00	1.00
2 prime	440.3	878.5	1763.1	2.00	2.00	2.00
3 third	523.6	1045.0	2095.6	2.38	2.38	2.38
4 fifth	662.3	1318.5	2649.2	3.01	3.00	3.01
5 octave	880.0	1755.9	3524.1	4.00	3.99	4.00

Figure 39.18. Carillon bell tower at Brigham Young University. (Photograph by M. A. Philbrick.)

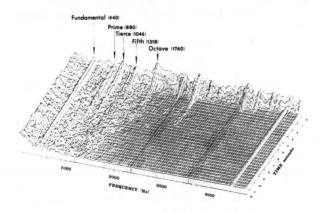

Figure 39.19. Time evolving bell spectrum. (Courtesy of I. G. Bassett.)

39.3 we discover several interesting facts. First, contrary to the usual assertions that bell pitch is determined by the strike tone (and prime mode), we see that the perceived pitch of these bells is in fact the pitch of the hum tone. Second, since there is no perceived pitch change as the bell is struck, the strike tone for these bells must be about the same as the hum tone. Since the second, fourth, and fifth modes have frequency ratios of almost exactly 2, 3, and 4 times the frequency of the fundamental (hum tone), the resulting perceived pitch will again be that of the hum tone. By comparing frequencies from the table, the three bells are seen to be quite accurately tuned to A3 = 220, A4 = 440, and A5 = 880 Hz, respectively.

Often, a fifth of a second or so after a bell is struck, a characteristic pulsating is heard. This variation in intensity is caused by beats due to slight differences in pitch between vibrating segments of the bell, probably resulting from nonuniformities in the mass distribution. This effect can be seen in Figure 39.19, along with the frequencies and decay rates of the various modes.

A carillon (Figure 39.18) is a set of at least two octaves of bells, very carefully tuned to chromatic intervals, and playable from a keyboard which permits control of expression through variation of touch. Although the carillon console resembles an organ console (with hand- and foot-operated levers rather than a keyboard and a pedalboard), the carillon action is more closely related to that of the piano. The console, although it appears to be clumsy, has been designed to permit the performer (without any assistance from electrical or pneumatic devices) to play, with sensitivity and virtuosity, bells weighing many tons. Because of the extremely high cost involved in manufacturing carillon bells and the large amount of space required for a carillon tower, chimes or electroacoustic devices have often been used to simulate bells. In Chapter 40 we will consider the electroacoustic method in more detail.

Handbells are often used in handbell choirs, where several are played simultaneously to create harmony. This can be done only because their upper partials are weak. The two principal modes of handbells have no circular nodes; one mode has four meridian nodes and the second has six meridian nodes. Some manufacturers tune the second mode frequency to three times that of the first. Others tune the second mode from 2.2 to 3.0 times that of the first.

Historical Notes

Although the perishable nature of membranes makes it difficult to determine the early history of drums, Mesopotamian artwork from about 3000 BC clearly depicts musicians playing these instruments. In

China, drums are traditionally said to have been invented in 2579 BC, and many types, shapes and sizes are still used there. Drums of many different styles and cultural uses have been developed over the centuries. Drums have flourished particularly in India and Africa, where they have traditionally been associated with ceremonies and magical power. The importance of drums in some primitive societies is demonstrated by the Banyankole people of Africa, whose drums are not only kept in a special shrine, but receive daily offerings of milk. The drums themselves actually own a large herd of cattle, the skins of which are used for making new drumheads.

Small pottery statues of women playing frame drums have been excavated from the ruins of Ur. Similar drums, with jangles attached, have been used by dancing women for centuries in the Middle Eastern countries. Eventually the instrument migrated to Europe where it became known as the tambourine. Today, 4000 years after Ur, the tambourine is still being played by the women of Gypsy ensembles and by Salvation Army bands.

The shell drums most likely had their origin in hollow logs capped with a membrane, while a membrane-covered clay pot may have been the ancestor of the kettledrum. A small prototype was known to have existed in Persia about 600 AD; the larger type of kettledrum dates to 12th Century Mesopotamia. The still-extant representations of these early instruments all indicate that the bowls had a flat bottom, presumably a carry-over from the clay pot ancestor. Large kettledrums reached Europe during the 15th century. Because of their large size they were played by riders mounted on camels or horses. During the 16th century the kettledrums were used for cavalry regiments and occupied an important position in many royal courts. In 17th century Germany, no one below the rank of Baron was allowed to own a kettledrum unless he had taken it into battle. The German drummers of this time were renowned for a spectacular playing style—a secret method imparted only to initiates of the Imperial Guild of Trumpeters and Kettledrummers. Court composers had the benefit of experiencing this art and, realizing the musical potential of these instruments, began to include them in orchestral composition. By the 18th century, kettledrums were considered part of the orchestra. Bach and Handel regularly called for the use of a set of tuned kettledrums. During the 19th century, as composers like Beethoven made greater demands of the tympani and used them in new and exciting ways, the instruments rapidly evolved into their modern form.

A tradition no less glorious accompanies the side drum, or snare drum, long associated with foot soldiers. The snare drum is a direct descendent of the Medieval tabor, but of larger size. Early representations of the Tabor show the single feature responsible for the unique, snappy sound of the snare drum: the snares. These consist of a group of wire strings stretched across the lower drum head. When the upper head is struck, the snares vibrate against the lower head to give the characteristic sound. The snare drum was used only occasionally in orchestral music during the 18th century, but by the latter part of the 19th century it had become firmly entrenchd as a legitimate member of the or-

chestral family.

The tuned percussion instruments may have originated in the late Stone Age with a prehistoric "lithophone" discovered in Indo-China. The set of tuned stone slabs show the typical flaking technique, whereby the rocks have been altered for tuning purposes. The xylophone, which probably originated in Africa, has a rather obscure ancestry. It may be descended from the primitive "pounding tubes": hollow bamboo pipes of different length which are tapped against the ground or against the thigh to produce sound. In a more advanced version, tubes of different length are pounded on a tree trunk laid on the ground. The use of metal bars for tuned percussion instruments dates to about 900 AD. By the 16th century the glockenspiel was in regular use in Germany; it soon found a place in the embryonic symphony orchestra. The marimba, an offspring of the xylophone, developed in Central America from a primitive xylophone brought across the Atlantic during the slave trade. The vibraphone was invented about 1920 in the United States.

Although bronze cymbals were common in Mesopotamia and China from the third millennium BC, the cymbals in use today are of Turkish origin. For over 300 years some of the finest quality cymbals have been manufactured by the Armenian family Zildjian (literally "cymbal maker").

Primitive bells were probably first constructed of stone during the Stone Age; the first metallic bells date to the Bronze Age. Throughout most of recorded history (until recently), bells were seldom employed as music makers. Their function seems to have been for religious ceremonies, as signal devices, or strictly utilitarian. The earliest extant bells date to 11th century Italy and Germany. These early European bells characteristically have many out-of-tune partials. The carillon bells (originating in the Netherlands during the 15th century) utilized tuned partials and could be used to play melodies or harmony, provided certain dissonant bell intervals were avoided.

Exercises

1. When a membrane is struck in the center, which modes will not be present? When a plate is clamped in the center which modes are not possible?

2. Sketch the fifth vibration mode for a vibrating bar.

3. Describe how the restoring forces associated with a stiff bar compare to the restoring forces in a piano string. Describe the relative importance of tension and stiffness in each case. What bearing does this have on the stretching of the partials in each case?

4. What observations can be made about the frequencies of the higher modes relative to the frequency of the lowest mode for nonpercussive instruments? For percussive string instruments? For vibrating bars? For vibrating membranes?

5. How do inharmonic partials affect waveshapes as compared to waveshapes created by harmonic partials? How do sound qualities differ?

6. Are there more or fewer modes present within a given frequency range for the tone of a percussive string as compared to a tone with harmonic partials? For a

bar? For a membrane? By how much do they differ?

7. Through comparison of chime and bell spectra (Figures 39.12 and 39.19), make arguments as to why chimes can often be used to simulate bells.

8. Determine the frequencies and amplitudes of the first six modes for the chime tone shown in Figure 39.12 right after the tone began. Determine the amplitudes of the modes at a later time. Which mode decays first? Which lasts the longest?

9. Repeat Exercise 8 using the time-evolving spectra of other instruments.

10. Describe bell modes 3, 4, and 5 in terms of nodal lines.

11. How many different vibrating portions are there in the first five modes of a drumhead? Of a bar? Of a plate? Of a bell?

12. Name other percussion instruments.

Demonstrations

1. Suspend a cast iron skillet (or any metal pan) from a string. Try striking it at different points and observe different tonal qualities. Is it definite or indefinite pitch?

2. Perform spectral analyses of drum tones produced when the drumhead is struck at different points.

3. Repeat Demonstration 2 for other percussion instruments.

4. Clamp a circular or square metal plate at its center. Sprinkle salt or sand or glitter on it. Excite it with a cello bow drawn across its edge. Vary the position of bowing.

5. Repeat Demonstration 4 but use a piece of dry ice held against the edge of the plate to excite it.

Further Reading

Backus, Chapter 14
Benade, Chapters 5, 9
Hall, Chapter 8
Kinsler etal, Chapters 3, 4
Olson, Chapters 5, 6
Rossing, Chapter 13

Bassett, I. G. 1982. "Vibration and Sound of the Bass Drum," Percussionist **19** (3), 50-58.

Brindle, R. S. 1970. **Contemporary Percussion** (Oxford).

Fletcher, H., and I. Bassett. 1978. "Some Experiments with the Bass Drum," J. Acoust. Soc. Am. **64**, 1570-1576.

Grutzmacher, M., W. Kallenbach, and E. Nellessen. 1965-66. "Acoustical Investigations on Church Bells," Acustica **16**, 34-45.

Hardy, H. C., and J. E. Ancell. 1961. "Comparison of the Acoustical Performance of Calfskin and Plastic Drumheads," J. Acoust. Soc. Am. **33**, 1391-1395.

Rossing, T. D. 1976. "Acoustics of Percussion Instruments—Part I," The Physics Teacher **14**, 546-555.

Rossing, T. D. 1977. "Acoustics of Percussion Instruments—Part II," The Physics Teacher **15**, 278-288.

Rossing, T. D. 1982. "Chladni's Law for Vibrating Plates," Am. J. Phys. **50**, 271-274.

Rossing, T.D. 1982. "The Physics of Kettledrums," Sci. Am. **247** (Nov), 172-178.

Rossing, T.D. 1982. Seven coauthored articles on various percussion instruments, Percussionist **19** (3).

Rossing, T. D., and H. J. Sathoff. 1980. "Modes of Vibration and Sound Radiation from Tuned Handbells," J. Acoust. Soc. Am. **68**, 1600-1607.

Schad, C., and H. Warlimont. 1973. "Acoustic Investigation of the Influence of the Material on the Sound of Bells," Acustica **29**, 1-14.

Taylor, H. W. 1964. **The Art and Science of the Timpani** (Baker).

Waller, M. D. 1961. **Chladni Figures: A Study in Symmetry** (Bell).

Audiovisual

1. **Vibrations of a Drum** (Loop 80-3924, EFL)

2. **Vibrations of a Metal Plate** (Loop 80-3932, EFL)

40. Electronic Instruments

When discussing families of musical instruments in Chapter 31, it was noted that instruments could be classified as belonging to one of three families: mechanical, electronic, or electromechanical. Mechanical (or acoustical) instruments in which there is no intervention of electronics have been discussed in Chapters 32-39. In this chapter we will consider instruments in which electronics play a role. We will first consider electromechanical instruments in which there are mechanical vibrating elements. Tone modifiers will then be considered as a special class of electromechanical instruments. Other tone generators and synthesis methods that can be used in electronic instruments will be discussed. Electronic organs will be discussed as practical examples of various synthesis methods. The extensive use of electronics for the reproduction of sound will be considered in Part VII.

Electromechanical Instruments

The basic elements of an electromechanical instrument are shown in Figure 40.1. The instruments belonging to this group are characterized by a mechanical vibrator; the motion of the vibrator is converted into an electrical signal by means of an appropriate transducer. The "body" filter consists of electronic circuitry designed to serve some of the functions of missing body parts. The amplifier produces adequate signal strength, and the loudspeaker converts the electrical signal into an acoustical signal. There are interesting examples from bowed strings, plucked strings, struck strings, and percussion.

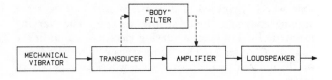

Figure 40.1. Elements of an electromechanical musical instrument.

An electronic violin is shown in Figure 40.2. The instrument is placed under the chin and bowed in a conventional manner. Electrical signals are produced when the metal strings vibrate in a magnetic field and then pass through the electronic body filter, which simulates the missing violin body. An early filtering scheme used many narrow filters spaced nonuniformly in frequency. A more recent scheme uses a single broad filter tuned near 3000 Hz. Some of its behavior was mentioned in Chapter 36. The electronic violin produces rather realistic tone quality. The strings behave somewhat differently than those of a normal violin, as they do not have to supply all of the radiated energy. The instrument is easier to play than the normal violin. When the loudspeakers are separate from the instrument, the ap-

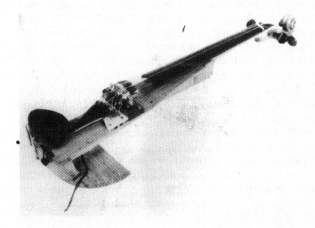

Figure 40.2. Electronic violin. (From Mathews and Kohut, 1973.)

parent position of the sound source is confusing as the player moves about.

The electric guitar (Figure 40.3) consists of six strings mounted on a heavy body that radiates little sound energy. The vibrations of the strings can be picked up at the bridge by contact microphones. Generally, however, electromagnetic pickups are used to sense the string motion and convert it into an electrical signal. The electromagnetic pickups are tiny coils of wire wrapped around an iron core. Several pickups may be placed at different points along each string. The position of the pickup determines how sensitive it is to a given mode. The electric guitar requires loudspeakers to radiate an appreciable amount of sound. Because of the amplification available, the instrument can compete in loudness with an ensemble of conventional instruments. The "body" filter may incorporate many different electronic tone controls. Tones produced by an electric guitar are typically long lasting because the string gives up very little energy to the body or to other sources of loss. Almost all of the radiated sound power is supplied electrically.

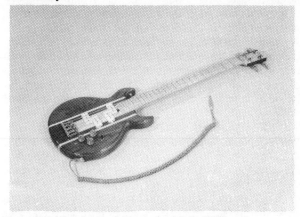

Figure 40.3. Electric guitar. (Courtesy of G. R. Williams.)

An early application of electronics to the piano was to remove the soundboard and use electronic pickups to capture and subsequently amplify the sound of the piano string. The sound produced by this method, however, had only the slightest resemblance to the tone of a piano. One reason is that the strings vibrate for a longer time when the soundboard is not present to dissipate the energy. A later development in electronic pianos utilized a vibrating reed struck by a hammer, with electronic pickup. The reed vibration decays in about the same time as a normal piano string, but the harmonic structure of the vibrating reed is quite different from that of the string. Thus, the resulting sound resembles a piano only in its decay characteristics. More recently, an electropiano has been developed which largely retains the conventional piano tone quality; it enables the piano to have a wider dynamic range, as well as new tonal capabilities. The strings are excited in the usual way (by striking), but the string vibration is transduced into an electrical signal, which is filtered in a manner similar to that of the soundboard before being amplified and radiated by loudspeakers. Such an "electronic soundboard" has several advantages and gives the performer much greater flexibility and control of the instrument.

The electronic violin, the electric guitar, and the electropiano all use strings as their primary vibrating elements, in a way similar to that of their conventional counterparts. Electronics are used to simulate conventional body properties. In the electronic carillon, small mechanical vibrators are used in place of any actual bell structure. This avoids the high cost of bells. The mechanical vibrators are in the form of bars or rods that have been tuned by proper shaping to have partials similar to those of a bell. Transducers are used to convert the bar or rod vibrations into electrical signals which are filtered, amplified, and radiated over loudspeakers.

Tone Modifiers

Any time a vibration is converted into an electrical signal, it can be electronically modified in many ways, including filtering and amplifying. All of the electromechanical instruments discussed above could also be classed as tone modifiers. Their primary classification, however, is not as tone modifiers because they are designed to have properties like those of their conventional counterparts. Consider the electronic violin, for example. It uses electronics as a means to simulate some part—the body in this case—of the conventional violin. A tone modifier may be considered an electromechanical instrument in which the emphasis is on tone modification and control rather than on simulation of a conventional instrument. The distinction is certainly not without ambiguity, but the classification is still of some use.

All instruments which permit electrical amplification might be classified as tone modifiers. However, usually some additional modification of tone is implied. An example of a tone modifier is the "electronic clarinet" illustrated in Figure 40.4. It consists of a conventional clarinet with a small microphone mounted to sense the pressure inside the clarinet mouthpiece and to convert it into an electrical signal. The tone modifier

part of this system consists of a frequency divider and a set of filters. The frequency divider makes it possible to obtain a tone one or more octaves lower than the tone played, so that a performer could provide self accompaniment with a simulated bass clarinet tone an octave lower. The filter provides a means for modifying the spectrum of the tone, thus enabling the production of different tonal qualities. If the frequency divider and filter are not used, the system can function as an amplifier for the clarinet tone. The resulting sound output will be the sound coming from the instrument directly plus the sound coming from the loudspeaker. This same approach is applicable to other wind instruments, such as the flute or trumpet.

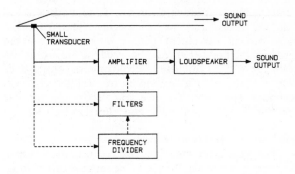

Figure 40.4. Tone modifier for clarinet.

Other Tone Generators

Instruments discussed to this point have mechanical tone generators (strings, reeds, etc.) and either mechanical or electrical filters. Early Hammond organs used a tone wheel (Figure 40.5) to generate electrical signals from a mechanical motion. The "toothed" tone wheel is rotated at a constant rate and causes a varying magnetic field at the pickup. The varying magnetic field induces a varying electrical signal in the coil (see Chapter 2). The frequencies generated depend on the number of teeth on a tone wheel and its rate of rotation.

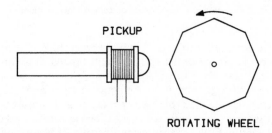

Figure 40.5. Rotating tone wheel with magnetic pickup used in Hammond organ.

At the present time the most extensive application of electronics in music is in "fully electronic" instruments as illustrated schematically in Figure 40.6. We see that the mechanical vibrator and transducer of Figure 40.1 have been replaced with an electronic

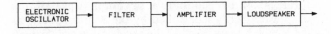

Figure 40.6. Elements of an electronic musical instrument.

oscillator in Figure 40.6. Otherwise the two "instruments" are essentially the same. We note that our "fully electronic" instrument requires an electrical-to-acoustical transducer —the loudspeaker—so that we can get sound from it. Electronic oscillators work in ways analogous to mechanical vibrators. The three important components of a mass and spring vibrator are mass, stiffness, and resistance; their electrical counterparts are inductance, capacitance, and electrical resistance, respectively. By creating electrical circuits with an appropriate combination of inductance, capacitance, and resistance, we can directly obtain oscillating electrical signals without having to use a transducer on a mechanical vibrator. Electronic oscillators and other electronic circuits are the building blocks for the various components of electronic instruments.

Synthesis Methods

With electronic building blocks at hand we must still decide how they are to be combined and used to produce the desired electrical signals. There are several basic synthesis methods in use, two of which are additive synthesis and subtractive synthesis. Some other methods will be discussed later. In **additive synthesis**, a complex signal is constructed by adding together various amounts of sinusoids of different frequencies. By the proper selection of frequencies and relative amplitudes, various tone colors can be achieved. (Recall that any complex musical tone can be constructed with frequency components of various amplitudes.) The additive synthesis process is represented in Figure 40.7. Additive synthesis has the primary advantage of flexibility. In principle, any complex signal conceivable can be synthesized by the additive method if a sufficient number of sine wave oscillators are used. For the most flexible synthesis, separate time variable frequencies and amplitude controls are used with each oscillator. The primary disadvantages of additive synthesis are the large number of oscillators required and the large amount of information needed to control them.

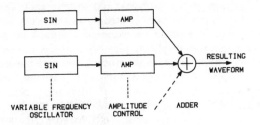

Figure 40.7. Elements of additive synthesis.

Subtractive synthesis starts with a tone which is rich in harmonics and modifies or attenuates these harmonics to produce the desired tone quality. The modifications are produced by various filters and other waveform modifiers. The original waveform (before filtering) is usually one which is easy to generate electronically, such as a sawtooth or a square wave. The subtractive synthesis process is illustrated in Figure 40.8. Two advantages of subtractive synthesis are the comparatively small number of components required, and the economy in information. In some ways, subtractive synthesis is more analogous to sound production in actual acoustical instruments. In the acoustical instrument, a vibrator (reed, string, etc.) produces an excitation rich in harmonics which is then selectively filtered by the body. In subractive synthesis a generator is used to produce waveforms rich in harmonics, which are filtered. However, electronically generated waveforms and electronic filtering are not identical with their acoustical counterparts. This leads to some lack of realism in the synthesis.

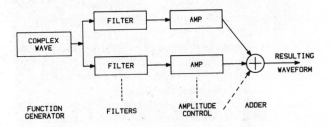

Figure 40.8. Elements of subtractive synthesis.

Several other synthesis methods have been motivated by the use of synthesizers and digital computers to be discussed in the next chapter. Sometimes even subtractive synthesis is considered too inefficient, and other less time consuming methods have been developed. One of these is **frequency modulation**, a process whereby the frequency of one wave is caused to vary by the instantaneous amplitude of another wave (see Chapter 41). Many additional frequency components are generated, especially when a complex wave is modulated. However, the additional frequency components generated cannot be easily specified beforehand. Even though the process is efficient, it lacks flexibility because there is no procedure that specifies what frequency modulation must to be used to achieve a particular result.

Another efficient synthesis method is that of **waveform distortion**, in which a waveform (such as a sinusoid) is distorted in some prescribed manner. A simple example of this would be the peak clipping of a sinusoid as described in Chapter 28. Many new high frequency components are generated in such processes. However, the method lacks flexibility, just as does frequency modulation, because there is no procedure that specifies what distortion must be used to achieve a given end.

One additional synthesis method is that of **waveform sequencing**, in which a waveform is stored in some form and then replayed on demand. This method is flexible and efficient, but requires a large amount of storage.

Electronic Organs

Electronic organs come in many sizes and styles, including small portable models, spinet models, church

291

models, and concert models. The smaller ones may have one or two abbreviated keyboards and a few foot pedals, while the largest ones may have three keyboards, a full pedal board, and numerous stops and controls. A typical church model organ is illustrated in Figure 40.9. Most electronic organs use subtractive synthesis; only a few use additive synthesis or waveform sequencing. As a key is depressed, an amplitude control must put an appropriate attack transient on the signal by amplitude modulating it. As the key is released, an appropriate decay transient must be supplied.

Figure 40.9. Baldwin model 645 classical electronic organ. (Courtesy of Baldwin Piano and Organ Co.)

The older Hammond organs using tone wheels provide an example of an instrument which uses additive synthesis. With this instrument, the performer has control over the relative amplitudes of nine independent harmonics, each having different frequencies but related to the fundamental. This enables the performer to reproduce the steady-state spectrum of many musical instrument tones, including the pipe organ. (The fact that the Hammond is unable to accurately reproduce such tones suggests the importance of factors other than steady-state spectra. Factors such as attack time, decay time, and ensemble must all be considered if a realistic representation of an actual musical instrument is desired.)

Subtractive synthesis is used in organs such as the Baldwin, Conn, Rogers, and others. Sawtooth, square wave, and triangular wave generators are used as sources. (Refer to Chapter 6 for details on waveforms and spectra.) A noise generator, producing a waveform with random consecutive values, is also used as a source. Low-pass, high-pass, and band-pass filters are used to produce desired spectra. As an example, suppose that we wish to synthesize an oboe-like sound on an electronic organ. It could be produced, according to

Figure 40.8, by choosing a saw-tooth wave of the desired frequency (controlled by the depressed key) as the complex wave, passing the saw-tooth wave through two band-pass filters (one peaking at 1000 Hz and one at 3000 Hz—see Chapter 33), and then adding the outputs.

Although considerable work has been done toward achieving realistic attack and decay times, the complex attack and decay transients of organ pipes cannot be duplicated inexpensively in electronic organs. Even though the steady state can be duplicated more effectively, it is not so important as the attack transient. One of the major deficiencies of electronic organs is the lack of an ensemble effect, as many models use only 12 tone generators for the entire instrument. A set of twelve master oscillators is used to generate the notes in a high frequency octave, which are then frequency divided to notes in the lower octaves. This results in all oscillators being mathematically in tune, with only small amounts of beating occurring (see Chapter 31). In comparing this to even a small pipe organ having hundreds of individual tonal sources, it is no wonder that the imitation is a rather poor one. Some organs use many individually-tuned oscillators which provide a more realistic ensemble effect, but the basic deficiencies of tone remain. Some of the more expensive electronic organs use several loudspeakers to produce a source with spatial properties more like those of pipe organs. The Allen organ uses a digital computer in a version of waveform sequencing in an attempt to produce sounds resembling those of a pipe organ. When the performer selects a desired tone quality, the sound is constructed from information stored in the computer. Although instruments such as these are a step toward removing some of the limitations on electronic simulation of pipe organ quality, most organists remain convinced that pipe organ quality has not been successfullly duplicated.

We have recently completed a preliminary study on whether organists can differentiate between electronic organ tones and pipe organ tones as presented by a tape recording. (Keep in mind that a pipe organ achieves a spatial effect which no number of well-placed speakers can duplicate, and so is placed at a disadvantage in a test of this kind.) The test involved tones from seven brand name electronic organs and nine different makes of pipe organ. The electronic instruments varied from new dealer demonstration models to instruments more than ten years old. The pipe organs varied from recently installed instruments to fifty-year old instruments. Only concert (or church) model electronic instruments were recorded. (Small electric entertainment organs do not make an attempt to duplicate pipe organ tones and were therefore of no interest to our study.) Some of the electronic instruments were situated in rooms with very little reverberation, as were some of the pipe organs. Likewise, many of the pipe organs and a number of their electronic counterparts were located in fairly reverberant churches. The test consisted of two parts; each part contained 100 tones (50 electronic and 50 pipe), with all families of organ tone (see Chapter 35) represented. Part one consisted of single notes (all C4); the listener was asked to identify them as being pipe or electronic. Part two consisted of a C-major triad (C4, E4, G4) which the listener was also asked to identify.

The test was administered to a dozen organists (students and faculty).

Although the number of subjects tested makes an insufficient basis for definitive statements, there are some interesting trends. In both tests organists could differentiate between pipe and electronic organs most of the time, but there was some degree of confusion (16-25%). The confusion factor was greater for the more expensive makes of electronic organ, and there was almost no confusion for the least expensive makes. Although the electronic reeds family of stops was the one most confused with pipe organ reeds, in no case did any electronic organ even begin to approach the 50% confusion level of guessing. 25% of the single electronic tones were called pipe tones, while 16% of the single pipe tones were called electronic. Apparently, although some organ stops on some makes of electronic organ do sound realistic, a real pipe does not sound like an imitation quite as often. For the major triad, the results were reversed—17% of the electronic tones were called pipe, while 23% of the pipe tones were confused with electronic tones. Even the more expensive electronics were confused less often on this test. One would think that the lack of real ensemble in the electronic organ tones would be apparent to the organist's trained ear. Why, then, were pipe organ tones, when played as a chord, confused with electronic tones more often?

While we cannot provide a definitive answer to the question at this time, we are prepared to venture a guess. We believe that one of the primary clues in differentiating pipe organ tones from electronic organ tones is the initial attack. When one listens to individual tones, the attack is very perceptible and, as we discovered in Chapter 31, the attack gives important information about the perceived tone. When a chord is played, there are three notes and as many as two dozen individual pipes sounding. Obviously, it is difficult to distinguish an individual pipe in such a case. We therefore conclude that, while the ensemble effect characteristic of pipe organs helps to make many of the electronic instruments sound "dead" in comparison, the same ensemble muddles an important acoustic clue (the attack of individual pipes) so that more of the pipe organs are confused with their electronic imitators.

Historical Notes

With the advent of the electronic age, the early years of the Twentieth Century saw the invention of a number of new electronic musical instruments. The first of these instruments to gain fame and some measure of general acceptance was the Theremin, invented in 1919 by Dr. Leon Theremin. The instrument consisted of two electronic oscillators and a loudspeaker contained in an upright box, to which were attached a horizontal loop antenna at the left side and a vertical rod antenna on the right. The player holds both hands in space; the proximity of the player's right hand to the rod determines pitch and the distance of the left hand from the loop determines loudness. Although this arrangement allows an adroit player to produce some very interesting effects, it is impossible to obtain distinct pitches without an intervening glissando. This disadvantage was eliminated in Les Ondes Martenot invented by a Frenchman, Maurice Martenot. Working in a wireless sta-

tion during World War I, Martenot noticed that occasionally radio tubes—then a new invention—would emit particularly clear and vibrant tones. After the war Martenot experimented with sound and electronics in order to control and utilize these sounds. By 1928 he had perfected the new keyboard instrument which bears his name. The Martenot may be played at the 7 octave keyboard, which gives discrete tones, or by a chromatic ribbon which (like the Theremin) yields continuous glissandi over 7 octaves. Further versatility is provided by several tone quality buttons and a loudness control. The loudness control may be activated by a key with the left hand, or by a lever if the left hand is needed on the keyboard. Although relatively few Martenots have been built, more than 350 composers have successfully used this interesting instrument in concertos, symphonic works, ballets, and film scores. The Ondes Martenot portended a revolution in the musical world, both in terms of the subtlety of expression and the new tone colors, which culminated with the invention of the synthesizer.

The Hammond organ, invented in 1929, was the first electronic instrument to be commercially successful. It's early start in a wide-open market and the distinctive sound of the Hammond were probably responsible for its popularity. The Hammond does not even pretend to imitate the sound of a pipe organ, but its low cost and relatively long lifetime have made it popular in many small churches. Its interesting tone colors and unusual effects make it useful for radio and T.V. programs; its fast attack makes it a favorite instrument for many popular music performers. In the decade following World War II, many electronic organs were developed with unusual methods of tone generation. Most declined into disuse; by the 1970's the industry had gone almost entirely to subtractive synthesis for tone generation. The one major exception is the Allen Organ Company which, since the late 1960's, has been totally committed to tone generation by waveform sequencing using digital computers. This is a very original method of tone generation, and the most sophicated technology in use today among the various manufacturers of electronic organs. However, sound of comparable quality can be produced by other high-priced electronic organs.

During the 1950's, many serious composers felt the need for new sonorities in the symphony orchestra. One method in vogue during this time was to use a tape recording, prepared in advance by the composer, as a source of unusual tonalities. Diversified sources of sound were used, including electronically-modified conventional instrument sounds, to produce new and different sounds. Through tape splicing and tape speed variation, a breadth of range and tonal complexity impossible with conventional instruments, was produced. With this technique, piano tones can be played two octaves lower while tympani tones can be played in the upper flute range. An interesting example of this technique is found in the "Rhapsodic Variations for Tape Recorder and Orchestra" by Luening and Ussachevsky, first performed in 1954.

Exercises

1. Why is the purely electronic generation of a piano tone difficult? (See Chapter 38.)

2. Suppose one wanted to synthesize bassoon-like sounds. Which electronic components would be used? Draw a diagram to indicate how they would be combined.

3. Draw a block diagram showing the combination of components that would be most useful for synthesizing flute-like sounds. (See Chapter 34.)

4. Suppose you are to synthesize an oboe-like tone as described in the text. However, you are to use additive synthesis rather than subtractive synthesis. How many oscillators will be needed for a tone of 500 Hz? How will the amplitude controls be adjusted?

5. Sketch a tone modifier for a flute. Where might the transducer be placed?

6. Figure 40.4 is a diagram of a tone modifier system for a clarinet. How could the frequency divider be used to change the sound? How could the filters be used to change the sound?

7. Suppose you are given the task of creating the "ideal" organ. How will you provide control of tone color, transients, and ensemble effects?

8. Discuss advantages and disadvantages of various synthesis methods.

Demonstrations

1. Visit several music stores and observe the tonal characteristics of the various electronic organs.

Further Reading

Backus, Chapter 15

Olson, Chapters 5, 10

Rossing, Chapters 26 - 28

Winckel, Chapter 9

Beauchamp, J. 1979. "Brass Tone Synthesis by Spectrum Evolution Matching with Nonlinear Functions," Computer Music J. 3 (2), 35-43.

Blesser, B.A., K. Baeder, and R. Zaorski. 1975. "A Real-Time Digital Computer for Simulating Audio Systems," J. Audio Eng. Soc. 23, 698-707.

Chowning, J.M. 1973. "The Synthesis of Complex Audio Spectra by Means of Frequency Modulation," J. Audio Soc. 21, 526-34.

Dorf, R. 1968. **Electronic Musical Instruments**, 3rd ed. (Radiofile).

Douglas, A. 1976. **The Electronic Musical Instrument Manual**, 6th ed. (Pitman).

Dudley, J.D., and W. J. Strong. 1975. "Analysis and Synthesis of Oboe Tones by the Linear Prediction Method," J. Acoust. Soc. Am. **58**, S131.

Mathews, M.V. 1982. "An Electronic Violin with a Singing Formant," J. Acoust. Soc. Am. **71**, S43.

Mathews, M.V., and J. Kohut. 1973. "Electronic Simulation of Violin Resonances," J. Acoust. Soc. Am. **53**, 1620-1626.

Mitsuhashi, Y. 1980. "Waveshape Parameter Modulation in Producing Complex Spectra," J. Audio Eng. Soc. **28**, 879-895.

Moorer, J.A. 1976. "The Synthesis of Complex Audio Spectra by Means of Discrete Summation Formulas," J. Audio Eng. Soc. **24**, 717-727.

Plitnik, G.R. 1975. "A Digital Computer Technique for the Production of Synthetic Oboe Tones," Frostburg J. of Math.

Pritts, R.A., and J.R. Ashley. 1975. "Live Electronic Music in Large Auditoriums," J. Audio Eng. Soc. **23**, 374-380.

Slaymaker, F.H. 1956. "Bells, Electronic Carillons, and Chimes," IRE Trans. Audio **AU-4**, 24-26.

Refer to Further Reading in Chapter 41.

Audiovisual

1. "Synthetic Oboe and Bassoon Tones" (GRP)

2. **Percussion, Mechanical, and Electronic Instruments** (3.75 ips, 2- track, 30 min, UMAVEC)

3. **Sounds—The Raw Materials of Music** (1.88 ips cassette, SILBUR)

4. Refer to Audiovisual in Chapter 41.

41. Synthesizers and Digital Computers

Electronics made its debut in the early twentieth century. Many of the electronic instruments discussed in the last chapter were fairly well-developed by mid-century, even though they have undergone changes and refinements since. Synthesizers and digital computers were introduced about mid-century and grew phenomenally in the 1970s. The future promises even more phenomenal growth for both of them. We will discuss various electronic components used in synthesizers, and then consider how they are organized in a synthesizer. We will then consider how digital electronics function and how they can be used in place of analog electronics for sound synthesis. Several additional applications of digital computers in music will then be considered. Brief historical notes will conclude the chapter.

Synthesizer Components

The electronic components used in synthesizers are basically those discussed in Chapter 40. (Electronic components used in synthesizers and in other electronic devices are often referred to as modules. We will use the terms modules, components, and building blocks somewhat interchangeably.) Signal generators are used to produce sine, sawtooth, square, triangular, and pulse waveforms (see Figure 6.12). Also, a noise generator may be used to produce non-periodic waveforms. Various filter modules (low pass, high pass, and band pass) are used to modify the signal coming from the source. In addition envelope generators are used to shape the overall amplitude of the signal, and modulators are used to modify the frequency and amplitude of the signal periodically. The signal is amplified and radiated as sound by a loudspeaker.

Although electronic organs and synthesizers use basically the same building blocks, they can be distinguished by the two different methods in which these components are used. An electronic organ has a number of predetermined capabilities built into it which can be used by the performer. Many notes can be generated at once; the instrument is used for live performance. A synthesizer has much greater flexibility in modifying and controlling the tonal capabilities. Until the mid to late seventies, synthesizers could produce only single-note sequences, and performances were created through repeated sound-on-sound recording. (In the early synthesizer recordings, as many as 40 tracks were used.) During the 1970's synthesizers were designed to produce several notes at one time; consequently their use in live performance has increased.

Perhaps some similarities and differences between electronic organs and synthesizers can be seen by contrasting the use of the electronic components in each. Let us suppose that a sawtooth wave is to be used as the source of sound in an electronic organ; the wave is to be filtered to achieve an oboe-like sound. The only control

the musician has over the final sound is in selection of the stop tab. The manufacturer has pre-wired both the waveform generator and the filter. If the musician does not like the sound he can select another stop, but he cannot change the tone quality of any given stop. When a key is depressed on the organ, a step voltage is produced, as shown in the upper part of Figure 41.1; when the key is released the voltage returns to its previous value. This voltage goes to an attack and decay generator (ADG), which produces predetermined attack and decay transients. A level voltage is maintained after the attack for as long as the key remains depressed. The voltage from the ADG multiplies the sawtooth wave from the generator to produce the output wave seen in Figure 41.1. The duration and shape of the attack and decay transients are also predetermined and not accessible to player control. The frequency sounded depends on which key is depressed and thus on which sawtooth generator is used.

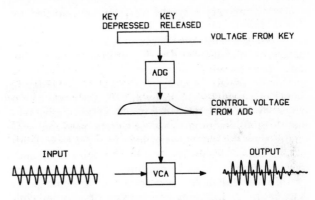

Figure 41.1. Amplitude modulation of a waveform with a time envelope. Signal from key goes to the attack and decay generator (ADG) which sends a signal to the voltage-controlled amplifier (VCA).

The emphasis in a synthesizer is on control flexibility; we need modules that can be easily controlled. Voltages are convenient to use for control signals, so a wide variety of voltage-controlled modules have been developed for use in synthesizers. A **voltage-controlled oscillator** (VCO) generates waveforms with the frequency controlled by the voltage sent to it—increasing the voltage raises the frequency and vice versa. The control voltage can come from a keyboard, from another VCO, or from any control source. A **voltage-controlled amplifier** (VCA) has a gain determined by an input control voltage which can come from any of many sources. Output from an ADG, discussed above, is often used as the input to a VCA. A **voltage controlled filter** (VCF) has filter characteristics, such as cut-off frequency, that can be controlled by an input voltage.

We now repeat the single organ tone example, but

this time on a synthesizer to illustrate some of the flexibilities possible. Suppose that the key voltage and ADG output are the same, but that we want an amplitude tremolo with the tone. We can use a VCO to generate a low frequency sinusoid, add it to the ADG output, and use their sum to control a VCA, as shown in Figure 41.2. The output waveform now shows amplitude modulation.

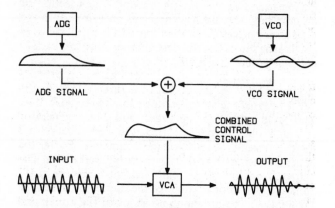

Figure 41.2. Amplitude modulation of a waveform with a time envelope. Signals from attack and decay generator (ADG) and voltage-controlled oscillator (VCO) are added to provide a control signal to the voltage-controlled amplifier (VCA).

Waveforms in a synthesizer can be modified by mixing, filtering, or modulation. When two waveforms are **mixed**, the values are simply added together at every instant, as in additive synthesis. For instance, the ADG signal and the VCO signal are added to produce the combined control signal in Figure 41.2. **Filtering** simplifies a complex sound by attenuating some of the frequency components, which results in subtractive synthesis. Four common types of filters used on synthesizers are the low pass and high pass filters with adjustable cut-off frequency, and the bandpass and band-reject (or notch) filters with adjustable center frequency. To **modulate** a waveform is to change it systematically, following the pattern of another controlling waveform. If the change is in amplitude, the result is amplitude modulation (AM), while a change of frequency is frequency modulation (FM). As an example of amplitude modulation, the input of Figure 41.2 is amplitude modulated by the combined control signal via the VCA to give the output waveform shown in the figure. A sine wave frequency modulated by a sine wave is shown in Figure 41.3. A square wave frequency modulated by a triangular wave is shown in Figure 41.4. Notice in Figure 41.4 that, as the triangular wave rises in amplitude, the square wave increases in frequency. The increasing frequency is observed as the pulses of the square wave become narrower and closer together.

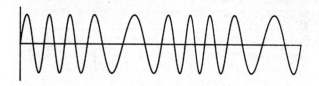

Figure 41.3. A sinusoid frequency modulated by another sinusoid.

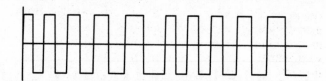

Figure 41.4. A square wave frequency modulated by a triangular wave.

Basically, the synthesizer operates entirely with voltages. Some modules are designed to be controlled by voltages (such as the VCO), and others are designed to produce control voltages. In fact the same voltage signals can be used for either control signals or output signals, the difference being in their function. A sawtooth waveform from an oscillator may be used as an output to be filtered and amplified, or it may be used to control another VCO generating a different waveform. A synthesizer, then, consists of signal-generating and signal-modifying modules, some of which are voltage-controlled by other signals. Each module is completely separate, with its own input and output points. The final signal from the synthesizer—called the audio signal because it ultimately becomes a sound wave—depends on how the modules are interconnected.

The possible combinations of voltage-controlled modules is limited only by imagination and resources. Frequency vibrato, spectrum vibrato, multiple decays, and so on are all possible. One of the primary concerns is how the great flexibility of these voltage-controlled modules can be organized and made accessible to the user. This will be considered in the next section.

Synthesizer Organization

A synthesizer consists basically of various generator, filter, modulator, mixer, and artificial reverberation modules. A multitrack tape recorder is usually incorporated into the system so that musical tone sequences can be stored and then recorded with additional tone sequences as they are generated. Sound is available to the listener through a power amplifier and loudspeaker (or headphones), as with other electronic instruments. Figure 41.5 illustrates a typical synthesizer organization scheme.

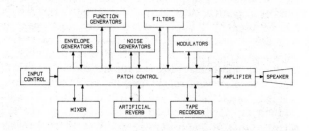

Figure 41.5. Typical synthesizer arrangement.

Means must be provided for making interconnections among the various modules in the synthesizer. This is accomplished via a patch panel, or its equivalent. In "plug and jack" patching, one end of a cord is plugged into a jack that is electrically connected to one module; the other end of the cord is plugged into a jack connected to another module. In this way the two modules are interconnected. This kind of patching is illustrated in Figure 41.6. Other kinds of patching done with matrix boards and switches accomplish essentially the same result, but in a "cleaner" way. Patching with cords and jacks may be the most versatile method, but other schemes are much more time-efficient.

Input signals to control the synthesizer can be derived in any number of ways. A keyboard is probably the most common device. In its simplest form it provides off-on voltages, as in organs. More sophisticated keyboards may have pressure sensitive keys, which increase the voltage output with key pressure to permit amplitude tremolo. Keys may be velocity-sensitive, to allow different attack transients, or sidewise displacement-sensitive to permit frequency vibrato. A joystick is an input control device much like a manual gear shift lever, except it can be set at any position. At least two control voltages can be obtained from a joystick, based on its horizontal and vertical positions. Touch-sensitive surfaces in which two-dimensional position and force are available, could be used in much the same way as a joystick. Early synthesizers used punched paper tape to provide control. Digital computers can be programmed to send control voltages to a synthesizer. Some small digital computers have synthesizer modules which can be plugged into and controlled by the computer, to synthesize music and speech. Virtually any device or method that can be used to generate a voltage can be used to control a synthesizer.

We now briefly summarize the function of each of the boxes in Figure 41.5. The envelope generators pro-vide attack and decay envelopes for controlling waveform amplitude. The function generators provide various waveforms at different frequencies. The noise generators provide white noise with equal energy in equal bandwidths, or pink noise with less high-frequency energy. The filters provide various kinds of spectrum shaping. The modulators provide variable amplification. The mixer allows us to add various amounts of different signals together. The artificial reverberation delays and attenuates the signal, and then adds it back to the original signal to simulate room reverberation. The tape recorder permits temporary storage of the tone sequence.

As an example, let us outline the procedures we might follow to create a two-voice piece of music. We choose the first voice to be percussive noise. The interconnections between modules are illustrated in Figure 41.7. Audio signals are shown by solid lines, and control signals are shown by dashed lines in the figure. Making the various connections between modules is called patching. Any given patch can be either for control or for audio, but not both. We patch the keyboard input to an envelope generator, the envelope generator and the noise generator to a modulator, the modulator output to a filter, and the filter output to the tape recorder. We set the envelope controls to produce a short attack and a long decay. We set the filter to bandpass the signal at a frequency of 1000 Hz. We play the note sequence on the keyboard. Voltages going from the keyboard to the envelope generator are given a sharp rise and slow decay; they modulate the noise, which is then filtered and recorded on tape.

Figure 41.6. Externally, patch-cord programmed synthesizer. Note the tape recorders included as part of the synthesizer setup. (Courtesy of Moog Music, Inc.)

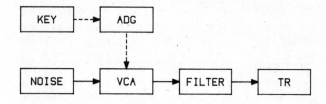

Figure 41.7. Interconnections between modules to generate percussive noise sounds.

We choose our second voice to be a bowed string with frequency vibrato, using a sawtooth wave as the source. We control the sawtooth generator with a very low frequency sinusoid to give vibrato, and set the envelope generator to produce a slow attack. We choose a bandpass filter set at about 3000 Hz. We patch the filter output and the tape recorder output to the mixer, and the mixer to the artificial reverberation and then to another tape track and the power amplifier. As we depress the keys, we hear our string tones mixed with the previously recorded percussive tones. The result is also recorded on tape for future reference.

Synthesizers were originally devised to produce music a line at a time, record it on one track of a multitrack tape, and then mix the results as many times as required to produce the desired complexity in the musical structure. The first synthesizers were limited to the control of one VCO at a time from the keyboard. However, recent synthesizers permit the control of several VCOs at a time, and ensembles of synthesizers can sometimes provide complete live performances. (Usually part of the "live" program is prerecorded onto magnetic tape.)

To be musically successful, a synthesizer should be designed to meet the following requirements: (1) The synthesizer should perform all of the basic generating and modifying operations of the classical studio, and provide resources consistent with the state of the art. (2) The design should contain no unnecessary inherent limitation. (3) The operation of the synthesizer should be straightforward, easy for a musician to understand, and should not require extensive technical knowledge. (4) Features which speed up the composition process and give the composer direct control should be stressed. (5) The instruments should be stable and precisely calibrated for repeatability and ease in working with a score. (6) The complete synthesizer should be as lightweight and portable as possible, and should present an aesthetically pleasing appearance. (7) Reliability should be achieved through the use of conservative design, quality components, and rugged construction techniques. (8) By using suitable control devices, the synthesizer should be readily adaptable to both live performance and programmed control.

Digital Synthesis

Most of the electronic organs and synthesizers discussed up to this point are analog devices, as their electrical signals are continuous. We want now to consider digital devices in which the signals are discrete and not continuous. The sounds we produce and hear are continuous. The electrical signals transduced from sound (via a microphone) and used to produce sound (via a loudspeaker) also are continuous. One may wonder why digital signals are of any interest.

Analog devices (e.g., waveform generators and filters) are inherently unstable, imprecise, and difficult to control. An oscillator set to produce a particular frequency will drift and need to be reset periodically. One cannot hope to turn a knob to the same dial setting and get precisely the same frequency from one time to another. Digital devices overcome these problems. A digital oscillator set to produce a given frequency will continue to produce the same frequency day after day, under variable environmental conditions. One can fully expect to get the same frequency each time that a digital dial is set to a particular value. Digital devices are also easy to control. A final compelling feature of digital devices is their continually decreasing cost as compared to analog devices. As digital electronic devices are made smaller and smaller, one can have more and more electronic functions in the same space and for about the same amount of money.

In digital devices, voltages are discrete and not continuous as in analog devices. One can consider a signal to be a series of numbers instead of a continuous voltage, as discussed in Chapter 6. It is possible to describe equivalances between analog signals and digital signals. Similarly, there is a digital counterpart for any analog device. This means that the previous discussion of analog synthesizers is also applicable to digital synthesizers. We can conceptually replace all analog components in a synthesizer with digital components, and have a digital synthesizer with its inherent advantages. There are, of course, many details to consider if one is to do this with actual devices.

Uses of Digital Computers

An analog synthesizer can be considered a special-purpose analog computer—an analog computer designed to accomplish a specific task. An electronic organ is a special-purpose analog computer in this sense. A digital synthesizer can be considered a special-purpose digital computer. There are also general-purpose digital computers. The distinction between special- and general-purpose computers is largely one of trade-offs. A special-purpose computer is designed to perform a restricted number of tasks very rapidly. A general purpose computer can be programmed to perform a wide variety of tasks, at the expense of a loss in speed. A general purpose digital computer could be used as a digital synthesizer. However, it may not be able to generate even one tonal sequence in real time.

Computers have been used to study and simulate a wide range of musical tone phenomena. For example, the computer has been used to analyze oboe and bassoon tones over a wide range of frequencies and dynamic markings. The results indicate that the spectral envelope concept, as shown in Figure 33.19, is valid and that the envelopes are fairly stable over the range of the instrument. Oboe and bassoon tones are then synthesized, using information about the fundamental frequency and the frequencies and amplitudes of the two peaks in the spectral envelope. The two peaks in the spectral envelope are modelled with two parallel filters, each having its own amplitude controls, as shown in Figure 40.8. The input is a pulse train having a fundamental frequency equal to that of the desired note. A pulse train has a spectrum with all harmonics having equal amplitude. When the pulse train is put into the two parallel filters, frequencies near the resonance peaks of the filters are passed through, while frequencies not near the peaks are attenuated. This results in a spectrum similar to that shown in Figure 33.19. The different dynamic markings are obtained by increasing or decreasing the amplitude of each resonant peak via the amplitude controls. The resulting tone qualities are clearly oboe-like and bassoon-like, although the notes do not sound completely natural.

The above scheme was used to produce an arrangement of Bach's Two-Part Invention in C Major for synthetic oboe and basson. The synthetic oboe was programmed to play the upper voice, while the synthetic bassoon played the lower voice. The result can best be described as "interesting." The problems which detracted from the realism are all amenable to reasonable solution. On the positive side, it was possible to produce the music with minimal effort on the part of the programmer. The following information was specified: the note to be played, its time value, the metronome marking, the dynamic marking, and the phrasing. After a separate recording of each voice was made, they were added together via tape recorder. On the negative side, there were slight round-off errors in the frequencies and time values of the notes, which resulted in the two voices being slightly out of tune and synchronization.

Music compilers have been developed that permit a "composer" to specify musical instrument characteristics, note characteristics, and note sequences. The instrument characteristics are supplied to the compiler, which simulates the instruments and "plays" the specified notes on them. The resulting sequences of numbers are then D/A converted and presented through a loudspeaker. Computers offer great flexibility for simulation of systems, but have the disadvantage of not performing in real time. The time required to calculate the musical waveforms is often many times longer than the duration of the waveform created, so they must be calculated, and stored, and then played all at once when the calculations are complete.

Computers can be used to control electronic organs, pipe organs, and synthesizers. There are several interesting features to having a computer attached to an organ. First of all, the computer can play music that humans are not capable of performing. Second, the computer can be used as an aid (e.g., for registration changes) for a live performer. Finally, the computer can be used to record (for later playback) a live performance.

Computers have also been used to compose music. Typically, this process involves the specification of rules stated in the form of the likelihood that something will occur, such as one note following another. These rules can apply to melody, harmony, note sequences, note durations, etc. Random numbers are generated by the computer and run through the rules of the system to generate note strings. In this fashion many different musical structures can be generated, but all of the structures will have the same rules in common. One of the earliest and best known examples of music composed by computer is the "ILLIAC Suite for String Quartet,"

first performed in 1959.

Historical Notes

The first true music synthesizer was probably the elaborate RCA Synthesizer constructed from 1949 to 1955 at the RCA Laboratories in Princeton, N.J. Its purpose was to produce tones with any combinations of frequency, intensity, growth, duration, decay, portamento, timbre, vibrato, and variation. Figure 41.8 is a schematic diagram of a larger and more elaborate version, the Mark II, which was completed in 1959 for the Columbia-Princeton Music Center. In this device, seven large racks of vacuum-tube electronic paraphenalia, such as oscillators and amplifiers, are controlled by a 40-channel paper tape reader.

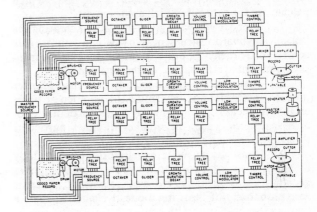

Figure 41.8. Schematic layout of RCA Music Synthesizer. (From Olson et al, 1969.)

Although some composers used the RCA Synthesizers, they were not the type of equipment available to most musicians. It remained for Robert A. Moog to develop and successfully market the first synthesizers which were small, relatively portable, and relatively inexpensive, a feature of prime importance to struggling composers. Developed in the late 1960s, these instruments first attracted attention because of the clever arrangements of popular songs by musician-composer W. Carlos, and his best selling album "Switched on Bach." The Moog synthesizer uses voltage-controlled elements, which has now become standard procedure in the industry. During the 1970s the synthesizer came into its own as many manufacturers rushed into production to meet the growing demand for these instruments. At the same time that more musicians wanted to play the synthesizer and more composers were writing music for it, new technology and stiff competition brought prices down to even lower levels. Today the budding musician can choose from the many models and types of synthesizers available from many manufacturers.

Exercises

1. Suppose one wanted to simulate trumpet sounds on a synthesizer. Which electronic components would be used? Draw a diagram to indicate how they might be combined.

2. Why must a D/A converter be used with a digital synthesizer in order to hear the sounds?

3. Describe how you might use the various func-

tions avilable on a synthesizer to create flute sounds or clarinet sounds. Illustrate each case with a diagram.

4. In Chapter 40 peak clipping of a sinusoid was discussed as one distortion useful in synthesizers. What harmonics are generated in the process? If one does "infinite" peak clipping what are the resulting waveform and spectrum?

5. Sketch the following waveforms: (a) a sawtooth wave amplitude modulated by a square wave; (b) a sawtooth wave frequency modulated by a square wave, (c) a square wave amplitude modulated by a square wave; and (d) a square wave frequency modulated by a square wave.

6. Sketch a figure similar to Figure 41.7 for the second voice of the example given in the text. Which patches are for control? Which are for audio?

7. Suppose you were given the task of creating the "ideal" keyboard instrument. How would you provide control of frequency, vibrato, tremolo, tone color, transients, ensemble effects, etc., to the performer?

8. Take a fairly simple melody (a Stephen Foster tune, for example) and determine the number of times each note occurs; the number of times particular two-note sequences occur; the number of times particular three-note sequences occur.

9. Construct a new "Stephen Foster-like" melody by using a set of two dice as your random number generator for determining two-note sequences. (There is zero probability of throwing a one, 1 chance in 36 of a two, and so on.) Start with some arbitrary note such as C. If there is a 50% chance of having an E follow the C and a 50% chance of having a G follow the C you might choose to have an E anytime you throw a 2, 4, 5, 6, or 10 with the dice (1/36 + 2/36 + 3/36 + 4/36 + 5/36 + 6/36 = 1/2) and a G for any other number thrown. Then once you have "thrown" an E or a G you can determine the next note by means of the probabilities from Exercise 8 in a similar fashion.

Demonstrations

1. Experiment with a synthesizer and try to simulate lip reed, mechanical reed, air reed, string, and percussion tones.

2. Invent some new musically interesting sounds with a synthesizer.

Further Reading

Alles, H.G. 1980. "Music Synthesis Using Real Time Digital Techniques," Proc. IEEE **68**, 436-449.

Gardner, M. 1978. "White and Brown Music, Fractal Curves, and One-over-f Fluctuations," Sc. Am. **238** (Apr), 16-28.

Hartman, W.M. 1975. "The Electronic Music Synthesizer and the Physics of Music," Am. J. Physics, **43** 755-763.

Hiller, L.A., and L.M. Isaacson. 1959. **Experimental Music** (McGraw- Hill).

Holmes, W.H., and J.P. Rist. 1980. "Real-Time Music Synthesis," Proc. 10th ICA, paper K-5.2.

Mathews, M.V. 1969. **The Technology of Computer Music** (MIT Press).

Moog, R. 1967. "Electronic Music—Its Composition and Performance," Electronics World.

Moorer, J.A. 1978. "How Does a Computer Make Music?" Computer Music J. **2** (1), 32-37.

Moorer, J.A. 1977. "Signal Processing Aspects of Computer Music—A Survey," Computer Music Journal **1** (1), 4-37.

Olson, H.F., H. Belar, and J. Timmens. 1969. "Electronic Music Synthesis," J. Acoust. Soc. Am. **32**, 311-319.

Refer to Further Reading in Chapter 40

Audiovisual

1. "Modulation" (REB)
2. "Musical Synthesizer Fundamentals" (REB)
3. "Bach's Two Part Invention in C for Synthetic Oboe and Bassoon" (GRP)
4. **Music from Mathematics** (Bell Labs Records #122227)
5. **Voice of the Computer** (Decca #DL710180)
6. **The Nonesuch Guide to Electronic Music** (Nonesuch #HC-73018)
7. **Switched-on Bach** (Columbia #MS7194)
8. **Switched On Bach II** (Columbia #KM 32659)
9. **A Clockwork Orange** (Warner Brothers #WB2573)
10. **1812 Overture** (London, Phase 4)
11. **Touch** (Columbia #MS7316)
12. **The Wild Bull** (Nonesuch #H-71208)
13. **Pictures at an Exhibition** (RCA Records)
14. Refer to Audiovisual in Chapter 40

VII. Electronic Reproduction of Music

Courtesy of Brigham Young University Archives. Photo by R. Madsen.

42. History and Overview of High Fidelity

Near the end of the Nineteenth Century, the almost simultaneous invention of the telephone and the phonograph began the historic development which was to culminate in the complex world of high fidelity that we know today. This present chapter will trace the development of devices for the recording and reproduction of music from Edison's original phonograph to the relatively recent development of quadraphonics. Along the way we will consider the general nature of high fidelity and modern stereophonic sound reproduction systems.

The Acoustic Phonograph

Mankind's interest in talking machines dates from almost the beginning of recorded history. The statue of Memnon, constructed at ancient Thebes about 1500 BC, was alleged to emit (due to hidden air chambers and an unknown mechanism) a vocal greeting to the sun at dawn. The medieval philosopher-scientist Robert Bacon constructed a mechanical talking head. In 1860 a Viennese inventor constructed a talking automaton, complete with rubber lips and tongue and a moveable mouth cavity. However, the true progenitor of the phonograph, was not these "talking machines," but the telegraph, a seemingly unrelated device.

The phonograph was invented by Thomas Edison as an unexpected offshoot of an invention never intended for the reproduction of sound. When Edison was young, he invented a machine to record and store on a rotating cylinder the "dots" and "dashes" emitted by a telegraph key. Years later, while experimenting with an improved version of the same device, he began to rotate the disc rapidly. Noticing that a musical tone was produced, he conceived the idea of embossing sound waves onto the disc. Preliminary experiments convinced him that the device would work, and so in 1877 the world's first phonograph was constructed. The invention combined simple elements which had been in existence for some time. Edison's original phonograph, shown in Figure 42.1, was an entirely mechanical device consisting of little more than a hand-rotated cylinder covered with tinfoil. Although the machine could only be used to repeat several spoken syllables, it did demonstrate that sound reproduction was a definite possibility. The sound track was a groove, with vertical hills and dales embossed on the foil at the bottom of the groove. The "record" could not be removed, stored, or duplicated. Edison lost interest in his experimental device and left it to others to commercialize it.

The first commercial sound reproduction device, the "graphophone," was devised in 1877 by Chichester Bell and Sumner Tainter. Wax cylinders were used as the recording medium instead of tinfoil and a clockwork mechanism was provided to move the needle across the cylinder at a more uniform speed. These innovations improved both the longevity of the cylinders and the

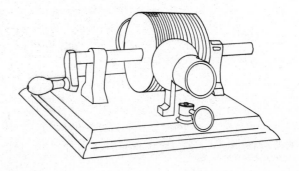

Figure 42.1. Sketch of Edison's original tinfoil phonograph. (After Read and Welch, 1976.)

quality of the reproduced sound. The basic elements of this system are diagrammed in Figure 42.2. When a sound enters the horn, a diaphragm at the small end of the horn vibrates the cutter, thus scratching a groove in the rotating wax cylinder, with "hills" and "valleys" representing the wave motion. To reproduce the sound, a playback needle followed the hills and valleys of the wax cylinder. The needle was attached to a diaphragm thus forcing it to vibrate, sending the reproduced sound out the large end of the horn. Figure 42.3 shows the recording process for the graphophone and Figure 42.4 illustrates the playback process. Although this system could handle only a limited range of frequencies, it was able to reproduce intelligible speech. Since the wax on the cylinder was relatively soft and deteriorated rapidly when played, the record had a somewhat limited lifetime.

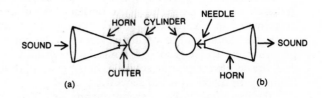

Figure 42.2. Graphophone recording/reproducing system: (a) recording mode; (b) playback mode.

Edison then improved his original concept and began marketing the "phonograph" as a competing brand. Because only a limited number of copies could be made of the wax cylinders, as many as 20 recording horns were used simultaneously. Thus, the unfortunate artist who wished to record his skills for posterity had to do so as many as 60 or 70 times in one day.

In 1888 a German-American, Emile Berliner, invented a whole new approach to sound reproduction. He substituted a flat disc, rotated on a turntable, for the

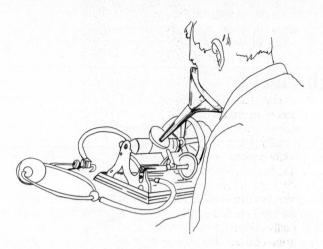

Figure 42.3. Sketch of Graphophone as used for recording. (After Read and Welch, 1976.)

Figure 42.4. Sketch of Graphophone as used for listening. (After Read and Welch, 1976.)

Figure 42.5. Sketch of hand-driven Gramophone. (After Read and Welch, 1976.)

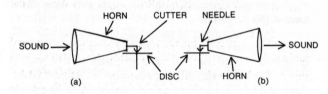

Figure 42.6. Berliner's disc recording device: (a) recording mode; (b) playback mode.

cylinder used by Edison and Bell/Tainter. Figure 42.5 shows Berliner's original hand-driven device, called the gramophone. The elements of the Berliner system are diagrammed in Figure 42.6. Sound entering the horn causes the diaphragm to oscillate, vibrating the cutter perpendicular to the groove. During the playback process, the needle follows the squiggles in the groove, thus vibrating the diaphragm and generating sound.

Berliner thought that both Edison and Bell/Tainter were wrong to rely on the "hill and dale" method of recording sound, as the needle might jump from one hill to another and miss some of the dales. In Berliner's invention the recording needle would vibrate from side to side, rather than up and down, producing a wiggly groove of constant depth on the disc of wax-coated zinc. On playback, the side-to-side motions yielded a louder, if not a higher quality, reproduction. Because of patents already in effect, however, Berliner had to devise a new way of creating the record groove. This he did by a process, developed after much laborious experimentation, using acid to etch the zinc disc after the recording needle had left its trace on the wax coating. Although the etching process left a grainy pattern which caused a great

deal of surface noise, Berliner's device had one immediate practical advantage over its rival; the playback needle automatically followed the helical groove, while the needle on the phonograph or graphophone cylinder player had to be cranked sideways while playing. Berliner's greatest triumph, however, came several years later. After more experimentation, he was able to perfect a method for mass-producing records—the same method which is still used today. The process involved plating the original disc with copper and nickel which, when removed, made a "reverse master disc (i.e., the grooves projected as ridges). When this master was pressed against a heated plastic disc, a replica of the original disc was created.

The final ingredient needed for Berliner to perfect his invention (which he patented as the gramophone) was a motor which would move the turntable at a constant speed. When the motor, devised by Eldridge Johnson, was finally added to the rest of Berliner's system, the Victor Talking Machine Co. was founded, and its trademark, the "Victrola" became widely known and used. Figure 42.7 shows the improved Berliner gramophone featuring Eldridge Johnson's spring-wound motor.

While Berliner's disc system was gaining popularity, the earlier cylinder devices were also making impressive gains. By the turn of the century the rivalry between disc and cylinder manufacturers was intense. The discs could be mass-produced and produced louder sound, but the cylinders had less surface noise, resulting in considerably better sound quality. Berliner had im-

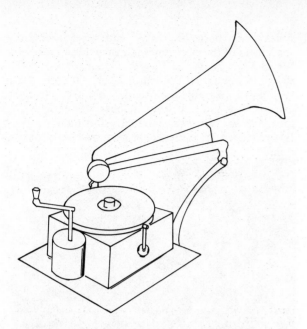

Figure 42.7. Sketch of Gramophone with spring wound motor. (After Read and Welch, 1976.)

proved the quality of his records by switching from hard rubber (which was very susceptible to warping and shrinkage during pressing) to a composite material invented by a button manufacturer (the same composition used for 78 rpm records up to the mid 1950's). However, the etching process remained his nemesis, since the acid would spread in all directions into the zinc from the smooth curve cut into the surface wax. The uneven etching created a master disc of poor quality, thus leading to poor-quality reproductions. Berliner knew that the etching process could be eliminated by making a wax electrically conductive and using electroplating technology to create his master discs, but the graphophone company owned the patent right to use wax masters. On the other hand, the graphophone people would have liked to use electroplating techniques to make a master cylinder, but the technology could only be applied to discs. Each company lacked some vital ingredient necessary to perfect its product. In 1901 the matter came to a head and the recording industry reached a temporary standstill due to these patent problems. Fortunately, each major interest had something to give while still needing something from the other, so in 1902 the impasse was resolved by an epoch-making patent pool. By making the basic patents available to all, the advantages of both systems were combined to the benefit of all. Berliner immediately abandoned his long-time zinc process and began recording on wax, while the graphophone companies switched from cylinders to the much more convenient discs. For the next several decades the lateral disc recording market was dominated by two forces, the American Graphophone Co. (Columbia Phonograph Co.) and the Victor Talking Machine Co. (Berliner Gramophone Co.). Thus began the golden age of acoustically-reproduced sound—which was to last until the advent of radio brought the next major advance in sound reproduction systems.

In 1906 the Victor Co. introduced the Victrola with an internal horn and a stylish cabinet similar to the in-

strument shown in Figure 42.8. Although it did not sound as good as the external horn models, it looked better in the home and was consequently readily accepted by the public. During the next two decades many advances were made in the acoustic recording and playback processes, but the systems suffered from several inherent limitations. To produce the acoustical power necessary to cut the wax during recording sessions, various subterfuges were employed. Singers usually stuck their heads inside the recording horn, while violins were equipped with special horns and diaphragms mounted directly on the instrument. Since the high frequencies did not record well, snare drums were particularly difficult to record and no placement of the horn yielded a satisfactory result. The solution eventually adopted was for the drummer to forego the snare drum entirely and to play on the large suitcase used to carry the drums. Recorded and played back on these primitive machines, this gave a plausible reproduction of the sound of a snare drum. When larger ensembles were recorded, the crowding of musicians around the recording horns and the unconventional arrangement of the performers necessitated by the insensitive recording mechanism made it difficult to achieve an inspired performance. This mechanical process of recording and reproducing music remained unchanged until 1925, the beginning of the electro-acoustic age.

Figure 42.8. Interior view of hand-cranked phonograph of about 1920. Note needle, diaphragm, and portion of horn acting as support arm. (Courtesy of G. Orvis photographed from G. R. Plitnik collection.)

Electronic Phonographs

The inauguration of regularly-scheduled radio broadcasting in 1919 marked the decline of the era of acoustic recording. By 1924 experiments in recording sound through the use of microphones were being conducted by AT&T engineers. In 1925 the financially ailing record companies, who had been losing sales as the radio audiences grew, released several records recorded using microphones, but intended for mechanical playback. The improved quality was immediately apparent. Sibilant sounds (such as /S/), which were previously difficult to record due to the many high frequencies, sounded clearer and more natural. There was also an overall volume increase which made some ensembles sound raucous and strident when played on the older equipment. To overcome these difficulties and boost sales, Victor began manufacturing a new mechanical phonograph, the Orthophonic Victrola, especially designed to play the new electrically-recorded discs. When Columbia followed suit, the smaller companies were crowded out in the competition. One of these, the Brunswick-Balke-Collender Co., decided to meet the competition head-on by being the first to manufacture an all-electric phonograph, the Brunswick-Panatrope. The Panatrope used a horseshoe magnet pick-up, a vacuum tube amplifier, and a small dynamic loudspeaker. While the volume was almost unlimited, the tone quality of the early models was not nearly so good as that of the Orthophonic Victrola. Later models produced better quality and soon other manufacturers abandoned acoustic reproduction entirely. In 1927 the Victor Talking Machine Co. merged with the Radio Corporation of America (which held all the important electronic patents) to form the RCA Victor Co. Thus was the inevitable merging of mechanical sound reproduction with electronics consummated. That this union has been fruitful is evidenced by the plethora of electronically-based derivatives.

A subtle philosophical change came with the new methods of electronically recording and reproducing sound. When sound reproduction was an entirely mechanical process, the goal was to re-create, as closely as was then possible, the original sound. With the advent of the electro-acoustic age, the capturing of sound began to be viewed as a synthesis, the projection of an illusion. The ailing phonograph industry conceded to the changing tastes of the radio-listening public and began to use some of the stock tricks of radio. Singing groups who could not sing without the assistance of a microphone gained popularity. This trend has continued unabated to the present time.

During the 1930's the recording industry settled on a standard record speed of 78 rpm, with grooves and stylus 3 mils wide (one mil is 0.001 inch). Due to the lateral motion on the disc surface, the grooves had to be spaced fairly wide apart so as not to interfere with each other. Consequently, a typical 12 inch record would play for only 7 to 12 minutes on one side. Above a frequency of 250 Hz, recording was done at constant "frequency-amplitude"—that is, the amplitude was decreased with increasing frequency. Below 250 Hz all frequencies were recorded at constant amplitude, due to the generally large amplitudes of bass tones. During the late 1930's the 250 Hz "crossover" point was moved upward to 500-600 Hz which meant that larger signals could be inscribed without bumping adjacent grooves on the record. The result was an increase in dynamic range with some loss in bass response, which could be compensated for by boosting the bass during playback. Soon, however, recording companies began arbitrarily raising the crossover point even higher in an effort to produce the loudest sounds possible. To further complicate matters, someone discovered that boosting the intensity of frequencies in the 3000 Hz region gave a more brilliant sound on the cheap phonographs then available. Unfortunately, each company was doctoring the sound in different and non-standard ways so that the consumer had to fiddle with the tone controls every time he changed records.

Although electronics had led to important improvements in sound reproduction, sound reproducing systems were basically "lo-fi" until after the rapid advancement of the electronics industry during the Second World War. A typical "lo-fi" system of the 1940s is shown in Figure 42.9. Recorded performances did not very closely resemble the original sound: the frequency response was not uniform, the bass was deficient, and the treble was grossly attenuated. After the war, true hi-fi systems became possibile, as discussed in the next section.

Figure 42.9. "Lo-fi" with built-in radio of about 1940. (Courtesy of G. Orvis photographed from G. R. Plitnik collection.)

High Fidelity Systems

Prior to the 1950's, the state-of-the-art "hi-fi" systems had a frequency range of about 200 to 5000 Hz with distortion less than 10%. In order to produce true hi-fi systems, deficiencies in the records, the electronics, and the loudspeaker systems had to be overcome. The range of frequencies which could be recorded on discs was extended upward by the invention, in 1949, of "microgroove" records. The microgroove, 0.7 mils rather than 3.0 mils wide, enabled engineers to slow the speed to 33.3 rpm without adversely affecting the high frequency response.

Consider a 12-inch record which rotates at 78 rpm, a speed equal to 1.3 revolutions per second. The circumference of a circle having a 12-inch diameter is 37.7 inches. The highest frequency which can be picked up by a stylus is limited by a wave length no longer than the stylus width. With a stylus width of 0.003 inches, the maximum number of waves which would fit on the groove at the outer edge of the record is 37.7 inches/0.003 inch = 12,600. Since the record makes 1.3 revolutions per second, this corresponds to a maximum frequency of 12,600 × 1.3 = 16,400 Hz. A similar calculation shows that by the time the needle reaches the inner-most grooves (where the diameter is only about 4 inches) the maximum frequency decreases to about 5500 Hz.

By reducing the groove width and stylus tip to 0.7 mil = 0.0007 inch, a 12-inch record can fit 53,900 waves on the outer-most groove. Even with the speed reduced to 33.3 rpm, the maximum frequency has been increased to 29,900 Hz. An additional bonus of the slower speed and closer spacing of the microgrooves is that the discs will play for up to 30 minutes on a side.

By 1954, the electronics and the loudspeakers had been improved sufficiently that true high-fidelity systems were available. These systems were capable of reproducing an extended frequency range with little distortion. Nevertheless, the sound was still somewhat lacking in realism, a condition that could not be corrected until the invention of stereophonic systems.

Although the term **high fidelity** (usually abbreviated **hi-fi**) has been in general use since the Second World War, it has never been satisfactorily defined. A definition often encountered is "close resemblance to an original sound when reproduced." Manufacturers have not hesitated to exploit such vague definitions for commercial profit by labeling many inferior products with the ubiquitous term hi-fi. Today the term has, for all practical purposes, become synonomous with "sound system". Further confusion has been generated by the general public's acceptance of the mere label as somehow magically improving the quality of generally mediocre equipment. Even so, the term hi-fi remains the only expression available to convey the ideal of all such systems: a reproduction which is completely faithful to the original sound. Furthermore, the foisting of the term hi-fi upon a gullible public should not obscure the very impressive progress, made by the audio industry in recent years. In fact, it is safe to say that the current "state of the art" is such that the degree of faithfulness to the original sound is limited only by your pocketbook. A practical definition of hi-fi for the potential enthusiast is "the best sound reproduction equipment available which is within your budget."

There are three different types of hi-fi enthusiast. The type of equipment deemed necessary for a particular enthusiast depends, to a great extent, into which group he fits. The first type is primarily a music lover who wishes to hear his favorite performances with minimal distortion. Since any reasonable quality of reproduction will satisfy this desire, slight distortions will go unnoticed. The second type is a technical perfectionist who is more concerned with the responses of his equipment than with the music itself. He may derive great satisfaction from listening to loud sounds with the least possible amount of distortion. The majority of hi-fi owners, however, do not fall into either camp. Rather, they represent the compromise position; they are not only interested in obtaining distortion-free reproductions but they also enjoy the music itself.

True faithfulness in a reproduced sound requires not only a distortion-free reproduction, but also a realistically-reproduced sound. During the years that the science of high-fidelity was developing, many attempts were made to achieve realistic monophonic reproductions. These attempts failed because there is no way that "acoustic space" can be perceived with a single-channel recording. The only way to achieve realism in a reproduction is to use two (or more) separately recorded and reproduced signals.

Stereophonic Systems

The term stereophonic is derived from the two Greek words: stereos, meaning "solid," and phone, meaning "sound." The composite word thus literally means "solid sound" or "three-dimensional sound." The term, stereophonic, coined by Alexander Graham Bell in 1880, was at that time actually a misnomer for binaural (or "two-eared") sound reproduction. The aim of binaural reproduction is to take the listener to the source of sound; true stereophonic reproduction attempts to bring the source of sound to the listener. When music is recorded binaurally, two closely-spaced directional microphones are mounted at the ear positions of a dummy head. The sound is then presented to a listener through earphones. Stereophonic recording utilizes two (or more) widely spaced microphones. Playback is through two loudspeakers. As we have seen in Chapter 15, we perceive sounds binaurally and use this "two-channel" information to localize sound or to listen selectively. When Bell experimented with binaural sound reproduction, he compared the spatial effect of the binaural reproduction with the apparent visual solidity seen in the stereopticon popular at that time. (The stereopticon uses two views of the same object photographed from slightly different positions.) The comparison, unfortunately, was not a good one. The eye is capable of quite remarkable accuracy in depth perception and in judging distance. The ear, however, has no comparable process which can be used to elicit depth sensations or to determine the distance of sound sources, although we do often judge the distance of a source by its relative loudness. Our visual perception of depth is largely determined by the degree to which our eyes converge on the object in order to bring it into focus. Since human ears cannot move to any appreciable degree, aural depth perception is nearly impossible. (However, some animals do possess movable ears and they may very well have the ability to determine the distance of a sound source.)

During the late 1940's and early 1950's, many experiments were conducted with "two-channel" (or stereo) discs. One early attempt utilized each side of the record for one of the two channels. A special tone arm was constructed with two needles, one on top of the record and one on the bottom, playing simultaneously. This sytem was doomed to failure for obvious reasons. In 1958 the first two-channel discs were released. Since monophonic records required only a lateral motion of the needle, the second channel was added by having the needle follow hills and dales on a plane perpendicular to the record surface. Thus, one channel was picked up with the conventional side-to-side motions of the stylus while the second channel was reproduced from the up-and-down motion. Since these two motions were perpendicular, they could be separated into two channels, even though picked up from the same stylus. These records, when played on a system with a stereo pickup, worked rather well. If played on a conventional mononphonic system, however, only the channel recorded as lateral motion was picked up. This resulted in some rather unusual-sounding records. A friend relates the time he heard an early stereo recording of a concerto for harpsichord and orchestra, played on monophonic equipment. After listening for several minutes it dawned on him that the harpsichord was missing! It had, unfortunately, been recorded on the up-and-down track.

The next innovation was to record the two channels on the side walls of a ''V'' shaped groove (see Chapter 44). The two channels could still be reproduced with a stereo pickup, but a monophonic needle (which only moves laterally) would pick up the sum of the two channels and nothing would be lost. The early stereo systems which used this approach, however, ran into another problem. The monaural needles and cartridges in use at the time were very non-flexible in the up-down mode, since they had been designed only to respond to lateral motion. When a Stereo record was played on a monophonic system, the needle tended to chew up the vertical parts of the recording, thus ruining the record. Finally, in the early 1960's, a monophonic sytem was developed with a stylus/cartridge much more flexible in the vertical direction, even though this motion was not used for sound reproduction. After the introduction of stereo, the sound reproducing industry settled into a stable period, lasting over a decade, which was shattered in the early 1970's by the introduction of the four-channel disc systems.

Quadraphonic Systems

Each major step in the one-hundred-year development of sound reproduction systems brought a new degree of realism. In the early years of this century the quality of reproduced sound was not important. Rather, people were enchanted by the idea of having a ''magic talking box.'' When the first electrical recordings appeared in 1925 people raved over the ''lifelike quality'' of the reproduction, in spite of the missing bass and treble. When hi-fi systems became popular after World War Two, people were very impressed with the quality of sound, even though their systems sound unbelievably crude to our ears today. With the advent of hi-fi stereophonic systems in the late 1950s, rather faithful reproductions of concert hall environments were at last made possible. Although stereo added a feeling of breadth to the sound, some felt that the effect still lacked depth; there was no feeling of being completely surrounded by sound. The next logical step in the development of total realism was the introduction of 4-channel sound: quadraphonics (from the Latin word quadri, meaning ''four''). For quadraphonic sound reproduction, four independent loudspeakers, each with its own power amplifier, are used. The loudspeakers are arranged with two in front of the listener and two behind. When the storage medium is tape, four independent tracks are used for recorded music (one track for each loudspeaker). Stereo records, however, have only two channels available and therefore the four channels required for quadraphonic sound had to be encoded into the two available channels. This was accomplished by two totally different, incompatible methods, involving different devices and different types of discs.

Quadraphonic systems started with high promise, but did not meet with consumer acceptance. J. R. Ashley coined the term ''quadrifizzle'' to describe four areas where the initial promise of quadraphonics fizzled out. The ''first fizzle'' concerns the idea that if two channels are better than one, four channels must be better than two. This fallacious reasoning resulted from the observation that a stereo system of 1960 sounded much better than a mono system of 1955. The stereophonic effect was undoubtedly one of the major reasons for this, but there were also other engineering developments (such as magnetic cartridges and good preamps) which improved the fidelity of stereo systems. Furthermore, certain other developments (such as acoustic suspension loudspeakers) brought the price of stereo systems down to that of a comparable monophonic system of five years earlier.

Two basic questions which consumers must ask themselves are: (1) What do we get for two more channels? and (2) What do the additional channels cost? For two additional channels we get a slight front-back effect (under good listening conditions), sometimes at the expense of stereo separation. Since there have been no accompanying major technological breakthroughs to reduce the cost of components, adding two more channels increases the cost of your system by a factor of 1.5 to 1.8.

The ''second fizzle'' is the utter failure of four-channel sound to better reproduce the listening environment of a concert hall in the average living room. The two reasons often given for the supposed superiority of four-channel sound over two-channel sound are (1) enhanced directional effects, and (2) added ambience. If we picture a listener facing two loudspeakers, with two identical speakers behind him, we think of being in a concert hall and hearing direct sound from the front speakers and reflected sounds from the rear speakers. This, however, is not how the music is usually recorded. When four independent channels (such as on four-track tape), recorded from four microphones located in the listening areas of a concert hall, are played back via independent loudspeakers the result is described by listeners as confusing. It is argued that a complete two-channel system (from microphone through loudspeaker) while not perfect, is optimum. When more channels are added, the results get worse, not better.

Ambience is, roughly speaking, equivalent to the reverberant sound field in an auditorium. If a concert is recorded with two microphones facing front and two aimed toward the rear of the auditorium, the ''rear'' mikes will record mostly reverberant sound. Since the volume of the reverberant sound is usually substantially lower than that of the direct sound, it is superfluous to require two extra channels to carry this limited amount of sound information. When the ambience information is added into the front channels, realistic reverberation effects are obtained.

The ''third fizzle'' has to do with the fact that the sale of electronic equipment is coupled to the availability of program material. Sales of four-channel equipment are slow, in spite of the blitz of advertising material which has appeared. The mono-to-stereo revolution came about because the public could hear clearly the superiority of stereo recordings over mono records. Stereophonic records and FM radio were developed in response to public demand for stereophonic sound. Quadraphonic systems, on the other hand, were developed before there was much public interest. Even at their peak, the selection of quadraphonic discs was small compared to the number of stereo discs available.

The "fourth fizzle" has been the failure to develop a quadraphonic record that has both the required channel separation and low distortion. Two quad record systems, matrix and CD-4, show these problems, respectively. The CD-4 system, because of the modulation process and wide bandwidth required, results in both bandwidth problems and distortion troubles. With the matrix systems, the two 20-kHz channels are retained without distortion, but a significant amount of crosstalk is created between channels. Thus, in choosing a system for quadraphonic records one is forced to choose between the lesser of two evils: crosstalk or distortion.

In conclusion, we can state that four-channel equipment costs approximately twice as much as stereo equipment, whereas the results are as often detrimental as beneficial to the overall quality of the reproduction.

Exercises

1. Why was the graphophone an improvement over Edison's phonograph? What feature of the original phonograph made it unusable as a device which could be exploited commercially?

2. Why were the original phonographs better for reproducing music than speech?

3. Explain how complex sounds, such as the human voice, can be represented by one signal on a phonograph recording. Refer to Chapter 11.

4. Compare the graphophone and the gramophone. How are they similar? Different?

5. Why did the introduction of electronic amplifiers in sound reproduction systems mean that smaller needles and lighter tone arms could be used?

6. Why was it easier to adapt tape recording technology to quadraphonic sound reproduction than it was to adapt recorded disc technology?

7. Explain why a single-channel hi-fi recording can never (no matter how many speakers are used) give a realistic reproduction of the original sound.

8. Which would sound more realistic as you move around a room, stereo loudspeakers or binaural earphones. Explain.

9. Why can more than two recording microphones often be used advantageously for stereo recording while more than two mikes for a binaural system would be superfluous? Explain.

10. Explain why a stereo system in 1960 did not cost substantially more than a comparable monophonic sytem in 1955.

11. Explain why ambience effects are better handled with two channels (although four recording microphones may be used) than with four.

Further Reading

Official Guide, Chapter 9

Olson, Chapter 9

Ashley, J. R. 1976. "Is Four Channel A Quadrifizzle?" IEEE Newsletter (Acoustics, Speech, and Signal Processing Society) **38**, 1-6.

Bauer, B., D. Gravereaux, and A. Gust. 1971. "A Compatible Stereo-Quadraphonic (SQ) Record System," J. Audio. Eng. Soc. **19**, 636-646.

Eargle, J. M. 1971. "Multichannel Stereo Matrix Systems: An Overview," J. Audio Eng. Soc. **19**, 552-559.

High Fidelity. 1977. Several articles on the history of high fidelity. High Fidelity **27** (Jan), 58-97.

Hodges, R. 1977. "Will Quadraphonics Rise Again?" Stereo Review (April), 33.

Inoue, T., N. Takahashi, and I. Owahi. 1971. "A Discrete Four-channel Disc and Its Reproducing System (CD-4 System)," J. Audio Eng. Soc. **19**, 576- 583.

Nakayama, T., and T. Minura 1971. "Subjective Assessment of Multichannel Reproduction," J. Audio Eng. Soc. **19**, 744-751.

Read, O., and W. Welch. 1976. **From Tin Foil to Stereo** (Howard Sams Co.).

Scheiber, P. 1971. "Four Channels and Compatibility," J. Audio Eng. Soc. **19**, 267-279.

Tager, P. G. 1967. "Some Features of Physical Structure of Acoustic Fields of Stereophonic Systems," Journal SMPTE **76**, 105-110.

Audiovisual

1. **Audio Archive** (8 min, color, 1983, Syracuse University Film Rental Center)

2. **The Recording Engineer: Sound's Great** (11 min, color, 1969)

3. **Phono-cylinders, Volumes 1 and 2**, ed. G. A. Blacker (Folkways Records, Albums FS 3886 and FS 3887, 1961)

43. Component Stereophonic Systems

The basic principle involved in all sound recording is to capture minute pressure variations in the air and to store them in a convenient form so they can be reproduced again at will. The goal of high-fidelity audio equipment manufacturers is to make this reconstituted sound as indistinguishable from the original as is technologically feasible. The fact that engineers have been successful in achieving this goal is apparent from listening tests where audiences were unable to differentiate between live music and a stereo reproduction both played behind a screen. In this chapter we consider the components of a modern stereophonic system. We begin with an overview of stereophonic systems. We then discuss the record player, the tape deck, the FM tuner, and finally the amplification stage. We will consider each component and its relevant specifications so that similar components by different manufacturers may be readily compared.

Overview of Stereophonic Systems

In the last chapter we discussed the meaning of the often misused term "high-fidelity," and the fact that there exists no clear line of demarcation between high-fidelity and "low-fidelity." Nevertheless, we can list some general requirements which are necessary for faithful sound reproduction. First, a uniform frequency response over a wide range must be possible. Second, low distortion is required so the reproduced sound is not "fuzzy" or unclear. Finally, noise produced by the system itself should be inaudible. When confronted by the plethora of stereo equipment now available, the problem faced by a prospective buyer is to translate these three abstractions into an actual selection of equipment. The remainder of this chapter is devoted to helping you do just that, by discussing individual components and their engineering specifications.

We begin with an overview of a modern high-fidelity stereophonic system. Such a system consists of three basic components: (1) a transducer to change the information stored on record or tape into an electrical signal, (2) a device to amplify and otherwise modify the electrical signal to a form suitable for (3) a second transducer, which transforms the electrical signal into sound. The first transducer may be a record player, a tape deck, or even a radio tuner. The second component usually is associated with pre-amplifiers, equalizers, tone controls, and power amplifiers. The final transducer may be earphones or loudspeakers. A diagram of these components, which we now consider in more detail, is given in Figure 43.1.

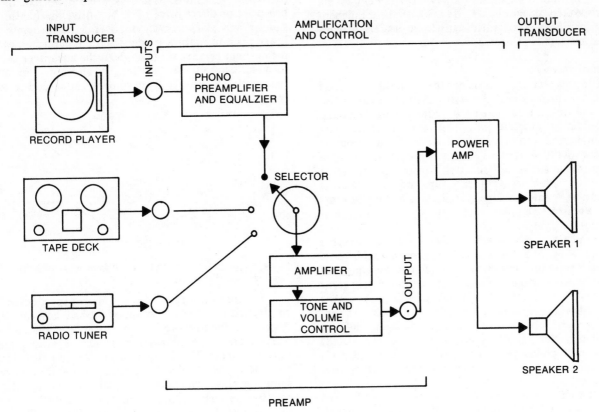

Figure 43.1. Diagram of a complete high-fidelity stereo sound system.

A record player consists of three components: a turntable, a tone arm, and a cartridge. The turntable rotates a record at a constant speed of 33.3 or 45 rpm (revolutions per minute). The tone arm pivots so as to guide the cartridge across the record and the cartridge converts the mechanical motion of the stylus into an electrical signal.

A tape deck retrieves magnetic patterns from a tape and converts them into electrical signals. (More detail on magnetic tapes will be presented in Chapter 44.) Because of the small signal output from the playback head, and because of the need for full equalization, the tape deck includes its own preamp and equalizer circuit.

A radio tuner picks up radio signals (via an antenna) and converts them into electrical signals which can be amplified in the stereo system. The two types of input signals are AM (amplitude modulation) and FM (frequency modulation). Since circuits for both types of reception can be provided for about the same price as either one alone, most inexpensive units sold are AM/FM combination tuners.

The preamp is electronic circuitry which takes the very small, varying voltage from the input device and makes it strong enough for the amplifier to use. Usually the preamp (when separate from the power amplifier) also contains all the control units and the appropriate equalization circuits. The preamp also contains the selector switch so that the record player, the tape deck, or the tuner can be selected.

The power amplifier is the device which takes the signals from the preamp (or from a tape deck or tuner) and amplifies them to a level sufficient to drive the loudspeakers.

The final component of any hi-fi system is the loudspeakers or earphones. Because of their independent significance we will discuss loudspeakers in considerable detail in Chapter 45.

Record Players

In this section we consider the turntable, the tone arm, the cartridge and the stylus. Although turntables are often taken for granted, extracting the signal from a record is an exacting job. First, the turntable must maintain the rotational speed without variation or "wow" and "flutter" may result. Second, the turntable must isolate the record and stylus from any unwanted vibration, which when picked up and amplified, is known as "rumble." Finally, the stylus and pickup cartridge must be well supported so the stylus can accurately trace the groove.

There are two different philosophies of turntable design. One, the "battleship" approach, achieves a smooth rotation through the use of massive construction. Because of the "flywheel effect," a massive turntable will rotate uniformly in spite of minor variations in the driving mechanism. The other approach is the exact opposite: the turntable is made as lightweight as possible so that variations may be effectively corrected. While each philosophy has its advocates, both methods are capable of yielding excellent results when well designed.

Two different types of motors are commonly used for turntable drive systems: induction motors and synchronous motors. **Induction motors**, found in the lower

cost units, rotate at a speed determined by the line voltage. So long as the voltage remains constant at 120 volts, the speed remains constant. However, the line voltage, especially at peak demand times, may drop below its rated value, causing a slight slowing of the turntable. More expensive turntables use **synchronous motors**, the speed of which is determined not by line voltage, but by the line frequency of 60 Hz. Because the line frequency is very carefully controlled and does not vary, the speed of these motors is quite constant. The most expensive turntables use hysteresis synchronous motors which, in addition to constant speed, produce fewer vibrations and noises.

The motor rotates the platter by means of one of three mechanisms. In a rim drive system (left part of Figure 43.2), the motor drive shaft turns a rubber idler wheel which rotates the turntable. Because this mechanism can readily transfer vibration from the motor to the turntable, and because the rubber wheel eventually develops flat spots which cause uneven speed, the rim drive is used only on the least expensive of turntables. A belt drive system (middle of Figure 43.2) is the most commonly used for two reasons. The flexible belt absorbs vibration and thus isolates the motor from the turntable and any slight changes of motor speed will be partially compensated for by the elastic belt. The direct drive system is found only on the most expensive turntables. The platter is mounted directly on the motor shaft (right part of Figure 43.2) and the speed of the motor is very accurately controlled. Although there are no frictional losses, and the platter reaches its exact speed soon after the motor is activated, some sophisticated engineering is required to minimize the motor vibrations. Which is to be preferred, then, belt drive or direct drive? The belt drive mechanism is simpler and gives very good overall performance in lower price turntables (under $200). For turntables in the medium price range ($200-500) the advantage shifts to direct drive. It will last longer, and other options, such as platter speed or pitch control, can be added easily.

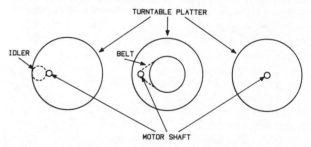

Figure 43.2. Turntable drive systems: rim drive (left), belt drive (middle), and direct drive (right).

A record must revolve at a constant speed and not be affected by spurious vibrations. Since any rotating device is prone to vibration, this must be minimized in one or more of the following ways: (1) the motors must run very smoothly, (2) all bearings and moving parts must be very carefully fitted, and (3) the drive motor should be suspended in elastic shock mounts. The three specifications which indicate how well a turntable meets these criteria are: rumble, wow and flutter, and speed accuracy. Rumble consists of any low frequency vibrations of the platter which are picked up by the cartridge. Rumble is usually specified as the level of "rumble noise" compared to a standard test tone level played on a record. The absolute minimum requirement for a hi-fi system is -35 dB meaning that the rumble level is 35 dB lower than that of the test tone. A good typical specification would be -45 dB, whereas a rumble of -60 dB would be inaudible.

If the turntable motor doesn't pull evenly, a rapid wavering of pitch, called **flutter** results. If the changes are quite slow (from 0.5 to 6 times per second), the effect is called **wow**. These terms refer to the same effect caused by imperfections in the drive system. They are specified as the percentage of speed change relative to the desired speed. On very good turntables this percentage should be below 0.1%, which is considered inaudible. An informal way of evaluating a turntable for constant speed is to listen to a record with long sustained piano chords. Any wow or flutter shows up immediately as an unsteadiness in the tone.

The **speed accuracy** of a turntable is its ability to maintain a constant speed over a period of several hours. It is given as a maximum change of speed expressed as a percentage. A good turntable should be accurate to 1%, but a person with a very sensitive ear may require 0.5%.

Because of the highly advanced state of the art in turntable design, audible rumble and flutter are no longer a problem—if they can be detected, the unit is probably defective. The places where audible differences in high quality turntables do occur are in their sensitivity to external vibration and to sound. Unfortunately, no standard specifications have been set for these problems. **Vibration sensitivity** is apparent as a potential problem when one considers that the phono cartridge responds to some curves on the groove wall of a record that are too small for a microscope to resolve. When this sensitive device is coupled to an amplification system which boosts the power by a factor of 10,000, the result is a very sensitive seismic detector. Any floor or wall vibrations which reach the record will be picked up and amplified. In order to help control this potential problem, the motor base should be isolated from the main frame by springs, and compliant rubber feet may be used under the turntable. A simple test for this problem is to unplug the turntable (so it will not rotate) and then carefully place the stylus in a record groove. Turn up the volume control and lightly tap the shelf on which the turntable rests. The weaker and briefer the resulting thump from the speakers, the better the isolation from vibration. **Acoustic sensitivity** is how readily the record (acting as a microphone) picks up sound in the room. To test for this, again stop the stylus in a record groove and record the phono output on tape when a radio is played in the room. The system will pick up the sound from the radio. The acoustic sensitivity is controlled by using a soft rubber platter mat under the record.

The tone arm must properly hold the cartridge in place and guide it across the record surface. For a high-fidelity system, the tone arm must (1) keep the cartridge as close to tangent to the groove as possible, (2) be in stable balance with the proper force exerted on the stylus, (3) pivot freely with low frictional drag, and (4) have no audible resonance. When a record is recorded,

the recording head moves radially across the disc, as shown by the dashed line of Figure 43.3. When a record is played, the tone arm generally swings in an arc about a pivot point—the stylus thus follows the curved solid line shown in Figure 43.3. Over most of the record, then, the stylus is angled slightly to the actual recorded groove. The departure from the true tangent to the groove is known as the **tracking error**, usually specified as degrees per inch. The pivot point for the tone arm is placed so that when the cartridge is halfway across a 12 inch record (i.e., four inches from the center of the record) the stylus will be exactly tangent to the groove and the error is 0° at this position. For every inch nearer or farther from the center, the tracking error accumulates. If, for example, the tracking error is specified to be 2 degrees/inch, the error is 2 degrees when the cartridge is 3 inches or 5 inches from the center, and 4 degrees when 2 inches or 6 inches from center. A high quality tone arm will have a tracking error of less than 2 degrees/inch or 4 degrees maximum. Tracking error can be reduced by using a long tone arm and by proper positioning of the pivot. For example, an 8-inch arm can be positioned so as to give a maximum tracking error of only 2.5 degrees (the use of a 16-inch arm would reduce the error to 1.5 degrees). A poorly mounted 8-inch arm, however, can given an error of some 18 degrees—obviously causing considerable distortion of the reproduced sound. The radial tone arm is a recent innovation which eliminates all tracking error. Instead of pivoting, a small motor drives the tone arm in toward the center of the record so that a perfect tangent is always maintained between the stylus and the groove. Although this device does eliminate tracking error, the typical error angle is usually so small that one wonders if it really makes any difference. Also, the best efforts of the manufacturer to eliminate tracking error by this means can be undone very easily if a careless hobbiest inadvertently mounts the cartridge at a slight angle.

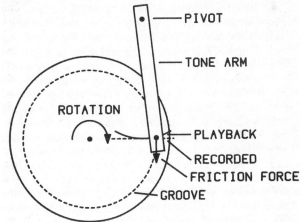

Figure 43.3. Tracking error between recording (dashed straight line) and playback (solid curved line). Frictional force tending to produce skating is shown by an arrow.

The **tracking force** is the downward force required (usually expressed in grams) to keep the stylus in the record groove and following the undulations. Lighter tracking forces mean less wear on the record and the stylus, but if the tracking force is too low the record may be damaged by groove skipping. The tone arm is balanced by an adjustable counterweight which can be set anywhere from 0 to 5 grams.

The tone arm's pivot point is a source of frictional drag. Manufacturers attempt to minimize frictional drag by one of several means: (1) the tone arm may rest on a single needle point, (2) miniature ball bearings may be used, or (3) knife edge pivots may be used. All three designs are capable of yielding excellent results. The maximum force, specified in milligrams, encountered by the tone arm as it pivots is the usual specification, although separate values may be specified for the horizontal and the vertical directions.

The effective mass of the tone arm is a specification which is not always available from the manufacturer, although it ought to be provided. The **effective mass** is the mass of the headshell and cartridge, the part of the arm closest to the stylus. The effective mass falls into one of three ranges: low (less than 10 grams), medium (10 to 16 grams), and high (above 16 grams). The hobbiest should know the effective mass of his tone arm before he attempts to match it with a cartridge. High mass tone arms should be used only with cartridges known to have low compliance, otherwise tone arm resonance may be a problem. Since a tone arm and cartridge have mass and rest on a springy suspension system, all the necessary ingredients for resonance are present. The system must be designed so the resonance frequency does not color the response. This is generally accomplished by keeping the resonance frequency below 16 Hz (the lowest audible frequency). If the resonance frequency is below 8 Hz, however, the system may be excited by motor rumble or record warps. It is therefore prudent to keep the tone arm resonance between 8 and 16 Hz. Another method of controlling tone arm resonance is to reduce the strength of the resonance by viscous damping at the pivot point, or by use of an accessory device such as a dust brush.

Skating is produced by the force which tends to pull the tone arm toward the center of the record. This force is caused by the friction between the stylus and the groove (Figure 43.3) acting at the odd angle at which the arm scans the disk. In the past this effect was ignored, but with the advent of high compliance cartridges the slightest imbalance will adversely affect the system performance. The stylus should contact both sides of the groove with equal force, but the skating force pushes the stylus against the inner groove wall which is the left channel side of the stereo system. This causes three problems. First, the extra force on the inner wall causes the left channel to wear faster than the right and causes the inner stylus surface to wear faster. Second, as the stylus is forced away from the outer wall, it does not make good contact with the wiggles in the groove, thus introducing some distortion in the right channel. Finally, the constant one sided force pushing the stylus off center may eventually cause a permanent kink in the stylus mechanism, thus unbalancing the stereo effect. For these reasons an antiskating device is often provided on better tone arms. The device pulls the tone arm (by means of small weights or springs) away from the inner wall of the groove so that the skating force is balanced. Since the skating force depends on the tracking force,

the antiskate mechanism is made adjustable from 0 to 5 grams, the same as that of the tracking force. It should always be adjusted to agree exactly with the setting of the tracking force.

Because high compliance cartridges are extremely sensitive to rough handling, many safety devices have been developed to prevent unintentional damage. One such device, an automatic positioning device, allows the audiophile to position the tone arm above the record at any starting point. The arm can then be lowered gently onto the disc without damage to stylus or record surface. Viscous fluid in the pivot bearings lets the arm float gently down toward the record in case the tone arm is accidently dropped over the record surface.

The actual transduction of mechanical energy into electrical energy occurs in the **cartridge**, a small unit located at the end of the tone arm. Transduction is accomplished in the following manner. A stylus, which is part of the cartridge, follows the groove of the rotating record and is caused to vibrate by the changing patterns on the walls of the groove. (We will consider the nature of the interaction between the record and the stylus in more detail later.) A transducer, also a part of the cartridge, is attached to the stylus and is caused to vibrate along with the stylus. The basic fidelity of a cartridge depends on how accurately the stylus tracks the record groove. When one considers the complex path the stylus must follow at high speeds without losing contact with the groove walls, this is nothing short of a technological miracle.

The stylus itself is usually sapphire or diamond. Despite the greater initial cost, diamond lasts approximately 20 times longer than sapphire and helps to increase the longevity of records. Two types of stylus are in common use: the conical stylus with a circular cross-section and the elliptical stylus having an elliptic cross section (see Figure 43.4). The elliptical stylus was developed as a means of reducing **tracing distortion**, defined as the failure of the playback stylus to trace exactly the path engraved by the recording stylus. Because a conical reproducing stylus has a circular cross-section it does not make contact with the same part of the groove wall produced by the chisel-shaped recording stylus (see Figure 43.4). Since tracing distortion can be minimized by making the radius of the playback stylus as small as possible, the elliptical stylus contact with the groove wall more accurately traces that produced by the recording stylus (see Figure 43.4). Also, the smaller contact area along the groove permits the elliptical stylus to negotiate narrower turns and thus yield a better high frequency response. These advantages, however, are offset by several disadvantages. Because of the smaller contact area, the elliptical stylus exerts greater pressure on the disc, which may cause records to wear faster. (An elliptical stylus set at a tracking force of 1 gm will produce the same record wear as a conical stylus set at 2 gms.) The elliptical stylus is more difficult to mount, more difficult to manufacture, and hence more expensive. Finally, the chance of having a problem with an elliptical stylus is increased because a slight unintentional rotation of the stylus will accelerate disk wear considerably.

There are two methods by which the mechanical motion of the stylus can be converted into an electrical

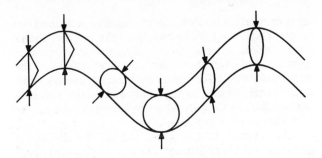

Figure 43.4. The elliptical stylus (right) makes contact with the record groove walls more like that of the cutting stylus (left) than does the conical stylus (center). (After Eargle, 1980.)

signal—piezoelectric or electromagnetic. The piezoelectric effect (see Chapter 6) causes a voltage to be generated across the surface of a crystal which is proportional to the force supplied by the motion of the stylus. The voltage is generally sufficient (0.1 to 1 volt) to be conducted directly to the amplifier without the need for preamplifiers or equalizing circuitry. Because piezoelectric cartridges are quite sensitive to excess heat, and because their high-frequency response is not very good, piezoelectric cartridges are only used for "lo-fi" systems. Electromagnetic cartridges contain a small magnet which produces a magnetic field surrounding a coil of wire. The stylus is attached to either the coil of wire or to the magnet. When the stylus vibrates as it moves in the record groove, its motion is passed on to the coil or to the magnet. In either case, the coil of wire moves relative to the magnetic field. When the conducting coil moves in relation to the magnetic field, a voltage is generated in the coil in accordance with Faraday's Law (see Chapter 2). The voltage generated (1 to 7 millivolts) is several orders of magnitude smaller than that generated with piezoelectric cartridges. Because of the small voltage, a wide frequency range can be reproduced without excessive distortion. However, the output requires boosting in the low-frequency range and attenuation of the high- frequency end of the spectrum. (This process, termed "equalization" will be discussed in more detail later.) Also, because of the very small cartridge voltages generated, at least one stage of preamplification is necessary.

We now consider the important specifications for cartridges. The **compliance** of a phono cartridge indicates how freely the stylus moves in the record groove. Compliance is defined as the distance the stylus moves (in cm) in response to an appflied force of one dyne. The higher the compliance (in units of cm/dyne), the more freely the stylus moves in the groove and the better the reproduction. A compliance specification of 15 × 10^{-6} cm/dyne means that when a force of 1 dyne pushes on the stylus, the stylus moves 15 millionths of a cm. (When comparing different cartridges, consider only the number before the 10^{-6}; the higher that number the more compliant the cartridge.) A compliance of 15 × 10^{-6} cm/dyne, is an adequate compliance. When the compliance is above 20 × 10^{-6} cm/dyne, the stylus is so pliable that it can be used satisfactorily only if the tone arm is capable of tracking the record at a very light tracking force (under 2 grams). Very high compliance cartridges (30 × 10^{-6} cm/dyne) should only be used on

tone arms that will track at less than a gram. With such systems, the record and stylus wear are minimized and a diamond tip stylus will last almost indefinitely.

The **dynamic mass** refers to the total weight of the moving parts of the cartridge. The lower this weight, the less the stylus is distracted by inertia due to the abrupt curves in the record groove. If the stylus shank is too light, however, another problem arises: the thin metal shaft becomes flexible and the motion of the diamond tip will not be accurately transferred to the moving coil or magnet, thus introducing some distortion. Since the weight of the diamond is determined by its size, an optimum compromise between the weight of the shank and its rigidity must be devised. Fortunately, modern metallurgy has produced some very stiff but lightweight alloys which are ideal for a stylus shank. Using these alloys, the dynamic mass can be reduced without noticeable distortion or high-frequency transmission losses. A dynamic mass specification of 1 milligram or less is excellent. By making the moving parts so light, the natural resonant frequency of the stylus is pushed well beyond the audible range (above 20 kHz), thus avoiding harsh resonance peaks. This helps in obtaining uniform frequency response from the cartridge.

The **frequency response** of the cartridge is a measure of how evenly the amplitudes of all the frequencies in the record groove are detected. This specification is usually written as a range of frequencies with a plus or minus rating. As an example of an excellent frequency response, 20-20,000 Hz ± 3 dB means that over the frequency range from 20 to 20,000 Hz the response doesn't exceed or fall below some average response by more than 3 dB. In general, the wider the frequency range and the smaller the + dB and -dB figures, the better the frequency response.

Stereo separation is a measure of the ability of the cartridge to keep the two signals from leaking across channels ("cross talk"). This is no mean feat when one considers that the two channels are picked up by one stylus. The stereo separation is specified as the separation (in dB) at 1000 Hz. A separation of 20 dB at 1000 Hz is typical, although high quality equipment may achieve a higher value. At higher and lower frequencies the separation is not so good. Consequently, some manufacturers specify the stereo separation at several frequencies across the audio range. The separation should be at least 15 dB at 10 kHz.

When the same input signal is fed to both channels, identical signals with the same amplitude should be picked up. The balance between the left and right channels for the same input signal at a specified frequency (usually 1000 Hz) is the **channel balance**. This specification should be within 1 dB, which means that the signals do not differ by more than 1 dB.

The **output voltage** is the voltage (in millivolts) generated by the cartridge when the stylus is tracking a 1000 Hz signal at a specified speed (usually 3.54 cm/s). This is not a measure of quality, but is provided so that the cartridge may be matched to the phono input sensitivity of the pre-amp, as will be discussed later.

A summary of specifications for an Empire 2000 E/111 cartridge are listed in Table 43.1.

Table 43.1. Empire 2000 E/111 cartridge specifications.

Frequency response	20-20,000 Hz ± 2dB
Tracking force range	0.75 - 1.5 gm
Stereo separation	20 dB 200 Hz 28 dB 50-15,000 Hz 20 dB 15-20 kHz
Stylus	0.2 × 0.7 mil elliptical
Effective tip mass	0.6 milligram
Compliance	20×10^{-6} cm/dyne
Tracking ability	32 cm/sec @ 1 kHz @ 1 gm
Channel balance	within 1 dB @ 1 kHz
Output	4.5 mV/channel @ 3.54 cm/sec

Tape Decks

Three different tape systems are in use today: open-reel, cassette, and cartridge. Open-reel machines are capable of the best fidelity, but they are also the most expensive. Better quality cassette machines are a close second in fidelity and sell for a considerably lower price. The cartridge (or 8 track) type of machine is used primarily in car stereo systems. Since fidelity is not a prime consideration in a noisy car, these units have not been developed to the same high quality standards as the cassette recorders. Since cartridge recorders serve no useful purpose in a home system, we will consider only open-reel (Figure 43.5) and cassette (Figure 43.6) machines.

The two types of tape decks are, at least in their basic principles, the same. The tape is magnetized by the recording head, with the amount of magnetization vary-ing in proportion to the electrical signal representing the sound. Thus, a magnetic replica of the original sound wave is stored on tape. During playback, the process is reversed: as the tape moves past the playback head, the stored magnetic information generates a varying electrical signal which, after being amplified, drives a loudspeaker. In Chapter 44 we will consider the nature of this process in more detail.

An open-reel tape deck has no amplifiers or speakers, but plays through the rest of the home hi-fi system. The tape is moved from a supply reel to a take-up reel by a drive system consisting of one or more motors, the capstan, and the pinch roller. The capstan is a thin steel shaft rotated by a motor. The pinch roller, a spring loaded rubber wheel, presses the tape against the capstan as shown in Figure 43.7. As the capstan rotates, the tape is pulled firmly across the tape heads at a constant speed. Open-reel decks come with one, two, or three drive motors. The simplest system, shown in Figure 43.8, consists of a single motor which drives the supply reel, the take up reel, and the capstan through an elaborate set of belts, pulleys, and linkages. The two motor units (Figure 43.9) employ one motor to drive the capstan and a second motor to drive the two reels. Three motor units differ from two motor units (Figure 43.9) in that they have separate motors for each reel. The reel motors drive the reels directly so no belt linkage is required. Since these systems do not require the elaborate set of linkages to get the rotating power where it's needed, they are the most reliable and they control tape speed most accurately. Naturally, they are also the most expensive units.

Less expensive tape decks may have only two heads: one for recording and playback and the second for erase. More expensive units have three heads to per-

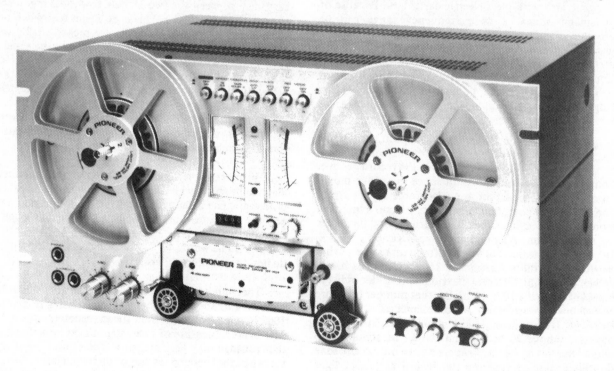

Figure 43.5. Open-reel tape deck. (Courtesy of Pioneer Electronics.)

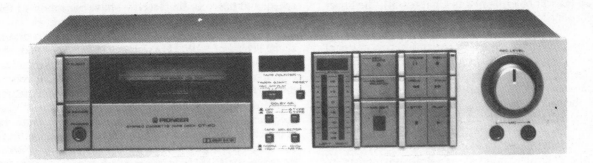

Figure 43.6. Cassette tape deck. (Courtesy of Pioneer Electronics.)

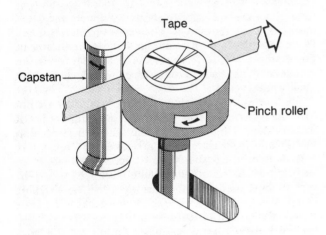

Figure 43.7. Capstan and pinch-roller assembly. (After Johnson, Walker, and Cutnell, 1981.)

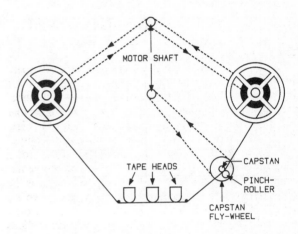

Figure 43.9. Two-motor tape drive system.

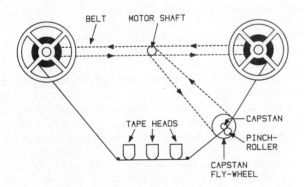

Figure 43.8. Single-motor tape drive system.

form the functions of erase, record, and playback. Units with separate record and playback heads have two advantages over units where these functions are combined into one head. First, the specialized heads are built with fewer compromises in design since each head serves only one purpose. Second, a separate playback head allows one to monitor the tape as it is being recorded so immediate corrections may be made if needed.

Cassette tape decks have both advantages and disadvantages compared to open-reel decks. On the plus side, cassette decks are smaller and more compact, tapes do not have to be threaded onto reels, and they cost considerably less than open-reel systems. On the negative side, cassette decks are not capable of fidelity quite comparable to that possible on open- reel decks.

The important specifications for tape decks are essentially the same as those we have defined for turntables or phono cartridges. The important tape deck specifications are: frequency response, S/N ratio, distortion, wow and flutter, stereo separation, and crosstalk. Unfortunately, these familiar terms are often applied with considerably less precision with tape decks than for other stereo equipment. The picture is further clouded by the fact that tape deck specifications are more difficult to establish because factors such as the type of tape used (iron oxide, chromium dioxide, etc.), the tape speed, and the recording level all affect the system's performance. A manufacturer may, for example, list an excellent frequency response for his equipment, but this specification may have been achieved by sacrificing the S/N ratio (of which he does not make an issue). We will discuss each of the above listed specifications for tape decks and indicate appropriate values for high quality equipment in the medium to medium-high price range.

The frequency response of a tape deck depends upon the tape speed and the recording level: the higher the speed of the tape, the better the high frequency

317

response (as will be explained in Chapter 44); the lower the recording level the better the frequency response, but the poorer the S/N ratio. The upper curves of Figure 43.10 show frequency responses for a signal recorded at 0 dB on the VU meter—the maximum level for recording without overdriving the system and causing tape saturation (see Chapter 44)—for the three different tape speeds of 1.88, 3.75, and 7.5 inches per second (ips). The fastest speed yields the best frequency response. The lower curves in Figure 43.10 show the results when the signal is recorded at -20 dB on the VU meter where all three speeds show a dramatic increase in the high-frequency response. The effect would be intermediate between these two cases if the signal were recorded at -10 dB. Not apparent on these graphs,

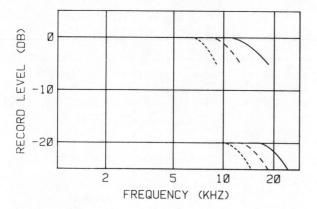

Figure 43.10. High frequency response for tape deck at three speeds: 7.5 ips (solid), 3.75 ips (large dashed), and 1.88 ips (small dashed). The upper curves are for recording at 0 dB and the lower curves are for recording at -20 dB.

however, is the progressive deterioration of the S/N ratio as the recording level is lowered. Essentially, the level at which the signal is being recorded is lowered, while the noise level remains the same, thus reducing the S/N ratio. Since cassette recorders have only one speed, 1.88 ips, the recording level is usually -20 dB. From the graphs we see that at this level a speed of 1.88 ips will yield approximately the same frequency response achieved on an open-reel machine at a speed of 7.5 ips recorded at 0 dB. Of course, the S/N for a cassette deck will be 20 dB below that of an open-reel machine.

The measured frequency response also depends upon the type of tape being used (e.g., chromium dioxide tapes yield a substantially improved frequency response), a variable seldom specified by the manufacturer. Since the frequency response of a tape deck depends upon the type of tape being used, the tape

speed, and the recording level, we have consolidated this information into Table 43.2 for easy reference.

The S/N ratio for a tape deck is the maximum signal level—of a 1000 Hz signal with no more than 3% distortion when recorded at a 0 dB level—of the deck as considered relative to the inherent noise level of the deck. Unless all of the information above (recording level, distortion, and frequency) is provided, the specification provided by a manufacturer should be read warily. Some manufactureres use a higher than standard reference signal (with 5% distortion) in order to make their S/N ratio appear better on paper. A reasonable S/N for a high quality open-reel or cassette deck would be 55 dB (weighted), using 0 dB as the reference level. The Dolby noise reduction system (see Chapter 44) will add another 5 to 10 dB to the S/N ratio. Since the S/N specifications are not entirely standardized, a potential buyer's best bet is to compare different tape decks by listening in a quiet environment. For example, record from disc to tape and listen to the amount of "hiss" added to the taped sound when a high quality tape is used.

Distortion figures given for tape recorders are not very meaningful since there are no standards of measurement. Even when the manufacturer states that the distortion measurements are taken at "maximum recording level," it should be realized that there is no standard definition of "maximum level." Even so, the type of tape used has not been specified. A listening comparison test between different tape decks is the most meaningful test for distortion.

Fortunately, specifications relating to the consistency of the tape speed, are amenable to definition and simple measurement. The wow and flutter in a high-quality tape deck should not exceed 0.15%.

Stereo separation and crosstalk, are similar whether applied to tape decks or phono cartridges. Both terms involve the leakage of the signal from one track to another track. **Stereo separation** is the number of dB that a signal in one channel is above the signal from the other channel which has "leaked over." Since the left and right channels are recorded on separate tracks (see Chapter 44), this problem is minimal for tape recorders. A cassette recorder can achieve a stereo separation of at least 40 dB, while open-reel machines typically show 50 dB or higher separation. (These are usually better separations than that which is present in the material being recorded.) **Crosstalk** is a more serious problem because it involves the leakage of information between two adjacent tracks. As we will see in Chapter 44, the adjacent tracks of a stereo tape deck contain the "other side" of the tape stored in the opposite direction. Any crosstalk will then appear as a garbled sound which will

Table 43.2. Frequency responses for open-reel and cassette tape decks using different audio tape at different recording levels and tape speeds.

Deck	Speed	Level	Tape	Frequency response
open-reel	7.5 ips	0 dB	low noise	50-13,000 Hz ± 3 dB
open-reel	7.5 ips	-10 dB	low noise	30-20,000 Hz ± 3 dB
open-reel	3.75 ips	0 dB	low noise	50-10,000 Hz ± 3 dB
open-reel	3.75 ips	-10 dB	low noise	40-14,000 Hz ± 3 dB
cassette	1.88 ips	0 dB	chromium dioxide	50-7,000 Hz ± 3 dB
cassette	1.88 ips	-20 dB	chromium dioxide	30-15,000 Hz ± 3 dB
cassette	1.88 ips	-20 dB	low noise	30-11,000 Hz ± 3 dB

interfere with the desired program. Thus, the specification for cross-talk for both open-reel and cassette machines should be 60 dB or higher.

FM Tuners

Radio waves carrying audio program material can be picked up by an appropriate antenna and converted into an electrical signal which is passed on to a tuner. Signals received by the tuner are either amplitude-modulated (AM) waves or frequency-modulated (FM) waves and it is the function of the tuner to demodulate them and produce an electrical signal bearing a more direct relationship to the audio signal. In AM broadcasting, the intensity of the carrier signal is constantly varied by the superimposed program material. In FM broadcasting, the intensity of the carrier wave is constant but its frequency is modulated by the information being broadcast. Let us now summarize the main differences in these systems. AM signals can be received over longer distances than FM, but with much more background noise. Although there are 20 times as many AM stations to choose from, only a restricted range of frequencies is broadcast (200-5000 Hz) and true high fidelity is impossible. Furthermore, in many localities it is difficult to accurately tune the desired station, and often the signal is garbled by competing stations (particularly at night). On the other hand, monophonic FM transmits frequencies from 30 Hz up to 15 kHz (almost to the limit of hearing), allowing high-fidelity reproduction. Also, the signals are low in noise and a station (once tuned accurately) remains in tune. The automatic frequency control (AFC) is a further refinement which holds the required station against "drift" effects. FM signals can also be used to transmit stereophonic signals effectively.

Often FM radio has been subject to the criticism that it is very wasteful to transmit only a relatively narrow range of audio frequencies over a bandwidth which could transmit a far greater frequency range. By means of multiplex techniques this ordinarily wasted capacity can be used to transmit two-channel stereo on a single channel. The simplest way to do this, if the frequency band available for transmission is from 0 to 53 kHz, would be to have one channel carried on the 0-15 kHz part of the band while the second channel would occupy the upper part of the band (23-53 kHz). Although this method is simple, it would be undesirable because someone who did not have a stereo FM tuner would receive only one of the two channels (the 0-15 kHz channel).

A practical FM stereo broadcasting system must be compatible with existing monophonic tuners. This has been accomplished in the following manner. First, the left (L) and right (R) signals are added to give a sum (L + R). The sum signal is broadcast on the 0-15 kHz channel, thus giving the "monophonic equivalent" of a stereo program. The R signal is also subtracted from the L signal to give an L-R "difference" signal. This difference signal is used to modulate a 38-kHz carrier signal; so the information is contained in the 23-53 kHz region, well beyond the range of audible frequencies. The stereo tuner detects the L + R and the L - R signals, then adds and subtracts them from each other.

In other words, two new signals are generated by algebraically manipulating the received channels. The operations yield the following: (L + R) + (L - R) = 2L and (L + R) - (L - R) = 2R. Note that this operation enables us to recover the original L and R channels as independent entities. Although this seems like a rather complex way to proceed, it is necessitated by the problem of having FM stereo broadcasting be compatible with existing monophonic equipment.

An FM tuner specification sheet contains the information necessary to evaluate how the tuner will perform under a variety of circumstances. These specifications are of two classes: those telling how well the tuner can select the desired station and those indicating how closely the output signals will resemble the original broadcast material. There are four important specifications (sensitivity, capture ratio, alternate channel selectivity, and AM rejection) in the first class and four (S/N ratio, total harmonic distortion, frequency response, and stereo separation) in the second. We will define each of these terms and present appropriate values for high-quality equipment.

Tuner sensitivity measures the minimum antenna signal which the tuner can transform into a satisfactory audio signal; the more sensitive the tuner, the better a distant or weak station can be received. Technically, **sensitivity** is the smallest input signal (expressed in microvolts) which can achieve 30 dB of "quieting" in the tuner. Quieting is the ability to suppress the background noise heard as a hiss between stations. A typical good specification for sensitivity might be 2 microvolts for 30 dB quieting. The lower the number of microvolts, the more sensitive the tuner—3 microvolts or less being a guideline for choosing good equipment. This will provide satisfactory reception up to a distance of 50 or 60 miles from the transmitter. Extrememly high sensitivity ratings (under 2 microvolts) are needed only in areas where there are not transmitting stations within 50 miles. Since 30 dB of quieting is still unacceptably noisy by hi-fi standards, some manufacturers specify a 50 dB quieting sensitivity. By this standard, quieting sensitivity should be no more than 9.7 microvolts for mono and less than 44 microvolts for stereo.

The **capture ratio** describes the ability of a tuner to suppress the weaker of two signals on the same transmission frequency while receiving the stronger. Technically it is defined as the number of dB by which a stronger signal must exceed a weaker signal (on the same broadcast frequency) for the audio output of the stronger to be 30 dB above the weaker. A good specification would be a 3.0 dB, or smaller, capture ratio.

In areas with many FM stations, alternate channel selectivity may be a more important consideration than the capture ratio. Channel selectivity is the ability of the tuner to suppress strong neighboring stations without affecting the desired receiving station. The **alternative channel selectivity** is defined as the number of dB by which the signal from an undesired station of different frequency can exceed that of a tuned-in station with the undesired audio program being 30 dB below the program of interest. A good value for alternate channel selectivity would be 60 dB or higher, with values of up to 100 dB being available.

AM rejection is a measure of the tuner's ability to suppress electrical interference of all types, including atmospheric static and man- made electrical noises such as fluorescent lamps or car engine ignitions. Some experts consider AM rejection to be one of the most important specifications for a tuner which will be used in an urban environment where the bouncing of signals from steel-frame buildings causes many spurious signals. A rating of -50 dB will give satisfactory AM suppression, but ratings of -60 dB are common on high-quality tuners.

The **signal-to-noise ratio** (abbreviated S/N) expresses the relative amount of interference with the desired signal. The S/N expresses the difference (in dB) between the desired signal level and background noise when the strength of the input radio signal is 1000 microvolts. Thus, if the tuner has a S/N of 60 dB, the noise will be 60 dB lower than the signal under the reception conditions of a 1000 microvolt input signal. A good value of the S/N for a stereo FM tuner is 60 dB or higher.

The tuner, like other electronic devices, produces a certain amount of unwanted higher harmonics of each frequency passing through the system. This is specified as the **total harmonic distortion** (THD). We will forego defining this important specification here, as it will be discusssed in some detail in a later section. The THD of a tuner should be specified for a 1000 microvolt signal carrying a 1 kHz audio signal, unless otherwise noted. For a stereo FM tuner, the THD should be 1.0% or less.

The frequency response of an FM tuner is defined in the same way as is the frequency response of a phono cartridge. The only pertinent difference is that the FM broadcast band is limited to audio frequencies between 30 Hz and 15 kHz. Any quality FM tuner should reproduce this entire range with a small level variation. A reasonable specification would be a frequency response from 30 Hz to 15 kHz ± 1 dB.

The final specification for tuners is **stereo separation**, which has the same meaning for FM stereo as for phono cartridges. For a high-quality FM tuner, the stereo separation should be 30 dB or more at 1 kHz, and at least 20 dB at 100 Hz and 10 kHz.

The Integrated Amplifer

In Figure 43.1 the selector switch, the pre-amplifier (preamp), the controls, and the power amplifier were shown as separate entities in a stereo system. In actuality, the selector and the controls are always part of the preamp unit, which is the ''hub'' of the sound system. The power amplifier may be a separate device, or it may be included with the preamp and controls as an integrated amplifier (Figure 43.11). If an FM tuner is also included, the unit is called a receiver. In this section we discuss the preamp and its associated controls, and the power amplifier. Important integrated amplifier specifications are defined in the next section.

The preamp unit serves several basic purposes. First, it enables the audiophile to select the desired input (phono, tape deck, or FM tuner). The selection control may also provide the electronic circuitry necessary to boost these different signal inputs to a common level (usually 0.5 volts). Although FM tuners and tape decks usually have their own built in preamps, the phono preamp is always part of the preamp unit.

The second function of the preamp is that of **equalization**, meaning that the audio signal has its amplitude increased or decreased in certain frequency ranges. For example, the amplitudes of audio signals to be FM broadcast are increased at frequencies above 1000 Hz, as shown by the solid curve in Figure 43.12. The equalization circuit in the preamp (or in the FM tuner) performs the opposite operation by attenuating all frequencies above 1000 Hz, as shown by the dashed curve in Figure 43.12. Since most noise in an FM broadcast band occurs in the higher frequencies, the stronger signals broadcast by pre-emphasizing the high frequencies increase the S/N ratio but, at the same time, create a distortion of the original sound. The de-emphasis of the equalization circuit restores the original sound by at-

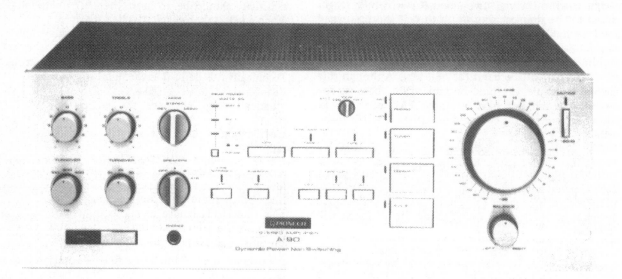

Figure 43.11. Integrated Amplifier. (Courtesy of Pioneer Electronics.)

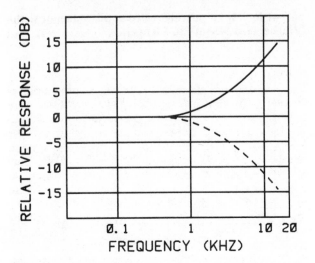

Figure 43.12. Pre-emphasis (solid curve) and equalization (dashed curve) for FM signals.

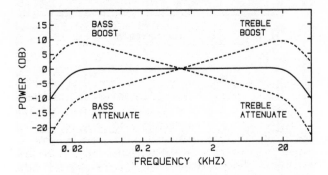

Figure 43.13. A flat amplifier response is shown by the solid curve when treble and bass controls are set to their neutral positions. The treble can be boosted or attenuated with the treble control and the bass can be boosted or attenuated by the bass control as shown by the dashed curves.

tenuating high frequencies, including noise, so that a favorable S/N ratio is maintained. A similar procedure, called RIAA equalization, is applicable to phono signals, while the Dolby system is a similar procedure applied to tape recording. Both of these will be discussed in more detail in the next chapter.

The third function of the preamp is to amplify the signals from the phono, tape, or tuner inputs. Power amplifiers usually require 7 volts input in order to achieve full power output. The preamp boosts any of the different possible input signals to the common level required by the power amplifier.

The final function is that of control. The most common preamp controls are volume controls, tone controls, a balance control, a loudness control, and filters. The volume controls, usually separate for each stereo channel, boost the signal level in the preamp. When a stronger signal is sent to the power amplifier, a louder sound will be produced by the loudspeakers. The tone controls consist of a bass control and a treble control for each channel. The bass control selectively boosts or attenuates the bass frequencies (below 800 Hz), while the treble control does the same for the treble frequencies (above 800 Hz). The maximum boost or cut in power from these controls is about 12 dB, as shown in Figure 43.13. The balance control is provided so that the sound from each stereo channel can be adjusted to the same level. This is usually done by setting the mode switch to mono, so that the same sound is coming out of each speaker, and then adjusting the balance control until these signals have the same loudness. When the mode switch is placed back in stereo mode, the channels will be balanced.

The loudness control (which is not identical to the volume control) is provided because one may wish to change the overall "level" of a program without having to adjust the tone controls. In Chapter 14 we learned that at low sound levels the ear is much less sensitive to

high and low frequencies. If one is listening to music from a hi-fi system at a fairly high level, and the volume controls are turned down, the recorded music will seem to lose its bass and its treble. The audiophile can boost both the treble and the bass control to re-establish the high fidelity, but the loudness control more simply achieves the same result. The loudness control may be used to reduce overall level, but as the level is decreased, the bass and treble are automatically increased.

Although filter switches are not a universal feature of pre-amps, they are a useful adjunct to any system. The two most common filters are the "scratch" or high frequency filter and the "rumble" or low frequency filter. The scratch filter is a low-pass filter which attenuates all high frequencies above the cut off frequency of about 4 kHz. Since scratches on records produce high frequency noise above 4 kHz, actuating this filter will remove such noise, along with all the other high frequency program material. The difference between a high frequency tone control and a scratch filter is that the tone control gradually reduces the high frequencies, while the filter strongly attenuates all frequencies above its cut-off. Figure 43.14 illustrates the effect of a scratch filter on a flat preamp response. (Compare this to the high frequency response shown in Figure 43.13 for treble attenuation.) The rumble filter is a high-pass filter with a cut-off frequency of about 100 Hz. Since turntable rumble is always low frequency, this filter will eliminate it, along with any 60 Hz hum picked up from the line voltage, and the bass end of program material. A comparison of the bass part of Figures 43.14 and 43.13 indicates the different effects of the rumble filter and the bass attenuation tone control. An additional filter found on more expensive units is the subsonic filter which removes those frequencies below 20 Hz. This filter will remove any low frequency vibrations which may have crept into the system, without removing any of the audible program material.

The power amplifier stage of an integrated amplifier boosts the relatively low power of the preamp to the high power output necessary to drive the

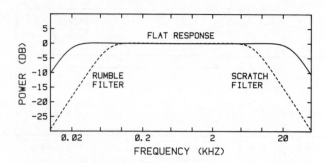

Figure 43.14. Flat amplifier response is shown by the solid curve and the effects of the rumble filter and the scratch filter are shown by the dashed curves.

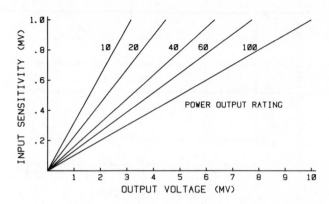

put, we look for a cartridge rated at this value, or slightly higher (e.g., 4.0 mV to 5.0 mV).

Figure 43.15. Graphs for determining phono cartridge output voltage for any given phono input sensitivity and power output rating (watts per channel).

loudspeakers. For the power to be transferred efficiently from the amplifier to the loudspeakers, the output impedance of the amplifier must be matched to the input impedance of the loudspeakers (usually 8 ohms). For this reason, most power amplifiers have several output impedances available (e.g., 4, 8 and 16 ohm) so that a variety of speakers of different impedances can be connected to the amplifier. It is also possible to connect combinations of speakers to the same output terminal.

Integrated Amplifier Specifications

The Institute of High Fidelity, (IHF) recommended in 1978 that 28 different standardized specifications be listed for amplifiers. We will define and discuss the seven specifications identified as being of primary importance: input signal sensitivity, maximum input signal, total harmonic distortion, frequency response, S/N ratio, continuous RMS power output, and dynamic headroom. We will also discuss several considered to be of secondary importance: intermodulation distortion, transient distortion, stereo separation, damping factor, and transient response.

The **input sensitivity** is the input voltage required to produce an output power of 1 watt at a frequency of 1000 Hz when the volume control is turned to its maximum position. The greater the sensitivity, the smaller the input signal required to produce an output power of 1 watt. This specification is not necessarily an indication of quality, but rather is a means of matching phono cartridges to preamp inputs, as will now be illustrated. Given a manufacturer's phono input sensitivity (in millivolts) and the amplifier's power output rating (in watts/channel), use the appropriate graph in Figure 43.15 to determine the required average ouput voltage (in millivolts) from a phono cartridge. This approach gives the minimum value for the phono cartridge output voltage so that the amplifier can deliver its maximum rated power when so required. As an example, consider an amplifier with a phono input sensitivity of 0.62 mV and a power output rating of 40 watts/channel. Locate 0.62 mV on the vertical axis of Figure 43.15. Draw a horizontal line from this point over to the slanted line labeled "40." At this point of intersection, drop a vertical line down to the axis labeled output voltage, and read the voltage, 3.9 mV in this case. Since this represents the minimum average phono cartridge out-

The **maximum phono input signal** is the largest input voltage (in millivolts) to an amplifier that will not cause distortion when the volume control is turned to a low level. When this maximum phono input signal is exceeded, the signal is peak clipped because the amplifier can no longer amplify to the level required (refer to Figure 15.6). Because of music's wide dynamic range, the maximum signal may be many times the nominal output, especially during loud musical climaxes. For a high-quality preamp, the maximum phono input rating should be at least 20 times the average output voltage.

Although at least three types of distortion plague audio equipment, only total harmonic distortion has been classified as a primary specification by the IHF. This type of distortion derives its name from the harmonics produced by any device whose input is a sinusoid. Harmonic distortion exists when an amplifier distorts the input signal by adding additional harmonics, or by increasing the strength of an instrument's existing harmonics. Since no stereo equipment can totally eliminate this problem, the goal of high-quality equipment is to minimize it. **Total harmonic distortion** (THD) is measured by using a sinusoid to drive the amplifier and then filtering the sinusoid from the output signal. Any frequencies remaining in the output are created by harmonic distortion. The strength of all harmonics present is measured and expressed as a percentage of the amplifier's output at a given power level. Figure 43.16 shows the THD as a function of the power output for a typical amplifier at a single test frequency. For this particular amplifier, the THD for a signal of any frequency remains less than 0.5% until a power of 50 watts is reached. At higher powers the THD increases rapidly. Curves such as the one in Figure 43.16 are used to determine the power rating of an amplifier. For the amplifier of this figure, the specifications could be listed as THD = 0.5% at full-rated power of 50 watts. If the manufacturer had decided to list this amplifier as a 60 watt amplifier, the THD would have to be given as THD = 0.8% at full-rated power. An acceptable value for the THD of a medium priced amplifier is THD = 0.5% or less at full-rated power. The lower the value the better the specifications; many medium and high priced amplifiers have THD of 0.1% or less.

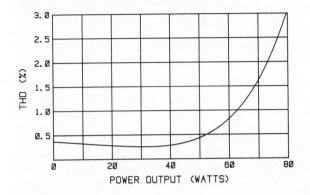

Figure 43.16. Total harmonic distortion (THD) versus power output for a power amplifier.

The frequency response of an amplifier, usually presented as a graph similar to Figure 43.17, is a measure of the output power of the amplifier (in dB) as a function of frequency, with a reference of 1 watt output at a frequency of 1000 Hz. If a graph is not provided, the specification may be written: frequency response from 20 - 20,000 Hz ± 2 dB. The wider the frequency range and the smaller the dB variation, the better the frequency response. If the plus and minus dB figures are not given, you may assume that the manufacturer has something to hide.

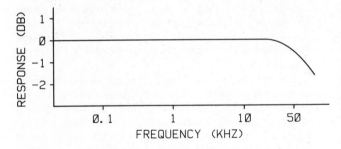

Figure 43.17. Power amplifier frequency response.

The S/N ratio for an amplifier is defined as the ratio of 1 watt of output power to the noise level produced by the amplifier. As discussed previously, the ratio is expressed in dB. For a quality amplifier, this ratio should be 70 dB or higher.

When the power ratings of amplifiers are discussed, confusion arises from two separate sources. First, some manufacturers state power specifications in a way such that inferior equipment "looks" better than it actually is. Second, the amount of power actually needed depends upon the efficiency of the loudspeakers, the type and size of the listening room, and the overall loudness desired. We will discuss this second set of factors in Chapter 45.

Let us consider the power specifications one is likely to encounter when purchasing an amplifier. Some manufactureres rate their equipment in terms of "music

power" or "dynamic power." Beware of this bogus, undefined specification designed to give an inflated power rating. The specification to look for is the **continuous average power output**, or RMS (root-mean-square) power output, defined as the average amount of power per channel that the amplifier can deliver to loudspeakers of specified input impedance on a steady basis, when both channels are driven, over the frequency range of 20 - 20,000 Hz, with less that 0.5% THD. Notice that this specification, in order to be complete, must specify each of the following items: the amount of power continuously available per channel, the input impedance of the speakers, the fact that both channels are being driven simultaneously, the frequency range, and the THD. The number of continuous watts per channel is specified so the consumer knows that the power rating is not based upon peak response, but rather upon a long-term average. This is the power available to each of the two channels. Even so, some manufacturers state the total power the amplifier can deliver (which is double the power/channel) in order to make their equipment appear to be more powerful. A reliable power measurement must also specify the input impedance of the load, since an amplifier will deliver more power to low impedance speakers. The tests must be performed with both channels activated; otherwise, if only one channel is being driven, the power rating will appear higher. Finally, as discussed previously, the frequency range and the THD are integrally related to the power output and must also be specified if the power rating is to have any real meaning. A typical specification for an amplifier of medium-high quality may appear as 20 watts (continuous) per channel, into 8 ohm speakers, both channels driven simultaneously, from 20 - 20,000 Hz with less than 0.5% THD at rated full power.

The common misconception about stereo amplifiers is that the power rating tells how loud a system will sound. This is true only in the same sense that the horsepower rating of a car tells you how fast the car will go. If you had two otherwise identical stereo systems, one having a 30-watt/channel amplifier and the other a 60-watt/channel amplifier, and both were played at top volume, the 60-watt system would sound ony slightly louder than the 30-watt system. The difference is in the "power reserve" for moments when needed. Amplifiers seldom operate at full power, but when the extra power is required, the power reserve will help keep the system from distorting. An amplifier is also capable of delivering more than its rated power without increased distortion for very short time periods (about 20 ms). A useful specification which measures an amplifier's ability to so perform is known as dynamic headroom. Technically, **dynamic headroom** is the number of dB of power above the rated continuous power an amplifier can deliver for short time periods without exceeding the rated THD. A typical specification for a quality amplifier would be 3 dB, which corresponds to a momentary doubling of the available power.

Although the next specifications are considered to be of secondary importance by the IHF, they provide useful information about an amplifier and are usually listed by the manufacturer. Even though the stated values of distortion may be small, small amounts of distortion, which may not register in our conscious mind, have a subliminal effect which, after a period of time, creates "listener fatigue." The type of distortion which seems to contribute most to listener fatigue is intermodulation distortion (IMD). IMD results from the fact that no amplifier can be made to be completely linear, and so sum and difference tones of the frequencies present are produced (see Chapter 15). Suppose, for example, that two sinusoids having frequencies of 60 and 1000 Hz are input to an even slightly nonlinear amplifier. The output will contain, in addition to these frequencies, their sum (1060 Hz), their difference (940 Hz), and other frequencies. (See Figure 15.6B for an example.) The fact that these new frequencies are usually not harmonically related to the original tones makes them particularly irritating. Since IMD cannot be totally eliminated, the best that the manufacturer can do is to keep it at a minimum.

Intermodulation distortion is defined by the following set of measurements. Two sinusoidal test tones of frequencies 60 Hz and 600 Hz (with an intensity ratio of 4:1) are input to the amplifier. Frequency filters at the output suppress these two frequencies; the remaining signals (produced by the IMD) are measured and expressed as a percentage of the total output. The manufacturer may list this specification at full-power output or at a 1-watt output level. For a quality amplifier, the IMD should be less than 1% at full power output.

A much subtler, but perhaps more important form of distortion is known as transient intermodulation distortion (TIM). Many audio experts are now convinced that the audible differences between similar amplifiers having identical power, THD, and IMD ratings are due primarily to TIM. This type of distortion seems to be prevalent in amplifiers with a large amount of negative feedback and a small time delay between input and output signals. When a high amplitude transient musical signal is input to the amplifier, the feedback necessary to reduce the amplitude of this transient arrives too late and peak clipping may occur momentarily. Naturally, the greater the dynamic headroom, the less the TIM produced under these conditions. To measure TIM distortion, the following procedure is used. A high frequency square wave is used to simulate the time-varying properties of a transient signal. The square wave is well suited for this purpose since it changes from a constant positive value to a constant negative value and back again during each cycle. If this square wave signal drives the amplifier into a nonlinear region during the steep rise and fall times, then other signals being amplified at the same time will be distorted. A test signal (Figure 43.18) is composed of a sinusoid having an amplitude one-quarter that of the square wave and a frequency about five times that of the square wave superimposed on the square wave. The intermodulation components of these two signals are measured and their combined amplitudes determined. TIM is the ratio of this amplitude to the original amplitude expressed as a percentage. A quality amplifier should keep TIM distortion below 0.5%, although studies have shown that TIM distortion percentages as low as 0.2% are aurally detectable.

The stereo separation of an amplifier is the amount

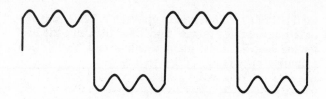

Figure 43.18. Test signal for measuring transient intermodulation distortion (TIM).

of left-channel signal leaking over to the right channel, and visa-versa. Separation is measured by sending a test tone into one channel and measuring the level of the signal present in the other channel. The difference between the two levels (expressed in dB) is the stereo separation, and the higher the rating, the better the separation. Even an amplifier of medium quality should be capable of a stereo separation of at least 30 dB in the mid-range frequencies (400 to 8000 Hz). Below 400 Hz the separation is not particularly important because diffraction effects spread the waves around the room. At very high frequencies (above 10 kHz) it becomes increasingly difficult to maintain stereo separation, but since program material and total energy at high frequencies is minimal, deterioration of separation in this range is not noticeable.

The **damping factor** describes an amplifier's ability to "dampen" or control unwanted residual movements of the speaker cone, resulting from inertia, after a signal is terminated. For example, when a "staccato" note terminates, the loudspeaker cone continues to vibrate slightly, creating an effect called "overhang." The amplifier can act as an "electrical shock absorber" to dampen these unwanted residual vibrations. The greater the ability of an amplifier to do so, the higher the damping factor, (expressed as a dimensionless number). Damping factors above 20 are generally considered adequate, but some amplifiers have dampling factors as high as 100.

Transients are any rapid changes in the amplitude of a signal, whether due to a crashing cymbal, the impact of piano hammers on strings, or the attack time of an oboe. Orchestral music is full of transients, and yet, transients are one of the most difficult things for an amplifier to reproduce accurately. When an amplifier doesn't respond rapidly enough to a transient in the signal, transient distortion inevitably results. When there is sufficient distortion of this nature, transient sounds may appear somewhat "blurred." The specification which purports to define an amplifier's ability to respond to transients is the **transient response**. Unfortunately, there is not a standard method of measuring transient distortion, so no objective measurement exists. Some manufacturers specify the "rise time" for their amplifiers; that is, the time (in microseconds) taken to reach some specified output level. Other manufacturers like to show pictures of oscilloscope tracings of the amplifier output when the input signal is a square wave. Since the square wave is the sharpest of all transient responses, the pictures give an indication of an amplifier's ability to respond, as shown in Figure 43.19. However, such pictures are subject to a considerable amount of interpretation, and will provide reliable comparisons only to trained personnel.

Figure 43.19. Amplifier transient measurement: (A) input square wave; (B) output from amplifier with poor frequency response; (C) output from amplifier with good frequency response.

Fortunately, there is a simple way to get some indication of an amplifier's transient response without having to resort to oscilloscope pictures. As a rough rule of thumb, the wider and flatter the frequency reponse of the amplifier and the higher the dynamic headroom, the better the transient response will be. This is why high-quality amplifiers often have a frequency response up to 50 kHz, even though we hear frequencies only up to about 18 kHz. These very high frequencies enable an amplifier to accurately reproduce signals with transients as steep as those of a square wave.

Exercises

1. High-fidelity sound systems may have a frequency range which extends beyond the range of human hearing. Is this extended range of any value in reproducing the sound?

2. What characteristics of the ear make bass and treble boost controls essential on modern hi-fi equipment? (Assume you want to listen to music which is balanced across the audible frequency range at both high-intensity and low-intensity levels.)

3. If the volume of a preamp is boosted so that output power is doubled, will the music sound twice as loud? Explain.

4. A stereo system produces a certain loudness when the power is turned to maximum. In order to double the possible sound level, the amplifier would have to be replaced with an amplifier having a greater power rating. How many times greater? Explain.

5. Explain why long sustained piano chords are useful in aurally testing for wow and flutter in a turntable.

6. Explain the difference between wow and flutter and speed accuracy for a turntable.

7. If the tracking error of a tone arm is given as 1.5 degrees/inch, compute the total tracking error angle at the outer edge of a 12 inch disc and at the inner edge of the same disc.

8. How would you rate the quality of the Empire cartridge from its specifications given in Table 43.1.?

9. Indicate the frequency response (or bandwidth) for normal hearing, a telephone, a hearing aid, a small tape recorder, a professional tape recorder, and a hi-fi amplifier.

10. When high-quality amateur tape recordings of music are made, the speed of the tape is usually 7.5 ips. Why would a lower speed not suffice?

11. Can better separation of stereo channels be achieved from a record or a tape? Why?

12. Explain why an equalization circuit is an essential part of modern hi-fi preamps.

13. Why is stereo AM radio broadcasting not practical?

14. The following specifications are for a Big Moose Hi-Fi Receiver. Frequency response: 30-30,000 Hz ± 25 dB; S/N ratio: 20 dB; Power output: 12 watts; Extras: single channel mixing. Evaluate the receiver from the specifications given. Would you buy this equipment?

15. If the frequency response of a piece of stereo equipment is given as 20-35,000 Hz, with no plus or minus dB figure, is the specification valid? Explain.

16. Explain the difference between tone controls and rumble and scratch filters on a preamp.

17. A typical power amplifier has the same input voltage and output voltage. If the input impedance is 10,000 ohms and the output impedance is 8 ohms, compute the power input (in watts) and the power output (in watts). (Hint: Power = (voltage)^2/impedance). If the input and output voltages are the same, why is this device called an amplifier?

18. Two 8-ohm loudspeakers are to be connected to the same amplifier output terminal. If the speakers are connected in parallel—that is, similar wires from each loudspeaker are connected to the same output terminal—what is their total impedance? Should they be connected to the 4-ohm, the 8-ohm or the 16-ohm output? Explain.

19. If the two 8-ohm speakers of Exercise 18 are connected in series, to which outputs (4-ohm, 8-ohm, or 16-ohm) should they be connected? Will this method or the method described in Exercise 18 give better results?

20. If the phono input sensitivity of a preamp is given as 8.5 mV, and the power output rating is 60 watts/channel, determine the phono cartridge output voltage which the cartridge should generate.

21. Explain why a 60-watt amplifier produces sounds only slightly louder than those of a 30-watt amplifier, all other factors being the same. Why then should you invest in an amplifier with more power?

22. Use Fourier's Theorem (see Chapter 6) to explain why an amplifier needs a frequency range up to 50 kHz in order to accurately reproduce a square wave with a good transient response.

Demonstrations

1. To see the importance of a wide frequency response for music, listen to a recording that has been bandlimited. This can be partially simulated by adjusting the controls on an equalizer. Listen to and compare music played over a quality hi-fi system, a telephone, a pocket transisitor radio, a typical clock radio, a "budget" hi-fi system, a hearing aid, and other systems.

2. The nonlinear response of the ear (see Chapter 15) can be demonstrated by comparing music played at a low level (40-50 dB) with that played at high levels (80-90 dB). While playing music with a large amount of bass at a high level change to a low level. Pay particular attention to the relative perceived bass at high and low sound levels.

3. Listen to music through reproduction systems that permit the following to be demonstrated: (a) undistorted music, (b) music at both high and low S/N, (c) low-pass filtered music, (d) high-pass filtered music, and (e) peak clipped music.

Further Reading

Aldred Chapters 9, 10
Eargle, Chapters 8-11

Johnson, Walker, and Cutnell, Chapters 2, 8, 10, 12, 13

Official Guide, Chapters 1, 2, 4-6, 10

Olson, Chapter 9

Traylor, Chapter 3, 7

Allison, R. **High Fidelity Systems** (Dover).

Feldman, 1979. "All About TIM Distortion," Radio/Electronics (June) 47- 49.

Fantel, H. 1976. **ABC's of Hi-Fi and Stereo** (Howard Sams Co).

Foster, E.J. 1977. "How to Judge Record-playing Equipment," High Fidelity (April), 60-68.

Hope, A. 1976. "Will Ambisound Shatter the Peace and Quiet of the Stereo Market?" New Scientist **69**, 222-224.

Institute of High Fidelity. 1978. "Standard Methods of Measurement for Audio Amplifiers," Publication A-202 (Institute of High Fidelity).

Maxwell, J. 1977. "Phono Cartridge Noise," Audio (March), 40-42.

Sands and Shunaman. 1969. **101 Questions & Answers about Hi-Fi and Stereo** (Howard Sams Co.).

Villchur, E. 1965. **Reproduction of Sound**, (Dover).

Audiovisual

1. **Frequency Response in Audio Amplifiers** (30 min, color, AIM)

2. **Introduction to Radio Receivers** (30 min, color, AIM)

3. **Waves, Modulation and Communication** (19 min, 1969, HRAW)

44. Recording Media

When we wish to save a certain set of musical sounds for later enjoyment, we must turn the sound into a form which can be stored conveniently. Perhaps the oldest way of doing so is by means of sheet music. Although the storage medium is very simple, an extremely complicated reproducing mechanism (the human performer) is required. Other media used extensively in the past to store music have included the player piano "rolls" and spiked drums in music boxes. Although these methods are simple and work amazingly well, they are not particularly convenient or readily accessible. Today four different media are commonly used to store almost any sound imaginable. The methods are mechanical, magnetic, optical, and digital. The mechanical method is used to produce plastic records; the magnetic method refers to tape recording tape; while the optical is used to store the sound track of movie film. Digital discs are the most recent innovation for storing music. In this chapter we will consider the first two media in some detail, and conclude with a discussion of the digital disc systems.

Disc Recording

The successful creation of a record requires four important links, each of which must maintain extremely high quality. These four links are (1) the recording machine, (2) the master disc, (3) the copying apparatus, and (4) the final product. We will consider these devices, as well as the magnetic pick-up cartridge in more detail.

The recording machine is a sophisticated, high-quality lathe which engraves a spiral track on an aluminum disc coated with cellulose acetate. This engraved disc, which must be absolutely free of any blemish, is known as the master. The recording head includes a steel stylus (see Figure 44.1) attached to a coil of wire located in a magnetic field (see Figure 44.2). When a varying voltage (the signal to be recorded) is introduced into the coil the stylus vibrates, tracing out a mechanical version of the time-varying input signal. The cutting edge of the stylus is kept very clean and sharp so that the groove walls will be smooth. As the cutting process proceeds, a highly flammable waste material, known as swarf, is formed. The swarf must be removed immediately; often this is accomplished by air suction. This recording process is diagrammed in Figure 44.3.

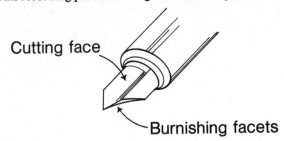

Figure 44.1. Cutting stylus.

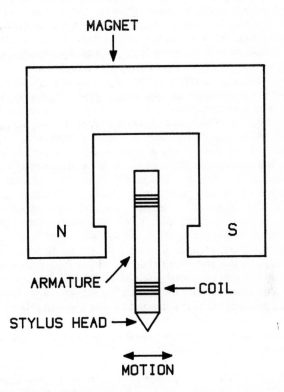

Figure 44.2. Cutting head.

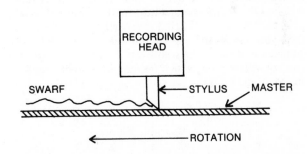

Figure 44.3. The disc recording process.

The depth of the cut made by the stylus must be very carefully controlled. Too deep a cut will yield an incorrect groove width, while too shallow a cut will not enable the reproducing head to track correctly. The correct depth is maintained by a feedback device which monitors the cutting process. In order to produce the fine groove spacing required of 33.3 rpm records, the cutting stylus is extremely small. The resulting groove width is 0.0064 cm with a depth of 0.0032 cm (see Figure 44.4). The number of grooves per inch averages out to about 250; for soft sounds the grooves are placed closely together, while for loud sounds they are spaced more widely. The space between grooves, called the land-width, is the area available for carrying the signal. Obviously, the groove excursion cannot exceed the land-width, but we do want the maximum excursion to be as great as possible to mask the surface noise. If the excursion is too great, however, the resulting stress will deform the material of the master. Since almost all of the large-amplitude components in music occur in the low-frequency range, the low frequencies are artificially attenuated in the recording process. Also, since most record playback noise occurs in the high frequencies, and since high-frequency tones tend to be under-modulated on a disc recording, the high frequencies are artificially boosted. The amount by which the bass is attenuated and the treble is boosted for disc recording has been standardized as discussed later. Obviously the playback preamp must perform the reverse process in order to reproduce the original sound.

After the master acetate disc has been cut, it is cleaned and dipped into a bath of silver nitrate, which causes a very thin layer of silver to be deposited. The silver makes the disc electrically conductive so that a nickel or copper plate can be made. A hammer and chisel are then used to remove the metal negative (called the father) from the master, which is destroyed in the process. Since the father is too valuable to be used for stamping out records, a mother is next made from the father in the same way that the father was produced from the master, except that the father is not destroyed in this process. Since the mother is a playable disc, flaws can be detected and corrected at this stage of the process. The mother is then used to generate further negatives, appropriately called sons, which are used to stamp out the vinyl records.

Since stereo records require two independent channels, the cutter must vibrate in two perpendicular directions, as is indicated in Figure 44.4. Each of the two walls of the groove (at 45 degrees to the perpendicular) has one waveform impressed upon it; the left or inner wall carries the left channel information, while the right or outer wall carries the right channel. During playback, the stylus mimics the original motions of the cutter. The complex motion of the stylus can then be resolved into two separate signals, which become the two channels of stereophonic sound.

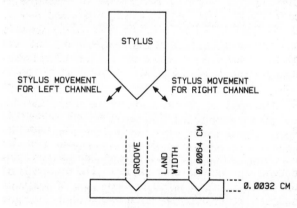

Figure 44.4. Movements of cutting stylus for two-channel recording (upper) and section of disc showing groove and landwidth (lower).

Let us consider how the signal may be retrieved from the groove. One method, used in many stereo systems, is to attach a stylus to an armature around which coils are wrapped, and suspend the armature in a magnetic field, as shown in Figure 44.5. When the stylus vibrates due to following a record groove (see Figure 44.6) an induced voltage is produced in the coil of wire (see Chapter 2) which closely approximates the pattern on the record. Stereo grooves having waves of different frequencies on their two sides are depicted in Figure 44.7 which shows (A) information on right channel only, (B) information on left channel only, and (C) information on both channels. Note the motion of the stylus tip in each of these cases.

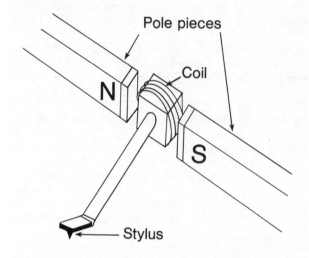

Figure 44.5. Magnetic phono pick-up. (After Rossing, 1980.)

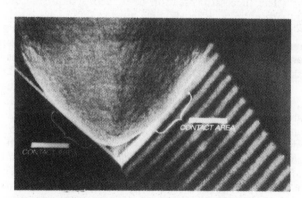

Figure 44.6. Stylus in "V-groove." (Source unknown.)

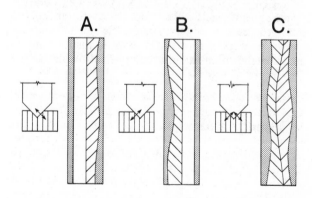

Figure 44.7. Sketches of stereo record grooves: (A) right channel recorded; (B) left channel recorded; (C) both channels recorded.

328

The frequency of a wave is encoded on a disc by the number of squiggles encountered by the stylus each second. The amplitude of the squiggles helps determine the amplitude of the electrical signal which also depends on the velocity of the stylus—the greater the velocity the larger the induced voltage. The velocity of the stylus depends on the frequency of the encoded wave and its amplitude. In a typical recording, the maximum velocity of the stylus is chosen to be about 5 cm/s for the largest signal on the disc. The maximum amplitude at any frequency (f) is given by (maximum velocity)/$2\pi f$. For signals having a frequency of 10 kHz, the maximum amplitude is about 0.00008 cm, while at a frequency of 1000 Hz the maximum amplitude is 0.0008 cm, and at 100 hz it is about 0.008 cm. Since frequency and amplitude are inversely proportional, the lower the frequency the greater the amplitude of the recorded signal, while even loud sounds at high frequencies will be recorded with relatively low amplitude. Since the bass signals require large stylus motion, the grooves have to be spaced farther apart on the disc, thus allowing for less recorded material. The high-frequency treble sounds are subject to obliteration by any noise due to the slight but unavoidable surface irregularities on the disc.

The solution to both of these problems is to attenuate the bass during recording and to boost the treble. The standard recording curve, called the RIAA (Recording Industries Association of America) recording curve is shown in Figure 44.8. By recording all discs with this bias, the bass excursions are reduced in amplitude, allowing more program material and less chance of consecutive grooves overlapping, and overall noise level is reduced by pre-emphasizing the weak treble sounds. The surface noise produced in the course of the needle-record contact is essentially "white noise," with equal energy per hertz. Thus, each higher octave,

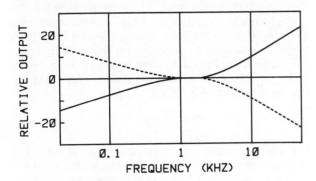

Figure 44.8. RIAA recording curve (solid) and RIAA playback equalization curve (dashed).

having twice the bandwidth, also has twice as much surface noise, making this a particular problem for the treble region. An untreated signal would show a progressive deterioration in the S/N ratio at high frequencies, as indicated in Figure 44.9A. Boosting the treble by 6 dB per octave keeps the S/N ratio constant, as shown in Figure 44.9B. To compensate for bass attenuation and treble pre-emphasis in recording, the opposite process must be effected in the preamp by electronic circuitry which boosts the bass and attenuates the treble, as

shown in Figure 44.8. This curve is known as the RIAA equalization curve and is an industry standard built into all phono preamps. The signal of Figure 44.9B, after being equalized during playback will appear as in Figure 44.9C. Notice that the level is flat, while the noise has been reduced to maintain a constant S/N ratio.

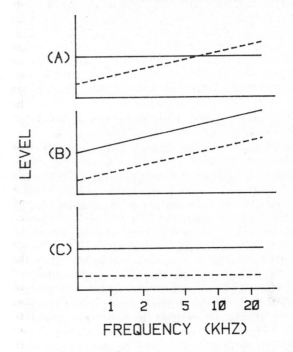

Figure 44.9. Effects of playback equalization showing relative levels of signal (solid) and noise (dashed): (A) untreated signal; (B) pre-emphasized signal; (C) signal after playback equalization

Tape Recording

Although a magnetic recording and playback device was invented in 1898, tape recorders did not really become practical until they were improved and developed by German scientists during World War II. Early devices utilized steel wire as the "tape." Although the quality was not bad compared with other machines of the day, the lack of suitable amplification and the inconvenience of the metal tapes curtailed development. By the late 1950's tape recorders were using electronic amplification for recording and playback, and the plastic based tape had been invented. The recent invention of small high quality cassette machines has made tape recorders a popular adjunct to any home hi-fi system.

Recording tape consists of a thin (5 to 50 micron) plastic base coated with finely ground magnetic materials in a binder which allows the magnetic particles to change their orientation. Figure 44.10 illustrates components of magnetic recording tape. The properties of the tape depend on which of two types of synthetic material, acetate or polyester (better known by the trade name of Mylar), is used as a base. Although many will argue the relative merits of acetate versus mylar tape, we will simply list some pros and cons of each type of tape.

Mylar is twice as strong as acetate, but it tends to stretch when pulled, as shown in Figure 44.11. Naturally, anything recorded on a stretched piece of mylar tape

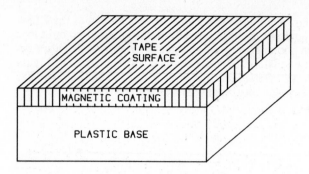

Figure 44.10. Components of magnetic recording tape.

will be totally ruined with no chance of recovering the lost information. Acetate tapes, on the other hand, will break without stretching. Since a broken tape can be spliced back together, many professional recording engineers contend that acetate is preferable since it can be patched, while a stretched Mylar tape is ruined beyond repair. Mylar is slightly more expensive, but it lasts considerably longer than acetate. In fact, Mylar-based tapes are among the most durable materials known to mankind, lasting longer than stone or steel. Acetate, on the other hand, gets dry and brittle with age; after 15 years or so it may disintegrate. Although mylar tapes can be identified by stretching, as shown in Figure 44.11, a non-destructive test is to look at the tape edge-on. If you can see light through the edge of the tape it is Mylar, whereas acetate tapes are opaque.

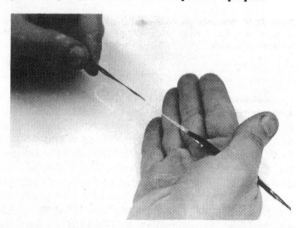

Figure 44.11. Stretched mylar tape showing how the magnetic coating has separated from the clear mylar base. (Courtesy of G. Orvis.)

Cassette and open-reel tapes come in several formats, depending upon their playing time. For cassette tapes, the number following the letter "C" is the total playing time of the tape; thus a C-60 tape plays 30 minutes on each side. Although C-120 tapes are available, a thinner tape is used making these tapes more likely to break or to jam. Open-reel tapes are gauged in mils. A standard 7-inch reel of 1.5 mil tape will run for 0.5 hours at 7.5 ips, and twice as long at 3.75 ips. Longer playing times are available with "extra play" tapes of 1 mil thickness or "double play" tapes of 0.75 mil thickness. Naturally, the thinner the tape, the greater the chance of having the tape stretch or break, and the greater the "print-through."

Tapes of comparable thickness come in three general varieties: top brands, house brands, and bargain tapes. The house brands cost less than the top brands not because the house brands are inferior, but because of more relaxed quality-control. The top brands have very high standards for their tapes, so a purchaser can be sure of always getting a high-quality tape. House brands will be good most of the time, but occasionally, one will be of poor quality and recording characteristics may vary slightly from tape to tape. The bargain, or "junk" tapes as they are known among audiophiles, are no bargain at all. If you are lucky, you will get reject computer tape, designed for very high-frequency use. Although the bass response is poor, the quality is not too bad for recording speech. If you are unlucky, however, you will get reject video tape which has the magnetic particles aligned in an up-down direction, rather than horizontally as in audio tapes. Although you can record on such tape, there will be a very audible "hiss." The magnetic coating of video tapes is also thinner than the coating used for audio tapes. The resulting saturation distorts all strong signals. Although you can purchase five reels of "bargain" tape (which usually comes in an unmarked white box) for the price of one reel of standard tape, if you want to be sure of a good recording, stay with standard brands.

Most open-reel tape decks use four tracks on tape one-quarter inch wide, as shown in Figure 44.12. Tracks 1 and 3 are the L and R channels for "side one," while tracks 2 and 4 are the respective L and R channels for "side two." Note that sides one and two are both on the same physical side of the tape, the side with the magnetic coating. To go from side one to side two, one simply turns the tape over. For side one, the two stereo heads line up with tracks 1 and 3, whereas for side two, the tape is turned over and the same two stereo heads line up with tracks 2 and 4. The reason for this staggered format is that when the early stereo tape recorders were made, the tape heads could not be made small enough so that adjacent tracks could be utilized. When the smaller cassette recorders came out in 1964, technology had progressed to the point that smaller heads could be made and mounted so that adjacent tracks could be used for each "side" of the tape, as shown in Figure 44.13, even though the cassette tape is only three sixteenths of an inch wide.

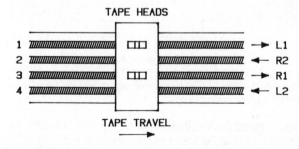

Figure 44.12. Track arrangement for open-reel stereo tape. The tape is "invisible" where it passes over the heads so that the head arrangement can be seen.

The magnetic coating on the audio tape's plastic base consists of a binder with finely ground iron oxide or chromium dioxide particles. For purposes of the present discussion we will assume the particles to be iron

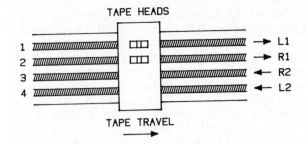

Figure 44.13. Track arrangement for cassette stereo tape. The tape is "invisible" where it passes over the heads.

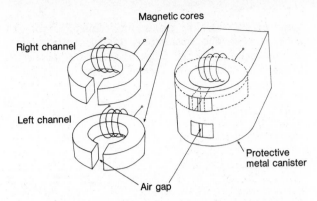

Figure 44.15. Stereo recording heads removed from cannister and in protective cannister. (After Johnson, Walker, and Cutnell, 1981.)

oxide. The iron oxide particles join into large groups of atoms called a magnetic domain, having a net magnetic field. When these magnetic domains are oriented in a random manner (Figure 44.14A), the material has no net magnetization. If a weak magnetic field is applied, some of the domains align with the field as shown in Figure 44.14B. As the strength of the magnetic field is increased, more and more of the domains align with the field until all domains are aligned in a condition called saturation as shown in Figure 44.14C. We may use this technique of "ordering" magnetic domains to store a signal on audio tape. A recording head (Figure 44.15) is used to align the magnetic domains according to a varying electrical signal, while a playback head converts the stored magnetic information back into a time-varying signal. This process is illustrated in Figure 44.16. Both frequency and amplitude information may be stored on magnetic tape. Amplitude is stored in terms of the number of aligned domains on any one region of the tape. A stronger recording signal creates a greater magnetic field which aligns more domains. The greater the number of aligned domains passing the playback head, the stronger the induced voltage, and the greater the amplitude of the signal. The greatest possible amplitude which can be stored on tape is limited when saturation occurs.

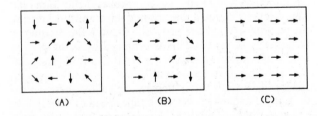

Figure 44.14. Magnetic domain alignments. (A) Random alignment of domains resulting in no net magnetic field. (B) Partial alignment of domains resulting in a weak stored magnetic field. (C) Complete alignment of domains resulting in a strong stored magnetic field.

Frequency is stored in terms of the distance between ordered regions on the tape. As the ordered magnetic domains on a recording tape pass the playback head, the induced voltage varies in time in a manner determined by the tape speed divided by the spacing between like-oriented domains. This spacing represents one wavelength of a stored signal. Consider a sinusoidal signal input to a recording head, as shown in Figure 44.16. When the signal is positive, a magnetic field which orients the domains to the right is created across

the gaps of the recording head. When the sinusoidal signal goes negative, the magnetic field changes direction and orients the domain alignment to the left. When the sinusoid becomes positive again, the domains are aligned again to the right. One wavelength of the sinusoid, then, is represented by two adjacent domains in a right-left arrangement.

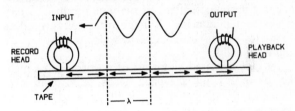

Figure 44.16. Magnetic recording and playback process. The input signal to the record head creates a magnetic field which is stored on the tape in the form of the alignment of magnetic domains (arrows). The stored field (arrows) induces a voltage in the output as it passes the playback head.

The size of the magnetic domains and the size of the gaps in the recording head limit the highest frequency which can be stored on tape. The problem of domain size can be illustrated with the following example. Suppose we wish to record a 15 kHz signal on a cassette player with tape speed of 1.88 ips. The wavelength on the tape is the tape speed divided by the frequency which gives 1.25×10^{-4} inches for this example. There must be at least two domains per wavelength so the domain size can be no larger than 0.63×10^{-4} inch. A modern cassette tape using chromium dioxide has a domain size on the order of 0.10×10^{-4} inch, considerably smaller than that required.

The size of the gap of the recording head limits the uppermost frequency to one whose wavelength is no smaller than twice the width of the gap. If the gap is so wide that a complete wavelength of a recorded signal fits between the pole pieces, there is no net magnetic field since the two domains offset each other as shown in Figure 44.17A. A maximum field between the poles is created when there is a single domain between the poles as indicated in Figure 44.17B. Since one magnetic domain corresponds to one-half wavelength, the highest frequency which can be recorded has a wavelength twice the length of the gap width. The gap for open-reel machines is about 2×10^{-4} inches. At a speed of 7.5 ips, the maximum possible frequency is 19 kHz, while at a speed of 3.75 ips, the maximum frequency drops to about 10 kHz. The considerably smaller gap in cassette machines allows the full range of audio frequencies to be reproduced at the low speed of 1.88 ips.

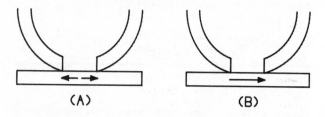

Figure 44.17. (A) A full wavelength of a magnetic alignment occupies the head gap. (B) A half wavelength of a magnetic alignment occupies the head gap.

The playback process is the reverse of the record process. Provided the tape travels at the same speed during playback as it traveled while being recorded, the magnetic patterns will recreate the original signal. As the magnetic patterns pass the playback head, the changing magnetic pattern of the tape produces a pattern in the gap inducing a voltage in the coil (according to Faraday's Law—see Chapter 2) which is a copy of the original signal.

A tape is erased by means of a third head which, like the record and playback heads, is also an electromagnet. A strong high amplitude signal of very high frequency (around 60 kHz) is used to randomize the magnetic domains. A considerably wider gap is used for erasure, however, since we want the domains to be randomized through-out the entire binder where iron oxide domains are located. The magnetic field produced by a wider gap can reach deeper into this layer so that all domains are randomized and no residual signal remains. It is quite easy to accidentally erase a tape. If the magnetic domains are given enough energy to break their binder orientation, they will reorient along any stray magnetic field present. One way this occurs is when the tape is heated by placing it near a heat vent; the greater the heating, the more domains will realign, and the greater the erasure.

Our discussion of the magnetic recording process has been intentionally oversimplified. A recording could be made by the process described to this point, but because this process is non-linear, the reproduction would be of poor quality. The nonlinearity arises from the relationship between the magnetization retained on a magnetic tape and the current in the record head coil used to produce the magnetization. The **transduction curve** in Figure 44.18 shows the relationship between residual magnetization on an audio tape and the "audio current" in the record head coil. Most of the transduction curve is nonlinear and thus distortion will result during recording. For example, suppose that two signals, I_1 and I_2 (where I_2 has twice the amplitude of I_1) are recorded. The residual magnetism B_2 is greater than twice the residual magnetism B_1 and the recorded signal is no longer an accurate representation of the original. However, portions of the transduction curve are linear. When all recording is done in a linear region—with a technique known as **biasing**—the response is linear. A

large amplitude signal of very high frequency (about 100 kHz) is added to the audio input signal. The purpose of this signal, known as the bias frequency, is to produce an average magnetic field at the middle of the linear part of the transduction curve. The superimposed audio signal produces small variations in the linear part of the curve resulting in a recorded signal that is not distorted. There is an optimum amplitude for the bias signal which depends on the type of tape being used. For this reason, a quality tape deck or cassette machine has bias switches in order to match the amplitude of the bias signal to the tape (e.g., standard, low noise, chromium dioxide).

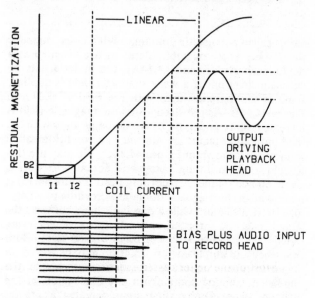

Figure 44.18. Transduction curve showing the relationship between the residual magnetization in the audio tape and the current in the record head coil. See text for further detail.

Like records and FM tuners, tapes require equalization during both recording and playback—otherwise both the very high and the very low frequencies would be severely attenuated. To compensate, the high and low frequencies must be boosted in amplitude during the recording/playback process, so that a flat frequency response is achieved. This is achieved by using a standard equalization known as the NAB (National Association of Broadcasters) equalization curves, shown in Figure 44.19. This figure depicts typical magnetic recording and playback curves for a tape speed of 1.88 ips. In order to obtain a good S/N ratio as well as to increase the bass, the treble is enhanced during recording while the bass level is increased during playback. The turnover point (marked X) moves up the scale as the tape speed is increased, having values of about 7000 Hz at 1.88 ips, 12,000 Hz at 3.75 ips, and 14,000 Hz at 7.5 ips. For this reason high-quality tape recordings of music are recorded at 7.5 ips while a higher-quality commercial tape made for producing a master disk is recorded at 15 ips.

When a magnetic tape is tightly wound on its spool a strongly magnetized section of the tape may slightly magnetize the tape on an adjacent layer. This phenomenon, known as "print-through," causes a "ghost sound" in the reproduction. Print-through can be reduced by using a heavier tape and by winding the tape less tightly. The rewind mechanism of tape

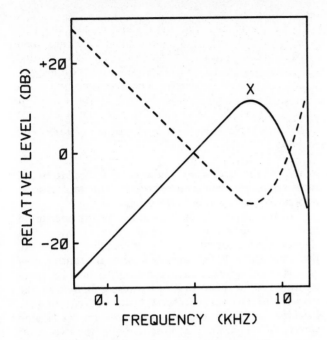

Figure 44.19. Equalization curves for recording (solid) and playback (dashed). The turnover point (X) occurs at about 7 kHz in this example.

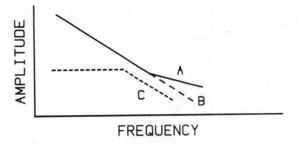

Figure 44.20. Dolby process. (A) Dolby signal with low amplitude signals boosted for recording. (B) Signals attenuated on playback. (C) Hiss reduced at high frequencies.

recorders winds the tape much more tightly than when the tape is played. Hence, open-reel tapes will suffer less print-through when stored on the take-up reel and rewound just before playing.

Most of the better cassette recorders, and many of the open reel machines feature a Dolby noise suppressor. The Dolby system reduces the hissing background inherent in all tape recording without narrowing the dynamic range of the music. The hissing sound is produced during playback and even blank tape produces a noticeable hiss. Because the head gap size is so small (only several times the size of the magnetic domains), only a limited number of domains are near the gap at any one time. When no signal is present, the domains are randomly aligned, but when the sample size is small, one direction of alignment will predominate. Since the domains are randomized, however, this alignment changes with time in a random manner. This process produces a ''signal'' of white noise, which we perceive as high frequency hissing.

The Dolby system is a two-step process. First, the input signal is carefully monitored. Whenever the amplitudes of the high frequencies are low, they are increased (up to 10 dB) so that the signal will remain substantially louder than the background noise. Only the high frequencies are boosted, since this is where the hissing is most bothersome. Second, when the tape is played back, the high frequencies are attenuated by the same amount. The result is that the softer sounds can be reproduced at their natural level without an undue amount of annoying hiss. Figure 44.20 outlines this process. The solid line represents the recorded signal. Note that the high frequencies have been boosted in amplitude, while the lower frequencies are unchanged. When, upon playback, the signal is restored to normal, the hiss at high frequencies is noticeably reduced as shown by the dotted line.

The Dolby process will function properly only when used both in recording and in playback. An ordinary tape played through a Dolby circuit will be missing high frequencies and hence sound dull and drab. On the other hand, a Dolby-recorded tape played without Dolby circuits will sound shrill because of the emphasized high frequencies. A Dolby tape may still be played on a recorder without the Dolby circuits, but the treble would have to be attenuated to get a reasonably natural sound.

Digital Audio

Storage of audio signals on discs and tapes is familiar to most people. Digital audio—a new form of recording, storage, and reproduction of audio signals—promises to have a profound influence on the future of recorded music. Digital audio should provide audiophiles improved quality and flexibility in audio as it becomes available on the mass consumer market. In this section we will discuss digital encoding and storage of audio signals, methods of retrieving digitally-stored information, and finally some pros and cons of digital audio.

Recording and reproduction with discs and tapes is done with continuous or analog signals. Digital recording requires that an analog voltage be represented by a series of numbers, and that these numbers be stored for later retrieval. The process of changing an analog signal into a digital or discrete signal is called **pulse code modulation** (PCM). It was described briefly in Chapter 6, but will be reviewed here. Suppose the audio signal is a sinusoid, such as the one illustrated in Figure 44.21A. An analog-to-digital converter is used to sample this signal at the points in time indicated by the vertical dashed lines. At each point in time, the sinusoid is sampled and the amplitude at that point is converted to and recorded as a number—a digital signal. Care must be taken to ensure that a type of distortion, called **aliasing** does not occur during the sampling process. Note that the sinusoid of Figure 44.21B, which is not present in the original signal, has the same set of digital samples as the waveform in A, and would appear in the output when the samples are reassembled. This is called aliasing distortion because the higher frequency wave is an ''alias'' (that is, false identity) of the original low-frequency wave. To avoid aliasing distortion, the sampling frequency must be at least twice as great as the highest frequency in the signal. The sampling rate of available digital systems is generally between 44,000 and 51,000 samples per second, which means that the highest usable frequency is about 20 kHz. To avoid

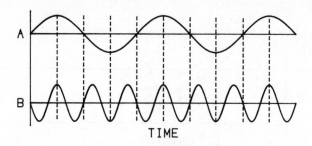

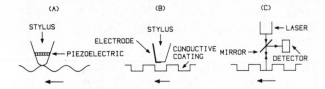

Figure 44.22. Digital audio playback schemes: (A) mechanical; (B) capacitive; (C) optical.

Figure 44.21. Sampling of two sinusoids illustrating aliasing. The samples are taken at the points on the two waves indicated by the dashed lines. Note that both sinusoids result in the same samples.

aliasing, all information above one-half the sampling frequency must be eliminated by using a low-pass filter.

After an analog signal has been LP-filtered to remove any frequencies above 20 kHz, it is sampled at a rate greater than 40 kHz by a sample and hold circuit. This circuit samples the input voltage every 25 microseconds or so and "holds" the voltage until it is time to take the next sample. The output from this circuit, still in analog form is fed to the ADC which quantizes the voltage to the nearest discrete number, which represents the amplitude at that point. This analog-to-digital conversion process is illustrated in Figure 6.13.

The resolution in amplitude depends on the number of bits per sample. The current generation of professional recorders uses 16 bits, which corresponds to 65,536 discrete values; home systems will use fewer. If the number of gradations is not large compared to the amplitude, quantization errors result, the effects of which can be seen in Figure 6.13. Amplitude changes of less than the amount of the digit-to-digit gradation of the converter cannot be encoded, resulting in quantization error. This error can be avoided by using a sufficient number of bits. After A/D conversion, the digital representation of the signal may be stored in a digital memory, on a digital tape or on a digital disc for later retrieval. The digital recording process would be as shown in Figure 6.13 with the output going to some digital storage medium.

Digital signals stored on a tape are represented by a series of off-on magnetic patterns rather than the continuous patterns of an audio tape and can be read by a digital tape reader. Digital signals are read from a disc by one of three methods as illustrated in Figure 44.22: mechanical, capacitive, or optical. The mechanical system utilizes a stylus traveling in a smooth groove without the undulations characteristic of a conventional disc. Instead, the stylus rides up and down over microscopic holes in the bottom of the groove. The "holes" and "non-holes" represent the zeros and ones of binary digital information. If the surface of the disc is made so as to be electrically conducting, and the stylus is a metal electrode, the system can sense changes of capacitance as the "holes" go by. In optical detectors, the "holes" are optically scanned with a laser. Optical discs are made dust and scratch-proof by means of a transparent layer over the recording surface.

After digital information has been retrieved from storage, it must be converted back into an analog signal. This process, the opposite of the encoding process, is illustrated in Figure 6.14. The numbers are D/A con-

verted and the "stair-step" signal is LP filtered to round the sharp corners and remove specious frequencies so as to provide reproduction of the original signal.

At this point, one may wonder why any one would bother storing a digital representation of an analog signal when for considerably lower cost they could store the "real thing." The answer is that no matter how good an analog device such as a recorder may be, there is always some distortion and noise inherent in the storage and retrieval process. If, instead of requiring the machine to record and reproduce a complex analog waveform we require that it handle only the two states of magnetization (on or off), there is less chance of error creeping in. A digital recorder with a fairly large amount of distortion and noise will still do a creditable job of storing zeroes and ones. The necessary capability of handling large amounts of data rapidly and without error in the reading and writing process is available from computer technology. Distortions arising from a digital approximation to an analog waveform become inaudible when the sampling rate is sufficiently high and the voltage resolution is sufficiently fine. The benefits of digital audio are evidenced by the impressive specifications of digital tape recorders: distortion at full recording level is less than 0.03%, flutter is unmeasurable, S/N ratio is greater than 96 dB, crosstalk is more than 90 dB down, and there is no discernable print-through. Furthermore, an unlimited number of copies of a master tape or of a master disc can be made with no loss of quality and each copy has the same high quality as the master.

Although the specifications of digital recorders are truly impressive, they do not perform perfectly, or even as well as might be expected. Audio specifications, which were developed to describe analog equipment, may not be entirely appropriate for digital equipment. The very low distortion figure (0.03%) is true only for the maximum signal level where the digital system performs best. This is opposite that of an analog device, where the distortion increases at higher levels. Distortion at low-levels in digital systems arises because as the signal amplitude gets smaller, the constant size gradations become larger relative to the signal. This quantization error can cause severe distortion near the bottom end of a digital recorder's dynamic range. In extreme cases, a sinusoid is "converted" into a type of square-wave by the quantization process. This effect, akin to amplifier clipping, introduces considerable distortion and produces a harsh, gritty sound. The subjective effect of this noise is considerably more unpleasant than ordinary tape hiss. Therefore, many digital recording machines add a very low level tape-hiss noise to mask this negative effect. Audio signals smaller than the smallest recordable gradations of a digital recorder are simply ignored, whereas with an analog recorder, signals which are below the system's noise level can still be heard.

Although the S/N ratio of a digital recorder is quoted as being 96 dB, the useful dynamic range may be somewhat less. When an analog recorder is working at the upper limit of its dynamic range and a loud burst drives it into a nonlinear range, the effect is gradual. A digital system, however, reaches its upper limit abruptly and any signal above that limit is completely clipped. This type of "hard" peak clipping produces considerably more distortion than the "soft" clipping of analog devices. The maximum usable digital recording level must be set somewhat below the actual maximum, thus reducing the effective dynamic range. Home digital tape recorders will probably use a 14-bit code to store amplitude, giving a maximum dynamic range of 84 dB, and a usable dynamic range that is considerably less.

Because digital discs or tapes must be capable of long uninterrupted playing times, the data must be packed very densely. Because of this tight packing, there is a high chance of missing bits or "dropouts." With an analog recorder, a dropout (caused by momentary loss of contact between tape head and tape) is a minor annoyance; in a digital recorder it is an unmitigated disaster. On some encoding devices, a dropout produces a burst of noise at maximum output level. In order to control this problem, manufacturers have included error correction circuits which predict what value the missing bit should have. If corrections fail, a muting circuit replaces the noise burst with a brief silence. Because of the problem with dropouts, professional digital recorders are used with two tape transports running at the same time, the second being a back-up machine to replace dropouts of the first.

Mixing and signal processing in the digital domain require some very expensive ancillary devices, as well as a person who understands how to program them. However, given these, digital tapes can be editted much more precisely than with the cut and splice methods used with analog tapes. Mixing also can be of better quality.

Two additional criticisms leveled at digital audio are (1) that the sound is harsh and "gritty" in the high frequency region and (2) that the sampling rates are too slow and the amplitude gradations too coarse to yield accurate reproductions. If these criticisms prove to be valid, it may be necessary to increase sampling rates to 100 kHz and to represent amplitude with more than 16 bits. Whatever the future prospects of digital audio may be, it seems safe to conclude that its influence on recorded music will be long-term, substantial, and beneficial.

Exercises

1. Why do low-frequency sounds have a greater amplitude than high-frequency sounds which seem to be equal in loudness?

2. Explain how the RIAA standards help to reduce surface noise and groove excursion.

3. Why does the erase head of a tape recorder have a wider gap than the record/playback head?

4. Why is an upper frequency limit inherent in tape-recorded music? Why does raising the tape speed increase this frequency limit?

5. Why is it not practical to reduce the size of the gap in the tape recording head to even smaller spacings than are in use today?

6. If you own a tape where a print-through is a particular problem why would it be wise not to rewind the tape after you finish playing it?

7. Compare and contrast records and tapes as sound storage media. What are the advantages and disadvantages of each?

8. Name and describe all the transducers used in the process of tape recording a sound and then playing it back.

9. Name and describe all the transducers involved in making a stereophonic record and then playing it back.

10. Based on the domain size for a chromium dioxide tape, compute the highest frequency which can be recorded when the tape speed is 1.88 ips.

11. Will the Dolby noise suppressor reduce the noise of old records when they are recorded for later playback? Why or why not?

12. When audio signals are stored on magnetic tape, how are the frequency and the amplitude stored? What magnetic properties represent frequency and amplitude respectively?

13. What are the advantages and disadvantages of using fast tape speeds?

14. What are the advantages and disadvantages of using wide tracks on tape?

15. What is tape biasing, and why is it necessary?

16. Explain the difference between tape biasing and equalization.

17. How are frequency and amplitude information stored on digital discs?

18. What are the advantages and disadvantages of digital discs as opposed to conventional discs?

19. Explain why digital tapes and records have no discernible wow and flutter.

20. Why is an LP filter necessary with digital devices? What should be the value of its cut-off frequency?

21. What is aliasing distortion, and how can it be eliminated?

22. What is quantization error, and how can it be controlled?

23. Is it wise to invest in a new turntable/tonearm/cartridge system now that digital audio is becoming available? Consider relative system costs, program material available, and cost of program material.

Further Reading

Aldred, Chapters 6, 8

Eargle, Chapters 8, 10, 11

Johnson, Walker, and Cutnell, Chapters 12, 13

Official Guide, Chapter 7

Traylor, Chapter 5

Bose, A. G. 1973. "Sound Recording and Reproduction. Part I: Devices, Measurement and Perception," Technology Review (June); "Part II: Spatial and Temporal Dimensions," Technology Review (Jul/Aug).

Feldman, L. 1982. "Equipment Profile: Technics

SV-P100 Digital Audio Cassette Recorder," Audio **66**, 46-50.

Kundert, W. R. 1978. "Everything You've Wanted to Know About Measurement Microphones," Sound and Vibration **12** (Mar), 10-23.

Meyer, E. B. 1982. "Digital," Stereo Review **47** (Apr), 56-59.

Neff R. 1981. "Digital Audio Players Push to Market," Electronics (Jan 13), 102-108.

Ranada, D. 1981. "Digital Audio, a Primer," Stereo Review (Feb), 63-65.

Rodgers, H., and L. Solomon. 1979. "A Close Look at Digital Audio," Popular Electronics (Sep), 39-44.

Rossing, T. D. 1980. "Physics and Psychophysics of High-Fidelity Sound, Part II," The Physics Teacher **18**, 278-289.

Weiler, H. D. 1956. **Tape Recorders and Tape Recording** (Radio Magazine, Inc.).

Woram, J. M. 1976. **The Recording Studio Handbook** (Sagamore).

Audiovisual

1. **The Soundman** (15 min, color, 1970, UEVA)

45. Electroacoustic Transducers

It is well known that in 1876 Alexander Graham Bell invented the telephone. Perhaps it is less widely realized that in order to transmit a telephone message he had to invent electroacoustic transducers—devices which change sound into electrical energy and vice versa. In this section we will be concerned with two basic types of electroacoustic transduction devices: the microphone and the loudspeaker.

Microphones

As we have seen in Chapter 6, microphones are transducers which transform acoustical energy into electrical energy. Since sound waves consist of tiny pressure variations in air, the molecules of air move from regions of greater density to regions of lesser density. The movement of air molecules as the pressure rises and falls is known as the particle velocity. Ideally, a microphone should respond to both pressure and particle velocity. In practice, most microphones respond only to the pressure variations. We will consider several types of pressure microphones and one type of velocity microphone.

In pressure microphones, a sound wave impinges on a diaphragm, causing the diaphragm to vibrate. The mechanical motion of the diaphragm produces the varying voltage in one of several possible ways. Let us review the three types of pressure microphones discussed in Chapter 6: the ceramic microphone, the condenser microphone, and the dynamic microphone.

A ceramic microphone generates a varying voltage by means of deformation of a piezoelectric material. The piezoelectric material generates a voltage proportional to the deforming force, which in turn depends on the pressure of the sound wave. The diaphragm is connected to one end of the ceramic as shown in Figure 6.2. Microphones of this type are small, are fairly inexpensive, and have a fairly uniform response from 20 Hz to 10 kHz. The ceramic microphone is widely used in P. A. systems and in hearing aids. The principal disadvantage of these microphones is that relatively small output voltages are obtained for normal acoustic pressures and consequently amplification is needed.

The condenser microphone depends on the variation in capacitance between a fixed plate and a tightly stretched metal diaphragm. The capacitance varies with the variation of the distance between the diaphragm and the plate. Since the electrical output varies directly with the changing capacitance, the output corresponds to the motion of the diaphragm. A diagram of this type of microphone is shown in Figure 6.3. The condenser microphone has a very high-quality response, but a high bias voltage (200 to 400 volts) must be supplied to make the microphone work. Also, a preamplifier is required in close proximity to the microphone to boost the power output. Condenser microphones are used exten-sively in hi-fi recording and for research purposes. Electret condenser microphones are now in wide use because they offer the advantages of a condenser microphone without the disadvantages—though they are slightly less sensitive.

A dynamic microphone has a coil of wire attached to the diaphragm as shown in Figure 6.4. The coil is free to move between the poles of a magnet. The electrons in the coil move with the coil and experience a force (as described by the magnetic force law) which produces a varying electric voltage across the ends of the coil. Dynamic microphones are capable of relatively high power output, are rugged, have good high frequency response, and are capable of a broad frequency response over a wide dynamic range. Since they are able to withstand the high intensity sound levels often associated with popular music, they are widely used for live performances and for recording sessions.

The electrical response in velocity microphones corresponds to the particle velocity in a sound wave rather than the varying pressure. The most common type of velocity microphone is the ribbon microphone, in which a light corrugated metallic ribbon is suspended in a magnetic field with both sides exposed to surrounding air. The ribbon is driven by the difference in sound pressure between the two sides, which corresponds to the velocity. The motion of the ribbon in the magnetic field produces a varying voltage corresponding to the original sound. This type of microphone has good low-frequency response which is quite uniform but the upper useful limit is usually about 9000 Hz. The response pattern of velocity microphones is bidirectional as compared to the omnidirectional response of pressure microphones. Since a velocity microphone responds equally to sounds in both directions it is useful for discriminating against undesirable sounds, balancing orchestral instruments, and picking up dialogue.

Loudspeakers

In this section we will consider the cone loudspeaker in some detail, followed by a brief discussion of horn loudspeakers. The next section deals with the baffling of loudspeakers in enclosures and the final section concerns loudspeaker specifications. Information about evaluating and purchasing loudspeakers will be presented in Chapter 46.

A loudspeaker is a transducer actuated by electrical energy that produces an acoustical waveform essentially proportional to the time-varying electrical input. The most common type of loudspeaker is the direct radiator dynamic loudspeaker. The dynamic loudspeaker is the inverse of the dynamic mike, both of which were described in Chapter 6. The varying input voltage is fed to the voice coil which is situated in a magnetic field produced by a permanent magnet. The varying voltage in the magnetic field causes the coil to experience a force and thus to move. The coil is attached to the loudspeaker diaphragm so that the motion of the coil drives the diaphragm, thus causing the air to vibrate and produce a sound. Because of the simplicity of construction, the small space requirements, and the fairly uniform frequency response, this type of speaker is almost universally used in radio, phonograph, and intercom systems.

A fundamental consideration for all loudspeakers is the amount of air that needs to be moved by the speaker. Figure 45.1 is a graph of sound power output (in watts) versus frequency. The slanted lines represent the amount of air (in cubic centimeters) that must be displaced by the speaker during each cycle. If the sound power output is to be 1 watt, we see that at 4000 Hz 0.1 cm^3 of air must be displaced, while at 400 Hz 10 cm^3 of air must be displaced. If the frequency is lowered to 40 Hz 1000 cm^3 of air must be displaced to generate 1 W of acoustical power. Thus, in order to reproduce low frequencies, large volumes of air must be moved, requiring a large transducer, many smaller transducers, or a small transducer capable of large excursions.

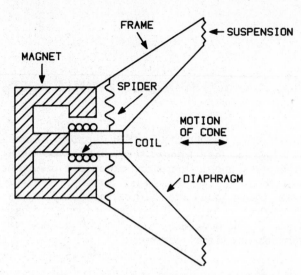

Figure 45.2. Schematic representation of a dynamic cone loudspeaker.

2. This force causes the coil to move in and out, thus driving the cone and causing the air to vibrate. In order to work properly, the voice coil must be centered exactly over the central magnet pole and be free to move in and out with no vertical excursions in the gap between the magnet poles. The **spider** is the support which allows the coil to move freely in the horizontal direction without allowing any vertical motion which would cause the coil to rub on the magnet poles. Figure 45.3 illustrates two different types of spiders. The gap clearance between the coil and the magnetic pole pieces is extremely small, on the order of 0.025 cm for a large speaker, and 0.008 cm for a small speaker. The speaker is always made so that the gap is as small as possible in order to increase the efficiency of the transducer, producing more sound power with a smaller magnet.

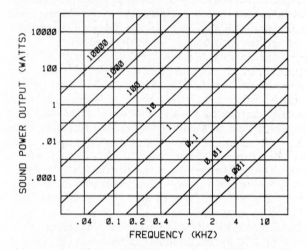

Figure 45.1. Sound power output versus frequency for different quantities of air displaced (labels on curves in cubic centimeters) per cycle. (After Olson, 1971.)

Figure 45.3. "Spiders" used to support loudspeaker coil and cone.

Figure 45.2 is a schematic representation of a dynamic cone loudspeaker. The four major elements are: (1) the diaphragm or cone, (2) the flexible suspension on the edge of the cone, (3) the permanent magnet, and (4) the voice coil. The entire assembly is held together by a metal frame. The cone, attached to the frame by the flexible suspension, is free to move in response to the input signal. The flexible suspension is a key element in the design of the loudspeaker. The permanent magnet provides a strong magnetic field surrounding the voice coil. The voice coil is a hollow cylinder with thin wire wrapped tightly around it. The coil is rigidly attached to the apex of the speaker cone so that when the coil is forced to vibrate the diaphragm follows. The wires from the voice coil are connected to the output terminals of the power amplifier.

When a time-varying voltage from the power amplifier exists in the voice coil, a time-varying force, directly proportional to the voltage, is produced on the coil via the electromagnetic effect discussed in Chapter

Should you remove the front grille from a loudspeaker enclosure you would see that the enclosure contains more than one loudspeaker, each of different size as shown in Figure 45.4. Since it is difficult to accurately reproduce the full audio range with one loudspeaker, two or three different loudspeakers are used to reproduce different parts of the frequency range. In a three-way system, the bass notes are produced by a large cone driver (typically 6 to 12 inches in diameter) called the **woofer**; the intermediate frequencies are produced by a **midrange** unit; and the high frequencies by a **tweeter**. Each loudspeaker produces sound only in its designated frequency range. The input signal, which covers the entire frequency range is subdivided by **crossover networks**, into the three separate ranges. The first crossover frequency is at 600 Hz with frequencies below 600 Hz being sent to the woofer. Frequencies above 600 Hz are sent to a second crossover network which separates the signal into frequencies above and

Figure 45.4. Two-way loudspeaker system with grille removed. (Courtesy of Pioneer Electronics.)

below the second crossover frequency of about 6000 Hz. (The actual crossover frequencies used vary from manufacturer to manufacturer and from system to system.) The response curve for a three-way system is shown in Figure 45.5. When multiple small loudspeakers are used to cover the same portion of the frequency range, no crossover network is required. These may be used to increase the bass without using a large speaker—two 6 inch speakers could be used instead of one ten inch. Two or three tweeters may be used to obtain better dispersion of high frequencies. A two-way system (Figure 45.4) uses a woofer and a tweeter, with a typical crossover frequency of about 1000 Hz. The response curve for such a system is shown in Figure 45.6.

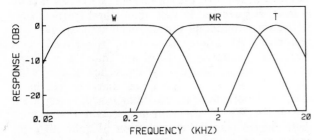

Figure 45.5. Idealized curves for three-way loudspeaker system showing woofer response (W) midrange response (MR) and tweeter response (T).

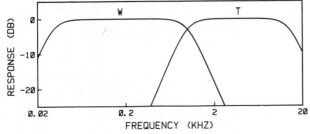

Figure 45.6. Idealized response curve for two-way loudspeaker system.

Woofers are relatively large (with diameters from 6 inches to 30 inches) because they must move enormous amounts of air in order to reproduce bass tones. The three ways in which a large volume of air can be moved by a woofer account for the different woofer designs. First, a woofer can have a large diameter, but a relatively small in and out excursion (see Figure 45.7). Second, several smaller speakers can each move a small volume of air, but with a large total quantity of air being moved. Finally, the diameter can be kept small, but provision can be made for greater cone excursions, so that more air can be moved. This is accomplished by speakers with very flexible suspensions capable of large in and out excursions. Bookshelf speakers are examples of small diameter, large-excursion loudspeakers (see Figure 45.4). A comparison of Figures 45.4 and 45.7 dramatically illustrates that size alone does not determine the bass frequency response of a speaker. Figure 45.7 is a photo of one of the first full-range loudspeaker systems developed by Bell Labs in the early 1930's. The enormous woofer on the bottom is a type of folded horn, while the horn "tweeters" on top provide dispersion of the high frequencies. Incredible as it may seem, a modern bookshelf speaker (see Figure 45.4) provides nearly the same frequency range as the "historic monster," but with less distortion and a more uniform frequency response. Admittedly though, the "monster" is capable of producing much greater sound power and at much greater efficiency.

Figure 45.7. Full-range loudspeaker system produced by Bell Telephone Laboratories in the 1930s. (Courtesy of Bell Laboratories Archives.)

Since a midrange unit must fill the gap between the woofer and the tweeter, some care is required to insure that the speakers are adequately matched to each other. First of all, the frequency ranges must be carefully matched, or a "hole" in the sound will result where the midrange and woofer responses do not overlap (see Figure 45.8A). Second, the midrange speaker must match the woofer and tweeter in efficiency or the effect will be that reflected in the response curves of Figures 45.8B or 45.8C. In Figure 45.8B the midrange is not sufficiently strong relative to the other speakers. As a result, the system will lack "presence." A vocalist, for instance, would seem to be singing from behind the accompanying instruments. In the situation reflected in Figure 45.8C, the midrange is overly strong and the middle frequencies dominate the sound. This is also highly undesirable. A well-designed three-way speaker system has response curves similar to those shown in Figure 45.5. The responses "cross" each other in frequency so that as one speaker "dies out" the next one is picking up. Also notice that the amplitudes of the three response curves are the same, indicating that the three speakers are evenly matched.

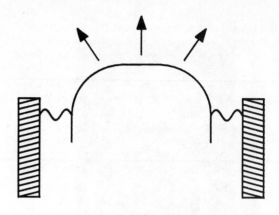

Figure 45.9. Dome tweeter used to produce high-frequency dispersion.

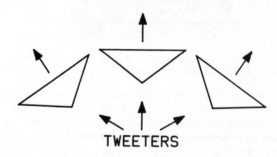

Figure 45.10. Three-tweeter array used to produce high-frequency dispersion.

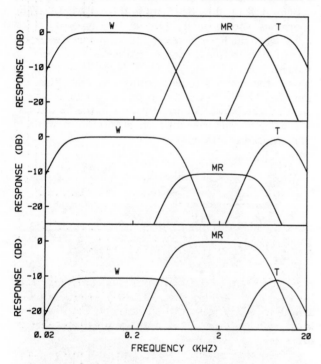

Figure 45.8. Idealized loudspeaker response curves showing: (A) frequency mismatch between woofer and midrange; (B) midrange too inefficient; (C) woofer and tweeter too inefficient.

Tweeters, ranging in size from 1 inch to 3 inches in diameter, must be extremely lightweight with a stiff supporting edge. Since the frequency response of a system is proportional to stiffness and inversely proportional to mass, the tweeter is designed to respond well to high frequencies. Three common types of tweeters being manufactured are: (1) the small flat cone, (2) the dome-shaped diaphragm, and (3) the horn tweeter. Since high-frequency sounds are highly directional, dome-shaped tweeters (see Figure 45.9) or horn tweeters help to disperse the sound. Another method often used to help increase the dispersion of high frequencies is that of using several tweeters, each pointed in a different direction, as shown in figure 45.10.

A loudspeaker diaphragm, having mass and stiffness, also has a resonance frequency. The heavier the diaphragm, the lower the frequency, and the stiffer the suspension, the higher the frequency. Although all loudspeakers have a resonance frequency, it is particularly important for woofers. A woofer driven by a constant amplitude input will yield a response curve (measured with a sound level meter) similar to the one of Figure 45.11. The sound output is maximum at the natural frequency of the speaker and decreases rapidly at lower frequencies. In fact, the natural frequency effectively determines the lowest tone that the speaker can reproduce. At lower frequencies, the sound level decreases too rapidly to be of any importance to the overall response. The natural frequency of a speaker can be lowered by using a larger diameter and more massive diaphragm or by providing it with a very loose suspension. We will see in the next section, that each of these choices leads to the selection of quite different enclosures for the loudspeakers.

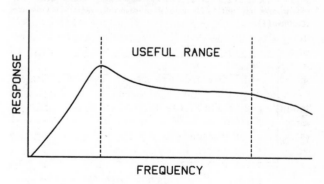

Figure 45.11. Loudspeaker response curve.

A horn loudspeaker consists of an electrically driven diaphragm mounted to a horn, as shown in Figure 45.12. The horn provides the coupling between the diaphragm and the air, increasing the low-frequency output. Direct radiator loudspeakers usually have quite low efficiencies, on the order of three to five percent, and are therefore not suitable for installations requiring large amounts of power, such as sound reinforcement systems. Horn loudspeakers can obtain efficiencies of 25 to 50 percent and consequently require much less power to drive than does a comparable direct radiator loudspeaker.

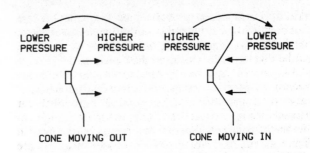

Figure 45.13. Destructive interference at low frequencies for unbaffled woofer.

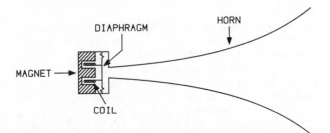

Figure 45.12. Elements of a horn loudspeaker.

The **cutoff frequency** of a horn is the lowest frequency that can be efficiently transmitted by the horn. The cutoff frequency is determined by the rate of the horn's flare and by the size of the horn's large end. When the wavelength of sound becomes several times the size of the horn opening, little sound is radiated. The cutoff frequency is the frequency of a wave having a wavelength four times the size of the opening.

Loudspeaker Enclosures and Baffles

We have discussed why several drivers of different size are needed in a loudspeaker to adequately reproduce the audible frequency range. We now consider the way in which a loudspeaker enclosure helps to determine the response. If the woofer were not enclosed in some manner, there would be a serious loss of bass response, as can be seen from Figure 45.13. When the diaphragm moves outward air is compressed in front of the speaker and rarefied behind the speaker. For low-frequency waves, the compressed air in front simply flows around the edge toward the region of rarefaction and the two tend to cancel each other. For high-frequency waves, the air does not have time to travel from front to back (or vice versa) before a new wave is created and thus there is little cancellation. The solution to the low-frequency problem is to mount a loudspeaker in a baffle to prevent this front-rear cancellation. Five of the most common mounting schemes are illustrated in Figure 45.14.

A baffle is any structure used to regulate or control the sound output from a loudspeaker. A flat baffle mounting, shown in Figure 45.14A, requires a relatively large baffle to obtain an adequate low-frequency response. For instance, to get a good bass response down to 70 Hz the baffle would have to be about 150 cm in diameter.

The open-back cabinet mounting, shown in Figure 45.14B, is widely used in radios, console phonographs, and television sets. The cabinet's resonance (usually oc-

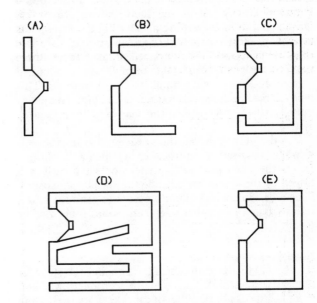

Figure 45.14. Common types of loudspeaker mounting and baffling systems: (A) flat baffle; (B) open-back cabinet; (C) bass-reflex cabinet; (D) acoustic labyrinth; (E) infinite baffle cabinet.

curring at 100 to 200 Hz) is used to enhance a weak bass, but this gives a characteristic "boomy" sound usually associated with speakers of this nature. Furthermore, low-frequency response below the cabinet resonance is decreased.

The phase-inverter or bass-reflex cabinet is completely enclosed (as shown in Figure 45.14C) with a small port, or opening, cut in the cabinet under the speaker (see Figure 45.4). The port inverts the phase of the waves; otherwise they would cancel. When this type of cabinet is designed properly it not only extends the low-frequency response, but it gives a more uniform output without the undesirable boomy quality of the open-back cabinet.

The acoustic labyrinth principle (see Figure 45.14D) is a variation of the bass-reflex cabinet, except that the path followed by the compressions and rarefactions from the rear of the speaker absorbs energy, not to reinforce the bass, but to prevent cancellation due to energy from the front. Because of the "folded up" path for the back wave, the total enclosure can be smaller in size than a completely enclosed system.

The completely enclosed cabinet, or infinite baffle mounting, as shown in Figure 45.14E, has no openings which communicate to the outside air. Since the cabinet is completely enclosed, the stiffness on the speaker is increased, which raises the resonance frequency. These considerations must be kept in mind when speakers of this type are designed, and it may be necessary to use a large cabinet in order to extend the bass range.

A popular speaker system for home use is the acoustic suspension design—a form of infinite baffle. Acoustic suspension systems, however, are different because they require very flexible speaker mountings. In free air the cone can undergo very large excursions with very little applied force. When these speakers are mounted in air-tight enclosures, the entrapped air "stiffens" the speaker suspension system. An acoustic suspension system produces a rather good bass response in a relatively small cabinet. These favorable characteristics, however, are achieved at the expense of efficiency; most acoustic suspension speakers are less than 1% efficient. When enclosed speaker cabinets are properly designed, appreciable outputs can be obtained even at 40 Hz for quite small (0.05 m³) cabinets. This type of speaker, however, has a reduced high-frequency output, which necessitates the use of a multispeaker system.

A recent innovation, the **passive radiator cabinet** is a type of bass-reflex loudspeaker in which a diaphragm covers the open port. This drone, or slave, driver is driven by the sound wave inside the speaker enclosure. Such a design enables manufacturers to produce a bass reflex cabinet of smaller size than would otherwise be possible.

Loudspeaker Specifications

An old truism maintains that a chain is no stronger than its weakest link. In the audio reproduction chain, there is little doubt that the loudspeakers are the "weak link" of the system. Loudspeakers also tend to be the most poorly defined components in the system. It is virtually impossible to compare two different speaker systems merely by studying the manufacturer's specifications. The only safe way for an audiophile to compare loudspeakers is to audition them in comparison tests—an important subject we will discuss in the next chapter. In this section we will consider speaker specifications, which provide important information to be used as background prior to auditions. Loudspeaker specifications fall into one of five categories: (1) those we don't need to know, (2) those we may like to know, but which have little bearing on the sound, (3) important ones usually provided, (4) important ones which need to be clarified, and (5) those which are usually not provided, but which should be provided.

The following specifications often quoted by speaker manufacturers, tell almost nothing about performance or quality, although some are important to an engineer designing the loudspeaker: (1) Magnet structure and weight. A heavier magnet, in and of itself, does not mean that better bass response will be achieved. (2) Voice coil diameter. A large voice coil does not ensure better bass response. (3) Free Air Resonance. (4) Crossover frequencies. The important point is not the specific crossover frequencies, but the matching of the different speakers at the crossover frequencies.

The following specifications while of general interest, tell us little about the quality of the sound produced: (1) Diameter of the transducers. Woofers come in a large range of sizes, but bigger is not necessrily better. The diameter of midrange speakers and tweeters is even less significant. (2) The number of transducers. A loudspeaker cabinet may contain anywhere from two to ten drivers, but the quality does not necessarily increase with the number of drivers. (3) Input impedance. The input impedance of a loudspeaker (4, 8, or 16 ohms) is specified so that you connect the speaker leads to the correct amplifier terminal, not as an indication of quality.

The following specifications are usually provided by the manufacturer. We will consider them in some detail since they provide important information about a loudspeaker.

(1) Minimum power rating. The minimum power rating of a loudspeaker system is the smallest amount of input electrical power from which the speaker can produce sound with low distortion at a reasonable listening level in an "average" room. If the rated power of the amplifier is less than the minimum power rating of the speaker, the sound level produced will be too low. To achieve normal listening conditions, the audiophile is forced to turn the volume of the amplifier up into the region where THD and IMD increase rapidly (see Figure 43.16). Because the amplifier may not have sufficient head room while being operated near or above its rated capacity, a loud burst of sound will drive it into the region of severe distortion. Since the amplifier cannot handle the input signal, it peak clips, producing harmonic components of the input signal (see Figure 15.6). Because most of the harmonics are in the high-frequency region, they are sent to the tweeter by the crossover network. If the distortion is very severe, the extra harmonics may overdrive the tweeter and damage it. Thus, there is more chance of damaging tweeters with an underpowered amplifier than with one adequately matched to the speaker. The minimum power rating is closely related to the efficiency of a speaker (to be discussed later) since a more efficient speaker produces the same sound output with less electrical power input.

(2) Maximum power rating. The maximum power rating is the greatest input electrical power the speaker can receive without damage to the drivers. If the amplifier delivers too much power, the speakers may be blown-out merely by turning up the volume control. Every speaker has a limit as to how much power it can handle without damage, but establishing that limit is not so easy. If the speaker is rated by "continuous power" (or RMS power), the rating will be very conservative, because the specification is in terms of how much power the speaker can handle when the input signal is a continuous sinusoid. The power-handling capacity of a loudspeaker under actual listening conditions is a complex function of the time distribution of the signal's amplitude and frequency. At very low frequencies, the limit of input power is determined primarily by mechanical considerations. If a woofer is driven past the end of its cone excursion capabilities, it may break or tear. For frequencies above 200 Hz, the input power limit is determined by thermal considerations because most of the input signal is converted into heat in the coils. When heat is created faster than it can be dissipated, the coils may be damaged by overheating. Loudspeakers can accept the high power peaks of music

when they are of short duration because the average power to the speakers is much lower than peak power. Most "music power" is produced at frequencies below 500 Hz and decreases rapidly with higher frequencies. A woofer, having a large voice coil, can usually dissipate at least 50 watts of power, while a small tweeter may be able to handle only 5 or 10 watts. When the woofer and the tweeter are combined into a system it should safely handle the full power output of a 200 W amplifier for typical musical material. Naturally, in order to be completely safe, it is prudent to install fuses, of the manufacturer's recommended rating, in your speakers. Generally, an amplifier's rated power should fall between the minimum and the maximum power ratings of the speakers. For example, for an amplifier rated at 40 watts/channel, the maximum power rating of the speakers should be around 50 watts, with a minimum of about 20 watts.

(3) Sensitivity is a measure of the sound pressure level (expressed in dB) which a speaker would produce at a standard distance (usually 1 meter) on the axis of the speaker, when driven with a specified input power (usually 1 watt). Thus a speaker rated at 90 dB (at 1 meter, for 1 watt input) will produce a sound level of 90 dB one meter in front of the speaker when it has an input of 1 watt of electrical power. A higher sensitivity does not indicate a higher quality speaker, but merely that you get more sound out for a given input level. One must be careful when comparing speaker sensitivities because different manufacturers use different input powers and different distances in the specification. As an aid in comparing loudspeaker sensitivities, consider that each doubling of the input power produces a 3 dB increase in the SPL. Thus, for each doubling of the reference input power subtract 3 dB from the given figures to find the sensitivity for 1 watt input. For example, if a specification states that the sensitivity is 93 dB at 1 meter for a 2 watt input, the sensitivity at 1 watt would be 90 dB. For each doubling of distance SPL is decreased by 6 dB while for each halving of distance SPL is increased by 6 dB. Thus, if the sensitivity is given as 82 dB at 2 meters for 1 watt input, the value at 1 meter would be 88 dB. Manufacturers should also state the frequency or frequencies at which the sensitivity is measured, because measurements at the most efficient frequency may yield a value 6 dB above that at frequencies where the speaker is least efficient, assuming a 3 dB tolerance in the frequency response.

The following important specifications are usually provided in some form by the manufacturer, but they need to be interpreted carefully. Otherwise, an audiophile may not obtain the information he thinks is being provided.

(1) Frequency response. The term "frequency response" has meaning only when a plus or minus dB figure is included (see Chapter 43). If only the frequency range is specified, it is reasonable to assume that the level has dropped by at least 10 dB at the two extremes. Occasionally a detailed frequency response curve, such as the one in Figure 45.15, will be given. However, the frequency response is usually given as something like "30 Hz to 17,000 Hz ± 4 dB" which may be misleading. Consider the response curves shown in Figure 45.15, both of which were made from the same

speaker system. The solid curve is the point-by-point response typical of most speaker systems, while the "stair-step" curve has been "averaged" over one-octave frequency bands. The argument goes that human hearing cannot distinguish amplitude peaks in a speaker system which are narrower than one-third octave, so the smoothed curve more nearly represents the way the speaker will be perceived. Whether or not this is true, a plus-or-minus dB figure taken from the smoothed curve will be considerably smaller than a similar figure from the detailed curve. Manufacturers do not usually specify when a smoothed curve has been used, but if response curves are available, look for speakers which do not show pronounced peaks or dips. A peak in the bass range, for instance, will make the speaker sound "boomy," while a dip in the mid-range will cause the speaker to lack presence.

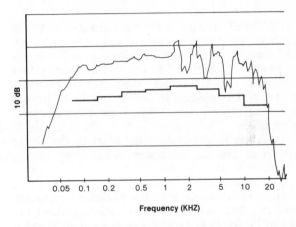

Figure 45.15. Loudspeaker on-axis frequency response curves. The solid line shows a swept sine wave response. The "stair-stepped" line shows one-octave band response. (Courtesy of J. Long.)

(2) Dispersion. Because high frequencies diffract less than low frequencies, high-frequency waves from a tweeter are heard primarily on the speaker axis or by room reflections. The "spread" of a sound wave emanating from a speaker is called "speaker dispersion." The specification which gives a loudspeaker's ability to disperse high-frequency sounds should contain three pieces of information: the angular dispersion in degrees, a dB level, and the frequency range to which this information applies. A typical spec may appear as: 100 degree dispersion, ± 6 dB between 60 Hz and 15,000 Hz. This information tells us that if we move along an arc anywhere within 50 degrees on either side of the speaker's central axis, the sound level remains constant, within 6 dB, over the frequency range specified. When dispersion is specified in degrees without information about the dB variation or the frequency range, the information is meaningless. A polar plot, such as the one shown in Figure 45.16, is the best way to present dispersion data, but such plots seldom appear on the specification sheet. Note that three different frequencies have been plotted, as their dispersions are not identical. For the speaker shown, whose axis is directed along the 0 degree line, we could claim a dispersion angle of 60 degrees (from 30° to 30°) ± 3 dB (since the SPL varies from 46 dB to 40 dB) over a frequency range to 19 kHz.

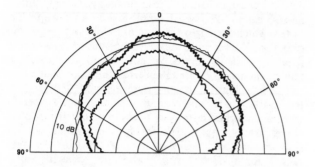

Figure 45.16. Polar plot of loudspeaker dispersion in three frequency bands. (Courtesy of J. Long.)

(3) The efficiencies of different loudspeakers may vary over an enormous range, although most are notoriously low. Efficiency (expressed as a percent) is defined as 100 times the ratio of the output sound power to the input electrical power. Most hi-fi loudspeakers have efficiencies ranging from 0.1% to 10%. If your speaker is 10% efficient, 90% of its input power is wasted as heat. It is important to know the efficiency of your speakers in order to match them to the power amplifier to produce the desired sound level. Unfortunately, most manufacturers provide the "recommended power input," which is a poor substitute for knowing the efficiency. However, since the efficiency is related to sensitivity, which is usually a specification provided, the graph of Figure 45.17 may be used to compute the efficiency from the given sensitivity. (This relationship is not exact because no account is taken of the off-axis output from the speaker. Nevertheless, for speakers of conventional design, the relationship yields a reasonably good approximation of efficiency.) Knowing the efficiency of a pair of speakers, we can work backwards to determine the amplifier power required per channel for the maximum desired acoustic output. In the next chapter we will discuss a procedure for relating the acoustic power output from a speaker to the resulting sound pressure level in a room.

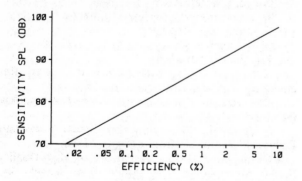

Figure 45.17. Sensitivity versus efficiency for a typical loudspeaker.

Although the following specifications are not generally provided by the manufacturer, they yield important information about a speaker system.

(1) Linearity of power response. Although power amplifiers are extremely linear in their amplitude response, the same is not true for loudspeakers. Often, as the input electrical signal increases, the sound output increases at a faster rate at high power levels and at a slower rate at low power levels, as illustrated in Figure 45.18. The overall effect of such a response curve is an increased dynamic range with slightly more sound produced at high power levels and less at low power levels. A typical linearity specification should appear (1) as a graph, or (2) as: linearity within 4 dB from 0.03 watts to 60 watts electrical input.

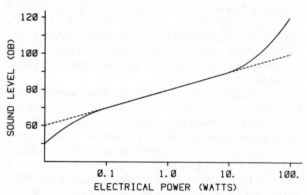

Figure 45.18. Loudspeaker output as a function of electrical power input is shown by the solid line. The dashed line shows a linear relationship for comparison.

(2) A specification giving the maximum attainable SPL at some specified distortion level is helpful to prospective buyers. If such were provided, it should appear in a form similar to the following: maximum SPL = 105 dB at 1 meter, with less than 2% THD and IMD from 100 Hz to 5 kHz.

(3) Distortion. These important specifications are almost never given for loudspeakers because the THD and IMD are always above 1%. Compared to the distortion ratings of amplifiers (usually around 0.1%), these distortions appear high. Still, there is no reason to hide these figures since distortion figures for loudspeakers will never be able to match amplifier distortion figures. When these figures are provided, they should appear in a form similar to: at any frequency THD less than 2% at SPL (1 meter) up to 85 dB, 3% at SPL 90 dB, 5% at SPL 95 dB.

A careful study of all of the above loudspeaker specifications should enable the prospective buyer to perform preliminary screening of possible loudspeakers. Naturally, the final decision should depend upon auditioning the speakers, a topic we consider in the next chapter.

Exercises

1. Describe how a dynamic microphone works (refer to Chapter 6).

2. Which type of microphone would be most useful for each of the following applications: (a) An accurate calibration of another microphone? (b) A hearing aid? (c) A tape recorder used to record speech? (d) A high-fidelity recording of an orchestra? (e) Recording dialogue on a radio program? (f) For use with dramatic or musical productions?

3. Do the size and style of a baffle used to enclose a

loudspeaker affect the bass or treble? Why?

4. Why must a woofer be relatively large? Why is a tweeter always small?

5. Why are tweeters always mounted on the front surface of a multi-speaker enclosure?

6. When you hear a hi-fi system playing next door, why do you hear mostly bass sounds?

7. Why do some multi-speaker units have several identical tweeters, each angled in a slightly different direction?

8. Contrast and compare the various speaker enclosures of Figure 45.14. Some items to consider are expense, efficiency, and bass response.

9. How will the SPL from a speaker with 10% efficiency connected to a 20 watt amplifier compare to a 1% efficient speaker connected to a 200 watt amplifier?

10. Why should loudspeakers always be mounted in a baffle?

11. What purpose does the horn serve on a horn-type speaker?

12. Why are exponential horns usually used for midrange speakers, and not for bass?

13. What would be the effect of putting a tube in the port of a bass reflex speaker?

14. Determine the efficiencies of speakers having sensitivities of 75 dB, 80 dB, and 90 dB.

15. Does a larger diameter ensure better bass response for a loudspeaker? Explain.

16. When comparing different loudspeakers, how much relative importance should you attach to the following: weight of permanent magnet, free air resonance, diameter of the woofer, and number of transducers?

17. If a certain loudspeaker has a maximum power rating of 60 watts, and a minimum power rating of 30 watts, what should be the power (per channel) of the amplifier?

18. A certain loudspeaker has a sensitivity rating of 98 dB at 1 meter for 4 watts input power. What would be the sensitivity at 1 meter for 1 watt input?

19. A speaker has a sensitivity rating of 98 dB at 4 meters for 1 watt input. What would be the sensitivity at 1 meter for 1 watt input?

20. A certain speaker has a sensitivity rating of 98 dB at 2 meters for 2 watts input power. What is the sensitivity at 1 meter for 1 watt input?

21. Response curves for four different speakers are shown in Figure 45.19. Assuming that these are all smoothed response curves, and that all other variables remain essentially the same, which speaker system has the "best" overall response? Why? Which speaker system will have a boomy bass? Why? Which system will lack presence? Why?

22. Which of the following speaker dispersion characteristics are best? Worst? Explain. Speaker A: $140° \pm 12$ dB from 60 Hz to 10 kHz. Speaker B: $120° \pm 8$ dB from 60 Hz to 14 kHz. Speaker C: $100° \pm 4$ dB from 60 Hz to 18 kHz.

23. Which of the following speaker power responses is most linear? Explain. Speaker A: ± 2 dB from 0.02 watt to 50 watt input. Speaker B: ± 6 dB from 0.01 watt to 60 watt input. Speaker C: ± 10 dB from 0.01 watt to 50 watt input.

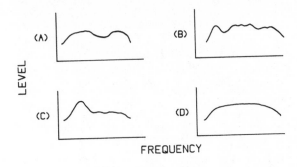

Figure 45.19. Idealized loudspeaker response curves for use in Exercise 21.

24. Give some probable reasons as to why the THD and IMD of loudspeakers are generally above 1% while that of amplifiers is typically less than 0.1%.

25. Why is a distortion rating for a loudspeaker without an SPL specification meaningless? What equivalent information is specified with the distortion specifications for amplifiers?

26. A horn loudspeaker has an opening of 0.6 m diameter. What is the wavelength of the cutoff frequency? Compute the cutoff frequency.

27. For a sound power output of 1 watt, how many cm^3 of air must be moved each second at frequencies of 40 Hz, 400 Hz, and 4000 Hz respectively?

28. Give several reasons why there are so many different loudspeaker designs on the market.

29. Name the four major components of a cone speaker and describe their function.

30. Explain how a small diameter woofer can have the same bass response as a woofer of larger diameter.

31. Explain why the resonance frequency of a combined woofer and baffle should be as low as possible.

32. How can one tell the difference between an acoustic suspension speaker system and a bass reflex system merely by looking at the sytem?

Further Reading
Aldred, Chapters 2, 4

Eargle, Chapters 5, 6

Johnson, Walker, and Cutnell, Chapter 6

Official Guide, Chapter 3

Olson, Chapter 9

Traylor, Chapters 2, 6

Allison, R. 1980. "Loudspeaker Power Requirements," Stereo Review (Aug), 62-65.

Dilavore, P., ed. 1975. **The Loudspeaker** (McGraw-Hill).

Feldman, L. 1973. "Speaker Specs, Facts and Fallacies," Radio- Electronics (Sept), 62-64.

Olson, H. F. 1976. "A History of High-Quality Studio Microphones," J. Audio Eng. Society 24, 798-807.

Rossing, T. D. 1980. "Physics and Psychophysics of High-fidelity Sound, Part III," The Physics Teacher **18**, 426-435.

46. The Complete Stereophonic System

A stereophonic sound reproduction system, like any other complex system is more than the sum of its individual components. The whole system consists of all the parts plus the manner in which the parts interact with each other and their environment. The components of a complete stereo system must be carefully matched to each other so as to avoid the "weak link" syndrome. A fact less widely recognized is that the audible part of the system (the loudspeakers) must be carefully matched to the listening environment. That is, the power output and the placement of the speakers have considerable effect on the resulting sound quality. In previous chapters we have considered the individual components of stereo systems. We now consider the system as a whole. In particular, we will present a procedure for selecting a stereo system, ways to estimate loudspeaker power requirements, suggestions for auditioning loudspeakers, considerations in designing a listening space, and comments on simulated environments.

Selecting a System

If you are contemplating the purchase of new hi-fi equipment, the very first step is to decide how much money you are willing to invest in the system. A fee of about $300 is the admission price to the world of hi-fi. We consider five groups of basic stereo systems differentiated by price, consisting of turntable and cartridge, stereo receiver, and two speakers. Additional components, naturally, will raise the price. The five groupings are: (1) $300 to $500; (2) $500 to $800; (3) $800 to $1500; (4) $1500 to $2400; and (5) above $2400.

The difference in system quality between these groups will now be explained. Some fine components are available in Group (1), the "bottom dollar" group, if you are not too discriminating and if you are very careful with your purchases. Group (2) will have more "life-like" speakers and a receiver with more power, resulting in fuller and more realistic sound. Most people will want equipment in Group (3), the "golden medium" group, as it offers the best return on investment. The sound is realistic and full with a fairly good bass response and very low distortion. Group (4) is where the really great systems begin, but the gains over Group (3) may be marginal, with considerably higher cost for slight improvements. Group (5), the "deluxe" group, is only for the wealthiest and most discriminating listener, with the additional gains beyond Group (4) being very marginal.

Once overall financial considerations are taken into account, the next step is to make a reasonable budget allocation. An optimum allocation for a basic four piece stereo component system is: 20-25% for turntable and cartridge; 30-40% for stereo receiver; and 35-50% for speakers (two). An optimum allocation for a six piece component system is: 15% for turntable and cartridge;

15% for tuner; 25% for open-reel tape deck; 20% for integrated amplifier; and 25% for speakers (two). Note that the highest percentage is recommended for the speakers, bearing in mind that these percentages are only guidelines. Since the speakers are the only part of the system which you actually hear, it makes sense to put your money into high quality speakers rather than into some aspect of the system which has only a nominal effect on the overall sound. Money invested in speakers may also save money elsewhere in the system. Buying high efficiency speakers for instance reduces the power requirement of your amplifier.

The next step is to determine the power requirements of your loudspeakers based on your listening space and your taste in loudness. Then, loudspeakers with the proper sensitivity that are within the budget allocation should be auditioned. Since auditioning the speakers is the most difficult job, we will describe it in a later section.

Once the speakers have been selected, and their power requirements are known, an amplifier with the correct power may be chosen. After an amplifier is selected, a matched turntable should be chosen. Last, but not least, you should connect all the components and listen very critically to the overall result. At this point it is helpful to listen to several albums with which you are familiar. At least some of the selections should be soft music, where deficiencies of quality may be more apparent.

Loudspeaker Power Requirements

The power requirements for a loudspeaker are directly related to the maximum SPL you wish to be able to produce. The decision as to the right maximum level is very subjective and depends on the following factors: the type of music to be played, the proximity and tolerance level of your neighbors, your personal preference, and the escalation of cost as the maximum SPL is increased.

If you wish to reproduce in your apartment the same levels you would experience at a live orchestra concert, you will need loudspeakers capable of generating an SPL of about 100 dB. However, the resulting loudness levels are likely to bring complaints from your neighbors. If you can be satisfied with a 97-dB level, the perceived loudness is only slightly lower but the power requirements are reduced to one-half. Each reduction in SPL of 3 dB results in cutting the power requirement in half. If, on the other hand, you enjoy rock-band levels of 120 dB you will need speakers which produce 100 times as much power.

The sound level created in a specific room by a particular pair of speakers depends upon the acoustical power radiated by the speakers, the size of the room, and the absorptive characteristics of the room. Large rooms with carpet, stuffed furniture, heavy drapes and large openings to other rooms require speakers with more power to reach a specified SPL than would be the case for a smaller or less absorptive room. The graph of Figure 46.1 is an attempt to assist the audiophile in relating his maximum SPL requirement to the loudspeaker power output. This graph relates the acoustic power produced by each of a pair of loudspeakers to the SPL in the reverberant field (at least 2.5 m from the loudspeakers) in an "average" large room (7 m × 4 m × 2.5 m) with "average" furnishings, and of reasonably solid construction. From the graph we can see that if a maximum SPL of 100 dB is to be produced, the acoustic power output from each speaker must be 0.20 W. (In the interests of safety, decorum, and economy we recommend a maximum SPL of 90 to 95 dB.) If the listening room is somewhat dead, the power should be increased by 50%. If the room is very dead and large, the power should be doubled. If the room is small and somewhat non-absorbing, decrease the power by 50%. If the room is very non-absorbing, cut the power in half. Once you know the power output required of the speakers, the necessary amplifier power is the speaker output power divided by the speaker efficiency. For example, a 95 dB level in an average room requires 0.063 acoustic watts from each speaker. If the speaker you select is 0.63% efficient, the electrical input power must be 10 watts/channel. If however, the room is highly absorptive, the input power from the amplifier should probably be 20 watts/channel. Loudspeaker efficiency can be estimated from its rated sensitivity via Figure 45.17.

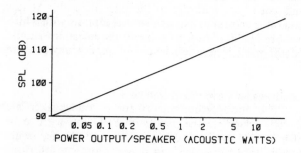

Figure 46.1. Sound pressure level in "average" room as a function of acoustical power output from each of two loudspeakers.

Auditioning Loudspeakers

Knowing the acoustic output power required of your loudspeakers, and knowing your loudspeaker budget, you are in a position to audition the speakers. To audition a loudspeaker is to do more than merely hear the sound it produces. Auditioning implies hearing plus a constant evaluation of the sound being produced. You should use one or two of your favorite albums, so that when comparing different speakers you are already very familiar with the program material.

Select as many speakers as possible with the proper power output and have them connected to a comparator for auditioning. The comparator, shown in Figure 46.2, is an electronic switching device allowing a listener to switch quickly from one speaker system to another. It is particularly important when comparing loudspeakers that the switching be rapid, as you are attempting to

compare different speakers at approximately the same time. Listen to them in a quiet room and compare the speaker systems two at a time. It is important to compare only two systems at any one time, as our auditory memory tends to be rather short and is easily fooled. Apply the following guidelines as you compare two systems until you have determined which you like better: (1) Be certain that each speaker has the same loudness, as we naturally favor louder sounds. (2) Keep the listening times the same (about 30 seconds), because we sometimes choose the sound we have heard the longer time. (3) Listen for the parts you like best in the music and judge which speaker reproduces these better. When you have selected the better system, compare it to another system and repeat the procedure until you have settled on one set of speakers. If possible, listen to this set for an extended time period—several hours in the hi-fi shop or at home for a trial listening period. "Tiring" of listening to the speakers for several hours is a sign of auditory fatigue caused by prolonged exposure to low levels of distortion. Should you experience auditory fatigue from listening to a particular set of loudspeakers, you would be wise to repeat the selection process. You have probably found your speaker system when you can listen to it for many hours without feeling nervous or "jumpy."

Figure 46.2. Comparator (upper) and left-channel set of speakers (lower) connected to comparator. The comparator controls eleven pairs of loudspeakers, eleven amplifiers, and eleven turntables so that any one of eleven pairs of speakers can be auditioned. (Courtesy of G. Orvis.)

Before a final decision is made regarding speakers, they should be checked for dispersion. If a dispersion plot is not provided by the manufacturer, perform the following simple test. Attach a single speaker to the output from the receiver. Turn on the FM radio and tune between two stations and place the muting switch in the "off" position. The "interstation static" that is heard through the speaker is a fairly good approximation to white noise. Starting on the speaker axis walk around the speaker while maintaining a constant distance from the speaker. As you walk, listen carefully until you detect an appreciable decrease in loudness for the higher frequencies. This angle, measured from the axis, determines the dispersion, as illustrated in Figure 46.3. If, for example, you move through an angle of 50° before the high frequencies begin to fade, the dispersion angle for high frequencies is 50° + 50° = 100°.

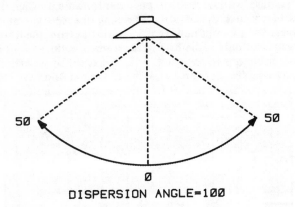

DISPERSION ANGLE=100

Figure 46.3. High-frequency dispersion of a loudspeaker.

Designing a Hi-Fi Room

Unfortunately, excellent stereo systems are often installed in inadequate or even deleterious environments with mediocre results because acoustical considerations are secondary. The position of the loudspeakers is usually determined by the contours of the room, the available space, and the overall visual effect. However, since only a fraction of the total sound produced by a loudspeaker travels directly to the listener, the interaction of the sound with the room boundaries and objects in the room is important. Furthermore, the acoustical effects of such interactions are predictable and controllable. Room shape, available space, surface characteristics, and speaker placement are important variables.

Consider the area of good listening in a rectangular room of dimensions 5.5×3.5 m as shown in Figure 46.4. The minimum recommended separation of the speakers is about 2.5 m. Note that the listening environment is inadequate if one is positioned too close to the speakers, too far from the speakers, or along the sides of the room. In the "typical" living room of this example, the speakers are positioned 2.5 m apart with each speaker 0.5 m from the nearest side wall. The area of good listening extends from a distance located 1.8 m from the center of the speakers to a distance 0.5 m from the back wall. No acoustical advantage would be realized by lengthening the room (say to 7 m) because a longer

room would only increase the area of poor listening. The next factor is the reflected sound which reaches your ears. The pattern of reflections affects your ability to create a stereo image as well as the overall tonal balance between channels.

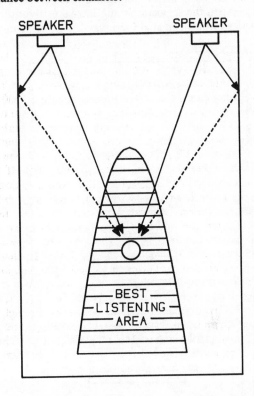

Figure 46.4. Loudspeaker placement in a rectangular room to provide stereo effect. The solid lines show possible paths of direct sound; the dashed lines show reflected sound. The shaded area away from the walls and at some distance from the loudspeakers is the best listening area for the perception of a stereo image. (After Olson.)

In the proper environment, a stereo system is capable of reproducing a panorama of sound localized so that one can judge the relative positions of performers. In order to achieve good stereo imaging, several conditions must be met. First, the speakers must be a matched pair that sound as much alike as possible. Second, the speakers must produce the same sound level and vibrate in phase. This can be checked by switching the amplifier to mono and adjusting the channel balance control to center the sound. Third, the listener should be positioned at an approximately equal distance from each speaker. Finally, the acoustical environments of the speakers must be matched insofar as possible. In the ideal case, this would be achieved by setting the speakers at identical distances from equivalent side walls with similar furnishings in their immediate vicinities. In practice, such symmetry may be impossible to achieve, so one avoids those placements in the worst geometrical assymetries and attempts to compensate for other, less than perfect, placements. If, for instance, the right speaker is situated next to a hard plaster wall and the left speaker is adjacent to heavy draperies, compensation can be made by placing sound deadening panels on the plaster wall in the vicinity of the speaker. In all cases, the speakers should be kept out of corners, unless they are specifically designed for such a location. Better stereo imaging occurs if the speakers are located 0.5 - 1.0 m from the side wall as illustrated in Figure 46.4. It may be advantageous to place sound absorbing materials on the side walls near the speakers (see Figure 46.5) when it is necessary to mount them close to side walls or when the speakers do not have a fairly uniform frequency response in all directions.

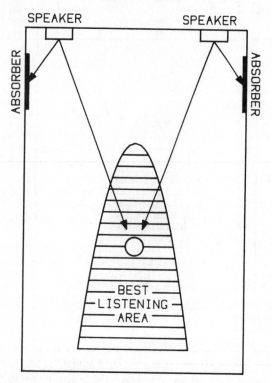

Figure 46.5. Placement of sound absorbing materials when loudspeakers are mounted close to walls.

Proper distribution of sound-absorbing materials in the room is important. Situations with one hard wall opposing an absorptive wall should be avoided. Reflecting surfaces should be alternated with areas of absorption on a given wall. In general, the wider the dispersion of a speaker, the more important it is to alternate areas of reflection with areas of absorption to achieve well-balanced sound.

The final factor in achieving high-quality sound is the avoidance of standing waves in rooms (see Chapter 10), which are to be avoided at all costs. A popular "hi-fi" myth claims that standing waves often provide a much-needed bass boost. In fact, standing waves do increase the mid-bass region somewhat, but reduce the deep-bass in precisely the region where the listener positions himself for the most advantageous imaging. The best way to avoid standing waves is to have a listening room with non-parallel walls and a cathedral ceiling. Since such conditions are rare, the next best step is to use large book cases or record shelves along the walls to break up standing wave patterns. It is also beneficial for the parallel floor and ceiling to have opposing acoustical characteristics. If for instance, a floor is covered by wall-to-wall carpeting, the ceiling should not be treated with acoustical tiles.

Reflected sound from side walls, ceiling, floor, and rear wall, added to the direct sound, result in constructive or destructive interference. The trick in positioning

loudspeakers is to place them at unequal distances from each of the surfaces so that destructive interference effects are minimized at different frequencies. Figure 46.6 illustrates the idea of boundary reinforcement in four different cases. In Case A, the speaker cabinet was placed directly on the floor with the cabinet against the rear wall and 0.7 m from a side wall. The woofer was 0.2 m above the floor and 0.3 m from the front wall, which produces reinforcement below 200 Hz, but a dip at 300 Hz. The strong cancellation at 500 Hz is due to floor reflections. The augmented bass and large mid-range dip, produce a "boomy" sound. For Case B, the woofer was 0.6 m above the floor. The floor reflection now occurs at the same frequency as the side wall reflection, thus producing a very noticeable cancellation at 200 Hz. The sound may be less boomy, but there is a noticeable "hole" in the mid-bass. In Case C, the woofer center was 1 m above the floor, resulting in more uniform bass response without loss of bass reinforcement. Case D demonstrates the effect of keeping the speaker at a height of 1 m, but moving it forward 1 m from the front wall. Not only is the bass reinforcement reduced, but a noticable dip is seen at 120 Hz because the woofer is now equidistant from the rear wall and the floor.

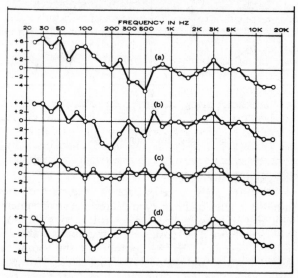

Figure 46.6. Illustration of loudspeaker response variations in a room due to different woofer positions. Refer to text for further detail. (From Mitchell, 1980.)

The following simple guidelines for speaker placement (see also references in Further Reading) are useful in compensating for boundary reflection effects.

(1) For conventional loudspeakers with drivers mounted on the front of the cabinet, the smoothest speaker response is obtained by locating the woofer at substantially unequal distances from the nearest three room surfaces. Carefully avoid placing the woofer equidistant from two or more boundaries.

(2) For two-way speakers, in which the woofer also handles the midrange frequencies up to 1,000 Hz or so, the speakers should not be placed on the floor or adjacent to a side wall because the close proximity of the reflecting surfaces will color the midrange output.

(3) A three-way system can be designed to take advantage of the boundary effects. Typically, this is done

by locating the woofer near the floor-wall intersection (to obtain strong bass reinforcement from the boundary reflections) and using the crossover to cut off the woofer's response around 400 Hz (to avoid a possible midrange dip from close-up reflections). The midrange driver is then located high on the cabinet, well away from any reflecting surfaces.

In summary, at least as much time should be spent analyzing and experimenting with speaker placement as is spent in selecting speakers. The reproducing capabilities of the best of speakers can too easily be negated by careless placement.

Simulated Environments

Any reproduced sound is heard in an environment different from that in which it originated. However, several systems have been devised that make it possible to capture much of the ambience of the original recording environment.

The diagram in Figure 46.7 is a speculative illustration of a type of environment control for listening to recorded sounds. This control would permit tailoring of the sound so that listening in a living room could be perceived as listening in a concert hall, a cathedral, etc. (see Chapter 21). The perception of sound depends on the reverberation time of the room and on relative proportions of direct and reflected sounds and their corresponding times of arrival. Suppose the living room is made as nearly anechoic as possible by covering the various surfaces with acoustic tile, drapery, and carpet. Four speakers are used, one mounted in each corner of the "anechoic living room." The character of the perceived listening room (whether concert hall, etc.) is determined by imposing various amounts of delay time and various losses in the separate channels. (Experiments indicate that four speakers are the minimum number necessary to create these effects, but more than four speakers would provide more realistic simulations of varied environments.) In other words, any desired sonic environment can be simulated by use of electronic modification and a sufficient number of loudspeakers. By controlling attentuations, delay times, etc. of signals sent to different loudspeakers, a large variety of listening environments may be simulated. However, these procedures are of limited usefulness at present because of the large amount of expensive equipment they require.

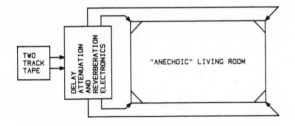

Figure 46.7. "Anechoic living room" with a loudspeaker in each corner and electronics designed to simulate various listening environments.

Exercises

1. If you had $1000 to spend on a basic stereo set, how much money would you allocate for the speakers, the amplifier, and the turntable/cartridge?

2. If you had $2000 to spend for a six-piece component system, how much money would you allocate for each of the components? Into which of the five categories would this system be placed?

3. If you had $2000 to spend for stereo equipment, would it be wiser to invest the money in a high-quality four-component system, or a lesser-quality six-component system? Explore the pros and cons of each approach.

4. List several reasons why the largest budget allocation should be for speakers.

5. What acoustic output power from your speakers is required to achieve 95 dB (in an average living room)? 105 dB?

6. For a speaker which is 10% efficient, compute the required input power for both cases in Exercise 5.

7. For a speaker which is 1% efficient, make the same calculations as in Exercise 6.

8. Do the calculations of Exercise 6 for a speaker that is 0.5% efficient in a highly absorptive room.

9. Summarize the recommended procedure for purchasing loudspeakers.

10. Describe how speaker placement in various locations within a room can produce different amounts of bass boost.

11. Which of the following speakers is most efficient. Speaker A: 200 watts in, 2 watts out; Speaker B: 20 watts in, 1.6 watts out; Speaker C: 2 watts in, 0.02 watts out.

12. 120 watts of electrical power are input into a speaker whose efficiency is 5%. How much of the input power is lost as heat? How much power is radiated as sound?

13. Describe the "dispersion test." Why is a speaker with wide angle dispersion characteristics desirable?

14. A room has dimensions of 3 × 4.8 m. Determine where the speakers should be placed in order to get the largest area of good listening. Should the speakers be placed on the shorter or the longer wall?

15. How far apart should stereo loudspeakers be placed for the best reproduction of the original sound field? What happens if the speakers are too close? Too far apart?

Further Reading

Johnson, Walker, and Cutnell, Chapter 6

Allison, R. 1965. **High Fidelity Systems** (Dover).

Mitchell, P. 1980. "Loudspeaker Placement," Stereo Review (Aug), 56-61.

Nakayama, T., and T. Minura. 1971. "Subjective Assessment of Multichannel Reproduction," J. Audio Eng. Soc. **19**, 744-751.

Olson, H. F. "A Review of Stereophonic Sound Research," RCA Publication 3J-3807 (Radio Corporation of America).

Rossing, T. D. 1981. "Physics and Psychophysics of High-Fidelity Sound, Part IV," The Physics Teacher **19**, 293-304.

Appendices

A1. Review of Elementary Math

Definitions

(1) Numerical constant: a quantity that always retains the same value. Example: speed of sound in air at T = 20°C.

(2) Arbitrary constants: numerical values assigned to a particular situation. Example: Assuming a constant speed of 20 mph, a man travels ten miles in one-half hour.

(3) Independent variable: variable for which the numerical value is chosen arbitrarily. (A variable is a quantity to which an unlimited number of values can be assigned.)

(4) Dependent variable: variable for which the numerical value depends on the value chosen for the dependent variable. Example: Given that a car is traveling at a constant speed of 20 mph, two variables are still involved, time and distance. Taking time as the independent variable, choose any arbitrary time interval, such as two hours. Then the dependent variable (distance) is determined to be 40 miles.

(5) Ratio: a quotient or indicated division, often expressed as a common fraction.

Review of Symbolic Relations and Equations

The math used in physics can be thought of as a set of symbolic relations used to show how "real things" relate to or depend on one another. For example, the distance traveled by an auto can be expressed with word symbols as the distance traveled is equal to the speed of travel multiplied by the time traveled. Alternatively, this relation can be expressed with mixed symbols as length = (speed) (time) or with other symbols, as ℓ = (v)(t).

We often deal with equations, which show some set of symbols equal to some other set. In many cases all symbols but one have known values. Then it becomes our task to calculate a value for the unknown symbol. To solve algebraic equations, a simple rule is to treat each side of the equation as one of a pair of identical twins. Each time something is done to one side of the equation (or twin) the same thing must be done to the other side of the equation (or twin). The object is to get the unknown quantity on one side of the equation and the known quantities on the other. Consider the following examples:

(1) $f = 1/T$ $T = 0.1$ $f = ?$

putting value of T in equation gives
$f = 1/0.1 = 10$

(2) $\ell = vt$ $\ell = 1000$ $t = 5$ $v = ?$

putting values of ℓ and t in equation gives $1000 = 5v$

dividing both sides of the equation by 5
gives $1000/5 = 200 = v$
so we find $v = 200$

Review of Square Roots and Logarithms

In addition to equations there are some special symbols that represent operations or special actions. $\sqrt{}$ is the square root symbol. It means to find a number which when multiplied by itself gives the number under the square root symbol. Study the following examples:

(1) $\sqrt{4} = 2$ $2 \times 2 = 4$
(2) $\sqrt{5} \approx 2.24$ $2.24 \times 2.24 \approx 5$

The logarithm (or "log") of a number is the power to which 10 must be raised to produce that number. The log of the product of two numbers is the sum of the logs of the two numbers. Study the following examples:

(1) $\log 10 = 1$ $10^1 = 10$
(2) $\log 1 = 0$ $10^0 = 1$
(3) $\log 100 = 2$ $10^2 = 100$
(4) $\log 0.1 = -1$ $10^{-1} = 0.1$
(5) $\log 100 = \log (10)(10) = \log 10 + \log 10 = 2$
(6) $\log 20 = \log (2)(10) = \log 2 + \log 10 = 1.3$

The Sine Function

The "sine" of an angle (where the angle is expressed in fractions of a cycle or degrees) is simply a number. The number can represent "real things," such as displacement, force, etc. The reason for using the "sine function" is that it can conveniently represent different kinds of waves. The sine for different angles can be generated by taking a stick L units in length, pinning one end at the point where the horizontal and vertical axes cross each other, rotating the stick through an angle θ, and measuring the displacement D of the other end of the stick from the horizontal axis, as shown in Figure A1.1. Note that the sine values will be the same for all rotations after the first so the table of sine values needs to go only from 0 to 1 cycle or 0 to 360 degrees. The sine of the angular rotation θ is defined as the displacement D divided by L or sine θ = D/L. Once a table of sine values has been put together we can look in the table for the sine value we want, rather than generating it with our rotating stick.

Exercises

1. If $v = f\lambda$, and you are given $\lambda = 3$ and $v = 21$, find f.
2. If $f = 6$ and $\lambda = 3$, find v.
3. If $f = 8$ and $v = 12$, find λ.
4. If $f = 1/T$, and you are given $T = 0.2$, find f.

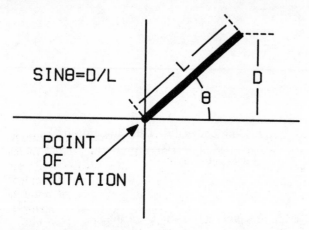

SINθ=D/L

POINT
OF
ROTATION

Figure A1.1. Rotating stick method for generating values of the sine function.

5. If f = 30, find T.

6. If d = vt, and you are given d = 30 and v = 10, find t.

7. If v = 2 and t = 5, find d.

8. If t = 100 and d = 2, find v.

9. If f = 0.16 $\sqrt{s/m}$ and you are given s = 90 and m = 10, find f.

10. If s = 360 and m = 10, find f.

11. Use Table A6.3 to determine the following: sine 0.5, sine 0, sine 0.25, sine 0.75, and sine 0.125.

12. Use Table A6.3 to determine the following: sine 0°, sin 90°, sine 162°, and sine 360°.

13. Why is the sine table only given for 0 to 1 cycle and 0 to 360 degrees?

14. Set up the "sine generator" of Figure A1.1 and compare values generated by it with those given in Table A6.3. Where do negative values come from?

15. Solve for the following by finding the number in the first column of Table A6.2 and its corresponding logarithm in the second column: log 100, log 1000, log 2, log 4, log 10, and log 1.

16. Solve for the following by finding the exponent in the second column of Table A6.2 and its corresponding power of ten in the first column: 10^0, 10^1, 10^6, 10^{-3}, 10^3, and 10^2.

17. Solve for the following by factoring the number into two or more smaller numbers, finding the logarithms of the smaller numbers in Table A6.2, and adding the logarithms: log 200, log 40, log 60000, log 4000, and log 9000.

18. The logarithm of a number specified by a single non-zero digit followed by a string of zeroes can be obtained by counting the zeroes and adding to this the logarithm of the digit. Why does this work? Apply this procedure to Exercise 17.

A2. Graphs and Graphical Analysis

It is frequently useful to observe how one part of a physical system changes with respect to some other part; that is, when one variable (the independent variable) is changed, how does another variable (the dependent variable) change? If a quantitative relationship exists between the variables, the relationship may be expressed (1) in an equation, (2) in a table, or (3) in a graph. As examples, we analyze some typical relationships found in the physical world. The relationships between the variables may be linear, non-linear, or oscillatory. In each example the information is expressed in equation form, in tabular form, and in graphical form. An example of graphical addition is also given.

Linear Relationship

The distance a car travels on a straight road when moving at a constant speed is given by $d = vt$, where d is distance, v is speed, and t is time. The distance travelled (in km) is tabulated in Table A2.1 at six different times (in hr) for a car moving with a speed of 40 km/hr. The values in the table are plotted in Figure A2.1. The plotted values are connected with a straight line to show the linear relationship between distance and time. The dependent variable (d in this example) changes by equal amounts when the independent variable (t in this example) changes by equal amounts in a linear relationship.

Table A2.1. Distance travelled at a constant speed of 40 km/hr.

Time (hr)	Distance (km)
0	0
1	40
2	80
3	120
4	160
5	200

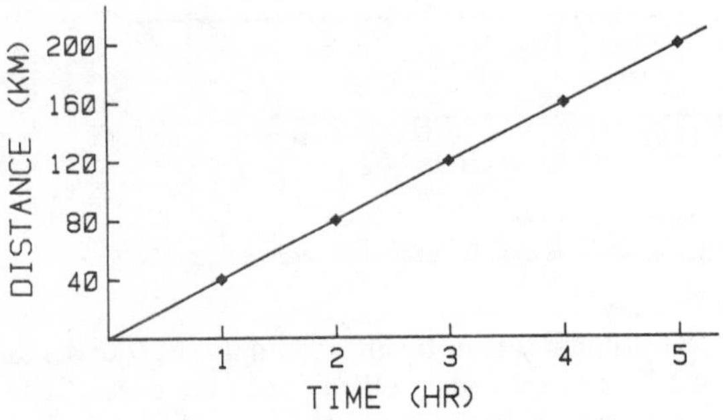

Figure A2.1. Plot of the linear relation d = 40t.

Nonlinear Relationship

The height of a rock above ground level when dropped from a 100 m high cliff is given by $h = 100 - 4.9 t^2$, where h is height (in m) and t is time (in s). The height is tabulated in Table A2.2 at different times. The values in the table are plotted in Figure A2.2. The plotted values are connected with a curved line to show the nonlinear relationship between height and time. The dependent variable (h in this example) changes by unequal amounts when the independent variable (t in this example) changes by equal amounts in a nonlinear relationship.

Table A2.2. Height of a rock dropped from a cliff.

Time (s)	Height (m)
0	100.0
0.5	98.8
1.0	95.1
1.5	89.0
2.0	80.4
2.5	69.4
3.0	55.9
3.5	40.0
4.0	21.6
4.5	0.8

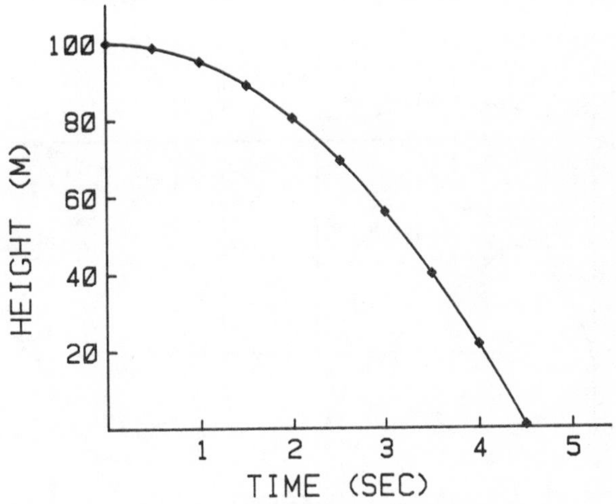

Figure A2.2. Plot of the nonlinear relation $d = 100 - 4.9t^2$.

Oscillatory Relationship

The displacement of a child in a swing is given by $d = 2.0 \sin(0.5 t)$, where d is displacement (in m) from the swing's resting position and t is time (in s). The displacement is tabulated in the second column of Table A2.3 at different times. (The argument of the sine function is assumed to be given in cycles.) The values from the table are plotted in the upper part of Figure A2.3. The plotted values are connected with a sinusoidal curve to show the oscillatory relationship between displacement and time. The dependent variable (d in this example) oscillates between positive and negative values as the independent variable (t in this example) increases.

Table. A2.3. Swing displacement, oscillator displacement, and their combined displacement.

Time (s)	Displacement (m)		
	Swing	Oscillator	Combined
0	0	0	0
0.2	1.18	0.31	1.49
0.4	1.90	0.59	2.49
0.6	1.90	0.81	2.71
0.8	1.18	0.95	2.13
1.0	0	1.00	1.00
1.2	-1.18	0.95	-0.23
1.4	-1.90	0.81	-1.09
1.6	-1.90	0.59	-1.31
1.8	-1.18	0.31	-0.87
2.0	0	0	0
2.2	1.18	-0.31	0.87
2.4	1.90	-0.59	1.31
2.6	1.90	-0.81	1.09
2.8	1.18	-0.95	0.23
3.0	0	-1.00	-1.00
3.2	-1.18	-0.95	-2.13
3.4	-1.90	-0.81	-2.71
3.6	-1.90	-0.59	-2.49
3.8	-1.18	-0.31	-1.49
4.0	0	0	0

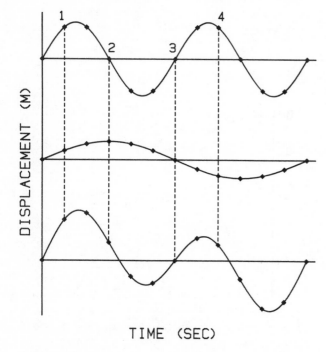

Figure A2.3. Plot of d = 2 sine (0.5t) in the upper curve, d = sine (0.25t) in the middle curve, and their sum in the lower curve.

Graphical Addition

It is often useful to add two simple oscillations to form a complex oscillation because this situation occurs in many everyday phenomena. Suppose we have an oscillator whose displacement, given by d = 1.0 sine (0.25t), is tabulated in the third column of Table A2.3 and plotted in the middle part of Figure A2.3. This oscillation can be added to the swing oscillation described in the previous section by adding the values in columns two and three of Table A2.3 to give the values in column four.

These two sinusoidal oscillations (upper and middle parts of Figure A2.3) can also be added graphically to give a complex oscillation (lower part of Figure A2.3). We will consider the addition of the numbered points to illustrate various features of graphical addition. Upper and middle points 1 are both positive; when they are added they produce the larger positive value plotted as point 1 in the lower graph.

Upper point 2 added to middle point 2 gives lower point 2; lower point 2 is equal to middle point 2 because upper point 2 is zero and adds nothing. Upper point 3 added to middle point 3 gives a lower point 3 of zero because both the upper and middle points have values of zero. Adding middle point 4 to upper point 4 gives a value smaller than upper point 4 because middle point 4 is negative. The result appears as lower point 4. Again, the points of the complex waveform (lower curve) are connected with curved lines to show the nature of the combined waves.

Exercises

1. Tabulate and graph values of v for the relation v = fλ as f is varied from 1 to 25 if λ = 4.

2. Tabulate and graph values of F for the relation F = ma as a is varied from 1 to 10 if m = 10.

3. Tabulate and graph values of f for the relation f = 1/T as T is varied from 0.1 to 10.

4. Tabulate and graph values of f for the relation f = 0.16 $\sqrt{(s/m)}$ as (s/m) is varied from 0 to 100.

5. Tabulate and graph values of dB for the relation dB = 20 log (p/20) as p is varied from 20 to 20,000.

6. Plot sine θ in Figure A2.4 for values of θ between 0 and 1 cycle.

7. Plot sine θ in Figure A2.4 for values of θ between 0° and 360°. How does this plot compare with that of Exercise 6?

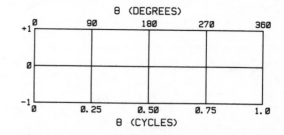

Figure A2.4. Grid to be used for plots of Exercises 6 and 7.

8. Graphically add the upper and middle curves in Figure A2.5 and plot the result in the lower part of the figure. (A ruler can be used to carry out the point-by-point additions.)

358

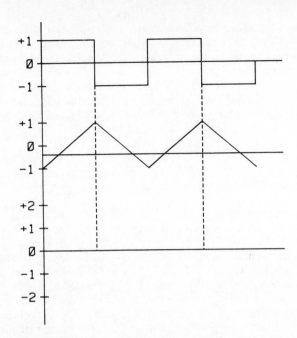

Figure A2.5. Upper and middle curves to be graphically added and plotted in the lower part of the figure as Exercise 8.

A3. Symbols, Quantities, and Units

In the hypotheses and abstractions of science it is necessary to express relationships between natural phenomena. Special symbols are typically used as a shorthand way of representing various physical quantites. Relationships expressed symbolically permit a great deal of economy as compared to writing everything out in words. Special symbols are not peculiar to science; they are used extensively in music, speech, and other facets of our lives. It is very helpful to use standardized symbols so that new symbols do not have to be learned every time a different person's writings are studied. However, although much standardization exists in science, uniformity of symbol usage is not complete.

When carrying out the meaurement step in a scientific method it is necessary to quantify the results of the various measurements. Standard units of measure should be employed so that one set of measurement results can be easily communicated to and interpreted by others. Combining the prefixes of Appendix 4 with standard unit symbols makes a whole range of standard units available. For example, ms (millisecond) is the symbol for one-thousandth of a second and cm (centimeter) is the symbol for one-hundredth of a meter. Scientists making similar observations can easily compare their experimental results with one another when standard units of measure are employed. (Think for a moment of the confusion that might develop in comparing various measurements of length of a particular object if each of ten different observers used as a unit of length the length of his shoe or index finger. In this case, you can see that the same object would be described in terms of ten length measures, each having a different numerical value.)

The definition of standard symbols and standard units of measure is quite arbitrary in many respects. Once the definitions have been made, however, a consistent use of the standards makes their utility value very significant.

Table A3.1. Symbols, quantities, and units.

Symbol	Quantity	Common Units
a	acceleration	m/s^2
d	displacement	m (meter)
f	frequency	Hz (cycle per second)
i	electric current	A (ampere)
ℓ	length	m (meter)
m	mass	kg (kilogram)
n	integer	—
p	pressure	Pa (pascal)
q	electric charge	C (coulomb)
s	stiffness	N/m
t	time	s (second)
v	velocity or speed	m/s
w	weight	N (newton)
x	any unknown quantity	
A	amplitude	various
D	mass density	kg/m^3, kg/m
E	energy or work	J (joule)
F	force or tension	N (newton)
K	constant	—
I	intensity	W/m^2
P	power	W (watt)
R	electric resistance	Ω (ohm)
S	surface area	m^2
T	period	s (second)
V	volume	m^3
Δ	indicates small change	—
λ	wavelength	m (meter)
ϕ	phase	cycle, degree
θ	angle	cycle, degree

Table A3.2. Symbols and quantities.

Symbol	Quantity
AC	absorption coefficient
ADC	analog-to-digital conversion
AFC	automatic frequency control
CB	critical band
DAC	digital-to-analog conversion
DRT	diagnostic rhyme test
GPE	gravitational potential energy
IMD	intermodulation distortion
JND	just noticeable difference
KE	kinetic energy
LL	loudness level
NC	noise criteria
PCM	pulse code modulation
PE	potential energy
PTS	permanent threshold shift
RMS	root-mean-square
RT	reverberation time
SHM	simple harmonic motion
SIL	sound intensity level
SL	sound level
S/N	signal-to-noise ratio
SPL	sound pressure level
STC	sound transmission class
TA	total absorption
THD	total harmonic distortion
TTS	temporary threshold shift
VCA	voltage-controlled amplifier
VCF	voltage-controlled filter
VCO	voltage-controlled oscillator

A4. Standard Prefixes

Many units are related by powers of ten in the metric system. New units of a more convenient "size" can be derived from standard metric units by the use of prefixes. Some standard prefixes are shown in Table A4.1. As an example, "kilo" used with a standard unit gives a new unit one thousand times the size of the original: a kilometer is one thousand meters. As a further example, "milli" used with a standard unit gives a new unit one thousandth the size of the original: a millisecond is one thousandth of a second.

Table A4.1. Standard prefixes with some examples.

Prefix	Symbol	Multiplication Factor	Example
giga	G	10^9	GHz
mega	M	10^6	MHz
kilo	k	10^3	km
centi	c	10^{-2}	cm
milli	m	10^{-3}	mm
micro	μ	10^{-6}	μm
nano	n	10^{-9}	nm

A5. List of Formulas

Several useful formulas are listed in the left column of Table A5.1. Quantities whose relationships are given by the formulas are listed in the right column of the table.

Table A5.1. Formulas and the quantities for which they show a relationship.

Formula	Quantities related by formula
$d = vt$	displacement, velocity, and time
$v = f\lambda$	velocity, frequency, and wavelength
$f = 1/T$	frequency and period
$F = ma$	force, mass, and acceleration
$f \approx 0.16 \; \sqrt{(s/m)}$	frequency, stiffness, and mass for a simple vibrator
$f_1 = f_s (v + v_1)/(v - v_s)$	frequency perceived by a listener and frequency produced by a source
$v = \sqrt{(F/\delta)}$	wave velocity, stretching force, and density for a string
$v = \sqrt{(1.4P/\delta)}$	wave velocity, pressure, and density for a gas
$E = Fd$	energy, force, and displacement
$P = E/t$	power, energy, and time
$I = P/S$	intensity, power, and surface area
$dB = 20 \log (p/20)$	pressure level and pressure
$dB = 10 \log (I/10^{-12})$	intensity level and intensity
$d = d_0 \sin (ft + \phi)$	displacement, displacement amplitude, frequency, time, and phase
$p = p_0 \sin (ft + \phi)$	pressure, pressure amplitude, frequency, time, and phase
$S = 4\pi d^2$	surface area of a spherical shell and distance from the center of the sphere
$v = 340 \text{ m/s}$	approximate speed of sound in air
$v = 34{,}000 \text{ cm/s}$	approximate speed of sound in air

A6. Tables of Functions

In the following tables an argument (n or θ) is given as a number in one column; the value of the function is given as a number in another column. In the case of the sine function table, the argument (θ) is given in three equivalent forms: degrees, radians, and cycle.

Table A6.1. Square roots

n	$\sqrt{n}$
1	1.00
2	1.41
3	1.73
4	2.00
5	2.24
6	2.45
7	2.65
8	2.83
9	3.00
10	3.16
11	3.32
12	3.46
13	3.61
14	3.74
15	3.87
16	4.00
17	4.12
18	4.24
19	4.36
20	4.47
21	4.58
22	4.69
23	4.80
24	4.90
25	5.00

Table A6.3. Sine functions

θ (degrees)	θ (radians)	θ (cycle)	sin (θ)
0	0	0	0.00
18	0.31	0.05	0.31
36	0.63	0.10	0.59
54	0.94	0.15	0.81
72	1.26	0.20	0.95
90	1.57	0.25	1.00
108	1.88	0.30	0.95
126	2.20	0.35	0.81
144	2.51	0.40	0.59
162	2.83	0.45	0.31
180	3.14	0.50	0.00
198	3.46	0.55	-0.31
216	3.77	0.60	-0.59
234	4.08	0.65	-0.81
252	4.40	0.70	-0.95
270	4.71	0.75	-1.00
288	5.03	0.80	-0.95
306	5.34	0.85	-0.81
324	5.65	0.90	-0.59
342	5.97	0.95	-0.31
360 (0)	6.28 (0)	1.00 (0)	0.00

Table A6.2. Logarithms

n	log n
0	—
1	0.00
2	0.30
3	0.48
4	0.60
5	0.70
6	0.78
7	0.85
8	0.90
9	0.95
10	1.00
20	1.30
50	1.70
100	2.00
1000	3.00

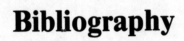

Bibliography

Publications

AAPT. 1972. **Apparatus for Teaching Physics** (American Association of Physics Teachers, Washington, D. C.).

Aldred, J. 1963. **Manual of Sound Recording** (Fountain Press).

Ashford, T. 1967. **The Physical Sciences**, 2nd ed. (Holt, Rinehart, Winston).

Backus, J. 1977. **The Acoustical Foundations of Music**, 2nd ed. (Norton).

Baines, A. 1969. **Musical Instruments through the Ages** (Penguin Books).

Ballif, J. R., and W. E. Dibble. 1972. **Physics: Fundamentals and Frontiers** (Wiley).

Baron, R. A. 1970. **The Tyranny of Noise** (St. Martin's Press).

Bartholomew, W. T. 1965. **Acoustics of Music** (Prentice-Hall).

Benade, A. H. 1976. **Fundamentals of Musical Acoustics** (Oxford University Press).

Berg, R. E., and D. G. Stork. 1982. **The Physics of Sound** (Prentice-Hall).

Beranek, L. L. 1962. **Music, Acoustics and Architecture** (Wiley).

Borden, G. J., and K. S. Harris. 1980. **Speech Science Primer** (Williams and Wilkins).

Bragdon, C. R. 1970. **Noise Pollution: The Unquiet Crisis** (University of Pennsylvania Press).

Bragg, S. W. 1968. **The World of Sound** (Dover).

Broch, J. T. 1971. **Acoustic Noise Measurements** (Bruel and Kjaer Instruments).

Chedd, G. 1970. **Sound** (Doubleday).

Crawford, F. S. 1968. **Waves, Berkeley Physics Course—Volume 3** (McGraw-Hill).

Culver, C. A. 1956. **Musical Acoustics**, 4th ed. (McGraw-Hill).

Doelle, L. L. 1972. **Environmental Acoustics** (McGraw-Hill).

Denes, P. B., and E. N. Pinson. 1973. **The Speech Chain** (Doubleday Anchor Books).

Durrant, J. D., and J. H. Lovrinic. 1977. **Bases of Hearing Science** (Williams and Wilkins).

Eargle, J. 1980. **Sound Recording**, 2nd ed. (Van Nostrand Reinhold).

Edge, R. D., ed. 1974. **Experiments on Physics and the Arts** (University of South Carolina).

Fant, G. M. 1960. **Acoustic Theory of Speech Production** (Mouton).

Flanagan, J. L. 1972. **Speech Analysis Synthesis and Perception**, 2nd ed. (Springer-Verlag).

Flanagan, J. L., and L. R. Rabiner. 1973. **Speech Synthesis** (Dowden, Hutchinson, and Ross).

Fletcher, H. 1953. **Speech and Hearing in Communication** (Van Nostrand).

Fletcher, N. H. 1976. **Physics and Music** (Heinmann Educational Australia).

Freier, G. D., and F. J. Anderson. 1981. **A Demonstration Handbook for Physics**, 2nd ed. (AAPT).

French, A. P. 1971. **Vibrations and Waves, MIT Introductory Physics Series** (Norton).

Hall, D. E. 1980. **Musical Acoustics** (Wadsworth).

Helmholtz, H. L. 1954. **On the Sensations of Tone** (Dover).

Hutchins, C. M., ed. 1975. **Musical Acoustics, Part I: Violin Family Components** (Dowden, Hutchinson, and Ross).

Hutchins, C. M., ed. 1975. **Musical Acoustics, Part II: Violin Family Functions** (Dowden, Hutchinson, and Ross).

Institute of High Fidelity. 1974. **Official Guide to High Fidelity** (Howard W. Sams).

Jeans, J. 1968. **Science and Music** (Dover).

Johnson, K. W., W. C. Walker, and J. D. Cutnell. 1981. **The Science of Hi-Fidelity**, 2nd ed. (Kendall-Hunt).

Josephs, J. J. 1973. **The Physics of Musical Sound** (Van Nostrand Momentum Book).

Kent, E. E., ed. 1977. **Musical Acoustics: Piano and Wind Instruments** (Dowden, Hutchinson, and Ross).

Kinsler, L. E., A. R. Frey, A. B. Copens, and J. V. Sanders. 1981. **Fundamentals of Acoustics** 3rd ed. (Wiley).

Knudsen, V. O., and C. M. Harris. 1950. **Acoustical Designing in Architecture** (Wiley).

Kock, W. E. 1971. **Seeing Sound** (Wiley Interscience).

Kock, W. E. 1965. **Sound Waves and Light Waves** (Doubleday).

Krauskopf, K., and A. Beiser. 1971. **Fundamentals of Physical Science**, 6th ed. (McGraw-Hill).

Kryter, K. D. 1970. **The Effects of Noise on Man** (Academic Press).

Kuttruff, H. 1973. **Room Acoustics** (Halsted-Wiley).

Ladefoged, P. 1975. **A Course in Phonetics** (Harcourt Brace Jovanovich).

Lass, N. I., ed. 1975. **Contemporary Issues in Experimental Phonetics** (Academic Press).

Meiners, H. F., ed. 1970. **Physics Demonstration Experiments, Volume I, Mechanics and Wave Motion** (The Ronald Press Co).

Miller, D.C. 1935. **Anecdotal History of the Science of Sound** (Macmillan).

Miller, G. A. 1951. **Language and Communication** (McGraw Hill).

Olson, H. F. 1967. **Music, Physics, and Engineering** (Dover).

Olson, H. F. 1978. **Modern Sound Reproduction** (Kreiger Publishing).

Peterson, A. P. G., and E. E. Gross, Jr. 1972.

Handbook of Noise Meausurement, 7th ed. (General Radio Co.).

Pierce, J. R., and E. E. David, Jr. 1958. **Man's World of Sound** (Doubleday).

Plomp, R. 1976. **Aspects of Tone Sensation** (Academic Press).

Plomp, R., and G. F. Smoorenburg, ed. 1970. **Frequency Analysis and Periodicity Detection in Hearing** (A. W. Sythoff).

Roederer, J. G. 1979. **Introduction to the Physics and Psychophysics of Music**, 2nd ed. (Springer-Verlag).

Rossing, T. D., ed. 1977. **Musical Acoustics** (American Association of Physics Teachers).

Rossing, T. D., ed. 1979. **Environmental Noise Control** (American Association of Physics Teachers.)

Rossing, T. D., 1982. **The Science of Sound** (Addison-Wesley).

Sabine, W. C. 1964. **Collected Papers on Acoustics** (Dover).

Sanders, D. A. 1977. **Auditory Perception of Speech** (Prentice-Hall).

Schafer, R. W., and J. D. Markel, ed. 1979. **Speech Analysis** (IEEE Press).

Sears, F. W., M. W. Zemansky, and Young. 1975. **College Physics** (Addison-Wesley).

Seashore, C. E. 1967. **Psychology of Music** (Dover).

Singh, S. S., and K. S. Singh. 1976. **Phonetics: Principles and Practices** (University Park Press).

Stevens, S. S., and H. Davis. 1938. **Hearing** (Wiley).

Stevens, S. S., ed. 1951. **Handbook of Experimental Psychology (Wiley).**

Stevens, S. S. and F. Warshofsky. 1956. **Sound and Hearing** (Time Science Series).

Sutton, R. **Demonstration Experiments in Physics** (McGraw-Hill).

Taylor, C. A. 1965. **The Physics of Musical Sounds** (American Elsevier).

Traylor, J. 1977. **Physics of Stereo/Quad Sound** (Iowa State University Press).

United States Gypsum. 1972. **Sound Control Construction**, 2nd ed. (United States Gypsum).

Van Bergiik, W. A., J. R. Pierce, and E. E. David. 1960. **Waves and the Ear** (Doubleday).

Von Bekesy, G. 1960. **Experiments in Hearing** (McGraw-Hill).

White, F. A. 1975. **Our Acoustic Environment** (Wiley).

White, H. E., and D. H. White. 1980. **Physics and Music** (Saunders).

Winckel, F. 1967. **Music, Sound and Sensation** (Dover).

Wood, A. 1974. **The Physics of Music** (Halsted Press).

Audiovisual Resources

AHP	Alfred Higgins Productions
AIM	Associated Instructional Media
BELL	Bell Telephone Companies
CF (COR)	Coronet Instructional Films
EBE (EBEC)	Encyclopedia Britannica Educational Corporation
EFL	Ealing Film Loop
EHF	Educational Horizons Films
EMCC	EMC Corporation
ES	Edmund Scientific Company
FLMFR	Filmfair Communications
FRSC (FRS)	Folkways Records and Services Corporation
GRP	G. R. Plitnik, Route 1, Box 100A, Mt. Savage, MD 21545 (Descriptive Acoustics Demonstration Tape)
GSF	Garden State/Novo Incorporated
HRAW	Holt, Rinehart and Winston
ILAVD	Institute of Laryngology and Voice Disorders
IOWA	Iowa State University
IPS	Introductory Physics Series
IU	Indiana University
JACBMC	Jacobs Manufacturing Company
MGH (MGHF)	McGraw-Hill Films
MLA	Modern Learning Aids Division of Ward's Natural Science
OPRINT	Out of print, but available in many film libraries
PAROX	Paramount-Oxford Films
RANK	J. Arthur Rank Organ
REB	Richard E. Berg, Physics Dept., Univ. of Maryland, College Park, MD 20742 (Set of Video Tapes)
SILBUR	Silver Burdett Company
SF	Sterling Educational Films
UEVA	Universal Educational and Visual Arts
UIOWA	University of Iowa
UKANMC	University of Kansas Medical Center
UMAVEC	University of Michigan Audio Visual Education Center
UMICH	University of Michigan
USNAC	U.S. National Audiovisual Center
WD	Walt Disney

Index

Notes